Lecture Notes in Computer Science 12002

Founding Editors

Gerhard Goos
Karlsruhe Institute of Technology, Karlsruhe, Germany
Juris Hartmanis
Cornell University, Ithaca, NY, USA

Editorial Board Members

Elisa Bertino
Purdue University, West Lafayette, IN, USA
Wen Gao
Peking University, Beijing, China
Bernhard Steffen
TU Dortmund University, Dortmund, Germany
Gerhard Woeginger
RWTH Aachen, Aachen, Germany
Moti Yung
Columbia University, New York, NY, USA

More information about this series at http://www.springer.com/series/7412

Jacques Blanc-Talon · Patrice Delmas ·
Wilfried Philips · Dan Popescu ·
Paul Scheunders (Eds.)

Advanced Concepts for Intelligent Vision Systems

20th International Conference, ACIVS 2020
Auckland, New Zealand, February 10–14, 2020
Proceedings

Springer

Editors
Jacques Blanc-Talon
DGA
Paris, France

Wilfried Philips ⓘ
Ghent University
Ghent, Belgium

Paul Scheunders
University of Antwerp
Wilrijk, Belgium

Patrice Delmas
University of Auckland
Auckland, New Zealand

Dan Popescu
CSIRO
Canberra, Australia

ISSN 0302-9743   ISSN 1611-3349   (electronic)
Lecture Notes in Computer Science
ISBN 978-3-030-40604-2   ISBN 978-3-030-40605-9   (eBook)
https://doi.org/10.1007/978-3-030-40605-9

LNCS Sublibrary: SL6 – Image Processing, Computer Vision, Pattern Recognition, and Graphics

This Springer imprint is published by the registered company Springer Nature Switzerland AG
The registered company address is: Gewerbestrasse 11, 6330 Cham, Switzerland

Preface

These proceedings gather the selected papers of the Advanced Concepts for Intelligent Vision Systems (ACIVS) conference which was held in Auckland, New Zealand, during February 10–14, 2020.

This event was the 20th ACIVS. Since the first event in Germany in 1999, ACIVS has become a larger and independent scientific conference. However, the seminal distinctive governance rules have been maintained:

- To update the conference scope on a yearly basis. While keeping a technical backbone (the classic low-level image processing techniques), we have introduced topics of interest such as deep learning, biometrics and compression, and watermarking. In addition, speakers usually give invited talks on hot issues;
- To remain a single-track conference in order to promote scientific exchanges among the audience;
- To grant oral presentations a duration of 25 minutes and published papers a length of 12 pages, which is significantly different from most other conferences.

The second and third items entail a complex management of the conference; in particular, the number of time slots is rather small. Although the selection between the two presentation formats is primarily determined by the need to compose a well-balanced program, papers presented during plenary and poster sessions enjoy the same importance and publication format.

The first item is strengthened by the notoriety of ACIVS, which has been growing over the years: official Springer records show a cumulated number of downloads of more than 550,000 for ACIVS 2005–2016 and more than 29,000 for ACIVS 2018 (the previous edition).

ACIVS attracted submissions from many different countries, mostly from Europe, but also from the rest of the world: Australia, Belgium, Brazil, China, Czech Republic, Finland, France, Greece, Hungary, India, Italy, Japan, New Zealand, South Korea, Poland, South Africa, Spain, Taiwan, the Netherlands, Tunisia, Turkey, the UK, and the USA.

From 78 submissions, 48 papers were selected for oral or poster presentation. The paper submission and review procedure was carried out electronically and a minimum of two reviewers were assigned to each paper. A large and energetic Program Committee, helped by additional reviewers, as listed on the following pages, completed the long and demanding reviewing process. We would like to thank all of them for their timely and high-quality reviews, achieved in quite a short time.

Finally, we would like to thank all the participants who trusted in our ability to organize this conference for the 20th time. We hope they attended a different and stimulating scientific event and that they enjoyed the atmosphere of the various ACIVS 2020 social events in the city of Auckland.

As mentioned, a conference like ACIVS would not be feasible without the concerted effort of many people and the support of various institutions. We are indebted to the local organizer, Patrice Delmas, for having smoothed all the harsh practical details of an event venue, and we hope to welcome them in the near future.

February 2020

Jacques Blanc-Talon
Patrice Delmas
Wilfried Philips
Dan Popescu
Paul Scheunders

Organization

ACIVS 2020 was organized by the University of Auckland, New Zealand.

Steering Committee

Jacques Blanc-Talon	DGA, France
Patrice Delmas	The University of Auckland, New Zealand
Wilfried Philips	Ghent University – imec, Belgium
Dan Popescu	CSIRO Data61, Australia
Paul Scheunders	University of Antwerp, Belgium

Organizing Committee

Patrice Delmas	The University of Auckland, New Zealand

Program Committee

Hamid Aghajan	Ghent University – imec, Belgium
Fabio Bellavia	University of Florence, Italy
Dominique Béréziat	Sorbonne Université Campus Pierre et Marie Curie, France
Janusz Bobulski	Czestochowa University of Technology, Poland
Philippe Bolon	University of Savoy Mont Blanc, France
Egor Bondarev	Technische Universiteit Eindhoven, The Netherlands
Don Bone	University of Technology Sydney, Australia
Salah Bourennane	École Centrale de Marseille, France
Catarina Brites	Instituto Superior Técnico, Portugal
Vittoria Bruni	Sapienza University of Rome, Italy
Giuseppe Cattaneo	University of Salerno, Italy
Jocelyn Chanussot	Université Grenoble Alpes, France
Patrice Delmas	The University of Auckland, New Zealand
Daniele Giusto	Università degli Studi di Cagliari, Italy
Monson Hayes	George Mason University, USA
Michael Hild	Osaka Electro-Communication University, Japan
Dimitris Iakovidis	University of Thessaly, Greece
Syed Islam	Edith Cowan University, Australia
Yuji Iwahori	Chubu University, Japan
Arto Kaarna	Lappeenranta University of Technology, Finland
Patrick Lambert	Polytech'Savoie, France
Sébastien Lefèvre	Université Bretagne-Sud, France
Céline Loscos	Université de Reims Champagne-Ardenne, France

Reviewers

Jocelyn Chanussot	Université Grenoble Alpes, France
Patrice Delmas	The University of Auckland, New Zealand
Alfonso Gastelum-Strozzi	Universidad Nacional Autónoma de México, Mexico
Trevor Gee	The University of Auckland, New Zealand
Georgy Gimel'farb	The University of Auckland, New Zealand
Daniele Giusto	Università degli Studi di Cagliari, Italy
Michael Hild	Osaka Electro-Communication University, Japan
Dimitris Iakovidis	University of Thessaly, Greece
Syed Islam	Edith Cowan University, Australia
Yuji Iwahori	Chubu University, Japan
Marquez Flores Jorge	Universidad Nacional Autónoma de México, Mexico
Arto Kaarna	Lappeenranta University of Technology, Finland
Patrick Lambert	Polytech'Savoie, France
Philippe Leclercq	The University of Auckland, New Zealand
Sébastien Lefèvre	Université Bretagne-Sud, France
Steven LeMoan	Massey University, New Zealand
Céline Loscos	Université de Reims Champagne-Ardenne, France
Gonzalo Pajares Martinsanz	Universidad Complutense de Madrid, Spain
Amar Mitiche	INRS, Canada
Jennifer Newman	Iowa State University, USA
Minh Nguyen	Auckland University of Technology, New Zealand
Jussi Parkkinen	University of Joensuu, Finland
Delmas Patrice	The University of Auckland, New Zealand
Dan Popescu	CSIRO Data61, Australia
Joris Roels	Ghent University, Belgium
Patrice Rondao Alface	Nokia Bell Labs, Belgium
Luis Salgado	Universidad Politécnica, Spain
Nel Samama	Télécom SudParis, France
Paul Scheunders	University of Antwerp, Belgium
Ionut Schiopu	Vrije Universiteit Brussel, Belgium
Guna Seetharaman	U.S. Naval Research Laboratory, USA
Syed Afaq Shah	Murdoch University, Australia
Ferdous Sohel	Murdoch University, Australia
Changming Sun	CSIRO, Australia
Tamas Sziranyi	University of Technology and Economics, Hungary
Attila Tanács	University of Szeged, Hungary
Sylvie Treuillet	Université d'Orléans, France
Florence Tupin	Télécom ParisTech, Université Paris-Saclay, France
Cesare Valenti	Università di Palermo, Italy
Marc Van Droogenbroeck	University of Liège, Belgium
Peter Veelaert	Ghent University – imec, Belgium
Nicole Vincent	Université Paris-Descartes, France
Domenico Vitulano	National Research Council, Italy
Damien Vivet	ISAE-SUPAERO, France

Contents

Biometrics and Identification

Image Analysis

Image Restauration, Compression and Watermarking

Deep Learning

Deep Learning-Based Techniques for Plant Diseases Recognition in Real-Field Scenarios

Alvaro Fuentes[1], Sook Yoon[2], and Dong Sun Park[3(✉)]

[1] Department of Electronics Engineering, Jeonbuk National University, Jeonju, South Korea
afuentes@jbnu.ac.kr
[2] Department of Computer Engineering, Mokpo National University, Muan, South Korea
syoon@mokpo.ac.kr
[3] Division of Electronics and Information Engineering, Jeonbuk National University, Jeonju, South Korea
dspark@jbnu.ac.kr

Abstract. Deep Learning has solved complicated applications with increasing accuracies over time. The recent interest in this technology, especially in its potential application in agriculture, has powered the growth of efficient systems to solve real problems, such as non-destructive methods for plant anomalies recognition. Despite the advances in the area, there remains a lack of performance in real-field scenarios. To deal with those issues, our research proposes an efficient solution that provides farmers with a technology that facilitates proper management of crops. We present two efficient techniques based on deep learning for plant disease recognition. The first method introduces a practical solution based on a deep meta-architecture and a feature extractor to recognize plant diseases and their location in the image. The second method addresses the problem of class imbalance and false positives through the introduction of a refinement function called Filter Bank. We validate the performance of our methods on our tomato plant diseases and pest dataset. We collected our own data and designed the annotation process. Qualitative and quantitative results show that despite the complexity of real-field scenarios, plant diseases are successfully recognized. The insights drawn from our research helps to better understand the strengths and limitations of plant diseases recognition.

Keywords: Deep learning · Plant diseases · Recognition · Real-field scenarios

1 Introduction

Plant diseases cause a lot of losses in the agriculture sector worldwide [1]. No matter boundaries, barriers or controlling techniques, plants can be severely affected by diseases and viruses. This is nowadays considered a big issue since it directly evolves to satisfy the demand for food which is related to human wellbeing [2]. Quantifying the impact of plant diseases on crops is a challenging problem in agriculture [3]. Once a plant is affected by any pathogen, it is often prone to get affected by more pathogens and cause significant losses in the whole crop [4].

© Springer Nature Switzerland AG 2020
J. Blanc-Talon et al. (Eds.): ACIVS 2020, LNCS 12002, pp. 3–14, 2020.
https://doi.org/10.1007/978-3-030-40605-9_1

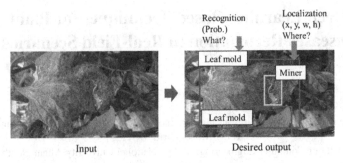

Fig. 1. Our proposed framework for plant disease recognition.

Deep learning has solved complicated tasks with increasing accuracies over the last years. The recent interest in this type of technology, and especially its potential application in agriculture, has fueled the development of efficient autonomous systems. Several methods [5–16] have addressed plant disease recognitions in different types of crops such as tomato, banana, strawberry, etc. Compared to traditional handcrafted feature-based methods, deep learning-based methods have shown that end-to-end approaches are a practical application for image classification and object detection with large scale datasets [17]. Although it is likely that these approaches are useful for identifying diseases in images, they present some limitations to provide sufficient characterization of the problem, especially in real-field scenarios.

We propose a set of new strategies that can successfully perform recognition of plant diseases and symptoms in real-field scenarios, while at the same time we provide some guidelines to develop better recognition techniques that can be adapted depending on the circumstances. In this work, we take a step towards deep learning to propose the following two frameworks: (1) A robust-meta-architecture for recognition of plant diseases and pests in real-field scenarios. (2) An efficient approach to address problems regarding class imbalance and false positives especially in applications with a limited volume of data. Figure 1 presents the general idea of our proposed approach for plant disease recognition.

Our cost-effective techniques can offer better possibilities to farmers in different parts of the world to manage crops properly while avoiding several losses caused by diseases and pests. Specifically, proper handling in correct timing refers to the early detection of symptoms of the plant and potential pathologies that can affect the entire crop. In that context, our proposed research involves challenging scenarios and difficulties such as variations in real-field scenarios, large inter- and intra-class variations, and class imbalance. In addition, looking to push forward our research, qualitative and quantitative experimental results presented in this work aim to reveal that our proposed systems can effectively recognize and locate plant anomalies while substantially achieving the expected descriptions and produce satisfactory prediction results in the complex task of tomato diseases and pest recognition in real-field scenarios. Furthermore, we expect that this research will serve as a reference guide to facilitate the design of more efficient systems since our application could be extended to other crops.

2 Related Works

An ideal scenario in agriculture involves high yields of high-quality products while holding production costs as low as possible. However, the scenario is not always that optimistic. Instead, farmers around the world, have to face several problems to manage their crops while avoiding losses mainly caused by diseases and pests.

Plants are affected by several types of pathologies, especially in tropical, subtropical, and temperate regions of the world [18]. The context of this problem is often related to factors such as temperature, humidity, excess of fertilizer, shade, light, variety of plants, etc. [19]. In terms of crop protection, traditional methods include the use of pesticides, which consequently results harmful to the environment, to the people as a potential risk for the farmers and consumers, and to the plants making them more vulnerable and resistant to chemicals. Based on those assumptions, we identify the following effects of our research:

- The estimation of the dimensionality of pathologies is relevant for rapid decisions in terms of production.
- A cost-effective method can help farmers to control the propagation of pathologies and thus avoiding economic losses.
- It influences the proper use of chemical compounds in a cost-effective way.

2.1 Techniques for Plant Disease Recognition

In practice, several techniques have been recently applied to identify plant diseases [20]. The problem has been addressed by using direct methods closely related to the chemical analysis of the infected area of the plant [21–23], or indirect methods employing physical techniques, such as imaging and spectroscopy [24, 25] to determine the properties of the plant and detect diseases based on the stress of a specific area. In that way, some important facts to consider should be, for example, cost-effectiveness, user-friendliness, sensitivity, and accuracy [26].

Plant anomalies identification has been studied by different means. Although previous methods show outstanding performance in the evaluated scenarios, they do not provide a highly accurate solution for the problems of real-field scenarios. Instead, their experiments are mainly conducted in a laboratory or using expensive techniques. Under those circumstances, our work focuses specifically on non-destructive methods to detect anomalies in plants. We study techniques based on images taken in a visible spectrum for the human eye or RGB images that show the symptoms of the plant.

2.2 Image-Based Plant Disease Recognition

Before that deep learning became popular in the field of computer vision, several methods based on handcrafted features had been widely applied specifically for image recognition. The facilities of deep learning have allowed researchers to design systems that can be trained and tested end-to-end. Due to the outstanding performance of Convolutional Neural Networks (CNNs) as a feature extractor in image recognition tasks, the idea has been also extended to applications in agriculture. Some works studied disease identification in different crops, such as apples [5], bananas [6], wheat [20], cucumber [7], among others.

The availability of accessible data has brought the opportunity to improve the accuracy of image-based disease classification. An important breakthrough is the Plant Village Dataset [27]. Recent works based on CNN for plant diseases classification use this dataset or part of it, to perform experiments. In [8], two CNN architectures AlexNet and GoogLeNet are used to identify 14 crop species and 26 diseases using images from [27]. In [9], 13 types of plant diseases out of healthy leaves are detected using an AlexNet CNN architecture. In terms of specific crops, research works have been also focused on plant disease classification for specific crops. [5] proposes an approach for apple leaf disease identification based on a combination of AlexNet and GoogLeNet architectures. In [10], the authors evaluate various CNN models to detect and diagnose images of plant diseases out of healthy leaves.

Although methods based on image classification have shown promising results in the task of plant disease identification, they present several limitations to provide a real characterization of the problem. Based on that, we identified the following facts:

- Challenges such as complex field conditions, infection stages, various pathologies in the same image, and surrounding objects, are not investigated.
- They mainly use images collected in a laboratory [27] and therefore do not deal with conditions presented in real scenarios.
- They are limited to the classification of single leaves with a homogeneous background, where the goal is to classify an image containing a sample into a fixed number of labels.

3 Proposed Deep Learning Techniques for Plant Diseases Recognition

Proper recognition of plant diseases in real-field scenarios involves several challenges and complexities which are mainly related to the conditions of the place. For instance:

- Multiple diseases in the same sample (leave, plant).
- Samples affected at various stages (early, middle, last).
- Location of the infection in the leaves (front or backside).
- Various patterns, colors, scales, etc.
- Surrounding objects in the scene.
- Illumination and background variation.

This section introduces the main contents of our research which is divided into the following parts: First, a deep meta-architecture for plant diseases and pests recognition. Second, a diagnosis framework to deal with false positives and class imbalance issues, called Refinement Filter Bank.

3.1 Meta-architecture for Plant Diseases and Pests Recognition

We propose a detector based on deep learning for plant disease recognition. The goal of this framework is to estimate multiple types of anomalies and their location in the image. This is a practical and applicable solution to detect the class and location of diseases in tomato plants, which in fact represents the main difference compared to traditional methods for plant disease classification.

Our work, as presented in Fig. 2, aims to identify ten types of diseases and pests that affect tomato plants using a deep meta-architecture as the main body of the system. The input image captured by a camera device with different resolutions and scales is fed to our system, which after being processed by our deep network (region-based framework) results in the class and localization of the infected area of the plant in the image.

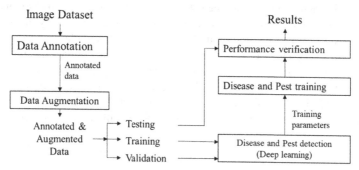

Fig. 2. Framework for plant diseases and symptoms recognition.

We extended the idea of object detection frameworks based on deep meta-architectures with feature extractors to find the class and location of diseases and pests in the image. For our research purpose, we have selected the Faster RCNN [28] as a meta-architecture and the VGG-16 network as a feature extractor.

Faster R-CNN. We extended the application of Faster R-CNN [28] and its Region Proposal Network (RPN) to estimate the class and location of regions that contain diseases in plants. The RPN is used to generate the proposals, including their class and bounding box coordinates. Then, for each proposal, we extract the features with an RoI Pooling layer and perform object classification and bounding-box regression to obtain the estimated targets.

To carry out the experiments, we have adapted the feature extractors to the conditions of our meta-architecture. For instance, in Faster R-CNN, each feature extractor includes the RPN and features are obtained from the "conv5" layer of a VGG-16 network. Our training objective is to reduce the loss between the ground-truth and the estimated results, as well as to minimize the presence of false positives in the final results by Non-Maximum Suppression (NMS). Each meta-architecture selects only candidates which are compared to their corresponding ground-truth.

3.2 Refinement Filter Bank for Plant Diseases and Pests Recognition

Although the task of tomato plant diseases and pest recognition can be effectively solved with satisfactory results by our approach introduced in the previous section. We consider that this task remains challenging due to the following conditions:

- The limited training data with significant unbalanced distribution makes the learning process more biased toward classes with more samples and variations.
- The discrepancy between the classes due to the inter- and intra-class variations results in a high number of false positives, which consequently limit the system to achieve higher accuracy in this complex recognition task.

To address the problems mentioned above, the system introduced in this section proposes a refinement diagnosis strategy. This approach consists of three main components: a primary diagnosis unit (Bounding Box Generator), a secondary diagnosis unit (CNN filter bank), and an integration unit. For each image and class category, the primary unit generates a set of bounding boxes with scores of a specific class instance and the coordinates that indicate the location of the target. Then, the secondary unit filters the confidence of each box by training CNN classifiers independently for each class to further verify their instance. Finally, the integration unit combines the results from the primary and secondary units. Figure 3 illustrates the overall proposed system.

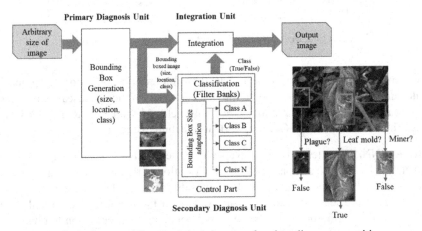

Fig. 3. Refinement Filter Bank-based system for plant disease recognition.

The input of the system is an image I of any arbitrary size. In the first part, the primary diagnosis unit (bounding box generation) proposes a set of boxes that contain the suspicious areas of the image. That is, for an input image I and object categories $C = \{1, 2, 3, \ldots, 10\}$, we want to extract the object proposals. The set of bounding boxes provide information such as the size, location and, class score.

The goal of the secondary diagnosis function is to verify if a bounding box is likely containing the correct target category or not. The inputs of the filter bank are the bounding boxes generated in the primary diagnosis unit. It works as a judging function, which decides whether a target belongs to the category as it was detected (True) or not (False). This framework works like a filter whose goal is to preserve bounding boxes with a higher recognition rate and eliminate the false positives and negatives from the list. An integration unit combines the results from the primary and secondary units.

4 Experimental Results

4.1 Tomato Plant Diseases and Pest Dataset

To validate the performance of our proposed frameworks, we have designed a procedure and collected our images specifically for our research purpose in real-field scenarios using different camera devices. We named our dataset as Tomato Plant Diseases and Pests dataset. Figure 4 shows a representation of some example images from our dataset. The categories and number of samples are shown in Table 1.

Fig. 4. Example images with their corresponding annotations from our Tomato Plant Diseases and Pest Dataset.

Table 1. Categories and number of samples in our Tomato Plant Diseases and Pest dataset.

Class	Number of images in the dataset[a]	Number of annotated samples (bounding boxes)[b]	Percentage of bounding box samples (%)
Leaf mold	1,350	11,922	24.06
Gray mold	335	2,768	5.57
Canker	309	2,648	5.33
Plague	296	2,570	5.17
Miner	339	2,946	10.63
Low temperature	55	477	0.96
Powdery mildew	40	338	0.68
Whitefly	49	404	0.81
Nutritional excess	50	426	0.85
Yellow leaf curl	3927	3927	7.90
Background [3]	2,177	18,899	43.54
Total	8,927	49,662	100

[a]Number of images in the dataset; [b]Number of annotated samples after data augmentation; [c]Category included in every image.

4.2 Quantitative Results

We performed experiments on our Tomato Plant Diseases and Pests dataset that includes ten types of annotated diseases and pest categories. We applied extensive data augmentation to avoid overfitting caused by the limitation of data. Our dataset has been divided into 80% training set, 10% validation set, and 10% testing set. The whole system represents a fast and effective solution that can generate outputs at 160 *ms* per image.

Meta-architecture for Plant Diseases Recognition. We implemented a deep meta-architecture and feature extractor. Table 2 presents the quantitative results of our detector. Faster R-CNN with VGG-16 shows a performance of mAP of 83%. In general, this framework generated satisfactory results, however, we can see a lack of performance in some categories. This issue has been further addressed by our framework based on a refinement filter bank.

Refinement Filter Bank for Plant Diseases Recognition. This study addresses the problem of false positives and class imbalance by implementing a Refinement Filter Bank framework. Table 2 shows the final results of our refinement filter bank-based system. The comparative values evidence a satisfactory improvement in all classes with respect to our previous results. The mAP demonstrates a difference of about 13%. This is, in fact, due to the implementation of the secondary diagnosis unit (CNN filter bank) that allows the system to filter misclassified samples focusing mainly on the reduction of false positives.

Table 2. Quantitative results of plant diseases recognition.

Class	Meta-architecture	Refinement filter bank	Difference in accuracy
Leaf mold	0.9060	0.9205	0.0145
Gray mold	0.7968	0.8910	0.0942
Canker	0.8569	0.9376	0.0807
Plague	0.8762	0.9710	0.0948
Miner	0.8046	0.9947	0.1901
Low temperature	0.7824	0.9821	0.1997
Powdery mildew	0.6556	0.9963	0.3407
Whitefly	0.8301	0.9929	0.1628
Nutritional excess	0.8971	0.9893	0.0922
Yellow leaf curl	0.8500	0.9500	0.1000
Mean AP (mAP)	0.8255	0.9625	0.1370

4.3 Qualitative Results

We evaluated the performance of bounding-box regression and class score for all categories in our Tomato Disease and Pest dataset. As shown in Fig. 5, our system is able to effectively detect the class and location of diseases and pests. We compared the estimated results with the ground-truth using an IoU > 0.5. We found that the best results are generated when the main part of the image consists of the target candidate, in contrast with images that include large background regions.

Our proposed framework shows several advantages compared to traditional methods, in particular, due to its robustness to deal for instance with objects of various scales, shapes, colors, etc. Moreover, our detector performs efficiently on images of real-field scenarios that include several variations.

Figure 6 illustrates some example results of the proposed framework on different types of diseases and pests that affect tomato plants.

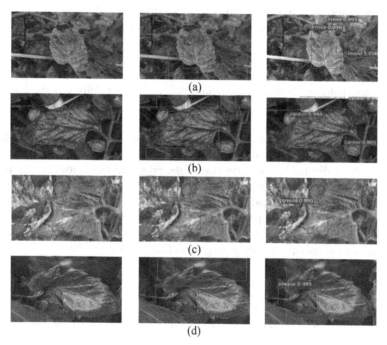

Fig. 5. Qualitative results generated by our proposed approach. A comparison between the ground-truth and the estimated results. (a) Leaf mold, (b) Canker, (c) Gray mold, (d) Plague.

Fig. 6. Example results of plant diseases and pests recognition in real-field scenarios.

5 Conclusion

In this work, we have proposed two efficient techniques based on deep learning that perform onto promising bounding boxes for plant diseases and pests recognition. Our research involves challenging scenarios and difficulties such as variations in real-field scenarios, large inter- and intra-class variations, and class imbalance. First, our meta-architecture for plant disease recognition introduced a practical and applicable solution for detecting the class and location of diseases in tomato plants. Experimental results demonstrated that our baseline meta-architecture can successfully achieve a mAP of 83%. Second, our diagnosis system for plant diseases and pest recognition based on a refinement filter bank presented a solution to deal with problems of class imbalance and false positives. Experimental results showed that using a secondary diagnosis function, we improved the results of the meta-architecture by a margin of 13%, achieving a mAP of 96.25%. We validated the performance of the proposed techniques on our tomato plant diseases and pest dataset that includes images of 10 different types of anomalies. Our cost-effective techniques offer better possibilities to farmers in different parts of the world to manage crops properly while avoiding several losses caused by diseases and pests. In addition, this technology can be adapted for large-scale cultivation, as it represents a tool to monitor breeding programs efficiently in real-time.

Acknowledgments. This work was supported in part by Basic Science Research Program through the National Research Foundation of Korea (NRF) funded by the Ministry of Education (No. 2019R1A6A1A09031717), and by the "Cooperative Research Program for Agriculture Science and Technology Development (Project No. PJ01389105)" Rural Development Administration, Republic of Korea.

References

1. Martinelli, F., et al.: Advanced methods of plant disease detection. a review. Agron. Sustain. Dev. **35**(1), 1–25 (2015). https://doi.org/10.1007/s13593-014-0246-1
2. Savary, S., Ficke, A., Aubertot, J.N., Hollier, C.: Crop losses due to diseases and their implications for global food production losses and food security. Food Secur. **4**, 519 (2012). https://doi.org/10.1007/s12571-012-0200-5

3. Donatelli, M., Magarey, R.D., Bregaglio, S., Willocquet, L., Whish, J.P.M., Savary, S.: Monitoring the impacts of pests and diseases on agricultural systems. Agric. Syst. **155**, 213–224 (2017). https://doi.org/10.1016/j.agsy.2017.01.019
4. Food and Agriculture Organization of the United Nations: "Averting Risks to the Food Chain," Rome (2009)
5. Liu, B., Zhang, Y., He, D., Li, Y.: Identification of apple leaf diseases based on deep convolutional neural networks. Sensors **10**, 11 (2018). https://doi.org/10.3390/sym10010011
6. Amara, J., Bouaziz, B., Algergawy, A.: A deep learning-based approach for banana leaf diseases classification. In: BTW 2017 - Workshopband, Lecture Notes in Informatics (LNI), Gesellschaft für Informatik (Bonn), pp. 79–88 (2017)
7. Kawasaki, Y., Uga, H., Kagiwada, S., Iyatomi, H.: Basic study of automated diagnosis of viral plant diseases using convolutional neural networks. In: Bebis, G., et al. (eds.) ISVC 2015. LNCS, vol. 9475, pp. 638–645. Springer, Cham (2015). https://doi.org/10.1007/978-3-319-27863-6_59
8. Mohanty, S.P., Hughes, D., Salathe, M.: Using deep learning for image-based plant diseases detection. Front. Plant Sci. **7**, 1419 (2016). https://doi.org/10.3389/fpls.2016.01419
9. Sladojevic, S., Arsenovic, M., Anderla, A., Culibrk, D., Stefanovic, D.: Deep neural networks based recognition of plant diseases by leaf image classification. Comput. Intell. Neurosci. **2016**, 3289801 (2016). https://doi.org/10.1155/2016/3289801
10. Ferentinos, K.P.: Deep learning models for plant disease detection and diagnosis. Comput. Electron. Agric. **145**, 311–318 (2018). https://doi.org/10.1016/j.compag.2018.01.009
11. Fuentes, A., Im, D.H., Yoon, S., Park, D.S.: Spectral analysis of CNN for tomato disease identification. In: Rutkowski, L., Korytkowski, M., Scherer, R., Tadeusiewicz, R., Zadeh, Lotfi A., Zurada, Jacek M. (eds.) ICAISC 2017. LNCS (LNAI), vol. 10245, pp. 40–51. Springer, Cham (2017). https://doi.org/10.1007/978-3-319-59063-9_4
12. Barbedo, J.G.A.: Impact of dataset size and variety on the effectiveness of deep learning and transfer learning for plant disease classification. Comput. Electron. Agric. **153**, 46–53 (2018). https://doi.org/10.1016/j.compag.2018.08.013
13. Fuentes, A., Yoon, S., Kim, S.C., Park, D.S.: A robust deep-learning-based detector for real-time tomato plant diseases and pests recognition. Sensors **17**, 2022 (2017). https://doi.org/10.3390/s17092022
14. Fuentes, A., Yoon, S., Lee, J., Park, D.S.: High-performance deep neural network-based tomato plant diseases and pests diagnosis system with refinement filter bank. Front. Plant Sci. **9**, 1162 (2018). https://doi.org/10.3389/fpls.2018.01162
15. Araus, J.L., Kefauver, S.C., Zaman-Allah, M., Olsen, M., Cairns, J.: Translating high-throughput phenotyping into genetic gain. Trends Plant Sci. **23**(5), 451–466 (2018). https://doi.org/10.1016/j.tplants.2018.02.001
16. Fuentes, A., Youngki, H., Lee, Y., Yoon, S., Park, D.S.: Characteristics of tomato diseases – a study for tomato plant. In: ISITC 2016 International Symposium on Information Technology Convergence, Shanghai, China, October 2016
17. Russakovsky, O., et.al.: ImageNet large scale visual recognition challenge. Int. J. Comput. Vis. **115**(3), 211–252 (2015)
18. Mabvakure, B., et al.: Ongoing geographical spread of tomato yellow leaf curl virus. Virology **498**, 257–264 (2016)
19. Heuvelink, E.: Tomatoes, Crop Production Science and Horticulture. CABI Publishing, Wallingford (2005)
20. Sankaran, S., Mishra, A., Ehsani, R.: A review of advanced techniques for detecting plant diseases. Comput. Electron. Agric. **72**, 1–13 (2010). https://doi.org/10.1016/j.compag.2010.02.007

21. Chaerani, R., Voorrips, R.E.: Tomato early blight (Alternaria solani): the pathogens, genetics, and breeding for resistance. J. Gen. Plant Pathol. **72**, 335–347 (2006). https://doi.org/10.1007/s10327-006-0299-3

22. Alvarez, A.M.: Integrated approaches for detection of plant pathogenic bacteria and diagnosis of bacterial diseases. Annu. Rev. Phytopathol. **42**, 339–366 (2004). https://doi.org/10.1146/annurev.phyto.42.040803.140329

23. Gutierrez-Aguirre, I., Mehle, N., Delic, D., Gruden, K., Mumford, R., Ravnikar, M.: Real-time quantitative PCR based sensitive detection and genotype discrimination of Pepino mosaic virus. J. Virol. Methods **162**, 46–55 (2009). https://doi.org/10.1016/j.jviromet.2009.07.008

24. Martinelli, F., et al.: Advanced methods of plant disease detection. a review. Agron. Sust. Dev. **35**, 1–25 (2015). https://doi.org/10.1007/s13593-014-0246-1

25. Bock, C.H., Poole, G.H., Parker, P.E., Gottwald, T.R.: Plant disease sensitivity estimated visually, by digital photography and image analysis, and by hyperspectral imaging. Crit. Rev. Plant Sci. **26**, 59–107 (2007). https://doi.org/10.1080/07352681003617285

26. Irudayaraj, J.: Pathogen sensors. Sensors **9**, 8610–8612 (2009). https://doi.org/10.3390/s91108610

27. Hughes, D., Salathe, M.: An open access repository of images on plant health to enable the development of mobile disease diagnostic. arXiv:1511 (2017)

28. Ren, S., He, K., Girschick, R., Sun, J.: Faster R-CNN: towards real-time object detection with region proposal networks. IEEE Trans. Pattern Anal. Mach. Intell. **39**, 1137–1149 (2016). https://doi.org/10.1109/TPAMI.2016.2577031

EpNet: A Deep Neural Network for Ear Detection in 3D Point Clouds

Md Mursalin[1(✉)] and Syed Mohammed Shamsul Islam[1,2]

[1] Edith Cowan University, Joondalup, WA 6027, Australia
{m.mursalin,syed.islam}@ecu.edu.au
[2] University of Western Australia, Crawley, WA 6009, Australia

Abstract. The human ear is full of distinctive features, and its rigidness to facial expressions and ageing has made it attractive to biometric research communities. Accurate and robust ear detection is one of the essential steps towards biometric systems, substantially affecting the efficiency of the entire identification system. Existing ear detection methods are prone to failure in the presence of typical day-to-day circumstances, such as partial occlusions due to hair or accessories, pose variations, and different lighting conditions. Recently, some researchers have proposed different state-of-the-art deep neural network architectures for ear detection in two-dimensional (2D) images. However, the ear detection directly from three-dimensional (3D) point clouds using deep neural networks is still an unexplored problem. In this work, we propose a deep neural network architecture named EpNet for 3D ear detection, which can detect ear directly from 3D point clouds. We also propose an automatic pipeline to annotate ears in the profile face images of UND J2 public data set. The experimental results on the public data show that our proposed method can be an effective solution for 3D ear detection.

Keywords: Ear biometric · Ear detection · 3D Pointcloud · Deep learning

1 Introduction

The ear is a magnificent organ of the human body that is generally used to detect, transmit and transduce sound. The outer shape of human ears contains distinguishing features that differ significantly among different subjects, even the ear of an identical twin is different from the other [1]. Researchers have shown that ear image analysis has numerous advantages over other biometric traits such as fingerprints, palmprints, iris, and face [2,3]. For instance, the acquisition technique is noninvasive, and the ear is not affected by expression variation. Furthermore, the structure of ears remains steady for a long age duration [4,5].

An essential task for ear biometrics is to detect ears from profile images. The 2D ear image-based techniques are regarded as the most popular for ear region localization as it involves less computations [6]. However, these 2D image-based techniques need to be performed in a constrained environment because

© Springer Nature Switzerland AG 2020
J. Blanc-Talon et al. (Eds.): ACIVS 2020, LNCS 12002, pp. 15–26, 2020.
https://doi.org/10.1007/978-3-030-40605-9_2

2D images are sensitive to changes in lighting conditions and pose variations. Furthermore, a 2D image can not differentiate between shapes and rotation angles. Recent developments in 3D imaging techniques overcome most of the limitations of 2D imaging [3]. Generally, a 3D scanner produces 3D data in the format of an unordered collection of points known as a point cloud.

The basic convolutional neural network (CNN) architectures require Euclidean structured input data formats such as multiview or 3D voxels for sharing weights and optimizing kernels. Since point clouds or meshes are a non-Euclidean data format, most of the work generally converts such data to Euclidean structured data before sending it to CNN architecture. Not only does representation conversion introduce unnecessarily voluminous data, but it also wraps natural invariances of the data, due to generation of quantization artefacts. To overcome this problem, we propose a deep neural architecture named EpNet, which is the modified version of the PointNet [7]. The EpNet is implemented directly on to 3D point clouds. An extensive review of the literature indicates that we are the first researchers applying a deep learning-based method for ear detection directly in 3D point clouds. The contribution of this work can be summarised as follows,

– A deep neural architecture for ear detection in 3D point clouds is proposed.
– A novel pipeline for automatic ear annotation from 3D profile images is introduced.

The rest of the paper is organized as follows: Sect. 2 briefly describes the related work, Sect. 3 explains the proposed methodology, Sect. 4 discusses the experimental results, and Sect. 5 draws the conclusion.

2   Related Work

Depending on the data type used, existing ear detection methods can be categorized as 2D, 3D, and the multimodal approach (using both 3D and the coregistered 2D image). An example of different data representation is shown in Fig. 1. In this paper, we mainly focus on 3D ear detection methods and explore the applicability of deep learning-based methods for ear detection.

Studies have been conducted for ear detection in profile image using 3D data. One of the pioneering work is presented by Chen et al. [8]. The authors extracted step edges from the 3D image and applied a modified iterative closest point (ICP) algorithm to detect helix and antihelix of the ear. However, their approach is sensitive to scale and pose variation. Zhou et al. [9] have proposed a method named histograms of categorized shapes (HCS) which used a 3D shape model combined with a support vector machine (SVM) classifier to detect ear from a 3D image. However, their approach fails to detect ear when prior knowledge about the given ear is not provided. Prakash et al. [10] introduced an edge connectivity graph to detect ear from 3D images. Resultingly, they were unable to handle the effect of off-plain rotation and performed poorly on the UND J2 dataset.

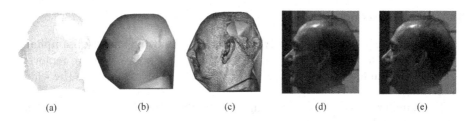

(a)   (b)   (c)   (d)   (e)

Fig. 1. Different representations of a sample profile image from UND J2 data set: (a) point cloud, (b) depth image, (c) 3D mesh, (d) 2D gray image, and (e) 2D colour image. (Color figure online)

A binarized mean curvature map-based method was presented by Pflug et al. [11] for detecting edges on 3D profile images. However, their algorithm failed to detect ear in the presence of occlusion. Lei et al. [12] proposed a novel ear tree-structured graph (ETG) method to detect ear from 3D profile images. Their method required manual annotations in the 2D depth images.

Recently, deep convolutional neural network-based approaches have been proposed for ear biometrics [13–17]. However, none of these methods used 3D images for ear detection.

3   Proposed Ear Detection Method

Our proposed ear detection method takes 3D point clouds as the input and outputs a set of 3D points that represents the ear location in the face image. Firstly, the input 3D point cloud is downsampled to N number of points. Each point has x, y, z coordinate values. The input dimension of each point is $I = C + P$, where C is the coordinate and P is the part id (here, face = 0 and ear = 1). We modify the PointNet network by eliminating some fully connected layers. The block diagram of our proposed method is shown in Fig. 2.

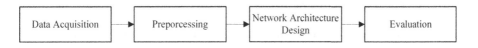

| Data Acquisition | → | Preporcessing | → | Network Architecture Design | → | Evaluation |

Fig. 2. Flowchart of our proposed method.

3.1   Data Acquisition

In this work, we collect 3D ear data named UND J2 from the University of Notre Dame [18]. This data set contains 1800 images from 415 different subjects with a resolution of 640×480. These include 681 images of 176 females and 1119 images of 239 males. All of these images were acquired using a laser scanner known as Minolta Vivid 900. The illumination conditions and poses are different among these images. The subjects are from different age groups with a variety of ethnic background. Additionally, some images contain occlusions with hair and earrings.

3.2   Preprocessing

Our preprocessing step consists of two parts: noise removing and downsampling. The raw profile face data of UND J2 data set contains noises that are removed using the median filter. A smoothing filter is then applied to eliminate the white noise.

The number of points in each image is 921,600. PointNet architecture computation depends on the number of points. Correspondingly, it is an essential step to downsample the data without losing the geometric shape of the object. We applied three different sampling techniques, including uniform box grid, non-uniform box grid and rand sampling technique. In this work, the non-uniform box grid filter is selected because it shows better sampling quality compared with the others. The sample output of different sampling technique is demonstrated in Fig. 3.

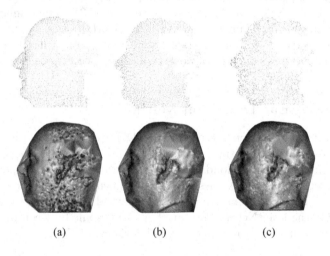

(a)   (b)   (c)

Fig. 3. Results of different downsampling techniques. Here, the first row is the point clouds and second row is the respective 3D mesh representation. Different techniques are: (a) box grid filter, (b) non-uniform box grid filter, and (c) random sampling.

3.3   EpNet Architecture

PointNet [7] is the first neural network that works directly on point clouds. The architecture of this network is simple, but it can efficiently extract discriminate features from input points. Firstly, the input points are passed through the transform network (T-net) and mapped into a feature vector. Then, a max pooling operator is used on this feature vectors to transform a permutation invariant global feature vector. Finally, the point feature vector and the global feature

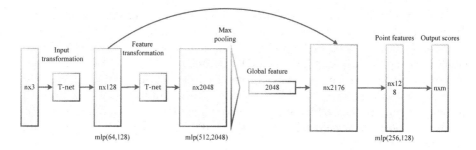

Fig. 4. The architecture of the proposed EpNet. T-net stands for transform network which makes the points invariant to permutation. Here, mlp represents the multi-layer perceptron and the number inside the mlp represents the number of layers.

vector are aggregated and mapped into an output vector using multilayer perceptron (MLP) networks. The output provides the scores for each point to a part id.

Our proposed EpNet architecture is based on PointNet part segmentation network [7]. The part segmentation network is designed for 16 different categories, where each category consists of numerous parts. The total number of parts for 16 categories are 50. In our ear detection problem, we have only one category which has only two parts: face (non-ear region) and ear. For this reason, we empirically decrease the number of MLP layers. So, our network is smaller compared with PointNet, which allows faster learning. In this work, we train the EpNet from scratch. The architecture of EpNet is shown in Fig. 4. The T-net stands for transform network, which offers the point sets invariant to permutation. We adopt the same transform network structure of PointNet. The final output shows the $n \times m$ scores for each of n points and m parts. In this paper, we apply the Adam optimizer [19] to train the EpNet.

3.4 Data Annotation

To the best of our knowledge, there is no annotated point cloud data for ear available in profile face images. As it requires significant controlled time—which increases the probability of operating errors related to operator visual fatigue, or susceptibility to distractions, or confusion during the annotating process—manual annotation is not viable for a large sample. Accordingly, we propose a novel technique to annotate the data automatically using the Basel face model (BFM) [20]. Firstly, we have generated 20,000 synthetic 3D faces using the BFM. The BFM was produced from geometric deformation of 100 male and 100 female faces. Thus, all of these generated images are different from each other. To make a similar view as the UND J2 data set, we then rotate the face point cloud by $-90°$ and delete the hidden points from the current viewpoint using the hidden point removal algorithm [21]. This procedure gives the left view of the synthetic image. All the generated images from BFM are hairless. So, we add random points to include hair occlusion in the 10,000 profile-faced point cloud. Next, we normalized the data to have values between 0 to 1. After normalization, we downsampled the data using non-uniform box grid filter sampling technique. The reason for downsampling is to reduce the computation for EpNet. Next, we labeled ears as the region of interest (ROI) in each 3D point cloud. Here, ground truth location of the ear region is known (because the face data is generated from the statistical model). Finally, we split the data into two groups: training and validation. We took 80% data for training and 20% data for validation. The number of epoch for training the network was 100 where the initial learning rate was 0.001, and the batch size was 12.

Algorithm 1: Fixing the over or under segmentation

Result: Annotated Ear

Calculate the difference $(diff)$ between the initial ground truth and prediction

if $diff <15\%$ then

| Increase the y_{min} and y_{max} using the mean value from correctly annotated ear;

end

if $diff >30\%$ then

| Decrease the y_{min} and y_{max} using the mean value from correctly annotated ear;

end

We have tested all 3D images from the UND J2 data set on the trained model. As our problem of ear detection is a binary class (points belong to non-ear is 0 and points belong to ear is 1) segmentation problem, and the significant portion of the points belongs to the non-ear, so initially we annotate all points in the point cloud as a non-ear. Thus, our ground truth values are all 0. After the testing with all images, we calculated the difference between the prediction and the initial ground truth, where the prediction contains both 0 (non-ear) and 1 (ear), and the initial ground truth includes only 0. In our observation, we found that if the difference is between 15–30%, then the ear points are localized accurately. Apart from the correct localization, we have also seen some over-segmented (difference more than the 30%) and under-segmented (difference less than the 15%) images. In our experiment, we found that all of the incorrect segmentations are in y direction. To correct the under-segmentation, we increase the y_{min} and y_{max} of the ear region with the mean values from the correctly segmented images. The over-segmented images are fixed by decreasing the value of y_{min} and y_{max} according to the mean values from the correctly segmented images (see Algorithm 1). The whole pipeline of data annotation is shown in Fig. 5.

3.5 Evaluation Metrics

In this work, we report four different standard evaluation metrics, including accuracy, intersection over union, precision and recall for object detection. All of

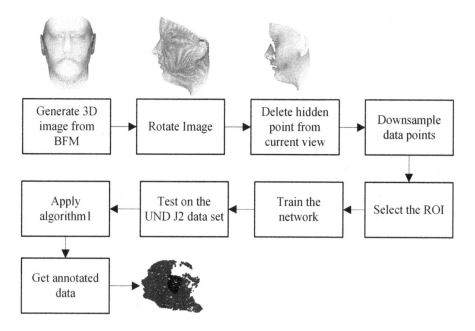

Fig. 5. Flow diagram of data annotation on the UND J2 data set.

these measurements are computed using the annotated ground truth. We use five different terms including true positive (TP), true negative (TN), false positive (FP), false negative (FN) and *Total points*. Here, TP represents the number of points correctly classified as part of an ear. TN represents the number points classified as non-ear points. FP represents the number of ear points classified as non-ear points. FN represents the number of non-ear points classified as ear points. *Total points* represents the number of points exist in a given test image.

The first measurement accuracy is calculated using the following equation,

$$Accuracy = \frac{TP + TN}{Total\ points} \tag{1}$$

The accuracy measurement shows the segmentation quality. However, this result is mostly influenced by the non-ear points that cover the majority portion of test images. Thus, even if most points are categorized as belonging to the non-ear class, our accuracy measurement is shown to have high values.

The next measurement, intersection over union (IoU) is computed as follows,

$$IoU = \frac{TP}{TP + FP + FN} \tag{2}$$

The IoU shows the ratio between the number of points present in both the detected ear areas and the ground truth, and the number of points in the union of the detected and annotated ear areas.

The precision and recall is calculated using the following equations,

$$Precision = \frac{TP}{TP + FP} \tag{3}$$

$$Recall = \frac{TP}{TP + FN} \tag{4}$$

The precision shows the percentage of correctly detected ear points and the ground truth ear points. This metric indicates how many ear points are identified from the actual ear points. The recall shows the percentage of correctly detected ear points from the predicted ear points. Detection accuracy is not defined generally. Different studies provide different measurement for calculating the detection rate. In this work, we used the IoU for defining the detection accuracy.

4   Results and Discussion

The input point cloud data is downsampled to 4096 points. This number is selected experimentally to retain the least visibility of the overall shape of point clouds. We choose 1100 data for training and 200 data for validation. For testing, we use 500 data which are not used during the training process. The network is trained using tensorflow where the number of the epoch is 100, and the batch size is 12. The test result is shown in Table 1.

Table 1. The average results on the UND J2 3D data set. The standard deviation is also given in each average results. All of these values are calculated using 500 test images.

Accuracy (%)	IoU (%)	Precision (%)	Recall (%)
93.09 ± 2.67	63.45 ± 11.14	75.56 ± 12.07	80.44 ± 10.30

Table 2. The average performance metrics between our method and the PED-CED [17]. The standard deviation is also given in each average results. Although the methods use different data set for evaluation, this comparison is to show the overall picture of the performance metrics.

Performance metrics	PED-CED [17]	Point-point (Ours)
IoU (%)	48.31 ± 23.01	63.45 ± 11.14
Precision (%)	60.83 ± 25.97	75.56 ± 12.07
Recall (%)	75.86 ± 33.11	80.44 ± 10.30

Table 3. Comparison of detection rate between our proposed method and other state-of-the-art methods.

Authors	Manual intervention	Detection accuracy (%)
Chen et al. [8]	No	92.6
Prakash et al. [10]	No	99.38
Yan et al. [18]	Yes	>97.6
Zhou et al. [9]	Yes	100
Pflug et al. [11]	Yes	95.65
Proposed method	No	100

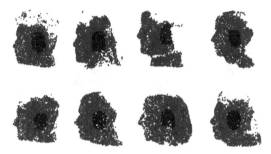

Fig. 6. Examples of our detection results on the UND J2 data set. (Color figure online)

The accuracy value is 93.09%, which is higher than other evaluation metrics, because both ear points, and non-ear points are contributed in the calculation. In the test image, the significant portion of the points as 81.16% are occupied by non-ear points while ear points occupy only 18.84% points. Accordingly, the accuracy metrics show the number of points is correctly detected in the given test image. However, it does not show the number of points that belong to ears as classified accurately (Table 2).

The IoU metrics show the actual performance of the detection, which is not affected by the points distribution of ear and non-ear classes. The false positive and false negative values have a direct impact on IoU measurement. The mean IoU is reported as 63.45%. This result can be improved by training with more diverse images because the typical errors occur due to the occlusions and pose variations. The reported precision in this work is 75.56%, which represents the percentage of detected points from the ground truth ear points. The recall value is 80.44%, that means 19.56% non-ear points are detected as ear points.

We also show some qualitative sample images of our test results in Fig. 6. Here, the blue colour represents the detected ear region. The qualitative results of the IoU values are illustrated in Fig. 7. Accordingly, the first column is the ground truth, the second column is the predicted results, and the third column is the difference between ground truth and prediction. The first row shows the output of the lowest IoU as 40.53%. We can see that most of the ear region is

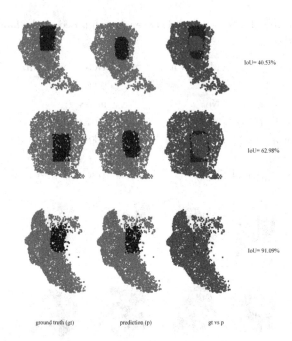

IoU= 40.53%

IoU= 62.98%

IoU= 91.09%

ground truth (gt)   prediction (p)   gt vs p

Fig. 7. The qualitative evaluation of our detection results using different IoU values. The first column is ground truth (gt), the second column is prediction (p), and the third column is the difference between ground truth and prediction. In the first two columns, the purple colour represents ear points, and the 3rd column red colour means the miss-match. (Color figure online)

detected accurately, where only the miss-match observed in the outer top and bottom area. The second row shows the output of IoU value as 62.98%. Here, the significant portion of the ear region detected correctly, although few miss-match is observed in the outer boundary region. The third row shows the output of the highest IoU as 91.09%, where most of the ear region is detected accurately. In this experiment, we have found that IoU of 40% can be used to detect the ear region. So, we consider the detection rate is 100% if the IoU is greater than or equal to 40%.

We also evaluate the performance between EpNet and PointNet++. The validation accuracy shows only 0.64% increase in case of PointNet++. However, the computation is higher in PointNet++, and it does not show significant improvement in our experiment.

It is essential to mention that in the existing literature, there is no standard evaluation method proposed for ear detection. Most of the work in this area reports different criteria for detection rate. Thus, direct comparison is not suggested as shown in Table 3. Here, we consider only those papers that used 3D images for ear detection.

5   Conclusion

In this paper, we present a method based on the deep neural network named EpNet for ear detection. To the best of our knowledge, this is the first study which applies the deep neural network-based approach for ear detection directly from 3D point cloud data. We have tested our trained model on the largest publicly available profile data set named UND J2. The experimental results show the effectiveness of our proposed EpNet for ear detection in 3D point cloud data.

References

1. Nejati, H., Zhang, L., Sim, T., Martinez-Marroquin, E., Dong, G.: Wonder ears: identification of identical twins from ear images. In: Proceedings of the 21st International Conference on Pattern Recognition (ICPR2012), pp. 1201–1204, November 2012
2. Burge, M., Burger, W.: Ear biometrics. In: Jain, A.K., Bolle, R., Pankanti, S. (eds.) Biometrics, pp. 273–285. Springer, Boston (1996). https://doi.org/10.1007/0-306-47044-6_13
3. Islam, S., Bennamoun, M., Owens, R.A., Davies, R.: A review of recent advances in 3D ear-and expression-invariant face biometrics. ACM Comput. Surv. (CSUR) **44**(3), 14 (2012)
4. Tiwari, S., Jain, S., Chandel, S.S., Kumar, S., Kumar, S.: Comparison of adult and newborn ear images for biometric recognition. In: Proceedings of 2016 Fourth International Conference on Parallel, Distributed and Grid Computing (PDGC), pp. 421–426, December 2016
5. Chen, H., Bhanu, B.: Contour matching for 3D ear recognition. In: 2005 Seventh IEEE Workshops on Applications of Computer Vision (WACV/MOTION'05)-Volume 1, vol. 1, pp. 123–128. IEEE (2005)
6. Emeršič, Ž., Štruc, V., Peer, P.: Ear recognition: more than a survey. Neurocomputing **255**, 26–39 (2017)
7. Qi, C.R., Su, H., Mo, K., Guibas, L.J.: Pointnet: deep learning on point sets for 3D classification and segmentation. In: Proceedings of the IEEE Conference on Computer Vision and Pattern Recognition, pp. 652–660 (2017)
8. Chen, H., Bhanu, B.: Shape model-based 3D ear detection from side face range images. In: 2005 IEEE Computer Society Conference on Computer Vision and Pattern Recognition (CVPR 2005)-Workshops, pp. 122–122. IEEE (2005)
9. Zhou, J., Cadavid, S., Abdel-Mottaleb, M.: Histograms of categorized shapes for 3D ear detection. In: 2010 Fourth IEEE International Conference on Biometrics: Theory, Applications and Systems (BTAS), pp. 1–6. IEEE (2010)
10. Prakash, S., Gupta, P.: An efficient technique for ear detection in 3D: invariant to rotation and scale. In: 2012 5th IAPR International Conference on Biometrics (ICB), pp. 97–102. IEEE (2012)
11. Pflug, A., Winterstein, A., Busch, C.: Ear detection in 3D profile images based on surface curvature. In: 2012 Eighth International Conference on Intelligent Information Hiding and Multimedia Signal Processing, pp. 1–6. IEEE (2012)
12. Lei, J., You, X., Abdel-Mottaleb, M.: Automatic ear landmark localization, segmentation, and pose classification in range images. IEEE Trans. Syst. Man Cybern.: Syst. **46**(2), 165–176 (2016)

13. Cintas, C., Delrieux, C., Navarro, P., Quinto-Sanchez, M., Pazos, B., Gonzalez-Jose, R.: Automatic ear detection and segmentation over partially occluded profile face images. J. Comput. Sci. Technol. **19**, 81–90 (2019)
14. Zhang, Y., Zhichun, M., Yuan, L., Chen, Y.: Ear verification under uncontrolled conditions with convolutional neural networks. IET Biometrics **7**(3), 185–198 (2018)
15. Wang, S., Du, Y., Huang, Z.: Ear detection using fully convolutional networks. In: Proceedings of the 2nd International Conference on Robotics, Control and Automation, pp. 50–55. ACM (2017)
16. Moniruzzaman, M.D., Islam, S.: Automatic ear detection using deep learning. In: Proceedings of the International Conference on Machine Learning and Data Engineering. iCMLDE2017 (2017)
17. Emeršič, Ž., Gabriel, L.L., Štruc, V., Peer, P.: Convolutional encoder-decoder networks for pixel-wise ear detection and segmentation. IET Biometrics **7**(3), 175–184 (2018)
18. Yan, P., Bowyer, K.W.: Biometric recognition using 3D ear shape. IEEE Trans. Pattern Anal. Mach. Intell. **29**(8), 1297–1308 (2007)
19. Kingma, D.P., Ba, J.: Adam: A method for stochastic optimization. arXiv preprint arXiv:1412.6980 (2014)
20. Paysan, P., Knothe, R., Amberg, B., Romdhani, S., Vetter, T.: A 3D face model for pose and illumination invariant face recognition. In: 2009 Sixth IEEE International Conference on Advanced Video and Signal Based Surveillance, pp. 296–301. IEEE (2009)
21. Katz, S., Tal, A., Basri, R.: Direct visibility of point sets. In: ACM Transactions on Graphics (TOG), vol. 26, p. 24. ACM (2007)

Fire Segmentation in Still Images

Jozef Mlích$^{(\boxtimes)}$ ⓘ, Karel Koplík ⓘ, Michal Hradiš ⓘ, and Pavel Zemčík ⓘ

Department of Computer Graphics and Multimedia, Faculty of Information
Technology, Brno University of Technology, Božetěchova 2, Brno, Czech Republic
{imlich,ikoplik,ihradis,zemcik}@fit.vutbr.cz
https://www.fit.vut.cz

Abstract. In this paper, we propose a novel approach to fire localization
in images based on a state of the art semantic segmentation method
DeepLabV3. We compiled a data set of 1775 images containing fire
from various sources for which we created polygon annotations. The
data set is augmented with hard non-fire images from SUN397 data
set. The segmentation method trained on our data set achieved results
better than state of the art results on BowFire data set. We believe
the created data set(http://www.fit.vutbr.cz/research/view_pub.php.cs?
id=12124) will facilitate further development of fire detection and segmentation methods, and that the methods should be based on general
purpose segmentation networks.

Keywords: Fire detection · Semantic segmentation · Deep learning ·
Neural networks · Emergency situation analysis

1 Introduction

An estimated 350 million surveillance cameras were operated worldwide as of
2016 compared with approximately 160 million in 2012. According to 2011 Freedom of Information Act requests, the total number of private and local government operated CCTV cameras was around 1.85 million over the entirety of
the UK. These days, it has become quite common that security cameras can be

Fig. 1. Example of fire segmentation

© Springer Nature Switzerland AG 2020
J. Blanc-Talon et al. (Eds.): ACIVS 2020, LNCS 12002, pp. 27–37, 2020.
https://doi.org/10.1007/978-3-030-40605-9_3

found almost anywhere. The ability to detect fire in early stage is important. If the existing cameras could be used for the detection task, it would be very beneficial for saving human lives and for protecting private property in case of an unexpected fire.

In some applications, it could be sufficient to detect presence of fire, but we decided to approach the task as image segmentation problem as it provides richer information and presence of fire can be still decided by simple post processing. The precise location of a fire in an image can be useful in many cases such as in omni-directional outdoor cameras and in autonomous fire-extinguishing systems. An example of precise fire localisation in complex scene by our approach is shown in Fig. 1.

The fire detection was previously addressed as a segmentation problem by task specific methods [5,7] and the method of Muhammad et al. [12] provides fire segmentation by thresholding hand picked activations of fire detection convolutional network. We believe that fire segmentation is very similar to other semantic segmentation problems such as traffic or indoor scene segmentation which are heavily studied. The state of the art methods in these tasks are based on fully convolutional neural networks [4,16] and suppose that the same methods should be effective for fire detection as well. Specifically, we have decided to follow up on work of Chen et al. [4] (DeepLabV3).

Section 2 presents existing data sets and algorithms for fire detection and segmentation in images and video. Section 3 describes a novel data set for fire segmentation: how it was compiled and what is characteristic about it. Section 4 describes DeepLabV3 and how we use it for fire segmentation. The evaluation methods and the achieved results are presented in Sect. 5 and then summarized in Conclusions - Sect. 6.

2   Related Work

Flame and smoke detection has been studied and a number of data sets for this task is available [1,2,5,8,9,14]. A detailed and systematic summary of older hand designed and simple machine learning methods can be found in [3]. Recent color based methods include [5–7]. Current state of the art in fire detection is represented by convolutional networks [12].

Existing fire detection data sets are summarized in Table 1. FlickrFire data set [1] contains high number of diverse images and it can be used for training and testing of fire detection methods. However, it contains only image level annotations which makes it unsuitable for segmentation and it is not available for download anymore. BowFire data set [5] can be used for evaluation of segmentation methods, but the training part consist of small cropped image regions which makes it unsuitable for modern segmentation methods. Other data sets [2,8,9,14] each contain large number of video frames from small number videos, which cover only small variation of possible fire scenes.

Contemporary semantic segmentation methods relay heavily on the quality and size of training data set. For example, Microsoft COCO [11] contains 200k

Table 1. Overview of fire and smoke detection data sets

Dataset	Type	Frames	Videos	Res. [Mpx]	Annot.	Fire	Smoke
FlickrFire [1]	Images	2000	0	0.2–13	Per image	50%	0
BowFire [5] train.	Crops	240	0	0.0025	Per crop	33%	33%
BowFire [5] eval.	Images	225	0	0.2–1.3	Mask	52%	0
FiSmo [2]	Videos	79,815	27	0.2	Per frame	70%	0
FiSmo: Rescuer [2]	Videos	29,895	61	0.1–2	Per frame	68%	0
FurgFire [14]	Videos	28,000	23	0.2–2	Bounding b.	50%	0
Firesense [9]	Videos	40,184	49	0.06–0.4	Per video	20%	25%
VisiFire [8]	Videos	36,025	40	0.1–0.4	Per video	12%	63%
Our dataset	Images	6,368	150	0.2 (– 8)	Polygons	27%	7%

images, ADE20k [16] for Places Challenge 2017 contains 20k images, and SUN397 data set [15] 100k images.

Our data set contains a large number of images with associated polygon annotations. The images covers most of the variability of possible fire scenes and it should be suitable to train contemporary semantic segmentation methods.

Historically, fire detection techniques were evolving from color thresholding trough texture classification. Those approaches are represented by [5,7].

The *BoWFire* method, published in [5], was, as the authors explain it, developed in reaction to algorithms with the following two step structure: First, the visual features (colour and brightness) are extracted from individual video frames, creating lots of false positives. Second, the temporal features are extracted and approximated to a mathematical/rule-based model, eliminating most of the false positive error. BoWFire is an acronym for *best of the both worlds of fire detection*; in other words, the goal is to eliminate the need to use videos and fix the imprecise colour classification by adding texture classification. Needless to say, texture classification had been used in video-based methods too, however, the BoWFire method achieved decent results with just images.

In *Fire Color Mapping–Based Segmentation, de Souza et al.* [6], fire detection with its custom probabilistic color-mapping method is presented whilst also analyzing entropy. The *BoWFire* data set for training and testing is also shown. In this approach, multiple colour models were used to see which combination of colour channels is the most effective for the task. The entropy analysis is based on fire regions being more irregular/unstable.

Nowadays, convolutional networks approach and semantic segmentation are used instead. Fully Convolutional networks consist of only convolutional and pooling layers, without the need of fully connected layers. The strightforward approach was to use a stack of same-sized convolutional layers to map the input image to the output one. This is computationally expensive. This problem is solved by an encoder-decoder architecture. The encoder is a typical convolutional network such as AlexNet or ResNet and the decoder consists of deconvolutional and up-sampling layers. The goal of downsampling steps is to capture semantic/contextual information while the goal of upsampling is to recover spatial information.

In [16], Zhou et al. propose their semantic segmentation network design called *Cascade Segmentation Module*. Their goal was to be able to parse a given scene in an image and then automatically remove custom content. For this purpose they also created their own data set, ADE20K, annotated by an "expert annotator". The Cascade Segmentation Module has three macro classes: background (containing classes like sky, road), foreground objects (e.g. car) and object parts (e.g. car wheels). This way, they can classify each group in separate streams and then combine the output. From the final output they can determine if a pixel belongs to background or foreground and is an object or object part, i.e. they achieve a multi-level segmentation.

Muhammad et al. [12] used deep neural network to detect and localize fire in video. It provides fire segmentation by thresholding hand picked activations of fire detection convolutional network. They proposed their own CNN architecture based on the SqueezeNet and AlexNet. The goal was to keep the computational and spacial requirements to a minimum while maintaining the accuracy. Eventually, the size of their model was reduced to 3 MB. Even with this size, they out-performed the state-of-the-art algorithms at the time.

DeepLabV3 [4], the Winner in 2016 ILSVRC Scene Parsing Challenge, is based on Atrous convolution, which allows effectively enlarge the field of view of filters to incorporate multi-scale context, in the framework of both cascaded modules and spatial pyramid pooling. In particular, the module consists of Atrous convolution with various rates and batch normalization layers which was important to train as well.

Figure 2 shows Atrous convolution with kernel size 3×3 and different dilation rates. Standard convolution corresponds to Atrous convolution with dilation rate 1. Employing large value of dilation rate enlarges the model's field-of-view, enabling object encoding at multiple scales.

3 Proposed Data Set

Our goal was to find the best parameters for existing segmentation tools to make them usable in real-world conditions. For the segmentation, we used state-of-the-art method for which we had compiled a specialized data set. The data set was then iteratively expanded and adjusted for the given domain.

To be able to train DeepLabV3 semantic segmentation model for fire, we had to prepare a large number of pictures that each have per-pixel annotation for flames and/or smoke. We have compiled our data set from multiple sources. We combined images from existing data sets, namely FlickFire, Firesense, and Visifire described in previous section, and unified and extended their annotations as needed. Next, we have searched for images and video sequences by keywords fire or flame using common search engines. We have included also 150 video sequences. We annotated only 6 frames on average per video, as more frames would be highly correlated and would not increase the information contained

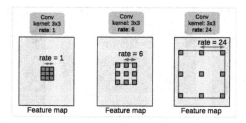

Fig. 2. Atrous convolution with kernel size 3 × 3. Taken from [4]

in the data set. We manually created annotations using LabelMe annotation tool [13].

The annotations are polygons created manually using mouse. We did not strive for pixel precision as fire and smoke do not have precise boundaries and it can be even transparent. In general this transparency leads multiple overlapping object classes; however, in our data set we focus on segmentation of fire class only. As consequence we can use standard tools such as a DeepLabV3 which use segmentation maps as ground truth. Examples of annotations from our data set are shown in Fig. 5.

In order to limit false positive detections, large amount of non-trivial negative data is usually needed (hard negative examples). We obtained such images from the SUN397 data set [15]. This set contains images of 397 scene categories which are annotated with object polygons, unfortunately fire and smoke classes included. We selected hard negative images from SUN397 based on responses of an early version of our fire segmentation tool. We had to manually filter these images as some of them contained fire (SUN397 includes classes volcano, campsite, burial chamber, restaurant kitchen, etc.). The selected images depict challenging scenes such as sun rise, sun set, street lighting at night, bright reflections on restless water surface etc. Table 2 shows from which scene classes were the hard negative images mostly selected.

Figure 3 shows how false positive responses are reduced when the hard negative images are included in the training set.

The final data set contains 1775 fire images and 4611 non-fire images. The annotation consists of image masks and polygons (LabelMe XML). The data set contain images in original resolution, but our experiments were done on images down-sampled to resolution 512 × 512 px. The dataset contains some annotations of smoke as well, but we didn't evaluated results of its detection.

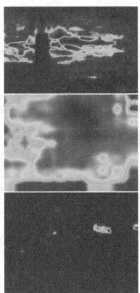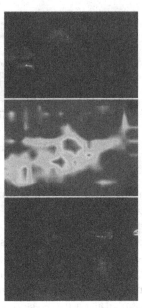

(a) Sample input images from SUN397 data set   (b) Probability of fire with basic training set   (c) Segmentation after hard negative mining

Fig. 3. Reduction of false positive responses with hard negative examples.

Table 2. Scenes from SUN397 which contain the largest number of hard negative images for fire segmentation.

# hard negatives	Class
59	c/cavern/indoor/
40	k/kitchen/
30	c/campsite/
28	a/airport_terminal/
26	p/pasture/
24	i/ice_skating_rink/outdoor/
24	c/clothing_store/
24	b/bathroom/
24	a/attic/

4 Proposed Network Structure

In our experiments, we used Deeplab-v3 re-implementation[1]. It relies on Tensor-Flow framework and Resnet models and it is able to import original Caffe-based networks trained by He et al. [10].

DeepLabV3 itself is augmenting data set by random horizontal flipping, cropping and padding, and changes of scale. Additionally, we augmented the data set in each training epoch by random changes in gain, saturation, hue, brightness, gamma, and by random rotations.

We were exploiting behaviour of DeepLabV3 with three different pre-trained models ResNet50, ResNet101, and ResNet152. These models have 50, 101, and 152 layers, respectively.

5 Experimental Results

For training, we compiled our own data set (see Sect. 3). For testing, we used the BoWFire data set from [5]. Results from other methods as they were disclosed in the aforementioned paper are also included in our results table (see Table 3). For more complete comparison, we have also included results from [6] and [12] (some values are missing due to not being mentioned in the source papers). The CNNFire [12] seems to be the currently best method in the world. We include both the stated values and approximated values read from their graph.

Table 3. Comparision of fire segmentation methods on BoWFire data set [5]

Paper	True positive rate	False positive rate	Precision	Accuracy	F-Measure	True negative rate (non-fire)
Color [5] ❏	77.00%	5.00%	62.00%		68.00%	
BowFire [5] ■	65.00%	4.50%	80.00%		72.00%	
Rutz [5] ◆	41.00%	1.00%	84.00%		55.00%	
Celik YCbCr [X7] ◗	62.25%	4.51%	90.20%	78.78%	68.33%	83.65%
Chen RGB [X5]	41.78%	2.27%	92.18%	69.67%	51.51%	65.79%
Chen RGB + HSI [X13]	10.60%	10.63%	32.44%	50.92%	13.86%	90.99%
Philips [X2] ▼	64.95%	1.65%	97.66%	81.60%	73.34%	79.00%
de Souza 2016 [6] ▲	86.02%	5.87%	93.67%	90.07%	87.23%	93.70%
CNNFire [12]	97.00%		86.00%		91.00%	
CNNFire T = 0.40 [12] ★	82.10%	2.50%				
CNNFire T = 0.45 [12] ★	85.50%	4.80%				
CNNFire T = 0.50 [12] ★	89.00%	7.50%				
DLv3 /w resnet v2 152	87.42%	1.60%	73.27%	97.79%	79.14%	99.45%

---

[1] Deeplab v3 TensorFlow re-implementation - https://github.com/rishizek/tensorflow-deeplab-v3.git.

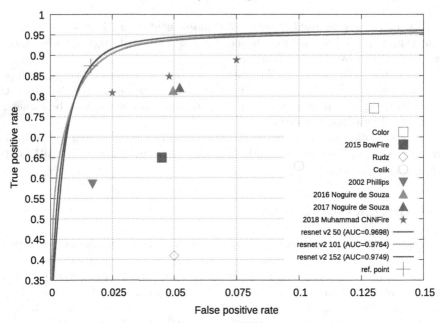

Fig. 4. Evaluation on BoWFire dataset [5]

' For comparison, the values of true positive rate, false positive rate, precision, accuracy, F-Measure, and true negative rate on non-fire data set are given in Table 3. Corresponding values in Fig. 4 and in Table 3 are marked with the same symbol.

True positive rate (TPR) against the false positive rate (FPR) for provided methods is plotted in Receiver operating characteristic space in Fig. 4. Our results are plotted as curve. Comparison with CNNFire shows 5–10 % improvement of TPR maintaining the same FPR.

Results achieved with different ResNet deepness are more less same. It suggests use of the fastest ResNet v2 50.

Examples of detection are shown at Fig. 5. You can see the input image, probability of fire class as colour map and also the colour map layered over the original image. The fourth column shows ground truth (fire in white and smoke in gray resp.). First and second sample comes from BowFire data set. Third and fourth image comes from flicker fire data set.

For evaluation, we also used our own data set and we have reached mean Intersection over Union (IoU) 70.51 % and Pixel Accuracy 98.78% on evaluation set. The details are summarized in Table 4.

Table 4. Evaluation on own data set

TPR	TNR	FPR	FNR	Precision	Accuracy	F-Measure	IoU
88%	0.8%	0.8%	12%	82%	99%	85%	70.51%

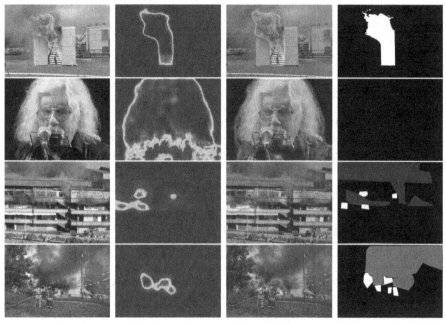

(a) Input image   (b) Fire probability   (c) Overlay on image   (d) Ground truth

Fig. 5. Examples of detection results of DeepLabV3 trained with our data set. First and second sample comes from BowFire data set. Second example is a hard negative. Third and fourth images come from flicker fire data set, these examples demonstrates annotations in our data set.

6   Conclusions

In our contribution, we focused on improving quality of fire detection by creating more robust and complex data set, which does not lack variability and also contains suitable negative set, and by exploiting state of the art semantic segmentation tools. The results, achieved by following aforementioned goals, are decisively outperforming the state of the art and set a higher standard.

To summarize the results in figures, we achieved 83.8% true-positive rate at 1.5% false positive rate and accuracy of 97.8% on BowFire data set. Of course, the "operational point" of the method can be adjusted to suit the application. We have achieved 88% True positive rate, 0.8% False positive rate and 99% Accuracy on our data set.

In our next efforts, we will try to further improve by applying semantic segmentation on videos (sequences of frames) with fire rather than single frames. We think this will be the way to eliminate even more false alarms while keeping the detection rate high.

Acknowledgements. The work presented in this paper was supported by the V3C - Visual Computing Competence Center, funded by the Technology Agency of the Czech Republic under the TE01020415 project.

References

1. Bedo, M.V.N., et al.: Techniques for effective and efficient fire detection from social media images. CoRR abs/1506.03844 (2015). http://arxiv.org/abs/1506.03844
2. Cazzolato, M.T., et al.: Fismo: a compilation of datasets from emergency situations for fire and smoke analysis. In: Proceedings of the Satellite Events (2017)
3. Çetin, A.E., et al.: Video fire detection – review. Digital Sig. Process. **23**(6), 1827 – 1843 (2013). https://doi.org/10.1016/j.dsp.2013.07.003, http://www.sciencedirect.com/science/article/pii/S1051200413001462
4. Chen, L., Papandreou, G., Schroff, F., Adam, H.: Rethinking atrous convolution for semantic image segmentation. CoRR abs/1706.05587 (2017). http://arxiv.org/abs/1706.05587
5. Chino, D.Y.T., Avalhais, L.P.S., Rodrigues, J.F., Traina, A.J.M.: Bowfire: detection of fire in still images by integrating pixel color and texture analysis. In: 2015 28th SIBGRAPI Conference on Graphics, Patterns and Images, pp. 95–102, August 2015. https://doi.org/10.1109/SIBGRAPI.2015.19
6. de Souza, B.M.N., Facon, J.: A fire color mapping-based segmentation: fire pixel segmentation approach. In: 2016 IEEE/ACS 13th International Conference of Computer Systems and Applications (AICCSA), pp. 1–8, November 2016. https://doi.org/10.1109/AICCSA.2016.7945741
7. de Souza, B.M.N., Facon, J., Menotti, D.: Colorness index strategy for pixel fire segmentation. In: 2017 International Joint Conference on Neural Networks (IJCNN), pp. 1057–1063, May 2017. https://doi.org/10.1109/IJCNN.2017.7965969
8. Dedeoglu, N., Toreyin, B.U., Gudukbay, U., Cetin, A.E.: Real-time fire and flame detection in video. In: Proceedings (ICASSP 2005). IEEE International Conference on Acoustics, Speech, and Signal Processing, vol. 2, pp. ii–669. IEEE (2005)
9. Grammalidis, N., Dimitropoulos, K., Cetin, E.: FIRESENSE database of videos for flame and smoke detection, July 2017. https://doi.org/10.5281/zenodo.836749, https://doi.org/10.5281/zenodo.836749
10. He, K., Zhang, X., Ren, S., Sun, J.: Deep residual learning for image recognition. CoRR abs/1512.03385 (2015). http://arxiv.org/abs/1512.03385
11. Lin, T., et al.: Microsoft COCO: common objects in context. CoRR abs/1405.0312 (2014). http://arxiv.org/abs/1405.0312
12. Muhammad, K., Ahmad, J., Lv, Z., Bellavista, P., Yang, P., Baik, S.W.: Efficient deep CNN-based fire detection and localization in video surveillance applications. IEEE Trans. Syst. Man Cybern.: Syst. 1–16 (2018). https://doi.org/10.1109/TSMC.2018.2830099
13. Russell, B.C., Torralba, A., Murphy, K.P., Freeman, W.T.: LabelMe: a database and web-based tool for image annotation. Int. J. Comput. Vis. **77**(1), 157–173 (2008). https://doi.org/10.1007/s11263-007-0090-8

14. Steffens, C.R., Botelho, S.S.D.C., Rodrigues, R.N.: A texture driven approach for visible spectrum fire detection on mobile robots. In: 2016 XIII Latin American Robotics Symposium and IV Brazilian Robotics Symposium (LARS/SBR), pp. 257–262, October 2016. https://doi.org/10.1109/LARS-SBR.2016.50

15. Xiao, J., Hays, J., Ehinger, K.A., Oliva, A., Torralba, A.: Sun database: large-scale scene recognition from abbey to zoo. In: 2010 IEEE Computer Society Conference on Computer Vision and Pattern Recognition, pp. 3485–3492, June 2010. https://doi.org/10.1109/CVPR.2010.5539970

16. Zhou, B., Zhao, H., Puig, X., Fidler, S., Barriuso, A., Torralba, A.: Scene parsing through ADE20K dataset. In: Proceedings of the IEEE Conference on Computer Vision and Pattern Recognition (2017)

Region Proposal Oriented Approach for Domain Adaptive Object Detection

Hiba Alqasir[(✉)], Damien Muselet, and Christophe Ducottet

Université de Lyon, UJM-Saint-Etienne, CNRS, IOGS, Laboratoire Hubert Curien
UMR5516, 42023 Saint-Etienne, France
h.alqasir@univ-st-etienne.fr

Abstract. Faster R-CNN has become a standard model in deep-learning
based object detection. However, in many cases, few annotations are
available for images in the application domain referred as the target
domain whereas full annotations are available for closely related public or synthetic datasets referred as source domains. Thus, a domain
adaptation is needed to be able to train a model performing well in
the target domain with few or no annotations in this target domain.
In this work, we address this domain adaptation problem in the context of object detection in the case where no annotations are available
in the target domain. Most existing approaches consider adaptation at
both global and instance level but without adapting the region proposal
sub-network leading to a residual domain shift. After a detailed analysis of the classical Faster R-CNN detector, we show that adapting the
region proposal sub-network is crucial and propose an original way to
do it. We run experiments in two different application contexts, namely
autonomous driving and ski-lift video surveillance, and show that our
adaptation scheme clearly outperforms the previous solution.

Keywords: Object detection · Domain adaptation · Deep learning ·
Faster R-CNN

1 Introduction

Object detection in images refers to the task of automatically finding all instances
of given object categories outputting, for each instance, a bounding box and
the object category. Recently, approaches based on deep Convolutional Neural
Networks (CNNs) have invaded the field thanks to both their efficiency and their
outstanding performances [17,18]. To address a given computer vision problem,
these methods require large training datasets with instance-level annotations.
However, for most real world applications, few annotations are available due
to the lack of image sources, copyright issues or annotation cost. To overcome
this problem, a current trend consists in training the network on a large public
annotated dataset (source domain), while adapting the network features to the

---

Partially funded by MIVAO, a french FUI project.

J. Blanc-Talon et al. (Eds.): ACIVS 2020, LNCS 12002, pp. 38–50, 2020.
https://doi.org/10.1007/978-3-030-40605-9_4

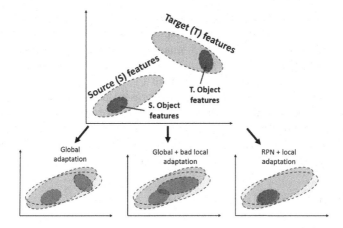

Fig. 1. Illustration of global, local and RPN adaptation (see text for details).

tested dataset (target domain). This approach is called domain adaptation [14, 23]. If no annotations are available in the target domain, the domain adaptation is referred as unsupervised.

In this context, the case of autonomous driving has been extensively addressed and a variety of datasets exists covering different urban scenes situations, illumination and weather conditions [4,6]. In this paper, we are particularly interested in unsupervised domain adaptation in a ski lift video surveillance scenario. The purpose is to detect dangerous situations during chairlift boarding by detecting relevant objects of the scene (e.g. safety bar, people, chairlift carrier). Instance level annotations are available for a reference chairlift where the chair model, the perspective view or boarding system may be different from the target chairlift [2].

Surprisingly, few works explicitly address the problem of unsupervised domain adaptation for object detection. Most approaches study the supervised case basically by fine-tuning a model pre-trained on the source dataset with few annotated images from the target domain, eventually freezing some layers to concentrate the training on the last layers [8,22]. Other recent approaches try to reduce the domain shift by transforming the source domain to make it close to the target one using style transfer [12]. The most significant contribution of domain adaptive object detection was proposed by [3]. Following [5], they added adversarial training components in the classical Faster R-CNN detector, in order to adapt both globally and locally the detector. Given the features from the two domains and considering the subset of features specific to the object (Fig. 1), a global adaptation, as illustrated in the bottom left of Fig. 1, may not match source and target object features. Thus, Chen et al. [3] also propose to adapt the features pulled from the regions returned by the Region Proposal Network (RPN). We argue that, since the RPN is trained on the source domain, the proposals from the target images may be wrongly detected and the local features used for the adaptation may be outside the target object features set (bottom center in Fig. 1). In this paper, we propose to adapt the RPN in order to ensure the features extracted from the target images to overlap with the source object features. A local adaptation through adversarial learning will thus better align source and domain features (bottom right in Fig. 1).

Our contributions are threefold: (1) We present a new viewpoint about the domain shift problem in object detection. (2) We propose to adapt the RPN as a global feature adaptation and integrate this new adaptation module in Faster R-CNN. (3) We run extensive experiments in two different applications contexts: autonomous driving and ski lift video surveillance.

2   Related Work

Object Detection. The first approaches proposed in the context of CNN were based on the region pooling principle [8,21]. In R-CNN [8], candidate regions detected by selective search were represented by a subset of pooled features and evaluated by an instance classifier. This two-stages principle was further refined in Faster R-CNN [18] with a common CNN backbone to extract the whole image features and two different sub-pipelines: the first one called RPN to generate proposals of regions which are likely to contain objects and the second one which is basically a classification and regression network aiming to refine the location and size of the object and to find its class. Besides these two-stages approaches, one-stage approaches directly predict box location, size and class in a single pipeline either by using anchor boxes with different aspect ratios [13] or by solving a regression problem on the feature grid [17]. Interested readers can refer to the review of recent advances in object detection in [1]. Since Faster R-CNN [18] provides very accurate results and has been largely studied, we propose to consider this network as a baseline in this paper.

Domain Adaptation. Unsupervised domain adaptation is needed when we want to learn a predictor in a target domain without any annotated training samples in this domain [14,23]. Obviously, annotations are available in a source domain which is supposed to be close to the target one. Two main types of methods have been proposed in this context. The first one is to try to match the feature distribution in the source and target domains either by finding a transformation between the domains [15] or by directly adapting the features [10]. One noticeable example is the gradient reversal layer approach proposed by Ganin et al. [5] that attempts to match source and target feature distributions. They propose to jointly optimize the class predictor and the source-target domain disparity by back-propagation. The second type of methods relies on Generative Adversarial Networks (GANs) [11]. The principle is to generate annotated synthetic target images from the source images and to learn (or fine-tune) the network on these synthetic target data [12].

Domain Adaptation for Object Detection. Few works consider domain adaptation for object detection particularly in the unsupervised setting. [16]

proposes class-specific subspace alignment to adapt RCNN [8] and [3] uses adversarial training inspired by [5] to adjust features at two different levels of a Faster R-CNN architecture. The adaptation at image level intends to eliminate the domain distribution discrepancy at the output of the backbone network while the instance level adaptation concerns the features which are pooled from a Region of Interest (RoI), before the final category classifiers. Following the same adversarial training approach, Saito et al. [19] argue that a global matching may hurt performance for large domain shifts. They thus propose to combine a strong alignment of local features and a weak alignment of global ones. To the best of our knowledge, none of the previous works considers the adaptation of the region proposal sub network of Faster R-CNN. They are then sensitive to any shift in the distribution of object bounding boxes between source and target domains.

In this work, we propose to incorporate two adversarial domain adaptation modules in Faster R-CNN: the first one at RPN-level to address the source-target domain shift of features of the region proposal module and the second one at instance-level to adapt the RoI-pooled features used in the final classification module.

3   Our Approach

In order to explain our adaptation scheme, we have to explain in details the workflow of Faster R-CNN [7], summarized in Fig. 2. Then, we present our approach to adapt this detector between different domains.

3.1   Faster R-CNN

Faster R-CNN is basically composed of two convolutional blocks called C_1 and C_2, providing two feature maps F_1 and F_2, respectively (cf. Fig. 2). Based on F_2, the RPN predicts a set of box positions used to crop the F_1 feature map using the RoI pooling layer (called RP layer, hereafter).

It is worth mentioning that the gradient can not be back-propagated through the RP layer towards the RPN, because this step is not differentiable. The authors of Faster R-CNN resort to an alternating training to cope with this problem [7]. It is crucial to understand this point when one wants to apply domain adaptation to Faster R-CNN. It means that we can not just plug a domain adaptation module after the last layers of Faster R-CNN (namely F_{3i}) and adapt in one shot the classification layers and the convolution blocks C_1 and C_2.

Back to the workflow of Faster R-CNN, the outputs F_{1i}, $i = 1, ..., N_p$, of the RP layer are cropped and resized parts of the F_1 feature map. N_p is the number of proposals returned by the RPN. The feature maps F_{1i} are then sent to shared fully connected layers FC_3 whose outputs F_{3i} are used to take the final decision of class and location.

From this workflow, we note that the classification and regression layers take as inputs either F_2 or F_{3i}, which are the key feature maps of the detector. In

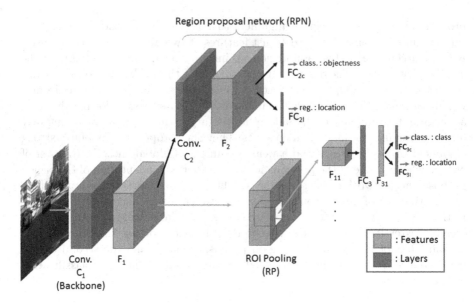

Fig. 2. Faster R-CNN Workflow.

the next section, we present how these feature maps can be adapted between the two domains.

3.2 Adapting Faster R-CNN

Let us consider a source domain $\mathcal{S}$ with $N_\mathcal{S}$ images $\{I_i^\mathcal{S}\}$, $i = 1, ..., N_\mathcal{S}$, each containing $n_i^\mathcal{S}$ objects, located at the positions $l_{ij}^\mathcal{S}$ and associated with the classes $c_{ij}^\mathcal{S}$, $j = 1, ..., n_i^\mathcal{S}$. Likewise, we denote $\mathcal{T}$ a target domain constituted of $N_\mathcal{T}$ target images $\{I_i^\mathcal{T}\}$, $i = 1, ..., N_\mathcal{T}$, each containing $n_i^\mathcal{T}$ objects, located at the positions $l_{ij}^\mathcal{T}$ and associated with the classes $c_{ij}^\mathcal{T}$, $j = 1, ..., n_i^\mathcal{T}$.

If the two domains are different (cameras, viewpoints, weather conditions, ...), there exists a domain shift between the joint distributions $P(I^\mathcal{S}, l^\mathcal{S}, c^\mathcal{S})$ and $P(I^\mathcal{T}, l^\mathcal{T}, c^\mathcal{T})$. In this case, we can not train the detector on the source data and obtain good results on the target data, without adaptation. The aim of domain adaptation is to decrease this distribution discrepancy so that $P(I^\mathcal{S}, l^\mathcal{S}, c^\mathcal{S}) \approx P(I^\mathcal{T}, l^\mathcal{T}, c^\mathcal{T})$. In the context of unsupervised domain adaptation, the labels (locations and classes) of the target data are not available and this is not an easy task to decrease the joint distribution discrepancy. By applying the Bayes' rule on the joint distribution, we obtain, for the source domain:

$$P(I^\mathcal{S}, l^\mathcal{S}, c^\mathcal{S}) = P(l^\mathcal{S}, c^\mathcal{S} | I^\mathcal{S}) P(I^\mathcal{S}) \tag{1}$$

Most of the domain adaptation approaches assume a covariate shift, which means that the shift between the source and target joint distributions is caused by the marginal distributions $P(I)$, while the conditional distributions $P(l, c|I)$ are constant across domains, i.e. $P(l^S, c^S|I^S) = P(l^T, c^T|I^T)$. Under this assumption, in order to decrease the joint distribution discrepancy, we have just to decrease the marginal distribution shift, so that $P(I^S) \approx P(I^T)$. In order to change the marginal distributions of the images, the classical approaches apply a transform T on the image features, so that $P(T(I^S)) \approx P(T(I^T))$. Usually, the transform T is a part of a convolution neural network.

In this paper, we propose to consider and adapt different feature maps extracted from the images. By looking at Fig. 2, we note that two feature maps are used as input for classification and regression layers, namely the F_2 feature map and the F_{3i} feature vectors. So, in order to adapt the detector to the source domain, we have to adapt the marginal distributions of F_2 and F_{3i}, so that $P(F_2^S) \approx P(F_2^T)$ and $P(F_{3i}^S) \approx P(F_{3i}^T)$.

In order to enforce these distributions to be closer, we propose to resort to an adversarial domain adaptation approach [5] called GRL for gradient reversal layer. Note that any other adversarial domain adaptation algorithms could have been used, we just use this one for a fair comparison with DA-Faster [3]. When plugged on a feature map F_k, the idea of GRL is to minimize the discrepancy between the feature distributions over the source and target domains $P(F_k^S)$ and $P(F_k^T)$ [5]. If the GRL is able to perfectly overlap these two distributions, we can conclude that the features extracted at this point of the network (F_k) are domain invariant and so can be applied either on the source or target domain with equivalent accuracies.

From the previous analysis, it is obvious that two GRL modules should be inserted in the detector: one after the feature map F_2 and one after the feature vector F_{3i}. It is worth mentioning that, when we plug a GRL module to a feature map, we back-propagate the (reverse-)gradient until the first layer of the C_1 convolutional block. Thus, the main advantage of our approach is that the reversal gradients are back-propagated through all the layers of the detector. Consequently, the backbone, the RPN and the local features are all adapted (see Fig. 3).

Formally, at training time, the total loss corresponding to a given training image $I_k \in I^S \cup I^T$ from domain $d_k \in \{S, T\}$ is given by:

$$L = L_{Fst} - \lambda \sum_{i,j} L_H \left(FC_{2a}(F_2^{i,j}(I_k)), d_k \right) - \lambda \sum_{i=1}^{N_p} L_H \left(FC_{3a}(F_{3i}(I_k)), d_k \right) \quad (2)$$

where L_{Fst} denotes the original Faster R-CNN loss activated only if $I_k \in I^S$, L_H denotes the cross-entropy loss, λ denotes the trade-off parameter to balance Faster R-CNN loss and domain adaptation losses, FC_{2a} and FC_{3a} denote the fully connected predictors for domain adaptation, $F_2^{i,j}(I_k)$ denotes the feature vector at location (i, j) of feature map F_2 for image I_k, and $F_{3i}(I_k)$ denotes the feature vector corresponding to the proposal region i of image I_k.

We note that the recent domain adaptive detection approaches [3,19] have not tried to adapt the RPN layer and we think that this is a strong weakness of

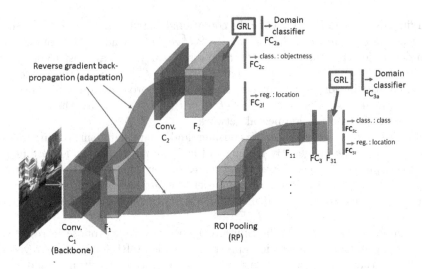

Fig. 3. Our domain adaptation for Faster R-CNN. See text for details.

these approaches. Indeed, as mentioned in [3] (called DA-Faster hereafter), the image-level adaptation is enforcing the F_1 target and source feature distributions to be closer but it is very hard to perfectly align them. This is one of the reasons why DA-Faster approach also applies instance level adaptation. But, it is clear in Fig. 2, that if F_1 features are not well adapted between the domains, the output of the RPN will also be different between the domains and consequently, the locations where the boxes F_{1i} are cropped from F_1 will be domain dependent. Therefore, the instance-level adaptation on F_{3i} features will not help to adapt the object detector between domains, since it will work on local features which are not equivalent between the domains (see Fig. 1).

4   Experiments

4.1   Experiment Setup

To evaluate the efficiency of our approach we conduct experiments in two contexts: autonomous driving and video surveillance of ski lifts. In each case, we train on a source dataset and test on a target dataset from a different domain. During training, likewise the other domain adaptive approaches, we use also images from the target domain, but without any label, while the source dataset images are provided with their bounding boxes instance annotations.

The baseline is Faster R-CNN model trained only on the source dataset. As mentioned earlier, our solution is inspired from DA-Faster [3] but our contribution is in the analysis of the domain shift in Faster R-CNN, conducting to the solution that the domain adaptation module (GRL) should be plugged at the RPN level. Consequently, the aim of these experiments is to compare DA-Faster with our approach in order to check the validity of our contribution in practice.

Thus, adapting the RPN in other solutions such as [19], or using other adaptation modules than GRL such as [14,23] might have provided better results, but it is out of the scope of this paper. Thus, for all the experiments, we compare our approach with DA-Faster [3].

As mentioned in [19], the results provided by the authors of DA-Faster are unstable and Saito et al. proposed to re-implement their own code for DA-Faster, conducting to lower results than the original paper [3]. So likewise [19], we report the results of DA-Faster with the implementation provided by [9] with the same hyper-parameters as our solution (results denoted *DA-Faster* hereafter), as well as the results provided by the original paper [3] (denoted *DA-Faster**), when available on the considered dataset.

To evaluate object detection we report the mean Average Precision (mAP) with intersection over union (IoU) threshold at 0.5 (denoted AP50), the mAP with IoU threshold of 0.75 (AP75) and the mAP averaged over multiple IoU from 0.5 to 0.95 with a step size of 0.05 (APcoco). The network is trained in an end-to-end manner using back-propagation and the stochastic gradient descent (SGD) algorithm. As a standard practice, Faster R-CNN backbone is initialized with pre-trained weights on ImageNet classification. We use a learning rate of 0.001 for 50k iterations, and 0.0001 for the next 20k iterations. Each iteration has 2 mini-batches, one from source domain and the other from target domain. The trade-off parameter λ to balance Faster R-CNN loss and domain adaptation loss is set to 0.1 as in [3]. We use a momentum of 0.9 and a weight decay of 0.0005.

4.2 Autonomous Driving

In this context we evaluate the domain adaptive detectors for two domain shifts: weather conditions (foggy and not foggy) and acquisition conditions (different cameras, different viewpoints and different scenes).

Fig. 4. One image from each dataset: the Cityscapes dataset (left), its foggy version (center) and the KITTI dataset (right).

Cityscapes → Foggy Cityscapes. In the first experiment we use the Cityscapes dataset [4] as source domain. It is a urban scene dataset with 2975 training images and 500 validation images. The 1525 unlabeled images are not considered. For training the network, we are using the 2975 train images and do not consider the validation images. There are 8 categories with instance annotations in this dataset, namely *person, rider, car, truck, bus, train, motorcycle* and *bicycle*. The target domain is the Foggy Cityscapes [20] dataset generated by applying fog synthesis on the Cityscapes dataset to simulate fog on real scenes (see Fig. 4). Thus, the number of images and labels are exactly the same as for Cityscapes dataset. For testing the detection, we are using the 500 validation images from Foggy Cityscapes. The results are summarized in Table 1. First, we can note that, without domain adaptation, the results of Faster R-CNN are very bad, underlying the strong need of adapting the network in case of weather condition variations. Thus, DA-Faster improves the results over Faster R-CNN, but we note that our approach clearly outperforms DA-Faster on this dataset, showing that the RPN adaptation helps in adapting the detector in case of weather condition variations.

Table 1. Detection results on Foggy Cityscapes (trained on Cityscapes dataset). The AP50 is reported for each class as well as the average APcoco, AP50 and AP75 over all classes.

	person	rider	car	truck	bus	train	mcycle	bicycle	APcoco	AP50	AP75
Faster R-CNN	18.8	20.5	24.2	17.0	8.0	6.2	7.2	5.0	06.20	13.35	05.42
DA Faster R-CNN	27.3	35.7	44.1	20.3	35.2	8.9	16.2	23.6	12.28	26.41	10.02
DA Faster R-CNN*	25.0	31.0	40.5	22.1	35.3	20.2	**20.0**	27.1	–	27.6	–
Ours	**27.8**	**35.8**	**45.1**	**23.5**	**42.1**	**26.1**	18.0	**27.6**	**13.70**	**30.47**	**11.04**

Cityscapes → KITTI. In this experiment Cityscapes is the source domain, and KITTI [6] is the target domain (see Fig. 4). KITTI is a benchmark for autonomous driving which consists of 7481 training images. Since the test set is not annotated we use all the training images with their annotations at test time to evaluate the performance. Only one category (*car*) is annotated in KITTI, so we consider this single class for evaluation. The results are summarized in Table 2. Once again, we note that the domain adaptation helps improving Faster R-CNN results. We see also that our approach outperforms DA-Faster for all the criteria when using the same hyper-parameters. The results provided in [3] are better than ours for *AP*50, but note that the implementation and hyper parameters are different from our tests. The comparison is therefore not fair.

Table 2. Detection results in KITTI training set (trained in Cityscapes dataset) for one class (Car) detection.

	APcoco	AP50	AP75
Faster R-CNN	26.73	58.60	21.54
DA Faster R-CNN	27.51	60.38	22.67
DA Faster R-CNN*	–	**64.1**	–
Ours	**28.39**	61.32	**23.59**

4.3   Video Surveillance of Ski Lifts

The MIVAO research project was launched in collaboration with a french start-up Bluecime, based on the needs of ski lift operators to secure chairlifts. MIVAO aims to develop a computer vision system that acquires images from the boarding station of chairlifts, analyzes the important elements (people, chairlift carrier, safety bar, ...) and triggers an alarm in case of dangerous situations. In this paper, we tackle this problem as an object detection task trying to detect the safety bar in the image, considering that it has to be closed when the chairlift leaves the boarding station. Across the ski resorts, the viewpoint, the background, the carrier geometry and the camera may be different and domain adaptive detectors are required to install new systems without a fastidious and time-consuming step of manual annotation.

Chairlift Dataset. For this experiment, we have created a dataset with images from two different chairlifts, called hereafter chairlift 1 and chairlift 2. The dataset contains 3864 images from chairlift 1 and 4260 images from chairlift 2. Example images are provided in Fig. 5. We can note that the main differences between the two chairlifts are in the viewpoints which are slightly different and in the presence of a cluttered background in the chairlift 2.

Fig. 5. Example images from our chairlift dataset. The two left images are from chairlift 1 and the two right images are from chairlift 2. The box annotations (open: red and close: green) are provided for the two right images, for illustration. (Color figure online)

The images are centered on the chairlift and manually labeled with the position and the dimensions of the bounding box containing the safety bar. From this information, we have created instance annotations with two categories: open safety bar and close safety bar, as illustrated on the two right images from Fig. 5.

Evaluation. The results are provided in Table 3. By training the baseline Faster R-CNN using images from one chairlift and test it on images from another chairlift, the results were surprisingly very good in terms of AP50. This can be explained by the important size of the ground truth bounding boxes that have high chance to well overlap random bounding boxes with similar dimensions. Obviously, when looking at the more demanding criteria such as *APcoco* or *AP*75, the need of domain adaptation is evident for precise object detection. The results show that the two domain adaptive detectors (DA-Faster and ours) are equivalent for the adaptation from chairlift 1 to chairlift 2, but they also show that our adaptation is much better than DA-Faster for the adaptation from chairlift 2 to chairlift 1. It is difficult to explain why DA-Faster is less accurate in one direction ($ch2 \rightarrow ch1$) than in the other direction ($ch1 \rightarrow ch2$). One assumption could be that in DA-Faster, the RPN is better trained on chairlift 1 since in this case the background is less cluttered. Thus, when applying it on chairlift 2, the adaptation process tends to promote features from the foreground and both the proposal and the classification are good. On the contrary, if the RPN is trained on chairlift 2, it will rely on cluttered features which are removed with the global adaptation and thus, for DA-Faster, the proposals will be bad on chairlift 1, leading to an important residual domain shift in the results. On the contrary, in our method, since the RPN is directly adapted, the residual shift is lower (see Fig. 1 and the related explanation in Sect. 1).

Table 3. Detection results on the chairlift dataset. First, adaptation from chairlift 1 to chairlift 2, and second adaptation from chairlift 2 to chairlift 1.

	$ch1 \rightarrow ch2$			$ch2 \rightarrow ch1$		
	APcoco	AP50	AP75	AP	AP50	AP75
Faster R-CNN	30.34	99.49	0.30	36.56	98.98	9.86
DA Faster R-CNN	50.51	99.50	**33.4**	42.56	98.99	11.1
Ours	**50.93**	**99.99**	30.7	**48.83**	**99.00**	**45.6**

5   Conclusion

In this paper, we have tackled the problem of domain adaptation for object detection. After a detailed analysis of the complete workflow of the classical Faster R-CNN detector, we have proposed to adapt the features pulled from this network at two different levels: one adaptation at a global level in the Region Proposal Network and one adaptation at the local level for each bounding box returned by the RPN. We have shown that these two adaptations are complementary and provide very good detection results. We have tested our solution on two different applications, namely the autonomous driving and the chairlift security. As future works, we propose to test more accurate adaptation procedures such as the approaches presented in [14,23]. These methods could help in the learning step to reach stable solutions which is a strong weakness of the domain adaptive Faster R-CNN. Furthermore, it could be interesting to adapt the features at different depth of the network as recommended by [19].

References

1. Agarwal, S., Terrail, J.O.D., Jurie, F.: Recent advances in object detection in the age of deep convolutional neural networks. arXiv preprint arXiv:1809.03193 (2018)

2. Bascol, K., Emonet, R., Fromont, E., Debusschere, R.: Improving chairlift security with deep learning. In: Adams, N., Tucker, A., Weston, D. (eds.) IDA 2017. LNCS, vol. 10584, pp. 1–13. Springer, Cham (2017). https://doi.org/10.1007/978-3-319-68765-0_1

3. Chen, Y., Li, W., Sakaridis, C., Dai, D., Van Gool, L.: Domain adaptive faster R-CNN for object detection in the wild. In: Proceedings of the IEEE Conference on Computer Vision and Pattern Recognition, pp. 3339–3348 (2018)

4. Cordts, M., et al.: The cityscapes dataset for semantic urban scene understanding. In: Proceedings of the IEEE Conference on Computer Vision and Pattern Recognition, pp. 3213–3223 (2016)

5. Ganin, Y., Lempitsky, V.: Unsupervised domain adaptation by backpropagation. arXiv preprint arXiv:1409.7495 (2014)

6. Geiger, A., Lenz, P., Stiller, C., Urtasun, R.: Vision meets robotics: the KITTI dataset. Int. J. Robot. Res. **32**(11), 1231–1237 (2013)

7. Girshick, R.: Fast R-CNN. In: Proceedings of the IEEE International Conference on Computer Vision, pp. 1440–1448 (2015)

8. Girshick, R., Donahue, J., Darrell, T., Malik, J.: Rich feature hierarchies for accurate object detection and semantic segmentation. In: Proceedings of the IEEE Conference on Computer Vision and Pattern Recognition, pp. 580–587 (2014)

9. Girshick, R., Radosavovic, I., Gkioxari, G., Dollár, P., He, K.: Detectron (2018). https://github.com/facebookresearch/detectron

10. Glorot, X., Bordes, A., Bengio, Y.: Domain adaptation for large-scale sentiment classification: a deep learning approach. In: Proceedings of the 28th International Conference on Machine Learning (ICML 2011), pp. 513–520 (2011)

11. Goodfellow, I., et al.: Generative adversarial nets. In: Advances in Neural Information Processing Systems, pp. 2672–2680 (2014)

12. Inoue, N., Furuta, R., Yamasaki, T., Aizawa, K.: Cross-domain weakly-supervised object detection through progressive domain adaptation. In: Proceedings of the IEEE Conference on Computer Vision and Pattern Recognition, pp. 5001–5009 (2018)

13. Liu, W., et al.: SSD: single shot multibox detector. In: Leibe, B., Matas, J., Sebe, N., Welling, M. (eds.) ECCV 2016. LNCS, vol. 9905, pp. 21–37. Springer, Cham (2016). https://doi.org/10.1007/978-3-319-46448-0_2

14. Long, M., Cao, Z., Wang, J., Jordan, M.I.: Conditional adversarial domain adaptation. In: Advances in Neural Information Processing Systems, pp. 1640–1650 (2018)

15. Pan, S.J., Tsang, I.W., Kwok, J.T., Yang, Q.: Domain adaptation via transfer component analysis. IEEE Trans. Neural Networks **22**(2), 199–210 (2010)

16. Raj, A., Namboodiri, V.P., Tuytelaars, T.: Subspace alignment based domain adaptation for RCNN detector. arXiv preprint arXiv:1507.05578 (2015)

17. Redmon, J., Divvala, S., Girshick, R., Farhadi, A.: You only look once: unified, real-time object detection. In: Proceedings of the IEEE Conference on Computer Vision and Pattern Recognition, pp. 779–788 (2016)

18. Ren, S., He, K., Girshick, R., Sun, J.: Faster R-CNN: towards real-time object detection with region proposal networks. In: Advances in Neural Information Processing Systems, pp. 91–99 (2015)

19. Saito, K., Ushiku, Y., Harada, T., Saenko, K.: Strong-weak distribution alignment for adaptive object detection. In: Proceedings of the IEEE Conference on Computer Vision and Pattern Recognition (CVPR) (2019)

20. Sakaridis, C., Dai, D., Van Gool, L.: Semantic foggy scene understanding with synthetic data. Int. J. Comput. Vision **126**(9), 973–992 (2018)

21. Sermanet, P., Eigen, D., Zhang, X., Mathieu, M., Fergus, R., LeCun, Y.: OverFeat: integrated recognition, localization and detection using convolutional networks. arXiv preprint arXiv:1312.6229 (2013)
22. Singh, B., Davis, L.S.: An analysis of scale invariance in object detection snip. In: Proceedings of the IEEE Conference on Computer Vision and Pattern Recognition, pp. 3578–3587 (2018)
23. Tzeng, E., Hoffman, J., Saenko, K., Darrell, T.: Adversarial discriminative domain adaptation. In: Proceedings of the IEEE Conference on Computer Vision and Pattern Recognition, pp. 7167–7176 (2017)

Deep Convolutional Network-Based Framework for Melanoma Lesion Detection and Segmentation

Adekanmi Adegun and Serestina Viriri[(✉)]

School of Mathematics, Statistics and Computer Science,
University of KwaZulu-Natal, Durban 4000, South Africa
{218082884,viriris}@ukzn.ac.za

Abstract. Analysis of skin lesion images is very crucial in melanoma detection. Melanoma is a form of skin cancer with high mortality rate. Both semi and fully automated systems have been proposed in the recent past for analysis of skin lesions and detection of melanoma. These systems have however been restricted in performance due to the complex visual characteristics of the skin lesions. Skin lesions images are characterised with fuzzy borders, low contrast between lesions and the background, variability in size and resolution and with possible presence of noise and artefacts. In this work, an efficient deep learning framework has been proposed for melanoma lesion detection and segmentation. The proposed method performs pixel-wise classification of skin lesion images to identify melanoma pixels. The framework employs an end-to-end and pixel by pixels learning approach using Deep Convolutional Networks with softmax classifier. A novel framework which learns the complex visual characteristics of skin lesions via an encoder and decoder subnetworks that are connected through a series of skip pathways that brings the semantic level of the encoder feature maps closer to that of the decoder feature maps is hereby designed. This efficiently handles multi-size, multi-resolution and noisy skin lesion images. The proposed system was evaluated on both the ISBI 2018 and PH2 skin lesion datasets.

Keywords: Deep learning · Deep convolution network · Skin lesion · Segmentation · Softmax classifier · Melanoma

1 Introduction

Melanoma skin cancer is a rapidly increasing skin cancer incidence with a very high mortality rate [1]. This type of cancer can be easily treated and cured if detected and diagnosed early [2]. Manual screening of skin lesions for detecting melanoma is cumbersome with inaccurate results. Many Computer Aided Diagnosis (CAD) techniques have been proposed in the recent past for the analysis and segmentation of skin lesions [3]. Both fully and semi-automatic methods have been however limited in performance due to the granulate and complex

© Springer Nature Switzerland AG 2020
J. Blanc-Talon et al. (Eds.): ACIVS 2020, LNCS 12002, pp. 51–62, 2020.
https://doi.org/10.1007/978-3-030-40605-9_5

visual appearance of skin lesions images [3]. Skin lesions possess features such as colour variation, multi-size, multi-resolution, low contrast with background, fuzzy borders and inhomogeneous textures. They are also characterised with artefacts presence such as hair, oil, bubbles, air etc.

Computer vision approaches such as segmentation techniques have been recently utilized for detection and diagnosis of melanoma skin cancer [4]. This work explores some recently developed deep learning techniques for the segmentation and possible detection of melanoma lesions from dermoscopy images. The performance of these techniques also experiences some restrictions due to the [5] distinctive features of the skin lesions. In this work, a deep learning framework that tends to overcome these limitations has been proposed for the detection and segmentation of melanoma lesions. The proposed framework employs an end-to-end training of skin lesion images with melanoma disease labels [6] and then, performs pixel by pixel classification of the images using the Deep Convolutional Network.

This work devises a novel Deep Convolutional Network-based framework which learns the complex visual characteristics of skin lesions via an encoder and decoder sub-networks that are connected through a series of skip pathway for easy transmission. Original UNet model has been redesigned though the introduction of series of skip pathways for connecting the encoder section with the decoder section. Only short skip connection was used in UNET [7]. The system is able to overcome the challenges with complex visual characteristics of the skin lesions through learning general visual characteristics by the encoder networks and the lesion boundaries details through the decoder networks. The system performs pixel-wise classification using the softmax classifier. The proposed network utilizes smaller number of trainable parameters such as employing 3×3 filter size and limiting the encoder-decoder level to five to reduce computational resources and time for processing the lesions images. The system was evaluated on publicly available skin lesions image dataset of ISBI 2018 and PH2 with performance metrics such as Accuracy, Confusion matrix, Dice Coefficient, Sensitivity and Specificity. Each pixel of the skin lesion image can then be presented as either True positives (TP), True negatives (TN), False positives (FP) and False negatives (FN). Our Contributions can be summarized as follows:

- We propose a novel Deep Convolutional Network-based framework which learns the complex visual characteristics of skin lesions via an encoder and decoder sub-networks that are connected through a series of skip pathway for efficient skin lesion segmentation. The encoder sub-network learn the coarse appearance and localization information of the lesion images while the decoder sub-networks learn the lesion boundaries information. The skip pathways brings the semantic level of the encoder feature maps closer to that of the decoder feature maps for easy transmission and improved processing speed.

- We propose pixel-wise classification of skin lesions images for quick and easy melanoma detection via softmax function.

- We compare our results with the existing methods. The results show that our method is effective for segmenting challenging melanoma lesions with fuzzy boundaries and heterogeneous textures.

2   Related Works

Machine learning techniques have been applied for various tasks in the past [8]. Recently, machine learning techniques most especially deep learning methods have been used to perform segmentation and classification of medical images. Specifically deep learning approach have been used in the segmentation and classification of skin lesion images. Bi et al. [9] proposed a method for automated skin lesion segmentation and melanoma detection by introducing a probability-based step-wise integration to combine complementary segmentation results derived from individual class-specific learning models. The system achieved an average Dice coefficient of 85.66% on the ISBI 2017 Skin Lesion Challenge (SLC), 91.77% on the ISBI 2016 SLC and 92.10% on the PH2 datasets with corresponding Jaccard indices of 77.73%, 85.92% and 85.90%, respectively. The complexity level of the system architecture is still higher.

A segmentation methodology through full resolution convolutional networks (FrCN) was proposed by Al-masni et al. [10]. The proposed method directly learns the full resolution features of each individual pixel of the input data without the need for pre or post-processing operations such as artefact removal, low contrast adjustment, or further enhancement of the segmented skin lesion boundaries. The system was evaluated using two publicly available databases, the IEEE International Symposium on Biomedical Imaging (ISBI) 2017 Challenge and PH2 datasets. The proposed FrCN achieved a segmentation accuracy of 95.62% for benign cases, 90.78% for the melanoma cases, and 91.29% for the seborrheic keratosis cases in the ISBI 2017. Bi et al. [11] proposed leveraging fully convolutional networks (FCNs) to automatically segment the skin lesions.

He et al. [12] proposed a skin lesion segmentation network using a very deep dense deconvolution network based on dermoscopic images. The deep dense layer and generic multi-path Deep RefineNet were combined to improve the segmentation performance. The deep representation of all available layers was aggregated to form the global feature maps using skip connection. The dense deconvolution layer is leveraged to capture diverse appearance features via the contextual information and to smooth segmentation maps for final high-resolution output. Bi et al. [13] proposed semi-automated skin lesion segmentation method that incorporates fully convolutional networks (FCNs) with multi-scale integration to overcome problems with over- or under-segmentation with challenging skin lesions such as when a lesion is partially connected to the background or when image contrast is low. Esteva et al. [14] developed CNN architecture using GoogleNet Inception v3 that was pre-trained on approximately 1.28 million images for melanoma detection. The system was pre-trained on a suitable dataset using a set of more than 129,450 high quality skin images. The system was deconstructed down to the pixel level and trained with pattern and its diagnosis [14].

Goyal et al. [15] proposed an end-to-end solution using fully convolutional networks (FCNs) for multi-class semantic segmentation. The system automatically segmented the melanoma into keratoses and benign lesions. Ramachandram et al. [16] proposed an approach based on a Fully Convolutional Neural Network architecture which was trained end to end, from scratch, on a small dataset. The semantic segmentation architecture utilized combined the use of atrous convolutions to increase the effective field of view of the network's receptive field without increasing the number of parameters, network-in-network convolution layers to increase network capacity and super-resolution upsampling of predictions using subpixel. A deep learning framework consisting of two fully-convolutional residual networks (FCRN) was proposed to simultaneously produce the segmentation result and the coarse classification result of skin lesion [17]. A lesion index calculation unit (LICU) was then developed to refine the coarse classification results by calculating the distance heat-map. A straight-forward CNN was also proposed for the dermoscopic feature extraction task between melanoma and non-melanoma lesions. Lastly Ramachandram et al. [18] presented an image segmentation method based on deep hyper-column descriptors for the segmentation of several classes of benign and malignant skin lesions. The system focuses on the task of accurately segmenting benign and malignant skin lesions in dermoscopic images through a means of lesion quantification.

Our proposed system aims at lowering trainable parameters to reduce computational resources and time and make the system feasible for real-time medical diagnosis. Most of the systems discussed above employ larger and more complex deep learning architecture. Our proposed system is able to perform both segmentation and pixel-wise classification of melanoma lesion pixels using a moderate-size deep convolutional network.

3 Methodology

Our methodology employs a pre-processing step that crops, resize and re-samples images to ensure that both the training images and the ground truth conform to the same resolution and size. The processed images are then sent into Deep Convolutional network. The architectural diagram of Deep Convolutional Network and the whole framework of the system is described in Figs. 1 and 2 respectively. Our architecture is a deep-supervised encoder-decoder network where the encoder and decoder sub-networks are connected through a series of skip pathways which are dense convolution block with two number of convolution layers. The dense convolution block brings the semantic level of the encoder feature maps closer to that of the decoder feature maps. This increases the processing speed. The whole process of learning and training is accomplished in end-to-end and pixels by pixels. Each pixel from the training images is assigned with a pixel from the ground truth labels. The framework also ensures less train-able weights through the application of the pixel-wise classification with the reduced number of feature maps in the last convolutional block and limited to five numbers of the encoding-decoding levels. This ensures a medium-size with lower cost in terms of memory and computation time.

3.1 Dataset

Data Pre-processing. The skin lesions images acquired for processing are always characterised with variation in size and multi-scale and multi-resolution nature features of the skin lesion images. The first task in the pre-processing stage is to have all images in the same scale and resolution via cropping, resizing and re-sampling. In this work, we use relatively small image size of 160×224 as inputs as this affects the input feature map size. This size has been chosen to avoid image deformations and for better presentation. The images are then automatically normalized by computing the mean pixel value the standard deviation for data normalization.

Image Augmentation. The system applied elastic deformations through random displacements before augmenting the dataset. Elastic deformation utilizes local distortion and random affine transformation for high-quality output. These transformation takes place with random displacement. In addition, simple and random rotation is adopted in the augmentation process to increase the training dataset and to improve the performance.

3.2 Deep Convolutional Network

The framework is based on Deep Convolutional Network with an Encoder-Decoder sub-network. The network learns the important visual characteristics and the pixels' localization of the skin lesions through the encoder sub-networks. The decoder sub-network learns the spatial information recovery and the lesion boundaries for the melanoma lesions. Training of skin lesions images in this network is performed from end-to-end and pixels-wise using only the pixels from lesion images and corresponding ground truth labels as the system input. The network architecture is described in Fig. 1 below.

In the encoder network, we use five 3×3 convolutions blocks with each block composing 2 convolutional layers, 2×2 max-pooling layers for down-sampling, filter kernel and ReLU activation function. The convolution layers perform feature extraction and generates feature maps from the input image. Features maps are extracted from the input images by the convolutional layers. The ReLU activation function transforms the feature maps to enable training and learning of the image patterns. The transformed output is then sent to the next convolution layer as input. This is illustrated below:

$$F_m = conv2d(ReLU, K_f, I_m) \tag{1}$$

where F_m is the extracted feature map, I_m is the input image into the convolutional layer, K_f is the filter kernel and $ReLU$ is the activation function applied on the feature map for each layer. The down-sampling layers performs size reduction of the extracted feature maps to eliminate and reduce features redundancy and over-fitting and eventually minimises the overall computing

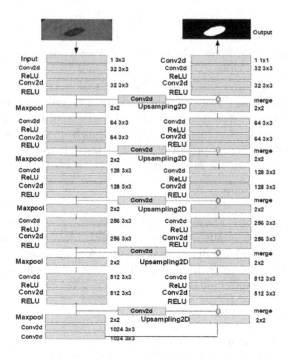

Fig. 1. Network architecture diagram for Deep Convolutional Network

processing time. This however causes reduction of the spatial features resolution of the input image. It utilizes the max-pooling function stated below.

$$layer = maxPooling2dLayer(poolSize) \tag{2}$$

The decoder learns the spatial features recovery information and the restores the original size of the feature map. In the decoder network, we use five 3×3 convolutions blocks with each block composing 2 convolutional layers, 2×2 up-sampling layers, filter kernel and ReLU activation function. The up-sampling layers with size 2×2 performs spatial feature recovery and boundaries localization.

$$y = upsample(x, n) \tag{3}$$

The skip connection in-between the encoder and the decoder has been replaced with series of skip pathway to ensure better segmentation. The encoded features are merged with the decoded features at a given spatial resolution.

3.3 Pixel-Wise Classification

The final decoder output with high dimensional feature representation is sent into a trainable soft-max classifier with dice loss function. This module identifies melanoma lesions through a pixel-wise classification of skin lesions images. The final feature maps are sent into the softmax module for pixels classification using the mathematical function stated below. The softmax classifier predicts the class for each pixel as melanoma or non-melanoma. In this research, we employ binary classification where n represents two number of classes and the output is a two-channel image of probabilities. The predicted segmentation therefore corresponds to the class with maximum probability at each pixel.

$$P(y = i|x) = \frac{e^{x^T w_i}}{\sum\limits_{n=1} e^{x^T w_n}} \tag{4}$$

where $x^T w$ represents the product of x and w, x is the feature map and w is the kernel operator.

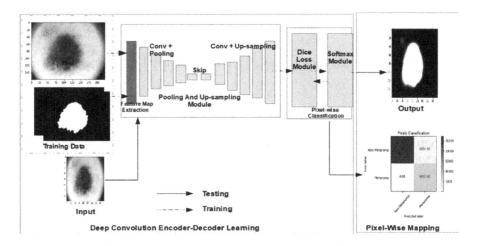

Fig. 2. The framework for the proposed system

4   Experimental Results

4.1   Dataset

The two well-established publicly available datasets used in the evaluation of the proposed segmentation method are from the ISIC challenge in skin lesion segmentation and PH2 data repository. PH2 [19] contains 200 skin lesion images with highest resolution of 765×574 pixels. This dataset was categorized into training and testing image set both comprising of images and ground truth labels respectively. The input dataset are skin lesion images in JPEG format while the ground truth are mask image in PNG format. ISBI 2018 [20] contains 2000 training images with the ground truth provided by expert clinicians. The proposed system is an expert system that has been subjected to test and generally acceptable. The image sizes possess highest resolution of 1022×767.

4.2   Evaluation Metrics

The most common and standard skin lesion segmentation evaluation metrics were used for comparison including: Dice Similarity Coefficient (DSC.), Sensitivity, Specificity, Precision and Accuracy. These metrics were used for evaluation of the model. They are illustrated below:

Dice Similarity Coefficient: It measures the similarity or overlap between the ground truth and the automatic segmentation. It is defined as

$$\mathbf{DSC} = \frac{2TP}{FP + 2TP + FN} \tag{5}$$

Sensitivity: It measures the proportion of those with positive values among those who are actually positive. It is also known as True Positive rate (TPR)

$$\mathbf{Sensitivity} = \frac{TP}{TP + FN} \tag{6}$$

Precision: This is the proportion of positive predicted values that are truly positive.

$$\mathbf{Precision} = \frac{TP}{TP + FP} \tag{7}$$

Accuracy: It measures the proportion of true results (both true positives and true negatives) among the total number of cases examined.

$$\mathbf{Accuracy} = \frac{TP + TN}{TP + TN + FP + FN} \tag{8}$$

Where FP is the number of false positive pixels, FN is the number of false negative pixels, TP is the number of true positive pixels and TN is the number of true negative pixels.

4.3   Results and Discussion

In the result, Figs. 5 and 6 show the performance of the proposed method on ISBI 2018 and PH2 skin lesion image datasets respectively. The results show improvement over existing methods as displayed in Table 1. The result give us accuracy and dice coefficient of 95% and 93% with the specificity and sensitivity of 95% and 93% respectively as shown in Table 1. This reflects in our predicted segmented output in the third row when compared with the ground truth result in the second row in Fig. 5.

Three images have been analysed pixel-wisely. The results are shown in Fig. 6 with the confusion matrix on the last column. The first image gives pixel classification result of 4060 pixels classified as melanoma correctly, 25135 pixels classified as non-melanoma correctly (this can be the background of the lesion or the healthy tissue) while 7 pixels are classified as non-melanoma in-correctly and 2639 pixels of non-melanoma classified as melanoma. The second image gives pixel classification result of 6957 pixels classified as melanoma correctly, 27195 pixels as non-melanoma classified correctly, 4 pixels of melanoma classified as non-melanoma and 1684 pixels of non-melanoma classified as melanoma. Lastly, the third image gives 5112 pixels classified as melanoma correctly, 29260 pixels classified as non-melanoma correctly while 153 pixels of melanoma are classified as non-melanoma and 1318 pixels of non-melanoma classified as melanoma.

Figure 3 below shows the dice coefficient and training loss curves. These curves identify that the loss reduces significantly as the dice coefficient increases. It clearly shows the relationship with the dice loss function relationship adopted in the model. The overall performance indicates that the dice coefficient which translates to the similarities between the predicted result output and the ground truth is very high with 93% of dice coefficient and less than 10% loss acquired. It can be inferred that the system performs efficiently.

Figure 4 shows the sensitivity and precision curves. These curves identify that the errors in classifying melanoma pixels as non-melanoma pixels reduces significantly as the sensitivity increase. Our sensitivity and precision shows that the number of melanoma pixels classified correctly as melanoma is high. It clearly shows the relationship between sensitivity and precision curves. The overall performance indicates that the sensitivity which translates to true positive rate is very high with value of 94% and the Precision also known positive predictive value also high with value of 93%.

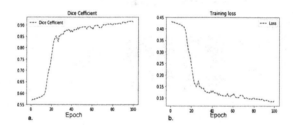

Fig. 3. (a) Dice coefficient curve (b) Training loss curve of the proposed method on the PH2 datasets

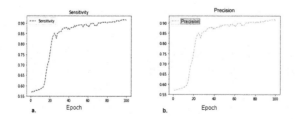

Fig. 4. (a) Sensitivity curve (b) Precision curve of the proposed method on the PH2 datasets

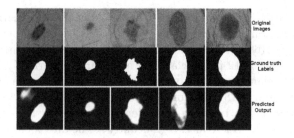

Fig. 5. Segmentation output of the proposed method on ISBI 2018 dataset with the third row showing the predicted results

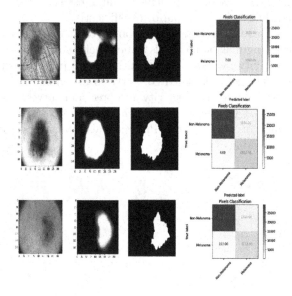

Fig. 6. Performance of the proposed method on PH2 dataset with the second column showing the segmented output and fourth column showing the confusion matrix result on each image. First column is the input image and the third column is the corresponding ground truth label.

Table 1. Segmentation performance (%) of the proposed model on PH2 datatest.

Techniques	Accuracy	Dice score	Sensitivity	Specificity
Proposed model	**95**	**93**	**93**	**95**
FrCN [10]	95	91	95	95
FCN [11]	94	90	95	94
mFCN [11]	96	91	96	95

5   Conclusion

This research explores Deep Convolutional Network-Based Framework for Melanoma Lesion Detection and Segmentation. The proposed system performs segmentation of melanoma lesions. The background of the images used have been given appropriate size for a reliable accuracy results. The system proposes pixel-wise classification of skin lesions for detection of melanoma cancer. The system aims at reducing deep learning architecture complexity in detecting melanoma. It also aims at developing an efficient system that can meet up with real time medical diagnosis task in diagnosing melanoma cancer. The proposed method is feasible for medical practices with the processing time for each dermoscopy image at averagely 5 s. The system was evaluated and tested on two publicly available database of ISBI 2018 and PH2. The results show that it out-performs some existing state-of-the-arts methods. Our results also show that the proposed system only under-performs in less than 10% of the image samples collected which could be caused by missing labels. Post-processing approach has been proposed for further refinements.

References

1. Apalla, Z., Lallas, A., Sotiriou, E., Lazaridou, E., Ioannides, D.: Epidemiological trends in skin cancer. Dermatol. Pract. Concept. **7**(2), 1 (2017)
2. Bi, L., Kim, J., Ahn, E., Feng, D.: Automatic skin lesion analysis using large-scale dermoscopy images and deep residual networks. arXiv preprint arXiv:1703.04197 (2017)
3. Masood, A., Al-Jumaily, A.A.: Computer aided diagnostic support system for skin cancer: a review of techniques and algorithms. Int. J. Biomed. Imaging **2013**, 1–23 (2013)
4. Mengistu, A.D., Alemayehu, D.M.: Computer vision for skin cancer diagnosis and recognition using RBF and SOM. Int. J. Image Process. (IJIP) **9**(6), 311–319 (2015)
5. Salido, J.A.A., Ruiz Jr., C.: Using deep learning to detect melanoma in dermoscopy images. Int. J. Mach. Learn. Comput. **8**(1), 61–68 (2018)
6. Khagi, B., Kwon, G.-R.: Pixel-label-based segmentation of cross-sectional brain MRI using simplified SegNet architecture-based CNN. J. Healthc. Eng. **2018**, 1–9 (2018)
7. Ronneberger, O., Fischer, P., Brox, T.: U-Net: convolutional networks for biomedical image segmentation. In: Navab, N., Hornegger, J., Wells, W.M., Frangi, A.F. (eds.) MICCAI 2015. LNCS, vol. 9351, pp. 234–241. Springer, Cham (2015). https://doi.org/10.1007/978-3-319-24574-4_28
8. Adegun, A.A., Akande, N.O., Ogundokun, R.O., Asani, E.O.: Image segmentation and classification of large scale satellite imagery for land use: a review of the state of the arts. Int. J. Civ. Eng. Technol. **9**(11), 1534–1541 (2018)
9. Bi, L., Kim, J., Ahn, E., Kumar, A., Feng, D., Fulham, M.: Step-wise integration of deep class-specific learning for dermoscopic image segmentation. Pattern Recogn. **85**, 78–89 (2019)
10. Al-Masni, M.A., Al-antari, M.A., Choi, M.-T., Han, S.-M., Kim, T.-S.: Skin lesion segmentation in dermoscopy images via deep full resolution convolutional networks. Comput. Methods Programs Biomed. **162**, 221–231 (2018)

11. Bi, L., Kim, J., Ahn, E., Kumar, A., Fulham, M., Feng, D.: Dermoscopic image segmentation via multistage fully convolutional networks. IEEE Trans. Biomed. Eng. **64**(9), 2065–2074 (2017)

12. He, X., Yu, Z., Wang, T., Lei, B., Shi, Y.: Dense deconvolution net: multi path fusion and dense deconvolution for high resolution skin lesion segmentation. Technol. Health Care **26**(S1), 307–316 (2018)

13. Bi, L., Kim, J., Ahn, E., Feng, D., Fulham, M.: Semi-automatic skin lesion segmentation via fully convolutional networks. In: 2017 IEEE 14th International Symposium on Biomedical Imaging (ISBI 2017), pp. 561–564. IEEE (2017)

14. Esteva, A., et al.: Dermatologist-level classification of skin cancer with deep neural networks. Nature **542**(7639), 115 (2017)

15. Goyal, M., Yap, M.H.: Multi-class semantic segmentation of skin lesions via fully convolutional networks. arXiv preprint arXiv:1711.10449 (2017)

16. Ramachandram, D., DeVries, T.: LesionSeg: semantic segmentation of skin lesions using deep convolutional neural network. arXiv preprint arXiv:1703.03372 (2017)

17. Li, Y., Shen, L.: Skin lesion analysis towards melanoma detection using deep learning network. Sensors **18**(2), 556 (2018)

18. Ramachandram, D., Taylor, G.W.: Skin lesion segmentation using deep hypercolumn descriptors. J. Comput. Vis. Imaging Syst. **3**(1), 1–3 (2017)

19. Mendonça, T., Ferreira, P.M., Marques, J.S., Marcal, A.R.S., Rozeira, J.: PH 2-A dermoscopic image database for research and benchmarking. In: 2013 35th Annual International Conference of the IEEE Engineering in Medicine and Biology Society (EMBC), pp. 5437–5440. IEEE (2013)

20. Codella, N.C.F., et al.: Skin lesion analysis toward melanoma detection: a challenge at the 2017 International Symposium on Biomedical Imaging (ISBI), hosted by the International Skin Imaging Collaboration (ISIC). In: 2018 IEEE 15th International Symposium on Biomedical Imaging (ISBI 2018), pp. 168–172. IEEE (2018)

A Novel Framework for Early Fire Detection Using Terrestrial and Aerial 360-Degree Images

Panagiotis Barmpoutis$^{(\boxtimes)}$ and Tania Stathaki

Department of Electrical and Electronic Engineering, Faculty of Engineering,
Imperial College London, London, UK
{p.barmpoutis,t.stathaki}@imperial.ac.uk

Abstract. In this paper, in order to contribute to the protection of the value and potential of forest ecosystems and global forest future we propose a novel fire detection framework, which combines recently introduced 360-degree remote sensing technology, multidimensional texture analysis and deep convolutional neural networks. Once 360-degree data are obtained, we convert the distorted 360-degree equirectangular projection format images to cubemap images. Subsequently, we divide the extracted cubemap images into blocks using two different sizes. This allows us to apply h-LDS multidimensional spatial texture analysis to larger size blocks and then, depending on the probability of fire existence, to smaller size blocks. Thus, we aim to accurately identify the candidate fire regions and simultaneously to reduce the computational time. Finally, the candidate fire regions are fed into a CNN network in order to distinguish between fire-coloured objects and fire. For evaluating the performance of the proposed framework, a dataset, namely "360-FIRE", consisting of 100 images with unlimited field of view that contain synthetic fire, was created. Experimental results demonstrate the potential of the proposed framework.

Keywords: Fire detection · Natural disasters · 360-degree images · Terrestrial · Aerial · CNN · Multidimensional analysis

1 Introduction

The environmental challenges the world faces nowadays have never been greater or more complex. Global areas that are covered by forests and urban woodlands, which comprise key parts of the global carbon cycle, are threatened by the impacts of climate change and the natural disasters that are intensified and accelerated by it. To this end, to address these impacts on people and nature, it is necessary to efficiently protect the forest ecosystems, by the occurrence of natural disasters maximizing the role of nature in absorbing and avoiding greenhouse gas emissions. Forest fires are one of the most harmful natural disasters affecting life around the world. It is worth mentioning that climate change and drier conditions have led to a marked increase of fire potential across Europe [1].

Thus, computer-based early fire warning systems that incorporate remote sensing technologies have attracted particular attention in the last years. These detection systems consist of visual cameras or multispectral/hyperspectral sensors while the main fire

© Springer Nature Switzerland AG 2020
J. Blanc-Talon et al. (Eds.): ACIVS 2020, LNCS 12002, pp. 63–74, 2020.
https://doi.org/10.1007/978-3-030-40605-9_6

detection challenge lies in the modelling and detection of the chaotic and complex nature of the fire phenomenon and the large variations of flame and smoke appearance in their representations. Detection techniques are based on various color spaces [2–4], spectral [4], spatial [5] and texture characteristics [6]. More recently, deep learning methods using a variety of algorithms such as YOLO [7], Faster R-CNN networks [8] and combination of fire representations in Grassmannian space and Faster R-CNN networks [9] have been designed, implemented and widely investigated. Among the most recent existing hazard-events detection approaches [9–11] the most successfully and commonly used base networks are the AlexNet [12], VGG16 [13], GoogLeNet [14] and ResNet101 [15] networks.

However, all the previously computer-based surveillance and monitoring systems for early detection of fire suffer from some limitations. Most of the frameworks taken to date use ground fixed, PTZ or human-controlled cameras with limited field of view. Other approaches require expensive and specialized aerial hardware with complex standard protocols for data collection and complex analysis methods, limiting their potential eventual widespread use by local authorities, forest agencies and experts [16]. Furthermore, the high levels of power, the long operation times and the high computational cost for the analysis that are required for the surveillance of wide areas do not allow the free of operation intervention (i.e. need to change UAVs' - unmanned aerial vehicles' - batteries) and under bad weather conditions (i.e. windy weather) appliance of these methods. Nevertheless, nowadays, 360-degree digital camera sensors become more and more popular and they can be installed in any place even in UAVs, hence proving to be a useful tool for the surveillance of wide areas [17].

In this paper, given the urgent priority around protecting the value and potential of forest ecosystems and global forest future, use of recently introduced 360-degree sensors and a novel computer-based approach for forest health surveillance and a better-coordinated global approach is proposed. More specifically, this paper makes the following contributions:

- We propose a new framework using terrestrial and aerial 360-degree digital camera sensors in an operationally and time efficient manner, aiming to overcome the limited field of view of state-of-the-art systems and human-controlled specified data capturing.
- A novel method is proposed for fire detection combining multidimensional texture analysis using Linear Dynamical Systems (LDS) and CNN ResNet101 network. Specifically, we first identify candidate fire regions of each image dividing the extracted cubemap images into two different size blocks (rectangular patches) and modelling them using LDS. Then we feed the candidate regions into the CNN network.
- To evaluate the efficiency of the proposed methodology, we created a dataset, namely "360-FIRE", consisting of 100 images of forest and urban areas that contain synthetic fire.

The rest of this paper is organized as follows: First, details of the proposed methodology are presented, followed by experimental results using the created dataset. Finally, some conclusions are drawn and future extensions are discussed.

2 Methodology

The framework of the proposed methodology is shown in Fig. 1. In this, recently introduced terrestrial and aerial 360-degree remote sensing systems are used in order to capture images with unlimited field of view. Once equirectangular images are acquired and due to the existence of distortions, they are converted to cubemap projection format images. Then, the extracted cubemap images are divided into blocks using two different sizes. Larger blocks are used in order to identify a high probability of existence of fire regions' and at the same time to reduce the computational time, whilst smaller blocks are used in order to accurately identify the candidate fire regions applying h-LDS multidimensional spatial texture analysis. Finally, the candidate fire regions are fed into a CNN *ResNet101* network for the classification to fire and non-fire regions.

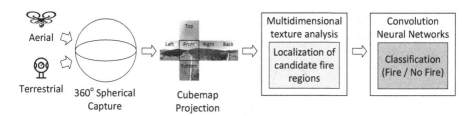

Fig. 1. The proposed methodology.

2.1 Introduction of Innovative Surveillance Schemes

To overcome the limited field of view data capturing and to achieve early and accurate detection of fire, two different formations, namely terrestrial and aerial 360-degree remote sensing systems are proposed. Terrestrial 360-degree digital cameras are ideal for areas with panoramic view while aerial 360-degree cameras are able to capture sphere images in areas where the installation of terrestrial cameras is not possible. It is worth mentioning that the required time to capture these types of data is estimated to be under a second using terrestrial cameras and under 30 s using a commercial UAV equipped with a digital camera.

Both of the proposed systems extract images in equirectangular projection (ERP) format. This native projection format is converted into cubemap projection (CMP) format in order to avoid false alarms due to the existence of distortions in the equirectangular images [18]. The CMP images consist of front, back, left, right, top and bottom images. This format is obtained by radially projecting points on the sphere to the six square faces of a cube enclosing the sphere (as illustrated in Fig. 2), and then unfolding the six faces. To this end, the spherical coordinates $p(theta, phi)$ are calculated using the normalized coordinates u and v of the equirectangular image:

$$theta = u * 2pi \tag{1}$$

$$phi = v * pi \tag{2}$$

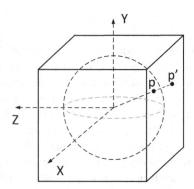

Fig. 2. Cubemap projection.

Then, the equivalent pixel coordinates $p'(x', y')$ for each square face of CMP are estimated as follows:

$$x' = \left(\frac{w}{2}\right)\left(\frac{theta}{pi} + 1\right) \tag{3}$$

$$y' = \left(\frac{h}{2}\right)\left(\frac{phi}{pi/2} + 1\right) \tag{4}$$

where w is the width, h is the height and *theta* and *phi* are the polar coordinates of the equirectangular image. Among all projection formats, CMP is widely used in the computer graphics community. Finally, as the top image in the proposed framework represents the sky in all cases, it was not taken into account for further processing.

2.2 Localization of Candidate Fire Regions Using Multidimensional Texture Analysis

For the localization of candidate fire regions, due to the need to achieve early detection of fire events, variant sizes of fires and different distances of fires from cameras, we propose a new modeling method through the division of 360-degree images into blocks using two different sizes (Fig. 3). Initially, we divide each region into larger size blocks with size n × n (we set n = 30) which are used in order to identify areas of higher probability of fire existence. Then, depending on the probability of fire existence these blocks are divided into four smaller size blocks (n = 15) in order to accurately detect the candidate fire regions. The goal of this, is two-fold: (a) Initially, we aim to reduce the computational time of the proposed framework and (b) to increase the reliability of the candidate fire regions localization procedure. Then, we apply multidimensional texture analysis in both block sizes through the higher order linear dynamical systems. Specifically, in the proposed approach, fire can be considered as a spatially-varying visual pattern dividing each individual block in 3 × 3 sub-patches and considering them as a multidimensional signal evolving in the spatial domain and modelling it through the following dynamical systems:

$$x(t + 1) = Ax(t) + Bv(t) \tag{5}$$

$$y(t) = \overline{y} + Cx(t) + w(t) \tag{6}$$

where $x \in \mathbb{R}^n$ is the hidden state process, $y \in \mathbb{R}^d$ is the observed data, $A \in \mathbb{R}^{n \times n}$ is the transition matrix of the hidden state and $C \in \mathbb{R}^{d \times n}$ is the mapping matrix of the hidden state to the output of the system. The quantities $w(t)$ and $Bv(t)$ are the measurement and process noise respectively, while $\overline{y} \in \mathbb{R}^d$ is the mean value of the observation data [19, 20]. Thus, this dynamical system models both the appearance and dynamics of the observation data, represented by C and A, respectively (Fig. 4). We then represent each sub-patch with a third-order tensor Y and apply a higher order Singular Value Decomposition to decompose the tensor:

$$Y = S \times_1 U_{(1)} \times_2 U_{(2)} \times_3 U_{(3)} \tag{7}$$

where $S \in \mathbb{R}^{n \times n \times c}$ is the core tensor, while $U_{(1)} \in \mathbb{R}^{n \times n}$, $U_{(2)} \in \mathbb{R}^{n \times n}$ and $U_{(3)} \in \mathbb{R}^{c \times c}$ are orthogonal matrices containing the orthonormal vectors spanning the column space of the matrix and $\times_j$ denotes the j-mode product between a tensor and a matrix. Since the columns of the mapping matrix C of the stochastic process need to be orthonormal, we can consider $C = U_{(3)}$ and

$$X = S \times_1 U_{(1)} \times_2 U_{(2)} \tag{8}$$

The transition matrix A can be estimated using least squares as follows:

$$A = X_2 X_1^T \left(X_1 X_1^T \right)^{-1} \tag{9}$$

where $X_1 = [x(2), x(3), \ldots, x(t)]$ and $X_2 = [x(1), x(2), \ldots, x(t-1)]$.

Assuming that the tuple $M = (A, C)$ describe each sub-patch, we estimate the finite observability matrix of each dynamical system, $O_m^T(M) = \left[C^T, (CA)^T, (CA^2)^T, \ldots, (CA^{m-1})^T \right]$ and we apply a Gram-Schmidt orthonormalization procedure [21], i.e., $O_m^T = GR$, in order to represent each descriptor as a Grassmannian point, $G \in \mathbb{R}^{m \times T \times 3}$ [5].

(a)	(b)

Fig. 3. Identification of candidate fire regions using: (a) larger size blocks and (b) smaller size blocks.

Finally, for the modelling of each fire candidate region, we apply VLAD encoding, which is considered as a simplified coding scheme of the earlier Fisher Vector (FV) representation and was shown to outperform histogram representations in bag of features approaches [22, 23]. More specifically, we consider a codebook, $\{m_i\}_{i=1}^r = \{m_1, m_2, \ldots, m_r\}$, with r visual words and local descriptors v, where each descriptor is associated to its nearest codeword $m_i = NN(v_j)$. The VLAD descriptor, V, is created by concatenating the r local difference vectors $\{u_i\}_{i=1}^r$ corresponding to differences $v_j - m_i$, with $m_i = NN(v_j)$, where v_j are the descriptors associated with codeword i, with $i = 1, \ldots, r$.

$$\overline{V} = \{u_i\}_{i=1}^r = \{u_1, \ldots, u_r\} \tag{10}$$

or

$$\overline{V} = \left\{ \sum_{\substack{v_j \, such \, that \\ m_1 = NN(v_j)}} (v_j - m_1), \ldots, \sum_{\substack{v_j \, such \, that \\ m_r = NN(v_j)}} (v_j - m_r) \right\} \tag{11}$$

while the final VLAD representation is determined by the L2-normalization of vector $\overline{V}$:

$$\overline{V}_{Euclidean} = \overline{V} / \|\overline{V}\|_2 \tag{12}$$

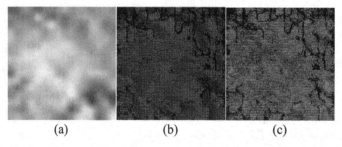

(a)   (b)   (c)

Fig. 4. Visualization of a: (a) candidate fire block, (b) transition matrix A and (c) mapping matrix C.

Finally, by ranking the similarities across all blocks, the majority rule of the s labels with the minimum distances is adopted in order to classify the examined block into candidate and non-candidate fire regions.

2.3 Fire Detection Using Convolution Neural Networks

After candidate fire regions are extracted, they are fed into a Convolution Neural Network (CNN). CNNs are one of the state-of-the-art deep learning approaches for hazard events detection. Although CNN architectures require large amount of training data, they are able to automatically learn very strong features. Thus, inspired by previous hazard events detection systems [9, 10] we chose to deploy ResNet101 feature extractor. With the advent of this model, researchers have developed deepened network structures that do not increase computational complexity. In the proposed methodology we use similar architecture to the original, aiming to train the parameters using fire focused images in order to solve the fire detection problem more effectively. Furthermore, we modify the number of neurons to two in the final layer of our architecture, enabling classification into fire and non-fire.

3   Experimental Results

Through the experimental evaluation we want to demonstrate the superiority of the proposed framework against other state of the art approaches and we aim to show that the proposed methodology improves the detection of fires.

To evaluate the efficiency of the proposed methodology, we created a dataset, namely "360-FIRE", consisting of 100 images of forest and urban areas that contain synthetic fire (90 of them consist of fire events and 10 of them are fireless 360-degree images). To the best of our knowledge and as 360-degree digital camera sensors are newly introduced type of cameras, there is not any dataset consisting of 360-degree images that contain fire. Thus, in order to create the dataset, we captured 360-degree images in different environments, and we used higher order SVD analysis (as shown in Eq. 9) in order to synthesize artificial flames [24]. Specifically, we represented synthetic video frames solving linear system equations and estimating the tensor generated at time k when the system is driven by random noise V [19]. Then the synthesized data was adapted to the 360-degree images and the size of fires was suitably adjusted with regards to the distance and the assumed start time of fire. For each captured image, we created 5 different fire events corresponding to different size of fires or fire locations. To evaluate the performance of the proposed methodology true positive, false negative, true negative, false positive rates and F-Score were used.

Fig. 5. CFDB and PASCAL VOC 2007 dataset training images containing actual fires and fire-coloured (non-fire) objects. (color figure online)

For the training of the proposed method for the localization of candidate fire regions using multidimensional texture analysis, we used the Corsican Fire Database (CFDB) that contain fire events and the PASCAL Visual Object Classes (VOC) 2007 dataset that contain fire-coloured objects (Fig. 5). Additionally, the CNN network was trained using the same datasets. Dataset sample images are shown in the Fig. 6. The implementation code of the proposed structure was written in Matlab and all calculations were performed on a 4 GB GPU and on a 6-core processor Intel Core i7-8750H, CPU 2.2 GHz. It is worth mentioning that in to have a fair comparison in our experiments we used the same training and testing set. In each case an image is labeled as a fire 360-degree image if it contains at least one fire image region.

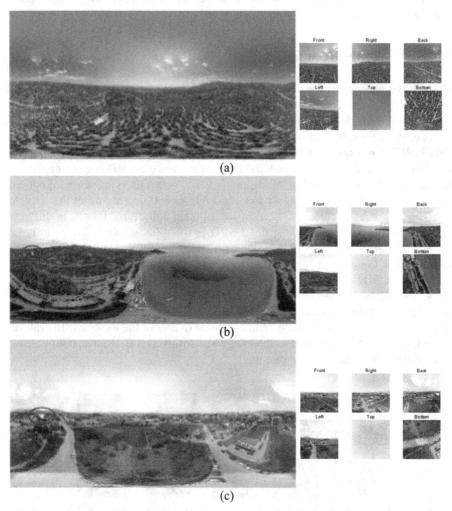

Fig. 6. "360-FIRE" dataset 360-degree images (left: equirectangular images, right: cubemap images) containing fires in different environments: (a) forest ecosystem, (b) forest ecosystem by the sea and (c) semi-urban environment.

Regarding the detection power of the proposed algorithm, we compared the proposed methodology to a spatial texture analysis method [20, 25] on its own as well as combined with a color localization method [5] and two Faster R-CNN [13, 15] architectures. As shown in Table 1, we compared the performance of the proposed approach firstly with a predefined color distribution for the localization of candidate blocks in combination

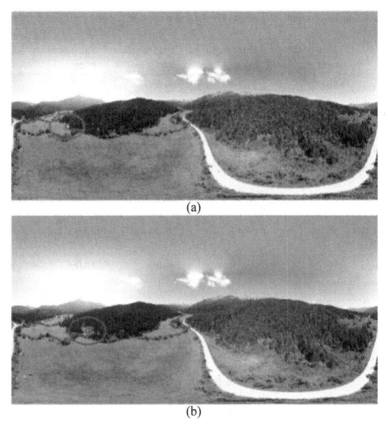

(a)

(b)

Fig. 7. Equirectangular images: (a) False negative of a long-distance fire, (b) True detection of a long-distance fire.

with the h-LDS approach for fire detection, secondly with the h-LDS approach playing the localization and detection role and finally with two Faster R-CNNs with VGG16 and Resnet101 base networks, achieving true positive rates 88.9%, 92.2%, 94.4% and 95.6% and F-score rates 77.7%, 83%, 88.1% and 88.7%, respectively. As depicted in Table 1, our method offers improved true positive rate (96.7%) and F-score (93.5%) compared to the second higher rates of 95.6% and 88.7% yield by the Faster R-CNN Resnet101 network. Furthermore, the proposed methodology increases true negative rates and reduces false negative and false positive rates.

Experimental results show that the proposed approach retains high true positive rates, while simultaneously significantly reducing false positives. This can be explained by the fact that the localization procedure eliminates many regions that are misclassified as fire or non-fire by the Faster R-CNN networks. Furthermore, it is obvious that the proposed methodology which includes dividing images into two different block sizes and extracting h-LDSs in order to estimate candidate fire regions is more efficient than the color analysis approaches, while CNN ResNet101 outperforms the discrimination ability of h-LDSs.

Table 1. Comparison results of various fire detection approaches using cubemap projection.

Method	True positive rates	False negative rates	True negative rates	False positive rates	F-score
Color & Spatial Texture Analysis (h-LDS)	88.9%	11.1%	60.0%	40.0%	77.7%
Spatial Texture Analysis (h-LDS)	92.2%	7.8%	70.0%	30.0%	83.0%
Faster R-CNN (VGG16)	94.4%	5.6%	80.0%	20.0%	88.1%
Faster R-CNN (ResNet101)	95.6%	4.4%	80.0%	20.0%	88.7%
Proposed	96.7%	3.3%	90.0%	10.0%	93.5%

Table 2. Comparison results of the proposed fire detection methodology using different projections.

Method	F-score
Equirectangular projection (ERP)	81.8%
Cubemap projection (CMP)	93.5%

However, in the proposed methodology some false negatives are noticed in small fires in long distance (Fig. 7a). In contrast, larger fires in shorter or same distance were accurately detected (Fig. 7b).

In Table 2, we present experimental results of the proposed methodology against the use of that equirectangular projection formats. More specifically, the use of cubemap projection formats achieves 11.7% higher detection rates.

Finally, we performed a computational speed test that is a crucial factor for early hazard events detection applications. Specifically, we estimated the time that is required for the processing of equirectangular images and cubemap images. Results show that the multi-thread processing of cubemaps requires 60% less computational time.

4 Conclusion

In this paper, we presented a novel framework that combines the newly introduced 360-degree digital camera sensors and modern signal and image processing techniques. This enables the near real-time environmental data acquisition, assessment, processing and analysis for the ultimate goals of ecosystem protection and forest and urban areas monitoring. The proposed framework will allow experts and scientists to achieve a better-coordinated global approach that will contribute to the limitation of negative impacts of climate change on forest ecosystems, air, and timber supplies.

In the future, a system for autonomous operation of UAV's which require one charge per day will be developed, in order to perform periodic flights every 30 min for the surveillance of wide areas. Additionally, in order to assess the effectiveness of the proposed we aim to extend our dataset using more data from a variety of urban, rural and forest areas. Furthermore, indoor images from buildings of cultural heritage will be captured and used. Finally, our goal is to install cameras at critical observation points for long periods in order to apply the proposed method in real hazard events.

References

1. European Environment Agency: Forest Fires (2019). https://www.eea.europa.eu/data-and-maps/. Accessed 12 June 2019
2. Töreyin, B.U., Dedeoğlu, Y., Güdükbay, U., Cetin, A.E.: Computer vision based method for real-time fire and flame detection. Pattern Recogn. Lett. **27**(1), 49–58 (2006)
3. Dimitropoulos, K., Tsalakanidou, F., Grammalidis, N.: Flame detection for video-based early fire warning systems and 3D visualization of fire propagation. In: 13th IASTED International Conference on Computer Graphics and Imaging, Crete, Greece (2012)
4. Grammalidis, N., et al.: A multi-sensor network for the protection of cultural heritage. In: 19th European Signal Processing Conference, pp. 889–893 (2011)
5. Barmpoutis, P., Dimitropoulos, K., Grammalidis, N.: Real time video fire detection using spatio-temporal consistency energy. In: 10th IEEE International Conference on Advanced Video and Signal Based Surveillance, pp. 365–370 (2013)
6. Dimitropoulos, K., Barmpoutis, P., Grammalidis, N.: Spatio-temporal flame modeling and dynamic texture analysis for automatic video-based fire detection. IEEE Trans. Circuits Syst. Video Technol. **25**(2), 339–351 (2014)
7. Shen, D., Chen, X., Nguyen, M., Yan, W.Q.: Flame detection using deep learning. In: 2018 4th International Conference on Control, Automation and Robotics, pp. 416–420 (2018)
8. Zhang, Q.X., Lin, G.H., Zhang, Y.M., Xu, G., Wang, J.J.: Wildland forest fire smoke detection based on faster R-CNN using synthetic smoke images. Procedia Eng. **211**, 441–446 (2018)
9. Barmpoutis, P., Dimitropoulos, K., Kaza, K., Grammalidis, N.: Fire detection from images using faster R-CNN and multidimensional texture analysis. In: ICASSP 2019-2019 IEEE International Conference on Acoustics, Speech and Signal Processing, pp. 8301–8305 (2019)
10. Giannakeris, P., Avgerinakis, K., Karakostas, A., Vrochidis, S., Kompatsiaris, I.: People and vehicles in danger-a fire and flood detection system in social media. In: 2018 IEEE 13th Image, Video, and Multidimensional Signal Processing Workshop, pp. 1–5 (2018)
11. Yang, L., Cervone, G.: Analysis of remote sensing imagery for disaster assessment using deep learning: a case study of flooding event. Soft Comput. **23**, 13393–13408 (2019)
12. Krizhevsky, A., Sutskever, I., Hinton, G.E.: Imagenet classification with deep convolutional neural networks. In: Advances in Neural Information Processing Systems, pp. 1097–1105 (2012)
13. Simonyan, K., Zisserman, A.: Very deep convolutional networks for large-scale image recognition. arXiv preprint arXiv:1409.1556 (2014)
14. Szegedy, C., et al.: Going deeper with convolutions. In: Proceedings of the IEEE Conference on Computer Vision and Pattern Recognition, pp. 1–9 (2014)

15. He, K., Zhang, X., Ren, S., Sun, J.: Deep residual learning for image recognition. In: Proceedings of the IEEE Conference on Computer Vision and Pattern Recognition, pp. 770–778 (2016)
16. Chowdary, V., Gupta, M.K.: Automatic forest fire detection and monitoring techniques: a survey. In: Singh, R., Choudhury, S., Gehlot, A. (eds.) Intelligent Communication, Control and Devices. Advances in Intelligent Systems and Computing, vol. 624, pp. 1111–1117. Springer, Singapore (2018). https://doi.org/10.1007/978-981-10-5903-2_116
17. Zia, O., Kim, J.H., Han, K., Lee, J.W.: 360° panorama generation using drone mounted fisheye cameras. In: Proceedings of the IEEE International Conference on Consumer Electronics, pp. 1–3, January 2019
18. Kim, J.H., et al.: U.S. Patent Application No. 15/433,505 (2018)
19. Doretto, G., Chiuso, A., Wu, Y.N., Soatto, S.: Dynamic textures. Int. J. Comput. Vision **51**(2), 91–109 (2003)
20. Dimitropoulos, K., Barmpoutis, P., Kitsikidis, A., Grammalidis, N.: Classification of multidimensional time-evolving data using histograms of Grassmannian points. IEEE Trans. Circuits Syst. Video Technol. **28**(4), 892–905 (2016)
21. Arfken, G.: Gram-schmidt orthogonalization. Math. Methods Phys. **3**, 516–520 (1985)
22. Jégou, H., Douze, M., Schmid, C., Pérez, P.: Aggregating local descriptors into a compact image representation. In: CVPR 2010-23rd IEEE Conference on Computer Vision & Pattern Recognition, pp. 3304–3311 (2010)
23. Kantorov, V., Laptev, I.: Efficient feature extraction, encoding and classification for action recognition. In: Proceedings of the IEEE Conference on Computer Vision and Pattern Recognition, pp. 2593–2600 (2014)
24. Costantini, R., Sbaiz, L., Susstrunk, S.: Higher order SVD analysis for dynamic texture synthesis. IEEE Trans. Image Process. **17**(1), 42–52 (2007)
25. Barmpoutis, P., Dimitropoulos, K., Barboutis, I., Grammalidis, N., Lefakis, P.: Wood species recognition through multidimensional texture analysis. Comput. Electron. Agric. **144**, 241–248 (2018)

Biomedical Image Analysis

Segmentation of Phase-Contrast MR Images for Aortic Pulse Wave Velocity Measurements

Danilo Babin[1]($\boxtimes$) (iD), Daniel Devos[2] (iD), Ljiljana Platiša[1] (iD),
Ljubomir Jovanov[1] (iD), Marija Habijan[3] (iD), Hrvoje Leventić[3] (iD),
and Wilfried Philips[1] (iD)

[1] imec-TELIN-IPI, Ghent University, Ghent, Belgium
danilo.babin@ugent.be
[2] University Hospital Ghent, Ghent University, Ghent, Belgium
[3] Faculty of Electrical Engineering, Computer Science and Information Technology,
University J. J. Strossmayer Osijek, Osijek, Croatia

Abstract. Aortic stiffness is an important diagnostic and prognostic parameter for many diseases, and is estimated by measuring the Pulse Wave Velocity (PWV) from Cardiac Magnetic Resonance (CMR) images. However, this process requires combinations of multiple sequences, which makes the acquisition long and processing tedious. We propose a method for aorta segmentation and centerline extraction from para-sagittal Phase-Contrast (PC) CMR images. The method uses the order of appearance of the blood flow in PC images to track the aortic centerline from the seed start position to the seed end position of the aorta. The only required user interaction involves selection of 2 input seed points for the start and end position of the aorta. We validate our results against the ground truth manually extracted centerlines from para-sagittal PC images and anatomical MR images. The resulting measurement values of both centerline length and PWV show high accuracy and low variability, which allows for use in clinical setting. The main advantage of our method is that it requires only velocity encoded PC image, while being able to process images encoded only in one direction.

Keywords: Image segmentation · Cardiac MRI · Pulse Wave Velocity

1 Introduction

Cardiovascular diseases are number 1 cause of death in the world today attributing to approximately 30% of all deaths. Aortic stiffness has proven to be an important parameter in estimating the overall cardiovascular health of patients [7], as well as diagnostic and prognostic parameter in many diseases, such as hypertension [5], Marfan syndrome [13], Turner syndrome [6], Metabolic syndrome [15], Diabetes [8], etc. The speed of pulse wave propagation, Pulse Wave Velocity (PWV), is used to estimate the aortic stiffness.

© Springer Nature Switzerland AG 2020
J. Blanc-Talon et al. (Eds.): ACIVS 2020, LNCS 12002, pp. 77–86, 2020.
https://doi.org/10.1007/978-3-030-40605-9_7

Measurement of PWV from MR images normally consists of 3 steps: (a) segmentation and centerline extraction of aorta (from anatomical MRI) for determining the aortic length and area of cross-sections, (b) computation of velocity curves from PC CMR images and (c) analysis of velocity curves for determining the pulse wave time propagation intervals between the aortic levels of interest. Segmentation and centerline extraction of aorta are usually performed on anatomical 3D MR images [2–4], while the calculation of velocity curves is done on Phase-Contrast MR images, which are velocity-encoded images taken in cross-sectional direction along the aortic centerline (and which are paired with anatomical MRI in order to perform segmentation of aortic cross-sections). Due to the combined use of different imaging modalities, measurement of MRI PWV can be tedious. Alternatively, para-sagittal MRI recording allows for an easier approach in PWV measurement, where a pair of anatomical and PC MRI images are recorded in para-sagittal orientation in a number of slice planes. In this fashion, the anatomical information is derived from the anatomical MR images and is used for masking the regions of interest in PC MRI, which simplifies the PWV calculation process. Once the region of the aorta is segmented from the anatomical images, velocity curves can be calculated by examining the average velocity over an aortic cross-section. One way of calculating time intervals between aortic levels is to use the maximum velocity change, as described in [10]. Approach of [9] uses cross-correlation to determine the time interval between flow waves at different locations along the aorta. A flow-sensitive 4D MRI PWV calculation method was proposed in [14]. Only few methods can perform the segmentation or centerline extraction directly from PC MR images. The method of [11] is graph-based with main emphasis on bifurcation detection, [1] performs tensor-based tracking of the aorta, [16] proposes semi-automatic level set PC MRI segmentation approach and [12] uses vector flow information to perform segmentation. However, none of these methods have been validated for PWV measurements.

In order to further simplify the PWV calculation workflow, we devise an aortic segmentation and centerline extraction method that works directly on PC MR images with velocity encoding in transversal direction. The method we present in this paper is semi-automatic, requiring only 2 user selected seed points that determine the start and the end position of the aorta. The benefit of our approach is that the anatomical MRI image is not needed in order to determine the aortic centerline and PWV measurement, which shortens execution times and simplifies user interaction. Also, it can be applied to a wider range of input images compared to the existing methods.

2   Method

We explain in this section our novel method for segmentation of para-sagittal PC MR images and the use of the segmentation result in centerline analysis and PWV calculation. The advantage of our approach is direct calculation on PC images without the need for accompanying anatomical images (i.e. only a single PC sequence is required to perform the processing), which leads to shorter processing times and lower memory usage. The method is composed of the following steps:

1. Extract a magnitude image from a PC image
2. Create a pulse wave propagation image
3. Segment aorta and extract its centerline
4. Measure PWV

The first step performs extraction of the magnitude of blood flow information from the PC image, which will be used a mask for the region of the aorta (it will contain high pixel values where prominent blood flow is detected). The magnitude image will be used to produce the pulse wave propagation image which will contain the propagation information of the blood flow over time in a single image slice. The pulse wave propagation image is used to extract the aortic centerline by "tracking" the values (indexes) of segmented regions in ascending order from ascending level to the abdominal level of the aorta. Finally, the centerline and the original PC velocity values are used to measure the PWV using the cross-correlation method for determining the time lapse between different positions along the aortic centerline. The detailed explanation of each of the steps follows.

2.1 Extraction of a Magnitude Image from a PC Image

Velocity encoding in PC images is governed by Velocity Encoding (VENC) value measured in cm/s, which is chosen to encompass the highest velocities that need to be recorded. In our data sets VENC value is set to 150 cm/s (spanning over 12 bits of gray pixel value), which means that the lowest possible pixel intensity represents the velocity of -150 cm/s and the highest pixel intensity represents the velocity of 150 cm/s. Hence, the mid-range between the maximum and the minimum possible gray pixel value corresponds to velocity value 0 cm/s. Let $\mathbf{p} = \{x, y, z\}$ be the 3D coordinates of a voxel in an image. In order to extract the magnitude of blood velocity $m(\mathbf{p}, i)$ at pixel $\mathbf{p}$ and time index i, we take the absolute value of subtraction of the original pixel gray value $g(\mathbf{p}, i)$ with the gray value corresponding to velocity 0 cm/s denoted as g_0:

$$m(\mathbf{p}, i) = |g(\mathbf{p}, i) - g_0| \tag{1}$$

The resulting magnitude image is shown in Fig. 1. Apart from the useful velocity magnitude information, the image also contains noisy pixels. These pixels are often isolated (resembling "salt and pepper" noise) and can be removed by morphological operations. The size of the structuring element (SE) is determined from the pixel spacing of the images and the expected size of the aorta. In order to avoid removing the useful signal, we perform first the dilation with circular SE of a predefined size, then the erosion with the SE of double the previous size, and finally another dilation with the original SE size. This sequence of morphological operators will remove much of the noise while still preserving the flow information regions.

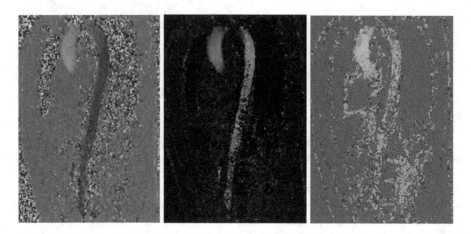

Fig. 1. Left: original PC image slice (time index 17 out of 40). Velocity encoding is done along the transversal axis, VENC = 150 cm/s. Center: extracted velocity magnitude image contrasts regions of high blood velocity. The absence of high velocity values in the aortic arch happens due to velocity encoding along the transversal axis. Right: pulse wave propagation image (color-coded grayscale image ranging from red to blue color) contains voxel gray values as time index of appearance of velocity above the preset value. (Color figure online)

2.2   Pulse Wave Propagation Image

In this subsection we will explain the principle behind our method for segmentation of relevant aortic regions in PC images. Since the PC images are acquired as multiple para-sagittal 2D in-time slices, we can expect that some part of the deviating aorta will be under-represented in the images (this is especially so for very tortuous aortas, where some parts are even not visible in any of the 2D slices). The under-represented parts contain some blood flow information which is often corrupted with noise, where in some cases the noise prevails over the useful signal. Hence, it is of high importance to be able to determine the regions with valid blood flow information and to discard the regions containing only noise. Therefore the goal is to create a 3D image (corresponding to the magnitude image) with voxel values indicating the earliest time of appearance of the pulse wave.

To achieve this, we refine the magnitude image by thresholding the velocity values. In order to remove the noisy and artifact pixels, we set the gray value threshold parameter g_t to promote only high velocities (velocities above a desired threshold velocity v_t). The actual threshold value is calculated from the VENC value of the PC MRI (in our case 150 cm/s) and the highest possible pixel intensity m_{max} (in our case m_{max} was $2^{11} - 1$, since the data was recorded in 12 bits):

$$g_t = m_{max}\, v_t\, /\, venc \tag{2}$$

Let I be the set of all time index values i for which a voxel with coordinates $\mathbf{p}$ in the magnitude image has a value larger than the given threshold value g_t:

$$I(\mathbf{p}) = \{i \in Z \mid m(\mathbf{p}, i) > g_t\}. \tag{3}$$

The pulse wave propagation image contains the time index values of arrival of the pulse wave with blood velocity above the specified threshold velocity:

$$f(\mathbf{p}) = \begin{cases} \min(I(\mathbf{p})), & I(\mathbf{p}) \neq \emptyset \\ 0, & I(\mathbf{p}) = \emptyset \end{cases}. \tag{4}$$

It should be noted that the original PC image and velocity magnitude image are 3D in time images (they can be considered as 4D images), while the pulse wave arrival image $f(\mathbf{p})$ is a 3D image where the time index information i is stored as voxel gray value (see Fig. 1).

2.3   Segmentation and Centerline Extraction

In order to create the segmentation and the centerline from the pulse wave propagation image, we perform morphological dilation of connected components per each time index value in the ascending time index order. This means that in the first iteration (time index 1), we extract all connected components with voxel value 1 and perform dilation with circular SE of size corresponding to the radius of the maximum inscribed circle fitting in the foreground region (non-zero value voxels) in the pulse wave propagation image. In this fashion we create larger connected components labeled with the time index values. The same principle is applied to all subsequent time index values, but with taking into account not to dilate over the connected components of lower time index values. Finally, for each time index we maintain only the connected components of a certain size, while we discard all the rest. The resulting image is shown in Fig. 2. The centerline is created by connecting all the centers of mass of each connected component and performing spline interpolation (see result in Fig. 2).

2.4   Pulse Wave Velocity Calculation

After extracting the aortic centerline, we mark the user defined start and end positions (which correspond to ascending and abdominal levels of the aorta) as locations for velocity curve calculations and PWV measurements. Alternatively, velocity curves can be calculated at any location along the centerline using the para-sagittal PC images (the most common locations are the ascending, descending, diaphragmal and abdominal level of the aorta). For each of the seed positions we calculate the average blood velocity as average grey value of pixels belonging to the region of the aorta (defined by the segmentation, as shown in Fig. 2) in the neighborhood defined as the maximum inscribed circle (analysis is per 2D slice because we only have 3 slice planes). This results in velocity curves for each of the positions with the number of samples equal to the number of images in the CMR time sequence. We use the cross-correlation method to determine the time shift between the velocity curves. Finally, the PWV is calculated from the measured lengths and velocity curve time shifts between the aortic positions.

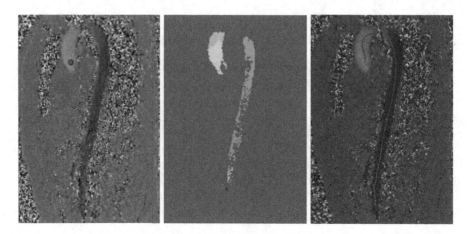

Fig. 2. Left: original Phase-Contrast image slice (time index 11 out of 40) with selected user seed points. Center: segmented regions of aorta corresponding to pulse wave propagation image that contrasts regions of high blood velocity (color-coded grayscale image ranging from red to blue color). The absence of high velocity values in the aortic arch happens due to velocity encoding along the transversal axis. Right: extracted aortic centerline. (Color figure online)

3   Results

The experiments were done on 10 healthy volunteers: 4 females and 6 males (age between 27 and 71 years old). Ethical committee approval was obtained for this study and all volunteers signed the written informed consent form. Images were recorded on Siemens Avanto Fit 1.5 T scanner as Phase-Contrast (PC) sequences with velocity encoding (VENC) value of 150 cm/s, and anatomical (Siemens trueFISP, i.e. Balanced-SSFP) images for ground truth centerline extraction. Para-sagittal PC sequences are recorded in 3 image planes. Each ECG triggered sequence (corresponding to one heart cycle) consists of 40 images.

Our proposed centerline extraction method for PC MR images was compared (in length) to manual centerline extraction on the same PC images and with manual centerline extraction on anatomical MR images (Siemens trueFISP, recorded either in transversal or sagittal orientation). The expert manual centerlines were drawn by placing 10 to 15 sample points along each aorta and using curve interpolation to produce the final ground truth centerlines. Results of length measurements on the whole aorta are shown in Table 1. The average mean absolute error over all cases is 12.7 mm, which is low in comparison to the length of the whole aorta, as confirmed by the average mean relative error value of 3.7%.

Taking into account that the main purpose for our proposed PC MRI segmentation and centerline extraction method is PWV calculation, we calculate the PWV to determine the applicability of our proposed method. The PWV is calculated using the same velocity curves for each PWV measurement approach

Table 1. Aortic lengths measured using the proposed method, manual centerline drawing on PC images and manual centerline drawing on anatomical images. Absolute error values of the proposed method were calculated against the manual measurements (as ground truth values).

Centerline length (mm)	1	2	3	4	5	6	7	8	9	10
Proposed method	429	452	412	452	441	434	471	333	445	341
Manual PC	422	470	403	502	461	432	455	358	454	365
Manual anatomical	421	462	387	476	436	422	455	333	454	326
Mean absolute error	7.5	14	17	33.3	12.5	7	16	12.5	9	19.5
Max absolute error	8	18	25	50	20	12	16	25	9	24
Mean relative error (%)	3.5	3	4.3	7.5	2.7	1.6	3.5	3.4	1.9	5.5

Table 2. Pulse Wave Velocity values calculated using length measurements of the examined centerline extraction methods. Absolute error values of the proposed method were calculated against the manual measurements (as ground truth values).

PWV (m/s)	1	2	3	4	5	6	7	8	9	10
Proposed method	6.1	7.1	4.7	9.4	4.1	4.9	4.8	4.6	4	2.9
Manual PC	6	7.5	4.6	10.7	4.3	4.9	4.6	4.9	4.1	3.1
Manual anatomical	6	7.4	4.4	10.1	4.1	4.8	4.6	4.6	4.1	2.7
Max absolute error	0.1	0.4	0.3	1.3	0.2	0.1	0.2	0.3	0.1	0.2

(in other words, the time intervals do not vary in PWV calculation for each approach, only the centerline extraction methods are different). Results of PWV measurements on the whole aorta are shown in Table 2. The average maximum absolute error amounts to 0.3 m/s, which is significantly lower than the clinically acceptable 1 m/s measurement error.

4 Discussion

The results of length measurements for our proposed method, the manual centerline drawing on PC image and manual centerline drawing on anatomical MRI are shown in Table 1. The results of centerline extraction for 3 cases are shown in Fig. 3, where the red line depicts the manually drawn centerline on the parasagittal PC image, while the blue centerline depicts the result of our proposed method. Visual inspection of the results shows that the centerline follows the ground truth very closely in linear segments of the aorta, but slightly shortens the curvature in the aortic arch. This is the reason for the average mean absolute error (over all cases) of 12.7 mm, although the average mean relative error is still quite low (3.7%). The method is robust to seed location selection as long as the selected seeds fall inside the region of the aorta.

The results of PWV measurements (Table 2) show that most of the calculated PWV values fall in range [2, 10] m/s (with one outlier), which is the expected

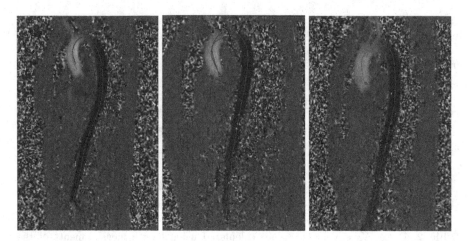

Fig. 3. Centerline extraction results. Manually extracted centerline is depicted in red, while the centerline extracted using our proposed method is blue. The proposed centerline extraction method follows closely the aortic centerline in the straight segments of the aorta and slightly differs in the curved segments (aortic arch). (Color figure online)

range of values in the healthy population. The calculated maximum absolute error values show that almost all values (except one) fall under $0.4\,m/s$, which is the error value that will not influence the diagnostic or prognostic relevance. The only case in which the error exceeds $1\,m/s$ is the measurement on a 71 year old male volunteer. The PC image of the volunteer exhibited lower blood velocity values, so the threshold velocity v_t (and consequently gray value threshold g_t) would need to be set to a lower value in order to correct the segmentation and centerline extraction. It has been shown that the older population exhibits stiffening of the aorta, which in turn creates higher PWV values. This is confirmed also in our test population, where the 71 year old male volunteer has the PWV value range [9.4, 10.7]. All other volunteers (all under 43 years of age) have normal range of PWV values. The para-sagittal PC MRI allows for PWV measurements on multiple places along the aorta. However, it should be noted that the PWV results per aortic segments will display higher variability (because of shorter aortic distances and pulse wave intervals).

5   Conclusion

We proposed in this paper a method for aortic region segmentation and centerline extraction from para-sagittal Phase-Contrast MR images with pulse wave velocity calculation as the end goal. The method works by tracking the pulse wave propagation from ascending level of the aorta down to the abdominal level of the aorta. The main advantage of our method is that it does not require any other images besides the Phase-Contrast image, which shortens the time required for image analysis and eliminates the need for additional image modality acquisition, leading to savings in time and cost while causing less patient discomfort. The method is semi-automatic requiring only 2 user input seed points for the start and end position of the aorta. The method was validated against the manual centerline extraction in para-sagittal PC images and anatomical MR images. Both centerline length measurements and PWV measurements on the whole aorta show high enough accuracy and low variability, which allows for use in clinical setting.

Acknowledgement. This work was supported by IWT Innovation Mandate spin-off project 130865: "WaVelocity: cardiovascular structure and flow analysis software" and by Croatian Science Foundation under the project UIP-2017-05-4968.

References

1. Azad, Y.J., Malsam, A., Ley, S., Rengier, F., Dillmann, R., Unterhinninghofen, R.: Tensor-based tracking of the aorta in phase-contrast MR images. In: Medical Imaging 2014: Image Processing, vol. 9034, p. 90340L. International Society for Optics and Photonics (2014)
2. Babin, D., Pižurica, A., Philips, W.: Robust segmentation methods for aortic pulse wave velocity measurement. In: IEEE EMBS Benelux Chapter, Annual symposium, Abstracts (2011)
3. Babin, D., Vansteenkiste, E., Pižurica, A., Philips, W.: Segmentation and length measurement of the abdominal blood vessels in 3-D MRI images. In: Proceedings of Annual International Conference of the IEEE Engineering in Medicine and Biology Society EMBC 2009, pp. 4399–4402 (2009)
4. Babin, D., Devos, D., Pižurica, A., Westenberg, J., Vansteenkiste, E., Philips, W.: Robust segmentation methods with an application to aortic pulse wave velocity calculation. Comput. Med. Imaging Graph. **38**(3), 179–189 (2014)
5. Brandts, A., et al.: Association of aortic arch pulse wave velocity with left ventricular mass and lacunar brain infarcts in hypertensive patients: assessment with MR imaging. Radiology **253**(3), 681–688 (2009)
6. Devos, D.G., et al.: Proximal aortic stiffening in turner patients may be present before dilation can be detected: a segmental functional MRI study. J. Cardiovasc. Magn. Reson. **19**(1), 27 (2017)
7. Devos, D.G., et al.: MR pulse wave velocity increases with age faster in the thoracic aorta than in the abdominal aorta. J. Magn. Reson. Imaging **41**(3), 765–772 (2015)
8. van Elderen, S., et al.: Cerebral perfusion and aortic stiffness are independent predictors of white matter brain atrophy in type 1 diabetic patients assessed with magnetic resonance imaging. Diabetes Care **34**(2), 459–463 (2011)
9. Fielden, S., Fornwalt, B., Jerosch-Herold, M., Eisner, R., Stillman, A., Oshinski, J.: A new method for the determination of aortic pulse wave velocity using cross-correlation on 2D PCMR velocity data. J. Magn. Reson. Imaging **27**(6), 1382–1387 (2008)
10. Giri, S., et al.: Automated and accurate measurement of aortic pulse wave velocity using magnetic resonance imaging. In: Computers in Cardiology, pp. 661–664, October 2007

11. Jeong, Y.J., Ley, S., Delles, M., Dillmann, R., Unterhinninghofen, R.: Graph-based bifurcation detection in phase-contrast MR images. In: Medical Imaging 2013: Image Processing, vol. 8669, p. 86691Z. International Society for Optics and Photonics (2013)

12. Jeong, Y.J., Ley, S., Dillmann, R., Unterhinninghofen, R.: Vessel centerline extraction in phase-contrast MR images using vector flow information. In: Medical Imaging 2012: Image Processing, vol. 8314, p. 83143H. International Society for Optics and Photonics (2012)

13. Kröner, E., et al.: Evaluation of sampling density on the accuracy of aortic pulse wave velocity from velocity-encoded MRI in patients with Marfan syndrome. J. Cardiovasc. Magn. Reson. 36(6), 1470–1476 (2012)

14. Markl, M., Wallis, W., Brendecke, S., Simon, J., Frydrychowicz, A., Harloff, A.: Estimation of global aortic pulse wave velocity by flow-sensitive 4D MRI. Magn. Reson. Med. 63(6), 1575–1582 (2010)

15. Roes, S., et al.: Assessment of aortic pulse wave velocity and cardiac diastolic function in subjects with and without the metabolic syndrome. Diabetes Care 31(7), 1442–1444 (2008)

16. Volonghi, P., et al.: Automatic extraction of three-dimensional thoracic aorta geometric model from phase contrast MRI for morphometric and hemodynamic characterization. Magn. Reson. Med. 75(2), 873–882 (2016)

On the Uncertainty of Retinal Artery-Vein Classification with Dense Fully-Convolutional Neural Networks

Azat Garifullin[1]([✉]) [iD], Lasse Lensu[1] [iD], and Hannu Uusitalo[2,3]

[1] LUT University, P.O. Box 20, 53851 Lappeenranta, Finland
{azat.garifullin,lasse.lensu}@lut.fi
[2] SILK, Department of Ophthalmology, Tampere University, ARVO F313,
33014 Tampere, Finland
hannu.uusitalo@tuni.fi
[3] Tays Eye Center, Tampere University Hospital, P.O. Box 2000,
33520 Tampere, Finland

Abstract. Retinal imaging is a valuable tool in diagnosing many eye diseases but offers opportunities to have a direct view to central nervous system and its blood vessels. The accurate measurement of the characteristics of retinal vessels allows not only analysis of retinal diseases but also many systemic diseases like diabetes and other cardiovascular or cerebrovascular diseases. This analysis benefits from precise blood vessel characterization. Automatic machine learning methods are typically trained in the supervised manner where a training set with ground truth data is available. Due to difficulties in precise pixelwise labeling, the question of the reliability of a trained model arises. This paper addresses this question using Bayesian deep learning and extends recent research on the uncertainty quantification of retinal vasculature and artery-vein classification. It is shown that state-of-the-art results can be achieved by using the trained model. An analysis of the predictions for cases where the class labels are unavailable is given.

Keywords: Bayesian deep learning · Blood vessels segmentation · Artery-vein classification

1 Introduction

A number of eye and systemic diseases influence the vasculature of the retina in different ways. The blood vessel characteristics in retinal images may provide visible evidence about numerous diseases such as hypertensive retinopathy, diabetic retinopathy, as well as other cardio- and cerebrovascular diseases [12]. The related characteristics include the shape and size of retinal vessels, arteriovenous ratio and arteriovenous crossing [14]. These characteristics may be obtained by using blood vessel segmentation masks produced by automatic machine learning techniques [5].

© Springer Nature Switzerland AG 2020
J. Blanc-Talon et al. (Eds.): ACIVS 2020, LNCS 12002, pp. 87–98, 2020.
https://doi.org/10.1007/978-3-030-40605-9_8

The topic of blood vessels segmentation is well studied by the community [1]. However, the artery-vein (AV) classification task remains challenging not only for machines, but also for humans. Despite the fact that discriminative features based on color and geometry are described, it is still difficult to distinguish arteries from veins [14] due to imperfect imaging conditions and limited visibility of the retinal blood vessels.

Recently, deep convolutional neural networks have become a common trend for retinal vasculature segmentation and AV classification because of the ability to automatically learn meaningful features. Welikala et al. [16] proposed a method based on a convolutional neural network (CNN) classifying arteries and veins in a patch-wise manner. The authors considered the problem as a multi-class classification task placing a softmax layer at the end of the network. The UK Biobank database was used from which 100 images were labeled and classification accuracy of 82.26% for arteries and veins was reported. Girard et al. [5] proposed to use a modified U-Net [15] with likelihood score propagation in the minimum spanning tree effectively utilizing information about the global vessel topology. The approach was tested on the DRIVE data set [8] and it achieved 94.93% accuracy for the AV classification. Badawi et al. [2] proposed to train a CNN with multiloss function consisting of pixelwise cross entropy loss and segment-level loss to overcome training issues appearing because of inconsistent thickness of blood vessels. The authors also created a new data set consisting of labeled subsets of EPIC and MESSIDOR [3] data sets and classification accuracy of 96.5% was reported. Hemelings et al. [7] applied the U-Net architecture for the task of AV classification stating the problem as a multi-class classification problem predicting labels for four classes (background, vein, artery, and unknown) with classification accuracy of 94.42% and 94.11% for arteries and veins, respectively. Zhang et al. [18] proposed cascade refined U-net which modifies the original model with multi-scale loss training and includes sub-networks for simultaneous AV and blood vessel segmentation. The authors achieved 97.27% arteriovenous classification accuracy evaluated on the automatically detected vessels.

In this work, a multi-label classification approach is considered with the uncertainty quantification experiments presented. Our approach is most similar to the method proposed by Zhang et al. [18] in a way how three-component loss is used. The main difference is that in this work, classification of arteries and veins are not conditioned on blood vessel predictions, but vessel labels are conditioned on arteries and veins. Using the multi-label classification approach, there is no need to separately model the AV crossings and background. To the best of authors' knowledge, this work is the first presenting uncertainty quantification experiments for the of AV classification. For the experiments, the RITE data set is utilized.

2   Data and Methods

2.1   DRIVE and RITE Data Sets

The DRIVE database is a common benchmark for the retinal blood vessel segmentation task [8]. It contains 20 train and 20 test images with two sets of manual blood vessel segmentations. The RITE data set [9] extends DRIVE with an AV reference standard containing four types of labels: arteries (red), veins (blue), overlapping (green), and uncertain vessels (white). An example test image is shown in Fig. 1.

2.2   AV Classification

Let f be a model with parameters $\boldsymbol{\theta}$ that maps an input image $\mathbf{x}$ to a map of logits with the same spatial dimensionality as the original image:

$$\hat{\mathbf{y}} = f\left(\mathbf{x}, \boldsymbol{\theta}\right). \tag{1}$$

Given predicted logits $\hat{\mathbf{y}} = [\hat{y}_{\text{artery}}\ \hat{y}_{\text{vein}}]$, probabilities of assigning labels to arteries and veins can be calculated as follows:

$$p_{\text{artery}} = \text{sigmoid}\left(\hat{y}_{\text{artery}}\right), \tag{2}$$

$$p_{\text{vein}} = \text{sigmoid}\left(\hat{y}_{\text{vein}}\right). \tag{3}$$

In the multi-label setup, the same pixel can be classified with both artery and vein labels, which is meaningful in the case of AV crossings. A vessel probability label can then be naturally inferred by a simple formula:

$$p_{\text{vessel}} = p_{\text{artery}} + p_{\text{vein}} - p_{\text{artery}}p_{\text{vein}}. \tag{4}$$

Since the data set contains the masks for both the AV classification and blood vessel segmentation, it is possible to state the following optimization problem

$$\hat{\boldsymbol{\theta}} = \arg\min_{\boldsymbol{\theta}} \left[\mathcal{L}_{\text{artery}}\left(\boldsymbol{\theta}\right) + \mathcal{L}_{\text{vein}}\left(\boldsymbol{\theta}\right) + \mathcal{L}_{\text{vessel}}\left(\boldsymbol{\theta}\right)\right], \tag{5}$$

where $\mathcal{L}$ denotes the binary cross entropy loss for the corresponding labels. This way even if the labels for arteries and veins are not given for uncertain vessel labels, it is possible to enforce a model to predict correct labels for the blood vessels.

2.3   Aleatoric and Epistemic Uncertainties

The approach described in the previous section gives only point estimates for the label probabilities and the model parameters are considered to be deterministic. In order to better capture imperfect data labeling and image noise, one can consider the model outputs and the parameters to be random variables. The first approach captures the heteroscedastic aleatoric uncertainty that depends

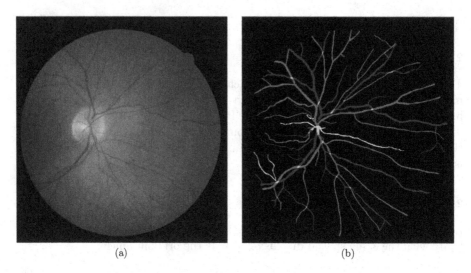

Fig. 1. The RITE data set: (a) An example test image and (b) corresponding artery-vein reference standard. (Color figure online)

on the input data, whereas the second represents the epistemic uncertainty that models a distribution of the learned parameters. More detailed explanations for the uncertainties can be found in [13] and [4]. In this work, a brief explanation for the AV classification task is given below.

Aleatoric uncertainty can be captured by modifying the original model to predict the mean and standard deviations of logits:

$$[\hat{\mathbf{y}}, \boldsymbol{\sigma}] = f(\mathbf{x}, \boldsymbol{\theta}). \tag{6}$$

In order to predict standard deviations, a second layer similar and parallel to the one used for logits is added to the output of the network. In order to ensure that the predicted standard deviations are positive, an additional absolute value activation is added to the output of the layer. The probabilities of the labels can then be calculated as follows:

$$\hat{\mathbf{p}} = \text{sigmoid}\left(\hat{\mathbf{y}} + \boldsymbol{\sigma} \odot \boldsymbol{\epsilon}\right), \quad \boldsymbol{\epsilon} \sim \mathcal{N}(\mathbf{0}, \mathbf{I}), \tag{7}$$

where $\odot$ stands for the Hadamard product and $\boldsymbol{\epsilon}$ are sampled during inference.

The main inference scheme for AV remains the same with the exception that instead of a point estimate, the model now yields N_A samples that are then used to calculate the loss (5). The final minimized loss is just an average over the predicted losses for each sample.

Epistemic uncertainty can be captured by considering the model parameters to be a random variable and considering the following posterior predictive:

$$p(\mathbf{y} \mid \mathbf{x}, \mathcal{D}) = \int p(\mathbf{y} \mid \mathbf{x}, \boldsymbol{\theta}) \, p(\boldsymbol{\theta} \mid \mathcal{D}) \, \mathrm{d}\boldsymbol{\theta}, \tag{8}$$

where $\mathcal{D}$ denotes a data set of input-output pairs. Typically, the parameter's posterior $p\left(\boldsymbol{\theta} \mid \mathcal{D}\right)$ for complex models such as deep neural networks is intractable and variational approximations are used. The posterior in (8) can be replaced by a simpler distribution $q\left(\boldsymbol{\theta}\right)$ and the training procedure can then be formulated as the minimization of the Kullback-Leibler divergence between the true posterior and the approximation.

In this work, the model f is parameterized as a dense fully-convolutional network (Dense-FCN) and Monte-Carlo dropout [4] is used for the variational approximation. The description of the utilized architecture is given below.

2.4 Architecture

The architecture utilized in this work is a Dense-FCN. It has been shown that Dense-FCNs have less parameters and may outperform other fully-convolutional network (FCN) architectures in a variety of different segmentation tasks [11]. Here we adapt the Dense-FCN architecture for the AV classification tasks.

The main building block of Dense-FCN is a dense convolutional block (DCB) where the input of each layer is a concatenation of the outputs of the previous layers. The block consists of repeating batch normalization (BN), ReLU, convolution and dropout $p = 0.5$ layers resulting in g feature maps (growth rate).

The main concept of Dense-FCN is similar to other encoder-decoder architectures in the sense that the input is first compressed to a hidden representation by the downsampling part, and then the segmentation masks are recovered by an upsampling part. The downsampling part consists of DCBs and downsampling transitions with skip connections to the upsampling part. The upsampling part consists of DCBs and upsampling transitions. An example of two blocks in downsampling and upsampling paths of a Dense-FCN is given in Fig. 2. The architectural parameters used are given below:

- Growth rate for all DCBs: $g = 16$.
- Downsampling path consists of five DCBs with depths $D_{\text{down}} = [4, 5, 7, 10, 12, 15]$.
- Upsampling also consists of five DCBs with depths $D_{\text{up}} = [12, 10, 7, 5, 4]$.
- The first and last convolution layers are the same as in Fig. 2.

2.5 Image Preprocessing

It was noticed in the experimental part of the work that simple preprocessing involving contrast enhancement and channel normalization improves the convergence and performance of the trained models. First, contrast-limited adaptive histogram equalization [19] with the clip limit of 2 and the grid size of 8×8 is applied and then each image channel is normalized to values between 0 and 255. The preprocessing scheme was used to reduce the effects of uneven illumination fields of the channel images (Fig. 3).

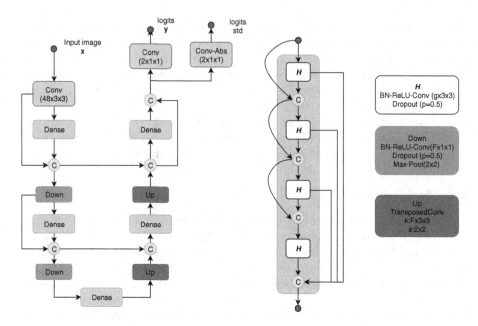

Fig. 2. Dense-FCN architecture: *Dense* stands for a DCB; C is a tensor concatenation; H is a block consisting of BN, ReLU and a convolutional layer with growth rate g; *Down* is a transition down block with F output feature maps; *Up* is a transition up with F output feature maps and 2×2 stride. *logits std* denotes standard deviations of logits.

2.6   Training Details

The Dense-FCN was pretrained for 200 epochs with 1000 steps per epoch on random patches 224×224 with the batch size equal to 5. Then it was fine-tuned for 50 epochs with 500 steps per epoch on full size images with the batch size equal to 1.

The weights were initialized using HeNormal [6]. In addition to dropout, $l2$ regularization with the weight decay factor 10^{-4} was used. As the optimizer, Adadelta [17] with the learning rate $l = 1$ and the decay rate $\rho = 0.95$ was used for both the pretraining and fine-tuning. The learning rate was dropped by a factor of 10 if the training loss was not decreased by 0.005 for 10 epochs. Data augmentation by using flipping, reflecting and rescaling (with scale rates 0.8 and 1.2) was applied in both cases. During the fine-tuning stage, the images were randomly padded to size 608×608 so that the size is divisible by 32 and could be properly compressed by the downsampling path. The parameter values were determined empirically based on initial experiments with the RITE database.

3 Experiments and Results

3.1 Training and Evaluation Strategies

Considering the given reference standard, the question arises of how to use the uncertain class labels and its effect on the final training results. Possible ways for utilizing this information are to consider these pixels to be arteries and veins simultaneously including uncertain (IU), or to exclude them from the training completely excluding uncertain (EU). In this work, a comparison of both training strategies is provided. The crossing labels are considered to be veins and arteries simultaneously. Both strategies are evaluated against the reference standard with excluded uncertain labels, and the vessels classification metrics are given by evaluating against the reference standard provided by the second expert.

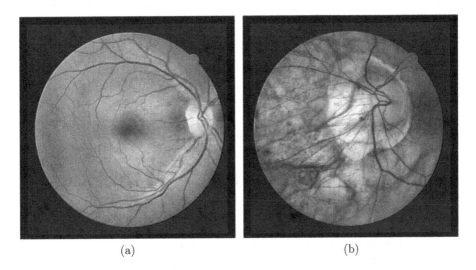

(a)   (b)

Fig. 3. Two examples of preprocessed RITE images.

Since the AV classification problem stated being multilabel, binary classification metrics were calculated for each class separately: area under receiver operating characteristic curve (ROC-AUC), accuracy, sensitivity and specificity.

During the inference stage, the model parameters are sampled 100 times and the number of inferred samples is $N_A = 50$. The final posterior predictive mean is calculated over all predicted samples, and the outputs aleatoric uncertainty U_A and epistemic uncertainty U_E are calculated as in [10]:

$$U_A = \mathbb{E}_q \left[\mathbb{V}_{p(\mathbf{y} \mid \mathbf{x}, \boldsymbol{\theta})} [\mathbf{y}] \right], \tag{9}$$

$$U_E = \mathbb{V}_q \left[\mathbb{E}_{p(\mathbf{y} \mid \mathbf{x}, \boldsymbol{\theta})} [\mathbf{y}] \right], \tag{10}$$

where $\mathbb{E}$ and $\mathbb{V}$ denote expectation and variance, respectively.

3.2 Experimental Results

The receiver operating characteristic (ROC) curves calculated after training with both strategies are shown in Fig. 4. The corresponding performance metrics are given in Tables 1 and 2. From the tables, it is clear that the AV classification performances are high, not far from the vessel pixel classification performance. Including uncertain labels into the training set leads to reduced classification accuracy for arteries and veins, but it slightly improves the performance of vessel classification. It is also clear that the Including uncertain strategy increases classification sensitivity, since the training procedure now takes all labeled vessels into account during the AV inference stage.

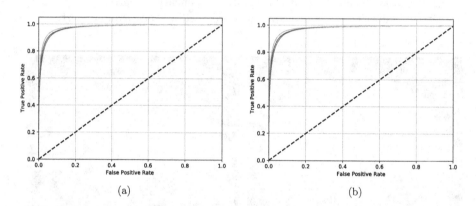

Fig. 4. ROC curves for arteries (red), veins (blue) and vessels (orange): (a) Excluding uncertain and (b) including uncertain strategies. (Color figure online)

Table 1. Evaluation results for the excluding uncertain strategy.

Label	ROC-AUC	Accuracy	Sensitivity	Specificity
Arteries	0.973	0.970	0.607	0.992
Veins	0.976	0.970	0.669	0.992
Vessels	0.980	0.960	0.749	0.989

The segmentation results for two example images from the test set are illustrated in Fig. 5. Comparing the results for the training strategies shows that the network trained with the EU strategy tends to be more discriminative for arteries and veins in the areas closer to the optic disc. The common issue for both strategies is the learned bias about the thin vessels being arteries and incapacity to capture connectivity patterns of the predicted segmentation masks inferring vein branches to be arteries.

The aforementioned problems can also be visualized as predicted epistemic and aleatoric uncertainties which are presented in Fig. 6 for the same images shown in Fig. 5. From the figure, it is clear that the epistemic uncertainty is larger near the optic disc where blood vessels cross. Further away from the optic disc it is concentrated mostly on the vessels' edges with a pattern similar to the one of the aleatoric uncertainty. Similar observations can be made from Fig. 7 where the uncertainties are compared for the two training strategies. The regions of highest uncertainty include vessel crossings and thin vessels even in the case correct classification.

3.3 Comparison with the State of the Art

The Table 3 shows a comparison of the proposed method with recently proposed methods. It is troublesome to directly compare the methods, since the evaluation methods and metrics used by different authors vary. The method proposed by Zhang et al. [18] is clearly superior compared to all the other methods, including the method studied in this work, but the authors use 5-fold cross-validation split, meaning that they have at least 32 images in the training set, whereas in this work the experiments were carried out using standard split with 20 images in the training set. Nevertheless, the performance obtained in this work is comparable with those recently published by Girard et al. [5] and Hemelings et al. [7].

Table 2. Evaluation results for the including uncertain strategy.

Label	ROC-AUC	Accuracy	Sensitivity	Specificity
Arteries	0.973	0.968	0.636	0.988
Veins	0.976	0.966	0.752	0.982
Vessels	0.981	0.961	0.797	0.984

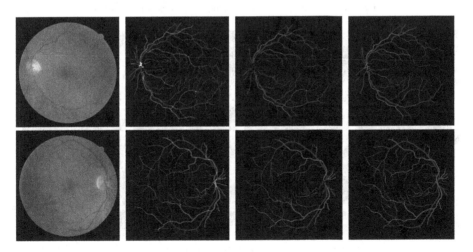

Fig. 5. Visualization of inference result: from left to right, the original image, reference standard, posterior predictive mean obtained with the excluding uncertain strategy with the including uncertain strategy.

Table 3. Comparison of evaluation results. The datasets are specified with splitting methods used by authors.

Method	Vessels accuracy	Arteries accuracy	Veins accuracy	Dataset
Girard et al. [5]	0.948	N/A	N/A	CT-DRIVE
Badawi et al. [2]	0.960	N/A	N/A	DRIVE (standard)
Hemelings et al. [7]	N/A	0.948	0.930	DRIVE (standard)
Zhang et al. [18]	N/A	0.977	0.975	DRIVE (5-fold CV)
This work	0.960	0.970	0.970	DRIVE (standard)

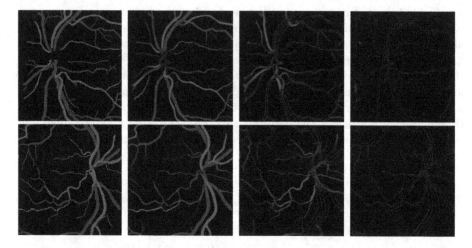

Fig. 6. Visualization of estimated uncertainty: from left to right, targets with removed uncertain labels and crossings, posterior predictive mean, epistemic uncertainty and aleatoric uncertainty. The results are obtained using the excluding uncertain strategy.

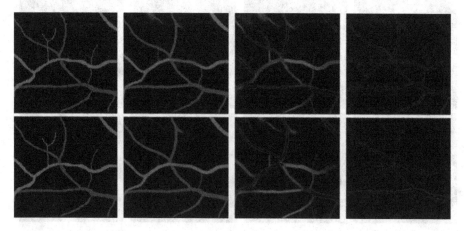

Fig. 7. Visualization of estimated uncertainty: from left to right, targets with removed uncertain labels and crossings, posterior predictive mean, epistemic uncertainty, and aleatoric uncertainty. The results are obtained using the excluding uncertain (top row) and including uncertain (bottom row) strategy.

4   Conclusion

In this work, multilabel classification of arteries and veins using a Bayesian fully-convolutional network was studied. It was shown that the misclassified areas on the images can be visualized using uncertainty estimates. The proposed approach is comparable with recent state-of-the-art approaches for blood vessel segmentation and AV classification methods.

The main topics for the future research are how to reduce the epistemic uncertainty and more careful study on the classification of uncertain labels in the RITE database. Retinal vasculature segmentation and AV classification methods typically include preprocessing procedures that affect the data. One of the opened questions, how different preprocessing techniques change the aleatoric uncertainty estimates. Other possible directions include differentiable end-to-end methods for modeling the connectivity and regularizations similar to [5] and [2].

References

1. Almotiri, J., Elleithy, K., Elleithy, A.: Retinal vessels segmentation techniques and algorithms: a survey. Appl. Sci. **8**(2), 155 (2018)
2. Badawi, S., Fraz, M.: Multiloss function based deep convolutional neural network for segmentation of retinal vasculature into arterioles and venules. BioMed Res. Int. **2019**, 1–17 (2019). https://doi.org/10.1155/2019/4747230
3. Decencière, E., et al.: Feedback on a publicly distributed image database: the messidor database. Image Anal. Stereol. **33**(3), 231–234 (2014)
4. Gal, Y., Ghahramani, Z.: Dropout as a Bayesian approximation: representing model uncertainty in deep learning. In: International Conference on Machine Learning, pp. 1050–1059 (2016)
5. Girard, F., Kavalec, C., Cheriet, F.: Joint segmentation and classification of retinal arteries/veins from fundus images. Artif. Intell. Med. **94**, 96–109 (2019). https://doi.org/10.1016/j.artmed.2019.02.004
6. He, K., Zhang, X., Ren, S., Sun, J.: Delving deep into rectifiers: surpassing human-level performance on ImageNet classification. In: Proceedings of the IEEE International Conference on Computer Vision, pp. 1026–1034 (2015)
7. Hemelings, R., Elen, B., Stalmans, I., Van Keer, K., De Boever, P., Blaschko, M.B.: Artery-vein segmentation in fundus images using a fully convolutional network. Comput. Med. Imaging Graph. **76**, 101636 (2019)
8. Hoover, A., Kouznetsova, V., Goldbaum, M.: Locating blood vessels in retinal images by piecewise threshold probing of a matched filter response. IEEE Trans. Med. Imaging **19**(3), 203–210 (2000)
9. Hu, Q., Abràmoff, M.D., Garvin, M.K.: Automated separation of binary overlapping trees in low-contrast color retinal images. In: Mori, K., Sakuma, I., Sato, Y., Barillot, C., Navab, N. (eds.) MICCAI 2013. LNCS, vol. 8150, pp. 436–443. Springer, Heidelberg (2013). https://doi.org/10.1007/978-3-642-40763-5_54
10. Hu, S., Worrall, D., Knegt, S., Veeling, B., Huisman, H., Welling, M.: Supervised uncertainty quantification for segmentation with multiple annotations. arXiv preprint arXiv:1907.01949 (2019)

11. Jégou, S., Drozdzal, M., Vazquez, D., Romero, A., Bengio, Y.: The one hundred layers tiramisu: fully convolutional DenseNets for semantic segmentation. In: 2017 IEEE Conference on Computer Vision and Pattern Recognition Workshops (CVPRW), pp. 1175–1183. IEEE (2017)
12. Jogi, R.: Basic Ophthalmology. Jaypee Brothers Medical Publishers, New Delhi (2008)
13. Kendall, A., Gal, Y.: What uncertainties do we need in Bayesian deep learning for computer vision? In: Advances in Neural Information Processing Systems, pp. 5574–5584 (2017)
14. Malek, J., Tourki, R.: Blood vessels extraction and classification into arteries and veins in retinal images. In: 10th International Multi-conferences on Systems, Signals Devices 2013 (SSD 2013), pp. 1–6, March 2013
15. Ronneberger, O., Fischer, P., Brox, T.: U-Net: convolutional networks for biomedical image segmentation. In: Navab, N., Hornegger, J., Wells, W.M., Frangi, A.F. (eds.) MICCAI 2015. LNCS, vol. 9351, pp. 234–241. Springer, Cham (2015). https://doi.org/10.1007/978-3-319-24574-4_28
16. Welikala, R., et al.: Automated arteriole and venule classification using deep learning for retinal images from the UK Biobank cohort. Comput. Biol. Med. **90**, 23–32 (2017). https://doi.org/10.1016/j.compbiomed.2017.09.005
17. Zeiler, M.D.: ADADELTA: an adaptive learning rate method. Technical report, December 2012. arXiv: 1212.5701 http://arxiv.org/abs/1212.5701
18. Zhang, S., et al.: Simultaneous arteriole and venule segmentation of dual-modal fundus images using a multi-task cascade network. IEEE Access **7**, 57561–57573 (2019). https://doi.org/10.1109/ACCESS.2019.2914319
19. Zuiderveld, K.: Contrast limited adaptive histogram equalization. In: Heckbert, P.S. (ed.) Graphics Gems IV, pp. 474–485. Academic Press Professional Inc., San Diego (1994). http://dl.acm.org/citation.cfm?id=180895.180940

Object Contour Refinement Using Instance Segmentation in Dental Images

Trung Van Pham[1,2(✉)], Yves Lucas[1], Sylvie Treuillet[1], and Laurent Debraux[2]

[1] PRISME, Polytech Orléans Site Galilée, 12 rue de Blois, BP 6744,
45067 Orléans cedex 2, France
van-trung.pham@etu.univ-orleans.fr,
{yves.lucas,sylvie.treuillet}@univ-orleans.fr
[2] Dental Monitoring, 75 rue de Tocqueville, 75017 Paris, France
{p.vantrung,l.debraux}@dental-monitoring.com

Abstract. A very accurate detection is required for fitting 3D dental model onto color images for tracking the milimetric displacement of each tooth along orthodontics treatment. Detecting the teeth boundaries with high accuracy on these images is a challenging task because of the various quality and high resolution of images. By training Mask R-CNN on a very large dataset of 170k images of patients' mouth taken with different mobile devices, we have a reliable teeth instance segmentation, but each tooth boundaries are not accurate enough for dental care monitoring. To address this problem, we propose an efficient method for object contour refinement using instance segmentation (CRIS). Instance segmentation provides high-level information on the location and the shape of the object to guide and refine locally the contour detection process. We evaluate CRIS method on a large dataset of 600 dental images. Our method improves significantly the efficiency of several state-of-the-art contour detectors: Canny (+32.0% in ODS F-score), gPb (+17.8%), Sketch Tokens (+17.3%), Structured Edge (+12.2%), Deep-Contour (+15.5%), HED (+2.9%), CEDN (+2.2%), RCF (+2.2%) and also the best result (ODS F-score of 0.819). Our CRIS method can be used with any contour detection algorithms to refine object contours. In that way, this approach is promising for other applications requiring very accurate contour detection.

Keywords: Contour detection · Instance segmentation · Object contour refinement

1 Introduction

A common issue in many medical applications concerns contours detection with a very high accuracy. In our case, we have a very large dataset of 170k of patients' mouth images taken at home with different mobile devices for dental care monitoring. A very accurate detection is required for fitting 3D dental model onto color images for tracking the milimetric displacement of each

© Springer Nature Switzerland AG 2020
J. Blanc-Talon et al. (Eds.): ACIVS 2020, LNCS 12002, pp. 99–107, 2020.
https://doi.org/10.1007/978-3-030-40605-9_9

tooth along orthodontics treatment, typically 0,25% of the diagonal image size. Processing these images is really a challenge due to the great difficulty to capture intra-oral scene leading to large variability in image quality.

In the past, teeth segmentation has been more investigated for X-ray imaging [14,15] than color images. With the development of mobile devices, teeth segmentation in color images has been recently proposed for biometrics applications [16], dental carries detection [17], or teeth color determination and 3D dental registration [18] or reconstruction with intra-oral CCD cameras [19].

Contour detection has a long and rich history in image processing. Before the recent development of deep learning, the traditional edge detectors are based on features extraction in images as gradient, brightness, color and texture information [1,2]. Next some machine learning based on carefully designed features from image patches were proposed [3,4]. Earlier deep learning based methods [5–7] continued by using convolutional layers to extract features from image patches. Recent approaches [9–11] developed end-to-end edge detection systems that are able to train and predict object contours in image-to-image fashion based on VGG-16 net [8]. On the other hand, last generation of instance segmentation by deep-learning, like Mask R-CNN, provides a reliable teeth detection but the boundaries are not accurate enough for dental care monitoring.

A specific concern for fitting 3D dental model is false positive in contour detection. Even deep learning based edge detectors do not provide contour with the required accuracy. The reason may be due to the lack of high-level information. We address this issue by using object instance segmentation to provide useful information on the location and the shape of teeth for refining the contour detection. The proposed approach can be appropriate for various computer vision application when high accuracy in contour segmentation is required. Next section presents the proposed pipeline. Experiments and results are discussed in Sect. 3, before a conclusion.

2   Proposed Method

The proposed pipeline, named Contour Refinement using Instance Segmentation (CRIS) is illustrated in Fig. 1. It consists in two main steps: (1) object instance refined segmentation and (2) contours refinement.

2.1   Object Instance Refined Segmentation

First, we apply Mask R-CNN [12] which appears as the state-of-the-art on several standard datasets. However, despite a learning from 170k dental images, there still remains issues in the accuracy of the contours given by this network. One issue is overlapping masks of neighboring teeth. If the overlap score of two instances is higher than a predefined threshold, the two instances are merged into one; otherwise, the boundaries of each mask is reduced by removing the intersection. By doing this, all the instance masks are individually separated (see the first image in Fig. 2).

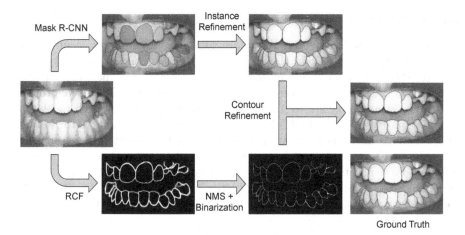

Fig. 1. Illustration of the proposed pipeline.

Then, the outline of each tooth is iteratively expanded inside a probable foreground mask (representing teeth). At each iteration, all the instances are simultaneously dilated by one pixel. If two masks become overlapped, their intersection will be removed from each one. This process stops when the maximum iteration number is reached or if there is no more space inside the foreground mask to expand instance boundaries (see the last image in Fig. 2). The probable foreground mask is obtained from a *quadmap*. By applying [13], we automatically segment the image into four types of regions (see the second image in Fig. 2): Certain Background (black); Unlikely Background (medium grey); Certain Teeth (dark grey); Probable Teeth (light grey). Background detection is based on Otsu's thresholding applied to channel a* (after color conversion from RGB to La*b*). Unlikely Background is the complementary part. Certain Teeth region is directly given by Mask R-CNN. Probable Teeth region is detected by limiting the Otsu's thresholding to the Unlikely Backgroung region. Finally, we get the limiting mask of probable foreground by merging Certain and Probable Teeth regions corresponding to dark and light grey (see the third image in Fig. 2).

Fig. 2. Object instance refined segmentation. From left to right: teeth instance obtained by Mask R-CNN, quadmap, foreground mask and refined instance contours.

2.2   Contour Refinement

The teeth boundaries provided by the previous step are not very accurate (see the first image in Fig. 3). On the other hand, the raw contours given by RCF detector [11] remain thick even after applying a standard non-maximal suppression (NMS) [1], as it can be observed on the second image in Fig. 3. We propose to mix the two approaches to get more accurate boundaries of each tooth. On thinned contours given by RCF+NMS, we select the closest points to the boundary of each instance (see the third image in Fig. 3). For unmatched contour part discarding a distance threshold, we detect end points (M_1, M_2). Then, if there exists also a connected contour between two points (R_1, R_2) at the same positions in RCF-based contours, we connect the contour between (M_1, M_2) by the shortest path between (R_1, R_2) using A star algorithm [22] (see the yellow path on the last image in Fig. 3).

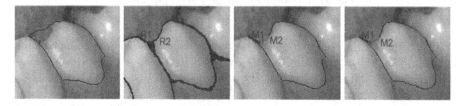

Fig. 3. Contour refinement. From left to right: refined instance boundary, RCF contours, matching points, connecting contours. (Color figure online)

3   Experiments

Dental Dataset. For Mask R-CNN training, we prepare a dataset of 170k images with manually labelled ground truth using our annotation tool which allows to manipulate the dental 3D models for fitting teeth masks on images as closely as possible (see two first rows in Fig. 4). For contour detectors training and evaluation, teeth contours were carefully annotated by experts on 600 images for ground truth for (see two last rows in Fig. 4). These images are taken by various mobile devices, with different resolutions up to 6 Mpixels. We use 300 images for training and 300 images for testing. For data augmentation, we first resize images to maximum width of 1600 pixels. Each image is then resized at scales of 0.25, 0.5 and divided equally into 4 parts at scales of 0.75, 1.0 before being flipped horizontally, vertically and both. By doing this, we augment the size of the training dataset by a factor of 40.

Implementation Details. To produce the raw contours, we compare several state-of-art detectors: Canny [1], gPb [2], Sketch Tokens [3], Structured Edges [4], DeepContour [5], HED [9], CEDN [10] and RCF [11]. To train the learning based approaches, we keep all settings and hyperparameters as default in the

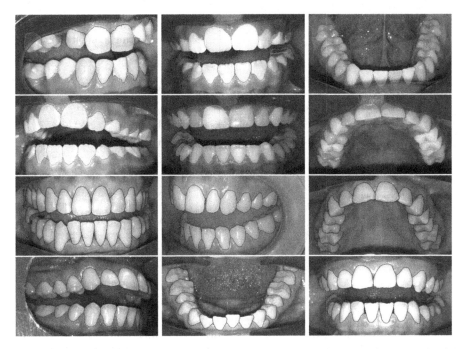

Fig. 4. Some ground truth teeth masks (two first rows) and contours (two last rows).

original implementation. For testing HED, CEDN and RCF, we scale images to the maximum width of 1024 pixels to avoid the GPU memory issue. The edgemap is then resized to their original size.

Evaluation Metrics. Contour detection accuracy is evaluated using the standard metric F-measure by applying an optimal fixed threshold for all images (optimal dataset scale, ODS) or applying an optimal threshold per-image (optimal image scale, OIS) [2]. The maximum tolerance allowed for matching contours is set to 0.25% of the diagonal image size.

Contour Detectors Comparison. Results are presented in Fig. 5 and Table 1. Figure 5 shows the precision-recall (PR) curves of ODS by varying the binarisation threshold. Note that CRIS-RCF is illustrated by one point as it produces directly the binary contours. In general, deep learning based detectors give better results. RCF achieves the best result in both ODS and OIS. The proposed pipeline for contour refinement using instance segmentation (CRIS-RCF) improves RCF scores.

Efficiency in Contour Refinement. We compare the ODS F-measures of original contour detectors facing different improvement strategies in Table 2: using mask filtering (i.e. selecting the connected contours which have intersecting parts with instances given by Mask R-CNN), using CR (i.e. CRIS without applying instance refinement step), or using CRIS. The proposed pipeline CRIS

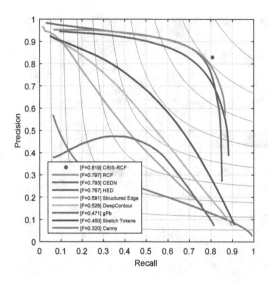

Table 1. Contour detection accuracy using standard metrics.

	ODS	OIS
Canny	.320	.333
gPb	.471	.488
Sketch tokens	.450	.457
Structured edge	.591	.599
DeepContour	.526	.528
HED	.767	.777
CEDN	.793	.796
RCF	.797	.807
CRIS-RCF	**.819**	**.819**

Fig. 5. PR curves for several contour detectors.

outperforms any contour detectors: the accuracy of traditional detectors can be significantly improved as for Canny (+32.0%), gPb (+17.8%), Sketch Tokens (+17.3%), Structured Edge (+12.2%), and even these of deep learning based detectors as for DeepContour (+15.5%), HED (+2.9%), CEDN (+2.2%), RCF (+2.2%). The improvement of the last deep learning methods (HED, CEDN, RCF) is much lesser degree because they already greatly reduce false positive contours. The proposed pipeline achieves also better improvement than using filtering or using CR step only.

The comparison between CR and CRIS emphasizes the importance of the object instance refinement step in our algorithm. When evaluating the performance of Mask R-CNN and refined instance contours, we observe that

Table 2. Efficiency comparison of some improvement methods.

	Original	Using filtering	Using CR	Using CRIS
Canny	.320	.558 (+23.8%)	.532 (+21.2%)	.640 (**+32.0%**)
gPb	.471	.473 (+0.2%)	.580 (+10.9%)	.649 (**+17.8%**)
Sketch tokens	.450	.474 (+2.4%)	.617 (+16.7%)	.623 (**+17.3%**)
Structured edge	.591	.636 (+4.5%)	.626 (+3.5%)	.713 (**+12.2%**)
DeepContour	.526	.606 (+8.0%)	.563 (+3.7%)	.681 (**+15.5%**)
HED	.767	.773 (+0.6%)	.747 (−2.0%)	.796 (**+2.9%**)
CEDN	.793	.799 (+0.6%)	.784 (−0.9%)	.815 (**+2.2%**)
RCF	.797	.802 (+0.5%)	.787 (−1.0%)	.819 (**+2.2%**)

the accuracy increases from ODS = .520 to ODS = .598 (+7.8%). This indicates that the improvement on object instance precision leads to the same effect on contour refinement. Using CR only can even weakly decreases the F-measure for the three last methods because of inaccurate instance boundaries given by Mask R-CNN.

Figure 6 displays some qualitative results of CRIS-RCF that demonstrate the efficiency of our refinement method in reducing the false positive contours. The two last columns show the results given by the proposed approach in case of improper contour detection given by RCF (second last column) or improper instance segmentation by Mask-CNN (last column). In the first case, our algorithm is not able to find the completed teeth boundary due to the missing contours of RCF, and in the second case, the instance contour is not close enough to the teeth boundary.

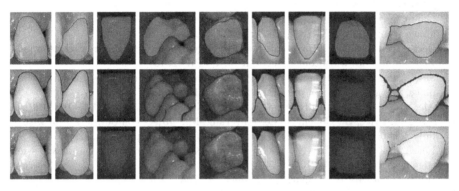

Fig. 6. Some qualitative results of our approach. From top to bottom: teeth instance masks obtained by Mask R-CNN, raw contours detected by RCF, CRIS-RCF refined contours.

4   Conclusions

We have developed an efficient method to refine contours using object instance segmentation. To our best knowledge, this is the first time a very large dental dataset, including 170k color images with ground truth teeth masks and 600 images with teeth contours, has been used. The experiments demonstrated that our refined contour detection achieves a much better accuracy in contour detection than most of the state-of-art contour detectors with ODS F-measure of 0.819. Our CRIS method can be used with any contour detection algorithms to refine object contours. In that way, this approach is promising for other applications requiring very accurate contour detection. The proposed refinement method could also improve dataset ground truth which is imperfectly annotated with polygons as it is a very time-consuming manual task. In the next months, we plan to apply it to our dental image database and to other computer vision standard datasets (e.g. PASCAL VOC [23]) as well. Developments are underway to validate a contour refinement approach by taking into account more classes in the instances including dental braces and the gum retractor.

References

1. Canny, J.: A computational approach to edge detection. PAMI **8**(6), 679–698 (1986)
2. Arbelaez, P., Maire, M., Fowlkes, C., Malik, J.: Contour detection and hierarchical image segmentation. PAMI **33**(5), 99–110 (2011)
3. Lim, J.J., Zitnick, C.L., Dollar, P.: Sketch tokens: a learned mid-level representation for contour and object detection. In: CVPR (2013)
4. Dollar, P., Zitnick, C.L.: Fast edge detection using structured forests. PAMI **37**(8), 1558–1570 (2015)
5. Shen, W., Wang, X., Wang, Y., Bai, X., Zhang, X.: DeepContour: a deep convolutional feature learned by positive-sharing loss for contour detection. In: CVPR (2015)
6. Ganin, Y., Lempitsky, V.: N4-fields: neural network nearest neighbor fields for image transforms. arXiv preprint arXiv:1406.6558 (2014)
7. Bertasius, G., Shi, J., Torresani, L.: DeepEdge: a multiscale bifurcated deep network for top-down contour detection. In: CVPR (2015)
8. Simonyan, K., Zisserman, A.: Very deep convolutional networks for large-scale image recognition. arXiv preprint arXiv:1409.1556 (2014)
9. Xie, S., Tu, Z.: Holistically-nested edge detection. In: ICCV, pp. 1395–1403 (2015)
10. Yang, J., Price, B., Cohen, S., Lee, H., Yang, M.: Object contour detection with a fully convolutional encoder-decoder network. In: CVPR, pp. 193–202 (2016)
11. Liu, Y., Cheng, M., Hu, X., Wang, K., Bai, X.: Richer convolutional features for edge detection. In: CVPR, pp. 5872–5881 (2017)
12. He, K., Gkioxari, G., Dollar, P., Girshick, R.: Mask R-CNN. In: ICCV (2017)
13. Rother, C., Kolmogorov, V., Blake, A.: "GrabCut": interactive foreground extraction using iterated graph cuts. In: SIGGRAPH (2004)
14. Said, E.H., Nassar, D.E.M., Fahmy, G., Ammar, H.H.: Teeth segmentation in digitized dental X-ray films using mathematical morphology. IEEE Trans. Inf. Forensics Secur. **1**, 178–189 (2006)
15. Shah, S., Abaza, A., Ross, A., Ammar, H.: Automatic tooth segmentation using active contour without edges. In: Biometrics Symposium: Special Session on Research at the Biometric Consortium Conference (2006)
16. Na, S.D., Lee, G., Lee, J.H., Kim, M.N.: Individual tooth region segmentation using modified watershed algorithm with morphological characteristic. Bio Med. Mater. Eng. **24**(6), 3303–3309 (2014)
17. Ghaedi, L., et al.: An automated dental caries detection and scoring system for optical images of tooth occlusal surface. In: EMBC (2014)
18. Destrez, R., Albouy, A., Treuillet, S., Lucas, Y.: Automatic registration of 3D dental mesh based on photographs of patient's mouth. Comput. Methods Biomech. Biomed. Eng. Imaging Vis. **7**, 604–615 (2018)
19. Abdelrahim, A.S., El-Melegy, M.T., Farag, A.A.: Realistic 3D reconstruction of the human teeth using shape from shading with shape priors. In: IEEE CVPRW (2012)
20. Otsu, N.: A threshold selection method from gray-level histograms. IEEE Trans. Syst. Man Cybern. **9**(1), 62–66 (1979)

21. Eduardo, S.L.G., Oliveira, M.M.: Shared sampling for real-time alpha matting. Comput. Graph. Forum **29**(2), 575–584 (2010)
22. Hart, P.E., Nilsson, N.J., Raphael, B.: A formal basis for the heuristic determination of minimum cost paths. IEEE Trans. Syst. Sci. Cybern. **4**(2), 100–107 (1968)
23. Everingham, M., Gool, L.J.V., Williams, C.K.I., Winn, J.M., Zisserman, A.: The Pascal visual object classes (VOC) challenge. IJVC **88**(2), 303–338 (2010)

Correction of Temperature Estimated from a Low-Cost Handheld Infrared Camera for Clinical Monitoring

Evelyn Gutierrez[1]([✉]), Benjamin Castañeda[1], and Sylvie Treuillet[2]

[1] Pontificia Universidad Católica del Perú, Av. Universitaria 1801, San Miguel, Lima, Perú
{egutierreza,castaneda.b}@pucp.edu.pe
[2] Laboratoire PRISME, Polytech, 12 rue de Blois, BP 6744, 45067 Orleans Cedex 2, France
sylvie.treuillet@univ-orleans.fr

Abstract. The use of low-cost cameras for medical applications has its advantages as it enables affordable and remote evaluations of health problems; however, the accuracy is a limiting factor to use them. Previous studies indicate that parameters from object position like distance camera-object and angle of view could be used to improve temperature estimation from thermal cameras. Nevertheless, most studies are focused on expensive thermal cameras with good accuracy. In this study, an innovative experimental setup is used to study the errors associated to temperature estimation from a low-cost infrared camera: FlirOne Gen3. In our experiments, the image acquisition is done from multiple point of view (distance camera-object and viewing angles) and by using a thermal camera manipulated by hand. Then, using a regression model, a correction is proposed and tested. The results show that our proposed correction improves the temperature estimation and enhance the thermal accuracy.

Keywords: Temperature correction · Low-cost infrared cameras · Clinical thermal imaging

1 Introduction

Thermographic has the potential as a noninvasive tool to be used in clinical monitoring as noted in some studies: [1–5]. However, accurate estimation of the temperature is sometimes required for using it in clinical settings. Professional infrared cameras have the highest accuracy (around ± 1 °C) but they are expensive and hard to manipulate.

In recent years, there have been improvements in IR imaging with portable and low-cost cameras as they are more convenient to be used for remote clinical monitoring [6, 7]. On the other hand, the disadvantage of low-cost cameras is the large errors. Technical specification of low-cost IR cameras like FlirOne Gen3 show accuracy of ± 3 °C or $\pm 5\%$.

Some studies exhibit the possibility to improve temperatures estimation from thermal cameras and recent ones consider low-cost cameras. Curran and al. proposed a correction by using a reference object with known temperature in the field of view of the thermal

© Springer Nature Switzerland AG 2020
J. Blanc-Talon et al. (Eds.): ACIVS 2020, LNCS 12002, pp. 108–116, 2020.
https://doi.org/10.1007/978-3-030-40605-9_10

camera [8]. The proposed correction worked well for various thermal cameras but no significant improvement was found for a low-cost IR camera like FlirOne.

It is known that the accuracy depends on many factors, including the distance of the object, the ambient temperature and the emissivity of the observed material. Theoretical studies [9, 10] and experimental results on human skin [11, 12] indicate that the emissivity decreases significantly when the viewing angle is greater than 60°, i.e. the angle between the optical axis of the camera and the normal to the surface, which leads to inaccurate temperature mapping.

Therefore, some works proposed correction for directional emissivity. Cheng and al. estimated a linear relationship from the calibration of the measurement error due to the viewing angle [13]. Zeise and Wagner used a non-linear least squares optimization minimizing the error of the measured output signals from theoretical models describing emissivity as a function of the viewing angle for known material emissivity class (metal/dielectric) [14]. Arnon and al. proposed a correction for clinical monitoring by thermal images based an empirical second-degree polynomial equation for calibrating the decreasing apparent temperature due to dependency of the skin emissivity on the viewing angle [15]. All the previous methods were tested and used with high end thermal cameras but not with low-cost IR cameras.

The distance between the camera and the object has also been considered. Ting and al. use a Kinect to obtain information on the depth of the image scene and proposed a correction based on it [16]. The correction was tested and accuracy was improved in a low-resolution thermal camera.

In the case of a handheld camera, the shooting protocol cannot be perfectly normalized and the point of view remains approximate. In this paper, we present an experimental setup using only a low-cost camera held in hand to study the variations of thermal measurements induced by the camera point of view. Then, we propose a correction based on a regression model to correct the temperature as a function of the viewing angle and the distance of the camera to object.

The document is organized as follows: Sect. 2 explains the experimental setup; Sect. 3 shows the analysis of absolute errors using a low-cost IR camera; Sect. 4 proposes a methodology for correction; and Sect. 5 presents the results after implementing the proposed correction.

2 Experimental Setup

The low-resolution camera used for the experiments is FlirOne Gen 3 (Flir Systems, Oregon, USA) which according to the specifications has accuracy ±3 °C or ±5%, thermal resolution 80 × 60 and thermal sensitivity is 150 mK.

To analyze the errors, an experimental setup where the temperature could be measured and controlled is required. This is achieved by utilizing a temperature-controlled stage and a controller with LCD touchscreen from LINKAM (Linkam Scientific Instruments Ltd, Waterfield, Epsom, United Kingdom). The temperature stage has a circular piece of metal in the middle which has 2 cm of diameter. The temperature could be set to any value between −196 °C and 125 °C; and the circular piece of metal will increase its temperature until reaching the desired degrees with an accuracy of less than ±0.1 °C.

Once the temperature is achieved, the device keeps it for around one hour. The metal piece had a reflective surface which was covered with carbon tape for eliminating reflections.

Additionally, four Aruco Markers are placed around the object of interest (piece of metal) to estimate the camera pose relative to the object. From the pose, we principally consider the distance of the camera from the object and the angle between the optical axis and the normal of the object surface. A label is placed on the side of the device to easily identify the set temperature in the experiments. The experimental setup is presented in Fig. 1.

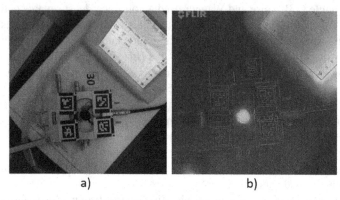

a)   b)

Fig. 1. Experimental set-up: (a) The temperature controlled stage and the Aruko markers; (b) RGB image edges overlaid on Infrared image.

Using this experimental setup and controlling the temperature from 20 to 40 °C, 365 thermal images were captured with FlirOne Gen3 camera connected to an iPad using handheld shooting and varying the point of view. The distance of the camera from the object was varied from 10 to 70 cm and the viewing angle from 0 to 60°. Some examples of image shooting are illustrated in Fig. 2 while Fig. 3 shows illustrates camera poses.

Fig. 2. Sample of images with different points of views.

The temperature has been measured in thermal images using a free package called "thermimage" within statistical software R based on the standard equations used in thermography and described in [17]. The estimation obtained is similar to the one used in

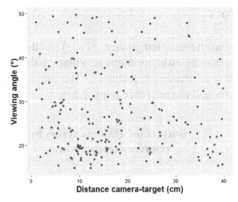

Fig. 3. Different camera poses regarding the distance between camera and target and the angular deviation between the optical axis and the target normal (viewing angle).

commercial Flir software: Tattersall demonstrate that the difference between the estimations from commercial Flir software and the free implementation has mean 0.03 °C and a standard deviation of 0.0312 °C [17].

The temperature estimation takes into account several parameters from the thermal camera which are named Plank constants. The only parameters that have been customized are the following: emissivity, ambient temperature and distance camera-object. For all images, the emissivity was specified to 0.98; ambient temperature to 22 °C; and the distance in centimeters was estimated from the four Aruco markers.

The temperature of the metal plate is measured in the thermal image by calculating the median value in a region of interest (ROI) of 6×6 pixels (T_{Est}). The region of interest is illustrated in Fig. 4.

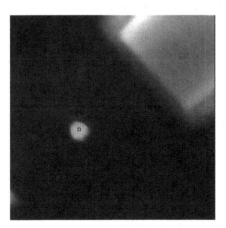

Fig. 4. Thermal image and the ROI of 6×6 pixels used to estimate the temperature of metal plate.

3   Analysis of Errors

The difference between the reference temperature (T_{Ref}) and the temperature estimation (T_{Est}) are used to calculate the estimation errors. Figure 5 shows the boxplots of the absolute error at different reference temperatures. The graph suggests a relationship between the absolute error and the reference temperature.

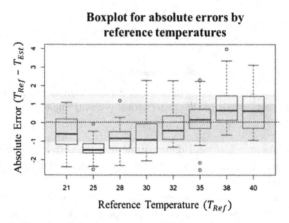

Fig. 5. Boxplot of absolute errors at different reference temperatures.

Similarly, to study the errors regarding the camera position, some plots of the absolute error against the distance and angles are presented in Fig. 6. On the left, the scatterplot shows the relationship between the estimation errors regarding the distance from the camera to the target and, on the right, regarding the viewing angles. A linear relationship can be assumed with the distance but it is not so evident for the angles.

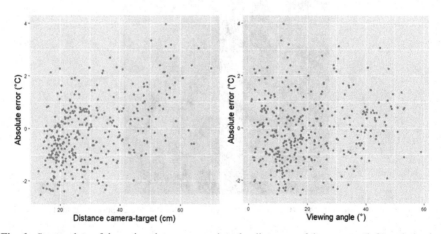

Fig. 6. Scatterplots of the estimation errors against the distances of the camera (left) and viewing angle (right)

Based on these observations, it seems possible to propose a correction of the measured temperature.

4 Proposed Model for Correction

As we want to use the correction for clinical purposes, the dataset used for the proposed correction contains only pictures with temperatures between 30 to 40°; distances between 10 and 50 cm; and angles between 0 to 40°. Using these criteria, a set of 190 images were finally selected.

A regression model is proposed to correct the estimated temperature. This model will be used to estimate reference temperature; however, T_{Ref} will not be used as the response variable. This variable is controlled in the experiments and therefore is not a random variable that could be fitted using regression model. Instead, T_{Est} is used as the response variable and for implementing this model, a methodology used commonly in chemistry for calibration of instruments is used. This methodology is called inverse prediction [18] and it allows to predict the values of the variable of interest: the reference temperature (T_{Ref}).

In order to define the variables used in the proposed model, a granular analysis of the estimated temperature against distance and angles has been done. In Fig. 7, the estimated temperature is seen as a function of the distances and angles. Each color represents the experiments in a specific controlled temperature. In this figure, a consistent linear relationship between the estimated temperature and the distance could be seen. In contrast, a weaker relationship with the angles is visualized with a second order polynomial relationship. Based on these results, a linear relationship with the distances and a second order polynomial relationship with the angles were considered for the structure of the proposed model.

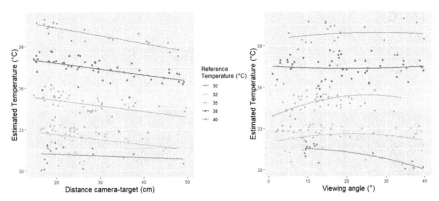

Fig. 7. Scatterplots of the estimated temperature against the distances from camera to object (left) and angle between camera plane and object plane (right).

For validating the proposed model, 5-fold cross validation was used and the Root Mean Squared Error (RMSE) was calculated. The average RMSE obtained in cross validation after using the proposed model is 0.855. In contrast, the RMSE of the initial temperature estimation was 0.969. This proves that using the proposed model structure, it is possible to reduce estimation errors.

Furthermore, the proposed model was fitted using the whole dataset to obtain the final estimates. Its structure is presented in Eq. (1), and estimated coefficients (β_i) are given in Table 1.

$$T_{Est} = \beta_0 + \beta_1 T_{Ref} + \beta_2 D + \beta_3 A + \beta_4 A^2 + \varepsilon, \tag{1}$$

Where D is the distance of the camera to the temperature controlled target and A is the viewing angle.

Table 1. Estimated beta coefficients for model described in Eq. (1)

Predictors	Estimates	Confidence interval	p-value
Intercept (β_0)	5.051	[3.998–6.988]	<0.001
Reference temp. (β_1)	0.874	[0.823–0.908]	<0.001
Distance (β_2)	−0.042	[−0.059−−0.030]	<0.001
Angle (β_3)	0.065	[0.004–0.102]	0.038
Angle squared (β_4)	−0.002	[−0.002−−0.0001]	0.038
Observations	190		
R2/adjusted R2	0.93/0.928		

Finally, Fig. 8 presents the boxplots without outliers for the absolute error before and after the proposed correction. Before correction, the errors were between −2.56 °C and 2.31 °C while after applying the proposed correction, the errors are between −1.44 °C and 1.66 °C.

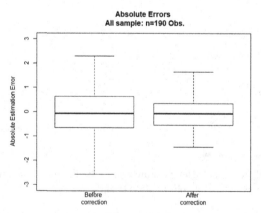

Fig. 8. Boxplot of the absolute errors before and after implementing the proposed correction.

5 Conclusions

Portable and low-cost cameras like FlirOne Gen 3 could have a low accuracy as described in their specifications. Moreover, in the use of clinical applications with handheld camera, the point of view cannot be controlled carefully.

The temperature accuracy depends on point of view and in particular, on the distance of the object and viewing angle. This study shows that it is possible to improve the temperature estimation by using a regression model and inverse prediction. As a result, this could enable improvements in low-cost infrared cameras for being used in clinical measurements.

Future work includes testing the proposed methodology for generalization using another low-cost thermal camera and a larger set of experiments varying environmental conditions (even if the temperature does not vary much indoors with air conditioning). The proposed correction will be used in thermal images of human skin. Finally, we plan to create 3D thermograms for wounds monitoring: corrected thermal information could be overlaid onto 3D models in order to provide accurate healing indexes.

Acknowledgments. This project has received funding from the European Union's Horizon 2020 research and innovation program under the Marie Sklodowska-Curie grant agreement #777661 (STANDUP project).

References

1. Colantonio, S., Pieri, G., Salvetti, O., Benvenuti, M., Barone, S., Carassale, L.: A method to integrate thermographic data and 3D shapes for Diabetic Foot Disease. In: Proceedings of the International Conference on Quantitative InfraRed Thermography (2006)
2. Serbu, G.: Infrared imaging of the diabetic foot. InfraMation **86**, 5–20 (2009)
3. Vilcahuaman, L., et al.: Detection of diabetic foot hyperthermia by infrared imaging. In: 36th Annual International Conference of the IEEE Engineering in Medicine and Biology Society (2014)
4. Fraiwan, L., Alkhodari, M., Ninan, J., Mustafa, B., Saleh, A., Ghazal, M.: Diabetic foot ulcer mobile detection system using smart phone thermal camera: a feasibility study. BioMed. Eng. OnLine **16**, 117 (2017)
5. Alametsä, J., Oikarainen, M., Perttunen, J., Viik, J., Vaalasti, A.: Thermal imaging in skin trauma evaluation: observations by CAT S60 mobile phone. Finnish J. eHealth eWelfare **10**, 192–199 (2018)
6. Bougrine, A., Harba, R., Canals, R., Ledee, R., Jabloun, M.: A joint snake and atlas-based segmentation of plantar foot thermal images. In: Seventh International Conference on Image Processing Theory, Tools and Applications, IPTA (2017)
7. Doremalen, R.V., Netten, J.V., Baal, J.V., Vollenbroek-Hutten, M., Heijden, F.V.D.: Validation of low-cost smartphone-based thermal camera for diabetic foot assessment. Diabetes Res. Clin. Pract. **149**, 132–139 (2019)
8. Curran, A., Klein, M., Hepokoski, M., Packard, C.: Improving the accuracy of infrared measurements of skin temperature. Extrem. Physiol. Med. **4**, A140 (2015)
9. Watmough, D.J., Fowler, P.W., Oliver, R.: The thermal scanning of a curved isothermal surface: implications for clinical thermography. Phys. Med. Biol. **15**, 1–8 (1970)

10. Hejazi, S., Spangler, R.: Theoretical modeling of skin emissivity. In: Proceedings of the Annual International Conference of the IEEE Engineering in Medicine and Biology Society (1992)
11. Lewis, D.W., et al.: Apparent temperature degradation in thermograms of human anatomy viewed obliquely. Radiology **106**(1), 95–99 (1973)
12. Ash, C.J., Gotti, E., Haik, C.H.: Thermography of the curved living skin surface. Mol. Med. **84**, 702–708 (1987)
13. Cheng, T.-Y. et al.: Curvature effect quantification for In-Vivo IR thermography. In: Biomedical and Biotechnology, vol. 2 (2012)
14. Zeise, B., Wagner, B.: Temperature correction and reflection removal in thermal images using 3D temperature mapping. In: Proceedings of the 13th International Conference on Informatics in Control, Automation and Robotics (2016)
15. Arnon, B., Oria, K., Arieli, Y.: Correction of the angular emissivity of human skin for clinical thermal imaging. Imaging Med. **9**(4), 103–108 (2017)
16. Ting, L.P.: Errors in thermographic camera measurement caused by known heat sources and depth based correction. Int. J. Autom. Smart Technol. **6**, 5–12 (2016)
17. Tattersall, G.J.: Thermimage: Thermal Image Analysis, 3 December 2017. http://doi.org/10.5281/zenodo.1069704. R package https://CRAN.R-project.org/package=Thermimage. Accessed 14 Sept 2019
18. Massart, L.M., Vandenginste, B.G.M., Buydens, L.M.C., De Jong, S., Lewi, P.J., Smeyers-Verbeke, J.: Handbook of Chemometrics and Qualimetrics: Part A. Elsevier, Amsterdam (1997)

Bayesian Feature Pyramid Networks for Automatic Multi-label Segmentation of Chest X-rays and Assessment of Cardio-Thoratic Ratio

Roman Solovyev[1], Iaroslav Melekhov[2,5], Timo Lesonen[3], Elias Vaattovaara[3], Osmo Tervonen[3,4], and Aleksei Tiulpin[3,4,5(✉)]

[1] IPPM RAS, Russian Academy of Sciences, Moscow, Russia
[2] Aalto University, Espoo, Finland
[3] Oulu University Hospital, Oulu, Finland
[4] University of Oulu, Oulu, Finland
aleksei.tiulpin@oulu.fi
[5] Ailean Technologies Oy, Oulu, Finland

Abstract. Cardiothoratic ratio (CTR) estimated from chest radiographs is a marker indicative of cardiomegaly, the presence of which is in the criteria for heart failure diagnosis. Existing methods for automatic assessment of CTR are driven by Deep Learning-based segmentation. However, these techniques produce only point estimates of CTR but clinical decision making typically assumes the uncertainty. In this paper, we propose a novel method for chest X-ray segmentation and CTR assessment in an automatic manner. In contrast to the previous art, we, for the first time, propose to estimate CTR with uncertainty bounds. Our method is based on Deep Convolutional Neural Network with Feature Pyramid Network (FPN) decoder. We propose two modifications of FPN: replace the batch normalization with instance normalization and inject the dropout which allows to obtain the Monte-Carlo estimates of the segmentation maps at test time. Finally, using the predicted segmentation mask samples, we estimate CTR with uncertainty. In our experiments we demonstrate that the proposed method generalizes well to three different test sets. Finally, we make the annotations produced by two radiologists for all our datasets publicly available.

1 Introduction

Heart failure (HF) is highly prevalent in different populations. As such, its prevalence varies from 1% to 14% of the population according to the data from Europe and the United States [10]. One clinical factor having impact on the diagnosis of HF is cardiomegaly, which is a condition affecting heart enlargement [10]. In clinical practice, assessment of cardiomegaly is trivial to a human

---

R. Solovyev and I. Melekhov—Equal contribution.

T. Lesonen and E. Vaattovaara—Equal contribution.

© Springer Nature Switzerland AG 2020
J. Blanc-Talon et al. (Eds.): ACIVS 2020, LNCS 12002, pp. 117–130, 2020.
https://doi.org/10.1007/978-3-030-40605-9_11

expert (radiologist) and typically done by a visual assessment. However, there are multiple clinical scenarios when the radiologist is not available, for example in emergency care or intensive care units.

Clinically accepted quantitative measure of cardiomegaly is cardiothoratic index (CTI) – a ratio of the heart's and the lungs' widths. In the literature, CTI is also often called a cardiothoratic ratio (CTR). CTR can be measured from chest radiographs which constitute over a half of radiographic imaging done in clinical practice [7].

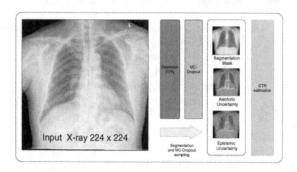

Fig. 1. Overview of the workflow proposed in this study. We use our proposed modification of Bayesian FPN and perform MC-dropout inference to obtain the multilabel segmentation masks (one for the heart and one for the lungs) as well as the aleotoric and the epistemic uncertainties per image.

Multiple recent studies have demonstrated promising results in assessing chest and other radiographs by applying Deep Learning (DL) [34,35,38]. These efforts indicate a possibility of reducing the amount of human labor needed for visual image assessments. Ultimately, this technology has potential to reduce the health care costs while keeping the same quality of diagnosis [27].

DL is a methodology of learning representations directly from data [29]. Typically, these representations (features) are learned with respect to the task, such as image classification or segmentation. The latter allows to classify image pixels individually and eventually obtain the locations and boundaries of the objects within an image. DL-based image segmentation was shown to be a core technique in assessing CTR from chest X-rays [9,22]. However, none of the existing CTR assessment or chest X-ray segmentation methods allow to obtain the model uncertainty which is crucial in clinical practice.

In this paper, we propose a robust Bayesian pipeline for CTR estimation, which segments the anatomical structures in the images and also outputs the model uncertainty. Our approach is based on Feature Pyramid Network (FPN) [23,30] with ResNet50 backbone [14] and instance normalization in the decoder [36]. For uncertainty estimation, we follow [20] and utilize Monte-Carlo (MC) dropout at test time. Schematically, the proposed approach is illustrated in Fig. 1.

The main contributions of this paper are summarized as follows: 1. We extend traditional DL-based methods for CTR estimation to Bayesian neural network which can predict pixel-wise class labels and uncertainty bounds from segmentation masks; 2. Compared to all the previous studies, we propose a challenging training dataset with diverse radiological findings annotated by a radiologist; 3. The model evaluation is performed on 3 widely-used public X-ray image datasets which were re-annotated in a similar way to our training dataset, but come from different scanners and hospitals; 4. To the best of our knowledge, this is the first work that uses Bayesian DL in both chest X-ray segmentation and CTR estimation domains. Our methodology allows to assess the uncertainty of the model at test time, thereby providing clinical value in potential applications; 5. Finally, we publicly release the annotations and the training dataset utilized in this study. We think that these data could set up a new, more challenging benchmark in chest X-ray segmentation.

2   Related Work

Chest X-ray Segmentation. The most relevant studies to ours are by Dong *et al.* [9], by Dai *et al.* [7] and also by Elsami *et al.* [11]. They introduced adversarial training to enforce the consistency between the predictions and the ground truth annotations. Both studies explore the same methodology while the former one is mainly focused on CTR estimation and uses the adversarial training for unsupervised domain adaptation (UDA), the latter is rather targeting segmentation of chest X-ray. The methods demonstrate that better generalization performance to unseen data can be achieved by using adversarial training.

Besides CTR estimation realm, there are other studies approaching the segmentation problem of chest X-ray images by applying DL. Arbabshirani *et al.* [1] and recent works [4,33] demonstrated remarkable performance in lungs segmentation. Furthermore, Wessel *et al.* [40] utilized mask R-CNN [13] to successfully localize, segment and label individual ribs.

From the segmentation field point of view in general, there exist multiple studies that use FPN as a decoder for image segmentation [21,25]. In particular, the study by Seferbekov *et al.* [30] explores a very similar architecture to ours and seems to be the first to demonstrate a combination of ImageNet-pretrained ResNet50 encoder with FPN decoder successfully applied to image segmentation.

In Bayesian segmentation, we note the study by Kendall *et al.* [20] that introduced the use of MC dropout for the uncertainty estimation in image segmentation. Furthermore, the recent study by [24] proposed to use a modification of DeepLab-v3+ [5] that allowed to achieve state-of-the-art segmentation results and at the same time obtain uncertainty estimates.

Limitations of the Existing Chest X-ray Segmentation Datasets. In this paragraph, we also tackle an important issue of existing annotations and images in CTR assessment and Chest X-ray segmentation realm. In particular, all the existing DL-based CTR estimation aforementioned methods have been trained on the datasets that *do not* include the true boundaries of the anatomical structures within the chest X-rays. While this does not have significant impact on CTR estimation in general, the absence of the true boundaries of heart and lungs limits the scope of applications that can be built using the automatic Chest X-ray segmentation (e.g. detection of plural effusion). Moreover, the existing datasets were originally from tuberculosis (TB) domain which also limits the reliable testing of the segmentation and CTR assessment models. We argue that *clinically applicable* methods must be trained and tested on the datasets that have diverse radiological findings.

3   Method

Overview. The proposed method is based on a combination of several state-of-the-art techniques for image segmentation. Specifically, our model leverages an encoder pre-trained on ImageNet [8], Feature Pyramid Networks-inspired decoder [23,30], instance normalization [36] and MC dropout at test time. The architecture of the proposed approach is illustrated in Fig. 2.

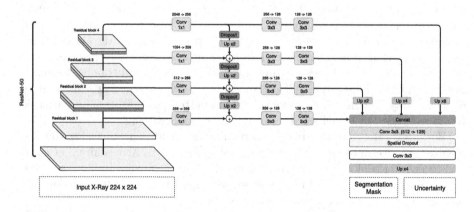

Fig. 2. FPN-based model architecture. In our model, we inserted the dropouts before the second, third and the fourth residual blocks (red). Besides, we also used the dropout in the FPN part of our model and used it before every upsampling layer in the feature pyramid (nearest neighbour). However, the dropout was not used before the upsampling layers that were followed by the concatenation of the feature maps. The decoder in our model used instance normalization in the convolutional blocks (yellow) and the final upsampling layer used a bi-linear interpolation. (Color figure online)

Backbone. We used a standard ResNet50 pre-trained on ImageNet [8]. We do not freeze the encoder during training and merely use it as is from the beginning. It is worth to note that our pre-trained encoder follows a Batch Normalization (BN) layer that learns the mean of the dataset during the training. Furthermore, we inserted the dropouts with a probability $p = 0.5$ before the second, third and the fourth residual blocks of ResNet50 (see Fig. 2).

Decoder. The decoder is a standard FPN. However, similarly to Seferbekov *et al.* [30], we do not use intermediate supervision and prediction at each layer of the feature pyramid as it is usually done for object detection [23]. As the sizes of the feature pyramid and the input image do not match, we use nearest neighbours upsampling. In contrast to [30], we replace each BN layer in the decoder by instance normalization (IN) layer at every level of the feature pyramid.

In addition to dropout units in the backbone, we also apply them after each 1×1 convolutional block of the feature pyramid as illustrated in Fig. 2. More specifically, the dropouts have been used in the decoder only before upsampling layers.

As the task of segmenting the chest X-ray structures is multi-label, rather than multi-class, the decoder has 2 outputs, where the first plane corresponds to the heart and the second one to the lungs. Before the final output layer we used a spatial dropout with a rate of 0.1.

Bayesian Segmentation Framework: Monte-Carlo Dropout. As mentioned previously, we leverage MC-dropout technique [12,20]. To capture the model's uncertainty, it is necessary to estimate the posterior distribution $p(\mathbf{W}|\mathbf{X}, \mathbf{Y})$ of the model's weights $\mathbf{W}$ given the train images $\mathbf{X}$ and the corresponding labels $\mathbf{Y}$. However, this distribution is intractable, therefore, its variational approximation $q(\mathbf{W})$ [24]. Gal and Ghahramani [12] have shown that training a neural network with dropout and standard cross-entropy loss function leads to minimization of Kullback-Leibler (KL) divergence between $q(\mathbf{W})$ and $p(\mathbf{W}|\mathbf{X}, \mathbf{Y})$:

$$\mathrm{KL}\left(q(\mathbf{W})\|p(\mathbf{W}|\mathbf{X}, \mathbf{Y})\right) \to \min_{\mathbf{W}}, \tag{1}$$

where $q(\mathbf{W})$ is chosen to be a Bernoulli distribution.

In our experiments, we enabled the dropout layers in both encoder and FPN, respectively. We then performed the sampling of T pixel-wise probability masks similarly to Kendall *et al.* [20]. Here, for every pixel $\mathbf{I}(i,j)$ of the input image $\mathbf{I}$, having T MC-dropout iterations, we generate the prediction $\hat{\mathbf{Y}}^{(t)}(i,j)$ at every t^{th} MC dropout iteration and eventually estimate $\mathbb{E}[\hat{\mathbf{Y}}(i,j)]$ to produce the segmentation masks $\mathbb{E}[\hat{\mathbf{Y}}_h]$ and $\mathbb{E}[\hat{\mathbf{Y}}_l]$ for the heart and the lungs, respectively.

Bayesian Segmentation Framework: Aleotoric and Epistemic Uncertainties. Besides the segmentation masks, the proposed framework also produced uncertainty estimates per pixel. As such, we computed both *aleotoric* and *epistemic* uncertainties. Briefly, the former one captures the inherent noise in the data (e.g. sensor noise) while the latter exhibits the model's uncertainty. Both of these uncertainties are important as aleotoric uncertainty allows to estimate the need of improving the sensor precision and the epistemic uncertainty enables to assess the need of larger training dataset [24]. Similarly to Mukohti and Gal [24], we approximated the aleotoric uncertainty for the test examples $\mathbf{x}$ given the train data $\mathcal{D}_{train}$ as a predictive entropy $\hat{\mathbb{H}}[y|\mathbf{x}, \mathcal{D}_{train}]$:

$$\hat{\mathbb{H}}[y|\mathbf{x}, \mathcal{D}_{train}] = -\sum_c \left(\frac{1}{T}\sum_t p(y = c|\mathbf{x}, \hat{w}_t)\right) \cdot \log\left(\frac{1}{T}\sum_t p(y = c|\mathbf{x}, \hat{w}_t)\right),$$

(2)

and the epistemic uncertainty was approximated as the mutual information between the predictive distribution and the posterior over the weights of the model:

$$\hat{\mathbb{I}}[y, \mathbf{W}|\mathbf{x}, \mathcal{D}_{train}] = \hat{\mathbb{H}}[y|\mathbf{x}, \mathcal{D}_{train}] + \frac{1}{T}\sum_{c,t} p(y = c|\mathbf{x}, \hat{w}_t)\log p(y = c|\mathbf{x}, \hat{w}_t). \quad (3)$$

In our experiments, we estimated both $\hat{\mathbb{H}}_{heart}[y|\mathbf{x}, \mathcal{D}_{train}]$ and $\hat{\mathbb{H}}_{lungs}[y|\mathbf{x}, \mathcal{D}_{train}]$ for each image (similarly, also $\hat{\mathbb{I}}_{heart}[y, \mathbf{W}|\mathbf{x}, \mathcal{D}_{train}]$ and $\hat{\mathbb{I}}_{lungs}[y, \mathbf{W}|\mathbf{x}, \mathcal{D}_{train}]$). For visualization purposes, we displayed the summed entropy and mutual information masks, respectively.

Training of the Model. During the training, we used the combination of binary cross-entropy (BCE) and soft-Jaccard loss (J) as done in various other studies [30]:

$$\sum_{c=l,h} \text{BCE}(\mathbf{W}, \mathbf{X}, \mathbf{Y}_c) - \text{J}(\mathbf{W}, \mathbf{X}, \mathbf{Y}_c) \rightarrow \min_{\mathbf{W}}, \quad (4)$$

where $\mathbf{W}$ are the model's weights, $\mathbf{X}$ are the images and $\mathbf{Y}_h$ and $\mathbf{Y}_l$ are the ground truth masks for the heart and the lungs, respectively.

During the training process, we use various data augmentation techniques to improve the robustness and decrease possible overfitting. In particular, we used random-sized crop and $\pm 5°$ rotation as our main data augmentations (with a probability of 0.5 per image). Noise addition, blur and and sharpening were used as secondary augmentations ($p = 0.05$ per image). Finally, elastic distortions, Random Brightness, JPEG compression and horizontal flips were used with a probability of 0.01 to regularize the images with hard cases.

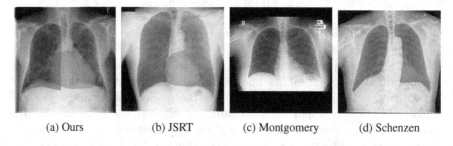

(a) Ours   (b) JSRT   (c) Montgomery   (d) Schenzen

Fig. 3. Original annotations in all the test datasets. In our experiments we re-annotated JSRT, Montgomery and Shenzhen datasets in a similar fashion to our dataset in order to have only the image distribution different, but the segmentation masks being annotated in exactly the same fashion.

All our models were trained with the image resolution of 224×224 pixels. To train the models, we used minibatch of 8, learning rate of $1e - 4$ and spatial dropout rate of 0.1. Our experiments were conducted in Keras [6] with Albumentations library [2] used for data augmentation.

4   Experiments and Results

4.1   Dataset

Our training set was derived from ChestXray14 dataset [39]. The original dataset included $112,120$ chest radiographs from $30,805$ patients, while the train data used in this study comprised 431 images randomly sampled from these data. For training, we used 294 images from 85 distinct patients. 38 images from 12 patients were used for validation and the remaining 99 images from 23 patients were eventually used as a test set.

The proposed multi-label dataset has the following findings and labels (the number of samples for each label is given in parenthesis): Cardiomegaly (27),

Table 1. Ablation study. IoU metric (higher is better) computed on the proposed dataset for Lungs (Table 1a) and Heart (Table 1b) obtained by different encoder-decoder architectures. For each decoder we indicate the best model as **bold** and the second best model as underscore. The best combination of Encoder + Decoder is highlighted as **gray** . We chose ResNet50 model as a backbone network and FPN decoder for further experiments. See Sect. 4.3 for more details.

Decoder \ Encoder	vgg16 [32]	vgg19	ResNet18 [14]	ResNet34	ResNet50	ResNet101	ResNet152	SE-ResNet18 [16]	SE-ResNet34	SE-ResNet50	DenseNet121 [17]	MobileNetV1 [15]	MobileNetV2 [28]
Unet [26]	0.906	**0.908**	0.895	0.897	0.902	0.895	0.906	0.846	0.843	0.908	0.892	0.907	0.874
FPN [23]	0.893	0.911	0.898	0.907	0.911	**0.913**	0.893	0.908	0.899	0.910	0.899	0.901	0.878
LinkNet [3]	0.905	0.907	0.861	0.893	0.860	0.861	0.875	0.874	0.821	0.904	0.858	0.880	0.874
PSPNet [42]	0.871	0.877	0.852	0.860	0.859	0.862	0.865	0.861	0.859	0.874	0.870	0.853	0.842

(a) IoU *Lungs*.

Decoder \ Encoder	vgg16 [32]	vgg19	ResNet18 [14]	ResNet34	ResNet50	ResNet101	ResNet152	SE-ResNet18 [16]	SE-ResNet34	SE-ResNet50	DenseNet121 [17]	MobileNetV1 [15]	MobileNetV2 [28]
Unet [26]	0.843	0.838	0.805	0.731	0.822	0.805	0.820	0.714	0.750	**0.848**	0.793	0.791	0.750
FPN [23]	0.814	0.786	0.799	0.836	**0.865**	0.863	0.806	0.819	0.806	0.814	0.849	0.787	0.766
LinkNet [3]	0.834	0.808	0.797	0.766	0.773	0.799	0.814	0.762	0.668	**0.839**	0.687	0.755	0.734
PSPNet [42]	0.764	**0.814**	0.740	0.776	0.717	0.712	0.745	0.703	0.741	0.779	0.781	0.654	0.661

(b) IoU *Heart*

Table 2. Selection comparison of different normalization techniques in decoders. Here, we did not experiment with PSPNet [42] decoder due to its low performance in backbone selection stage.

Decoder \ Normalization	batch (BN) [18]	group (GN) [41]	instance (IN) [36]
Unet [26]	0.902	0.886	0.911
LinkNet [3]	0.860	0.910	0.915
FPN [23]	**0.911**	**0.914**	**0.916**

(a) Lungs.

Decoder \ Normalization	batch (BN) [18]	group (GN) [41]	instance (IN) [36]
Unet [26]	0.822	0.822	0.870
LinkNet [3]	0.773	0.828	0.862
FPN [23]	**0.865**	**0.868**	**0.884**

(b) Heart.

Emphysema (30), Effusion (76), Hernia (9), Infiltration (72), Nodule (18), Atelectasis (51), Mass (37), Pneumothorax (34), Pneumonia (3), Pleural thickening (13), Fibrosis (14), Consolidation (15), Edema (3). There were 177 samples with no findings and the remaining images had at least one of the aforementioned radiological findings. The label distribution described above, makes our dataset more realistic and challenging for segmentation compared to the previous studies.

It is worth to mention that the original data were provided with the labels mined from the radiology reports, however the dataset did not have any segmentation masks. Our radiologist (radiologist A) annotated the train, validation and the test sets. An example of the annotations is presented in Fig. 3. Compared to the other existing datasets illustrated in the same figure, our annotations delineate the true anatomical contours which makes the segmentation more challenging.

4.2 Auxiliary Test Datasets

We evaluated our method on three auxiliary test sets each derived from three *independent public datasets*. Specifically, the JSRT dataset [31,37], Monthomery County X-ray set [19], and Shenzhen Hospital Chest X-ray dataset [19] have been used in our experiments. The original annotations for these datasets did not include the true boundaries of the lungs underlying the heart, or had missing annotations of the heart. To evaluate the performance of our method trained on our datasets with true lung boundaries, practicing radiologist B having the same experience to the radiologist A, annotated 50 random images for the test evaluation from each of the auxiliary test datasets. The comparison between the original annotations in the test datasets and our annotations is presented in Fig. 3.

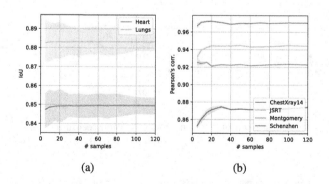

Fig. 4. Performance on the test sets vs. the number of MC dropout samples. Shown 95% intervals were computed with bootstrapping. Subplot (a): IoU per organ and subplot (b): CTR correlations with the ground truth per dataset.

Table 3. Quantitative results for each of the datasets. Here, we present the IoU for the Lungs and the Heart and also the Pearson's correlation between the ground truth CTR (computed from the manually annotated masks) and the predicted CTR (computed from the predicted segmentation masks).

Dataset	Heart		Lungs		Pearson's corr.
	IoU	Dice	IoU	Dice	
ChestXray14	0.87	0.93	0.92	0.96	0.87
JSRT	0.82	0.90	0.87	0.93	0.95
Schenzhen	0.84	0.91	0.87	0.93	0.97
Montgomery	0.86	0.92	0.87	0.93	0.92

4.3 Ablation Study

Backbone and Decoder. Latest advances in image segmentation demonstrate that transfer learning is helpful in image segmentation. As such, the use of encoders pre-trained on ImageNet [8] is a core technique in all the current state-of-the-art segmentation networks [21]. In our study, we investigated multiple pre-trained models with state-of-the-art decoders, namely U-Net [26], FPN [23], LinkNet [3] and PSPNet [42] (Table 1).

Our experiments demonstrate that in lung segmentation, ResNet101 and ResNet50 with FPN decoder yielded the best and second best results, respectively (Table 1a). In heart segmentation, ResNet50 backbone outperformed ResNet101 and both of the encoders with FPN decoder achieved best and second best results. Therefore, for simplicity and speed of computations, we selected ResNet50 with FPN to be our main configuration.

Normalization in the Decoder. Once the best configuration has been obtained, we assessed the influence of normalization in the decoder. In particular, we investigated whether replacement of batch normalization to group or instance normalization has any effect on the performance of our model. These experiments are presented in Table 2. The results demonstrate that instance normalization achieves better performance compared to other normalization types.

4.4 Test Set Results

Optimal Number of Monte-Carlo Samples. In our experiments, we assessed the influence of MC dropout onto the performance of our segmentation and CTR estimation pipeline. As such, we computed the aggregated IoU values for heart and lungs ground truth masks. Besides, we also computed the Pearson's correlation of CTR computed using the ground truth and the predicted masks for different number of MC samples. These results are visualized in Fig. 4. From this plot it can be seen that the optimal number of iterations on all datasets with respect to IoU and CTR correlations is close to 20. We use this number for our further evaluation of the developed method.

Quantitative Results. For the optimal number of MC samples (20 according to Fig. 4), we performed the quantitative evaluation of our model on all the test datasets, namely ChestXray14, JSRT, Shenzhen and Montgomery. In addition to the IoU, the Pearson's correlation coefficient between the CTR values for manually annotated masks and the predictions have been reported. These results are presented in Table 3.

Segmentation Examples. In Fig. 5, we visualized the examples of segmentation and the uncertainty estimates (both aleotoric and epistemic). The proposed method achieves accurate results demonstrating good segmentation performance in general. However, we note that our model does not predict the sharp corners of the lungs. Furthermore, from the epistemic uncertainty maps, it can be seen that our model is not confident in the bottom part of the lungs as they are typically very difficult to annotate since they can be hardly distinguished in the images.

5 Discussion

In this paper, we developed a novel approach for automatic segmentation of chest X-rays and assessment of CTR. Our approach is a modified FPN with ResNet50 backbone and MC dropout. In the extensive experimental evaluation, we found that the proposed configuration with instance normalization in the decoder yielded the best results compared to other investigated network configurations. Besides, it is worth to note that for the first time in CTR estimation realm, we proposed to assess it using Bayesian deep learning.

In this paper, we focused not only on developing state-of-the-art method for segmenting the chest X-rays, but also tackled the issue of annotation of these data and the availability of reliably annotated train and test data. As such, we proposed multiple new datasets that were annotated by radiologists and we plan to publicly release these data to facilitate further research.

Despite the advantages of our proposed method, this study has still some limitations. In particular, we did not experiment with training the models from scratch and used transfer learning. The second limitation of our study is that we did not compare our method to state-of-the-art unsupervised domain adaptation approaches [4,9,11]. However, this would require re-implementation of previously presented methods as our annotations for all the test set differ from all the previously published techniques. We leave this limitation for the future work. Another important limitation of our study is that the annotators of the test data differ. In particular, one radiologist (radiologist A) annotated the train and the test sets derived from ChestXray14 dataset [39] and another radiologist (radiologist B) annotated the images from JSRT [31], Montgomery and Shenzhen datasets [19]. While we think that this particular limitation has insignificant impact onto our results, we still plan assess the inter-rater agreement between the annotators of the data.

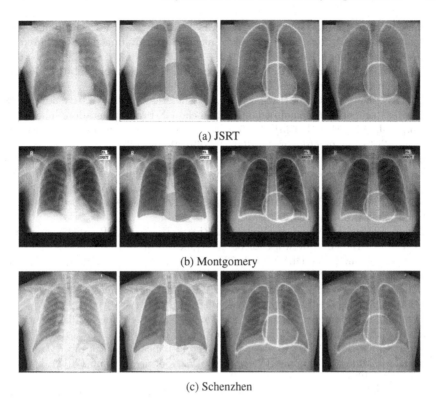

(a) JSRT

(b) Montgomery

(c) Schenzhen

Fig. 5. Examples of segmentation and uncertainty estimates for each of the test datasets (random examples are shown). Here, from left to right: original image, predicted segmentation mask, aleotoric and epistemic uncertainties.

To conclude, this paper introduced a novel, more challenging setting for segmenting organs in chest X-rays and proposed a Bayesian modification of FPN that allowed to estimate the CTR with the uncertainty bounds using MC-dropout. We think that the proposed approach has multiple applications in the clinical practice, as such, it could be useful for quantitative monitoring of CTR for patients with cardiomegaly in intensive care units. Another interesting application is the image quality assessment since our model is able to predict the aleotoric uncertainty for every test image. Finally, for the benefit of the community, we publicly release our dataset, implementation of our method and the pre-trained models at https://github.com/MIPT-Oulu/ChestBFPN.

References

1. Arbabshirani, M.R., Dallal, A.H., Agarwal, C., Patel, A., Moore, G.: Accurate segmentation of lung fields on chest radiographs using deep convolutional networks. In: Medical Imaging 2017: Image Processing (2017)

2. Buslaev, A., Parinov, A., Khvedchenya, E., Iglovikov, V.I., Kalinin, A.A.: Albumentations: fast and flexible image augmentations. arXiv preprint arXiv:1809.06839 (2018)
3. Chaurasia, A., Culurciello, E.: LinkNet: exploiting encoder representations for efficient semantic segmentation. CoRR abs/1707.03718 (2017)
4. Chen, C., Dou, Q., Chen, H., Heng, P.-A.: Semantic-aware generative adversarial nets for unsupervised domain adaptation in chest X-ray segmentation. In: Shi, Y., Suk, H.-I., Liu, M. (eds.) MLMI 2018. LNCS, vol. 11046, pp. 143–151. Springer, Cham (2018). https://doi.org/10.1007/978-3-030-00919-9_17
5. Chen, L.-C., Zhu, Y., Papandreou, G., Schroff, F., Adam, H.: Encoder-decoder with atrous separable convolution for semantic image segmentation. In: Ferrari, V., Hebert, M., Sminchisescu, C., Weiss, Y. (eds.) ECCV 2018. LNCS, vol. 11211, pp. 833–851. Springer, Cham (2018). https://doi.org/10.1007/978-3-030-01234-2_49
6. Chollet, F., et al.: Keras (2015)
7. Dai, W., Dong, N., Wang, Z., Liang, X., Zhang, H., Xing, E.P.: SCAN: structure correcting adversarial network for organ segmentation in chest X-rays. In: Stoyanov, D., et al. (eds.) DLMIA/ML-CDS -2018. LNCS, vol. 11045, pp. 263–273. Springer, Cham (2018). https://doi.org/10.1007/978-3-030-00889-5_30
8. Deng, J., Dong, W., Socher, R., Li, L.J., Li, K., Fei-Fei, L.: ImageNet: a large-scale hierarchical image database. In: Proceedings of CVPR (2009)
9. Dong, N., Kampffmeyer, M., Liang, X., Wang, Z., Dai, W., Xing, E.: Unsupervised domain adaptation for automatic estimation of cardiothoracic ratio. In: Frangi, A.F., Schnabel, J.A., Davatzikos, C., Alberola-López, C., Fichtinger, G. (eds.) MICCAI 2018. LNCS, vol. 11071, pp. 544–552. Springer, Cham (2018). https://doi.org/10.1007/978-3-030-00934-2_61
10. Dunlay, S.M., Roger, V.L., Redfield, M.M.: Epidemiology of heart failure with preserved ejection fraction. Nat. Rev. Cardiol. **14**, 591–602 (2017)
11. Eslami, M., Tabarestani, S., Albarqouni, S., Adeli, E., Navab, N., Adjouadi, M.: Image to images translation for multi-task organ segmentation and bone suppression in chest x-ray radiography. arXiv preprint arXiv:1906.10089 (2019)
12. Gal, Y., Ghahramani, Z.: Dropout as a Bayesian approximation: representing model uncertainty in deep learning. In: Proceedings of ICML (2016)
13. He, K., Gkioxari, G., Dollár, P., Girshick, R.: Mask R-CNN. In: Proceedings of ICCV (2017)
14. He, K., Zhang, X., Ren, S., Sun, J.: Deep residual learning for image recognition. In: Proceedings of CVPR (2016)
15. Howard, A.G., et al.: MobileNets: efficient convolutional neural networks for mobile vision applications. CoRR abs/1704.04861 (2017)
16. Hu, J., Shen, L., Sun, G.: Squeeze-and-excitation networks. In: Proceedings of CVPR (2018)
17. Huang, G., Liu, Z., van der Maaten, L., Weinberger, K.Q.: Densely connected convolutional networks. In: Proceedings of CVPR (2017)
18. Ioffe, S., Szegedy, C.: Batch normalization: accelerating deep network training by reducing internal covariate shift. In: Proceedings of ICML (2015)
19. Jaeger, S., Candemir, S., Antani, S., Wáng, Y.X.J., Lu, P.X., Thoma, G.: Two public chest x-ray datasets for computer-aided screening of pulmonary diseases. Quant. Imaging Med. Surg. **4**, 475–477 (2014)
20. Kendall, A., Badrinarayanan, V., Cipolla, R.: Bayesian SegNet: model uncertainty in deep convolutional encoder-decoder architectures for scene understanding. In: Proceedings of CVPR (2015)

21. Kirillov, A., Girshick, R., He, K., Dollár, P.: Panoptic feature pyramid networks. In: Proceedings of CVPR (2019)
22. Li, Z., et al.: Automatic cardiothoracic ratio calculation with deep learning. IEEE Access **7**, 37749–37756 (2019)
23. Lin, T.Y., Dollár, P., Girshick, R., He, K., Hariharan, B., Belongie, S.: Feature pyramid networks for object detection. In: Proceedings of CVPR (2017)
24. Mukhoti, J., Gal, Y.: Evaluating Bayesian deep learning methods for semantic segmentation. arXiv preprint arXiv:1811.12709 (2018)
25. Rakhlin, A., Shvets, A.A., Kalinin, A.A., Tiulpin, A., Iglovikov, V.I., Nikolenko, S.: Breast tumor cellularity assessment using deep neural networks. arXiv preprint arXiv:1905.01743 (2019)
26. Ronneberger, O., Fischer, P., Brox, T.: U-Net: convolutional networks for biomedical image segmentation. In: Navab, N., Hornegger, J., Wells, W.M., Frangi, A.F. (eds.) MICCAI 2015. LNCS, vol. 9351, pp. 234–241. Springer, Cham (2015). https://doi.org/10.1007/978-3-319-24574-4_28
27. Saba, L., et al.: The present and future of deep learning in radiology. Eur. J. Radiol. **114**, 14–24 (2019)
28. Sandler, M.B., Howard, A.G., Zhu, M., Zhmoginov, A., Chen, L.C.: MobileNetv 2: inverted residuals and linear bottlenecks. In: Proceedings of CVPR (2018)
29. Schmidhuber, J.: Deep learning in neural networks: an overview. Neural Netw. **61**, 85–117 (2015)
30. Seferbekov, S.S., Iglovikov, V., Buslaev, A., Shvets, A.: Feature pyramid network for multi-class land segmentation. In: Proceedings of CVPRW (2018)
31. Shiraishi, J., et al.: Development of a digital image database for chest radiographs with and without a lung nodule: receiver operating characteristic analysis of radiologists' detection of pulmonary nodules. Am. J. Roentgenol. **174**(1), 71–74 (2000)
32. Simonyan, K., Zisserman, A.: Very deep convolutional networks for large-scale image recognition. In: Proceedings of ICLR (2015)
33. Souza, J.C., Diniz, J.O.B., Ferreira, J.L., da Silva, G.L.F., Silva, A.C., de Paiva, A.C.: An automatic method for lung segmentation and reconstructionin chest X-ray using deep neural networks. Comput. Methods Programs Biomed. **177**, 285–296 (2019)
34. Tiulpin, A., et al.: Multimodal machine learning-based knee osteoarthritis progression prediction from plain radiographs and clinical data. arXiv preprint arXiv:1904.06236 (2019)
35. Tiulpin, A., Thevenot, J., Rahtu, E., Lehenkari, P., Saarakkala, S.: Automatic knee osteoarthritis diagnosis from plain radiographs: a deep learning-based approach. Sci. Rep. **8**, 1727 (2018)
36. Ulyanov, D., Vedaldi, A., Lempitsky, V.: Instance normalization: the missing ingredient for fast stylization. arXiv preprint arXiv:1607.08022 (2016)
37. Van Ginneken, B., Katsuragawa, S., ter Haar Romeny, B.M., Doi, K., Viergever, M.A.: Automatic detection of abnormalities in chest radiographs using localtexture analysis. IEEE Trans. Med. Imaging **21**, 139–149 (2002)
38. Wang, J., et al.: Grey matter age prediction as a biomarker for risk of dementia: a population-based study. BioRxiv (2019)
39. Wang, X., Peng, Y., Lu, L., Lu, Z., Bagheri, M., Summers, R.M.: ChestX-ray8: hospital-scale chest X-ray database and benchmarks on weakly-supervised classification and localization of common thorax diseases. In: Proceedings of CVPR (2017)

40. Wessel, J., Heinrich, M.P., von Berg, J., Franz, A., Saalbach, A.: Sequential rib labeling and segmentation in chest X-ray using Mask R-CNN. In: Proceedings of ICMIDL (2019)
41. Wu, Y., He, K.: Group normalization. In: Ferrari, V., Hebert, M., Sminchisescu, C., Weiss, Y. (eds.) ECCV 2018. LNCS, vol. 11217, pp. 3–19. Springer, Cham (2018). https://doi.org/10.1007/978-3-030-01261-8_1
42. Zhao, H., Shi, J., Qi, X., Wang, X., Jia, J.: Pyramid scene parsing network. In: Proceedings of CVPR (2017)

Deep-Learning for Tidemark Segmentation in Human Osteochondral Tissues Imaged with Micro-computed Tomography

Aleksei Tiulpin[1,2(✉)], Mikko Finnilä[1], Petri Lehenkari[1,2],
Heikki J. Nieminen[1,3,4], and Simo Saarakkala[1,2]

[1] University of Oulu, Oulu, Finland
`aleksei.tiulpin@oulu.fi`
[2] Oulu University Hospital, Oulu, Finland
[3] University of Helsinki, Helsinki, Finland
[4] Aalto University, Espoo, Finland

Abstract. Three-dimensional (3D) semi-quantitative grading of patho-logical features in articular cartilage (AC) offers significant improvements in basic research of osteoarthritis (OA). We have earlier developed the 3D protocol for imaging of AC and its structures which includes stain-ing of the sample with a contrast agent (phosphotungstic acid, PTA) and a consequent scanning with micro-computed tomography. Such a protocol was designed to provide X-ray attenuation contrast to visu-alize AC structure. However, at the same time, this protocol has one major disadvantage: the loss of contrast at the tidemark (calcified car-tilage interface, CCI). An accurate segmentation of CCI can be very important for understanding the etiology of OA and *ex-vivo* evaluation of tidemark condition at early OA stages. In this paper, we present the first application of Deep Learning to PTA-stained osteochondral sam-ples that allows to perform tidemark segmentation in a fully-automatic manner. Our method is based on U-Net trained using a combination of binary cross-entropy and soft-Jaccard loss. On cross-validation, this app-roach yielded intersection over the union of 0.59, 0.70, 0.79, 0.83 and 0.86 within 15 μm, 30 μm, 45 μm, 60 μm. and 75 μm padded zones around the tidemark, respectively. Our codes and the dataset that consisted of 35 PTA-stained human AC samples are made publicly available together with the segmentation masks to facilitate the development of biomedical image segmentation methods.

Keywords: Osteoarthritis · 3D histology · Deep Learning

1 Introduction

Osteoarthritis (OA) is a common field of interest in micro-computed tomogra-phy (μCT) research. OA is primarily characterized by progressive degeneration

© Springer Nature Switzerland AG 2020
J. Blanc-Talon et al. (Eds.): ACIVS 2020, LNCS 12002, pp. 131–138, 2020.
https://doi.org/10.1007/978-3-030-40605-9_12

of structure and composition articular cartilage (AC), along with the sclerotic changes in subchondral bone [4]. These changes in the microstructure of AC and subchondral bone can be visualized in three-dimensions (3D) using μCT. Conventionally, without any external X-ray contrast agents or sample processing protocols, only calcified tissue can be visualized. Thus, direct μCT imaging of soft tissues, such us AC, is not possible. To mitigate this limitation of X-ray imaging, several contrast agents have been introduced to provide X-ray attenuation contrast for the AC, such as phosphotungstic acid (PTA), CA4+ and others [6,10,14].

Specifically for OA, a novel *ex-vivo* μCT contrast method and a protocol to quantify collagen distribution in AC has recently been introduced along with the 3D grading system [10,11]. There, PTA was validated as a contrast agent, since it directly binds to collagen and significantly increases the attenuation contrast within the cartilage tissue [7,11]. However, despite the unique possibility to image soft tissues, PTA staining has one major drawback when it is used for osteochondral tissue: X-ray attenuation contrast at the tidemark (calcified cartilage interface; CCI) is lost due to the accumulation of PTA. Another drawback of the PTA staining is the occasional occurrence of non-enhancing regions, *i.e.* voids, at the CCI [10]. Both of these limitations and the typical examples of the PTA-stained samples analyzed in this study are illustrated in Fig. 1.

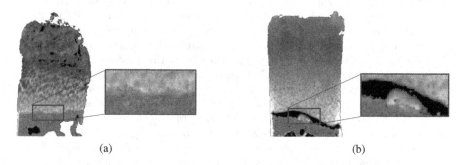

(a)   (b)

Fig. 1. Examples of the slices from $\varnothing = 2$ mm human osteochondral plugs imaged with contrast-enhanced μCT. (a) a typical sample showing the loss of the contrast at the CCI. (b) a typical non-enhancing region (void) which occurs with some samples.

An accurate analysis of CCI from PTA-stained μCT image stacks is of high importance in the evaluation of early OA-induced changes [8]. Two straightforward solutions exist: either to perform a manual annotation of this area, or, alternatively, perform double imaging – with and without PTA. However, both of these options are time consuming and could be avoided with the help of Machine Learning. In clinical OA research Machine Learning is has been applied to various tasks [2,12,15,19–21], however, its application in OA basic research so far has been limited [1].

Recently, one form of Machine Learning – Deep Learning (DL) has become a gold standard in medical image segmentation [17]. Fully-convolutional neural networks (CNN) have shown drastic improvements in the performance of the segmentation methods and decreased their computational time [17]. In particular, U-Net CNN architecture [16] allowed to significantly improve the bio-medical image segmentation.

In this study, we tackled the problem of automatic tidemark segmentation in PTA-stained osteochondral samples using Deep Learning. This study has the following contributions:

- We present a method based on Deep Learning that allows to perform assessment of tidemark in PTA-stained human osteochondral samples.
- We also present a data acquisition protocol based that allowed to obtain the segmentation masks without their explicit annotation by a human expert.
- In our experiments, we demonstrated the performance of popular U-Net architecture and assessed binary cross-entropy, focal and soft-Jaccard losses.
- Finally, we release our source code and the dataset with the ground truth masks for the benefit of the community.

2   Materials and Methods

Our imaging pipeline consisted of sample preparation, imaging, data pre-processing and, finally, image segmentation. The graphical illustration of this process is demonstrated in Fig. 2 and also in Fig. 3, respectively. The following sub-sections describe our methodology in details.

2.1   Samples Preparation and Imaging Protocol

We followed the institutional guidelines and regulations (Institutional ethics approval PPSHP 78/2013, The Northern Ostrobothnia Hospital District's ethical comittee) during sample extraction. The samples were obtained from $n = 20$ patients undergoing total knee arthroplasty surgery (informed consents obtained). At the preparation stage, the osteochondral plugs ($\varnothing = 2$ mm, depth ≈ 4 mm) were drilled from tibial and femoral condyles. These plugs were then frozen under $-80\,°$C. Before the imaging, we thawed the osteochondral plugs and fixed them in 10% neutral-buffered formalin for a minimum of 5 days. Subsequently, these plugs were wrapped into parafilm and orthodonic wax to avoid sample drying during the imaging process.

At first, we stained the samples with CA4+ contrast agent and imaged them using a μCT system (Bruker microCT Skyscan 1272, Kontich, Belgium; 45 kV, 222 μA, 3.2 μm voxel side length, 3050 ms, 2 frames/projection, 1200 projections, 0.25 mm aluminum filter) to be used in another study. After the imaging, CA4+ was washed out and the plugs were stained in PTA for 48 h before the second round of imaging with μCT using the same imaging settings.

Both CA4+ and PTA data were reconstructed using NRecon software of version 1.6.10.4; Bruker microCT, Kontich, Belgium. Eventually, these 3D stacks were co-registered using rigid intensity-based registration (mean squared error loss) with a subsequent manual adjustment. Subsequently, CA4+ stacks' intensities were thresholded to obtain the hard tissue masks used as segmentation ground truth. At the final step of the process, we graded each individual cartilage feature from PTA-stained samples according to the 3D histopathological grading system [10].

2.2 Data Pre-processing

Our imaging protocol allowed to obtain the 3D volumes of human cartilage and the mask annotations for the underlying mineralized tissues. The original size of the reconstructed samples ranges from 756×756 to 1008×1008 pixels in width and 884 to 2067 pixels in height (including the empty space around the sample). To harmonize the data and reduce its size, we firstly cut the bottom 30% of the scanned volume and performed a global contrast normalization of its intensities to $[0, 1]$ range. Subsequently, we performed a thresholding with a cut-off 0.1 and summed all the intensities of the obtained volume along the Z-axis. We used active contours method from OpenCV [3] to identify the largest closed contour in the obtained summed image and then identified its center of mass.

Having the center of mass of the sample in XY plane, we performed the cropping of the original volumes and the corresponding ground truth masks to the size of $448 \times 448 \times 768$ (XYZ) voxels. All the volumes and their masks were

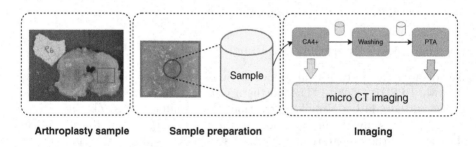

Fig. 2. Data acquisition pipeline: from sample preparation to imaging.

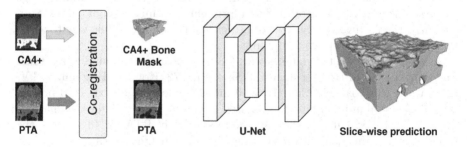

Fig. 3. Data processing pipeline. Here, we co-registered CA4+ and PTA samples and obtained the segmentation masks for hard tissues. These masks were used in training of our segmentation model.

then split into ZX and ZY slices to enlarge the dataset in slice-wise segmentation done by a U-Net-like Deep Neural Network [16].

2.3 Network Architecture

Our model is inspired by U-Net [16] with minor modifications. Here, we used 24 convolutional filters as the base width of our model and doubled this quantity every time after the max-pooling layer. The depth of the model was set to 6 and bilinear interpolation was used in the decoder of our model. Finally, every convolutional module of the model had two consequent blocks of convolution, batch normalization and ReLU layers.

2.4 Loss Function

In this study, we evaluated several loss functions. As such, we investigated Binary Cross-Entropy (BCE), soft-Jaccard loss $(1 - J$; J – soft-Jaccard index), focal loss and also a combination of BCE and soft jaccard losses. Instead of computing a direct sum of BCE and soft-Jaccard losses, Iglovikov et $al.$ [5] proposed to combine BCE and a negative of $\log J$:

$$L(\mathbf{w}, \mathbf{X}, \mathbf{y}) = \mathrm{BCE}(\mathbf{w}, \mathbf{X}, \mathbf{y}) - \log J(\mathbf{w}, \mathbf{X}, \mathbf{y}), \tag{1}$$

where $\mathbf{w}$ are the model's weights, $\mathbf{X}$ are the images and $\mathbf{y}$ are the ground truth segmentation masks. The idea behind combining the losses in this case, is such that the BCE penalizes the incorrect pixel-wise classification while the soft-Jaccard loss is rather rather optimizes the structural consistency between the predictions and the ground truth.

Similarly to [5], we also found that the loss in Eq. 1 yields better performance than when computing soft-Jaccard without a logarithm. In particular, the use of a log transform here is motivated by the following empirical observation: by the end of the network's training, the values of the total loss vary insignificantly when the soft-Jaccard loss value (or soft-Jaccard index is high) and the gradients, therefore, are not be strong. Such situation appears specifically in the calcified cartilage interface segmentation task – the network learns to well segment the hard tissues with high Jaccard coefficient in overall, but it is not able to accurately segment the edges which are required for a subsequent morphological analysis.

2.5 Evaluation Metric

As a main evaluation metric, we used Jaccard coefficient (intersection over the union, IoU). IoU was computed only at the area padded around the tidemark. In particular, we identified the location of the tidemark slice-by-slice and for every slice we created a padded region of $\pm P$ pixels. Such masks allowed to estimate the IoU only within the zone of the interest ignoring the other, non-relevant parts of the sample, e.g. bone. Besides the IoU, we also computed the complimentary metrics: Dice's and Volumetric similarity scores.

3   Experiments

3.1   Implementation Details

We implemented our models and training pipelines using PyTorch [13]. To augment our data, we applied random cropping, horizontal flip and random gamma-correction, varying value of gamma from 0.5 to 2. To make our model applicable to the real-life scenario when the black edges (air around the sample) are seen in the full sample, we first performed a padding to 800 × 800 pixels before random cropping. For the validation set, we used the original size of the images of 768 × 448. We used SOLT library [18] to perform data augmentation.

All our experiments were conducted with Adam optimizer [9], batch size of 32, learning rate of $1e-4$ and a weight decay of $1e-4$. For the focal loss, we used the standard hyperparameters: $\alpha = 0.25$ and $\gamma = 2$. All the experiments were done using group-5-fold stratified cross-validation, where the group division was performed by subject id and stratification was done using the previously mentioned 3D histopathological grades obtained for the calcified zone [10].

We assessed the results on sample-wise out-of-fold predictions. Here, we averaged the inference results for each sample's ZX and ZY slices and thresholded the obtained masks with the threshold of 0.3 for the combined loss and 0.5 for BCE and focal losses, respectively. The padding values P for computing the IoU were set to 15 μm, 30 μm, 45 μm, 60 μm, 75 μm, 90 μm, 105 μm, 120 μm, 135 μm and 150 μm.

3.2   Segmentation Performance

The performance of our network with different loss functions on cross-validation for IoU, Dice's and Volumetric similarity scores is presented in Fig. 4.

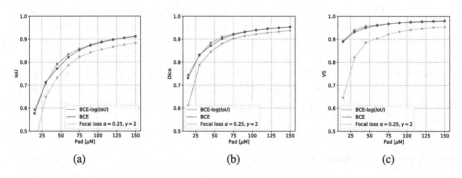

Fig. 4. Median values of performance metrics for different levels of padding around the tidemark. Here, subplots (a), (b) and (c) show the performance for IoU, Dice and Volumetric similarity scores, respectively.

More fine-grained assessment of the median values of the performance metrics and their standard deviations is presented in Table 1. From Fig. 4 and Table 1 it can be seen that for all the metrics, a combination of BCE and jaccard losses from Eq. 1 yields better performance in the close proximity to the tidemark.

Table 1. Median and standard deviation of IoU for different levels of tidemark padding.

Loss	Pad [μm]				
	15	30	45	60	75
BCE	0.57 ± 0.14	$\mathbf{0.71 \pm 0.11}$	0.77 ± 0.10	0.82 ± 0.09	0.85 ± 0.08
Focal	0.44 ± 0.19	0.65 ± 0.18	0.73 ± 0.15	0.79 ± 0.14	0.82 ± 0.12
BCE-log (Jaccard)	$\mathbf{0.59 \pm 0.13}$	0.70 ± 0.10	$\mathbf{0.79 \pm 0.08}$	$\mathbf{0.83 \pm 0.08}$	$\mathbf{0.86 \pm 0.07}$

4   Conclusion

In this study, we for the first time applied Deep Learning to μCT imaged osteochondral samples in order to segment the tidemark. The results presented in this paper are promising and indicate the possibility of accurate CCI segmentation even with a 2-dimensional method. Despite this, we believe that the presented results can further be improved. In particular, we think that a optimizing the segmentation of the tidemark directly with a volumetric model, e.g. 3D U-Net could yield better results. Finally, the future studies should also leverage other, surface-related metrics, e.g. hausdorff distance for more precise assessment of the segmentation results. The codes and the dataset are released on the project's GitHub page: https://github.com/MIPT-Oulu/mCTSegmentation.

Acknowledgements. This work was supported by Academy of Finland (grants 268378, 303786, 311586 and 314286), European Research Council under the European Union's Seventh Framework Programme (FP/2007-2013)/ERC Grant Agreement no. 336267, the strategic funding of the University of Oulu and KAUTE foundation. We would also like to acknowledge CSC IT Center for Science, Finland, for generous computational resources. Tuomas Frondelius is acknowledged for the initial experiments with the data and Santeri Rytky is acknowledged for the useful comments and proofreading of the paper.

References

1. Abidin, A.Z., Deng, B., Dsouza, A.M., Nagarajan, M.B., Coan, P., Wismüller, A.: Deep transfer learning for characterizing chondrocyte patterns in phase contrast X-ray computed tomography images of the human patellar cartilage. Comput. Biol. Med. **95**, 24–33 (2018)
2. Antony, J., McGuinness, K., O'Connor, N.E., Moran, K.: Quantifying radiographic knee osteoarthritis severity using deep convolutional neural networks. In: 2016 23rd International Conference on Pattern Recognition (ICPR), pp. 1195–1200. IEEE (2016)

3. Bradski, G.: The OpenCV library. Dr. Dobb's J. Softw. Tools **25**, 120–126 (2000)
4. Glyn-Jones, S., et al.: Osteoarthritis. Lancet **386**(9991), 376–387 (2015)
5. Iglovikov, V., Shvets, A.: TernausNet: U-net with VGG11 encoder pre-trained on imagenet for image segmentation. arXiv preprint arXiv:1801.05746 (2018)
6. Karhula, S.S., et al.: Micro-scale distribution of CA4+ in ex vivo human articular cartilage detected with contrast-enhanced micro-computed tomography imaging. Front. Phys. **5**, 38 (2017)
7. Karhula, S.S., et al.: Effects of articular cartilage constituents on phosphotungstic acid enhanced micro-computed tomography. PLoS ONE **12**(1), e0171075 (2017)
8. Kauppinen, S., et al.: 3D morphometric analysis of calcified cartilage properties using micro-computed tomography. Osteoarthritis Cartilage **27**(1), 172–180 (2019)
9. Kingma, D.P., Ba, J.: Adam: a method for stochastic optimization. arXiv preprint arXiv:1412.6980 (2014)
10. Nieminen, H., et al.: 3D histopathological grading of osteochondral tissue using contrast-enhanced micro-computed tomography. Osteoarthritis Cartilage **25**(10), 1680–1689 (2017)
11. Nieminen, H., et al.: Determining collagen distribution in articular cartilage using contrast-enhanced micro-computed tomography. Osteoarthritis Cartilage **23**(9), 1613–1621 (2015)
12. Norman, B., Pedoia, V., Majumdar, S.: Use of 2D U-Net convolutional neural networks for automated cartilage and meniscus segmentation of knee MR imaging data to determine relaxometry and morphometry. Radiology **288**(1), 177–185 (2018)
13. Paszke, A., et al.: Automatic differentiation in PyTorch (2017)
14. Pauwels, E., Van Loo, D., Cornillie, P., Brabant, L., Van Hoorebeke, L.: An exploratory study of contrast agents for soft tissue visualization by means of high resolution X-ray computed tomography imaging. J. Microsc. **250**(1), 21–31 (2013)
15. Pedoia, V., Majumdar, S., Link, T.M.: Segmentation of joint and musculoskeletal tissue in the study of arthritis. Magn. Reson. Mater. Phys. Biol. Med. **29**(2), 207–221 (2016)
16. Ronneberger, O., Fischer, P., Brox, T.: U-Net: convolutional networks for biomedical image segmentation. In: Navab, N., Hornegger, J., Wells, W.M., Frangi, A.F. (eds.) MICCAI 2015. LNCS, vol. 9351, pp. 234–241. Springer, Cham (2015). https://doi.org/10.1007/978-3-319-24574-4_28
17. Shen, D., Wu, G., Suk, H.I.: Deep learning in medical image analysis. Annu. Rev. Biomed. Eng. **19**, 221–248 (2017)
18. Tiulpin, A.: SOLT: streaming over lightweight transformations (2019). https://github.com/MIPT-Oulu/solt
19. Tiulpin, A., et al.: Multimodal machine learning-based knee osteoarthritis progression prediction from plain radiographs and clinical data. arXiv preprint arXiv:1904.06236 (2019)
20. Tiulpin, A., Thevenot, J., Rahtu, E., Lehenkari, P., Saarakkala, S.: Automatic knee osteoarthritis diagnosis from plain radiographs: a deep learning-based approach. Sci. Rep. **8**(1), 1727 (2018)
21. Tiulpin, A., Thevenot, J., Rahtu, E., Saarakkala, S.: A novel method for automatic localization of joint area on knee plain radiographs. In: Sharma, P., Bianchi, F.M. (eds.) SCIA 2017. LNCS, vol. 10270, pp. 290–301. Springer, Cham (2017). https://doi.org/10.1007/978-3-319-59129-2_25

Quadratic Tensor Anisotropy Measures for Reliable Curvilinear Pattern Detection

Mohsin Challoob[✉] and Yongsheng Gao

School of Engineering, Griffith University, Brisbane, Australia
mohsin.challoob@griffithuni.edu.au,
yongsheng.gao@griffith.edu.au

Abstract. A wide range of applications needs the analysis of biomedical images as a fundamental task to extract meaningful information and allow high throughput measurements. A new method for the detection of curve-like structures in biomedical images is presented by exploiting local phase vector and the structural anisotropy information at various directions. We introduce an oriented gaussian derivative quadrature filter not only for estimating the local phase vectors, which include line features, but also for its immunity to inhomogeneous intensity and its capability to enhance curved structures having various diameters, leading to more reliable hessian analysis. A novel measure function-based hessian tensor is proposed to detect curvilinear patterns by incorporating the anisotropic indices (coherence and linearity) of curved features, producing a uniform and strong response. Over multiple orientations, the responses are maximized to achieve a rotationally invariant response, and to detect target structures with different widths and illuminations. The evaluation of the proposed method on the extraction of retinal vessels and leaf venation patterns exhibits its superior performance against state-of-the-art methods.

Keywords: Curvilinear detection · Quadratic tensor anisotropy · Measure indices

1 Introduction

The analysis of biomedical imaging data plays a signification role in health care purposes and life sciences. In particular, the accurate enhancement, detection, quantification, and modelling of curve-like patterns is an essential step in bioimage informatics and medical quantification. For example, retinal blood vessels provide clinicians with information to diagnose, screen, and evaluate a wide range of pathologies such as diabetes and arteriosclerosis, and the automatic extraction of leaf veins is desirable for the classification of plant species [1–4]. However, the detection of curvilinear structures in biomedical images is a challenging task due to several difficulties including the changes in the intensity, the variability of features (width, length, tortuosity), inhomogeneous background, noise, and poor contrast between curved patterns and the background. A number of automatic detection approaches has been introduced to extract curve-like patterns from biomedical

© Springer Nature Switzerland AG 2020
J. Blanc-Talon et al. (Eds.): ACIVS 2020, LNCS 12002, pp. 139–150, 2020.
https://doi.org/10.1007/978-3-030-40605-9_13

images, but these methods suffer from many weaknesses. A phase congruency tensor (PCT) [2], which was introduced for detecting curvilinear structures in biomedical images, is sensitive to noise and unable to detect curved structures with different diameters, where the phase congruency cannot distinguish between lines, edges and mach bands and fuses them together. The morphological approaches [4, 6] apply an overly long line-shape structuring element rotated at several directions. This results in a difficulty in detecting high tortuosity patterns, produces a non-uniform response, and increases the sensitivity to background noises. An isotropic nonlinear filtering approach [5] applies circular masks to detect lines which are narrower than maximum width parameter and there is no bandpass filtering capability, leading to the deficiencies of the susceptibility to intensity inhomogeneity, structure width size, and non-uniform response across different curved morphologies. The studies [7] and [8] use hessian matrix to measure the response of vesselness and neuriteness, respectively, where this matrix is computed with the second order derivatives of gaussian. These hessian-based methods tend to suppress junctions and rounded structures, causing the discontinuity of curved network. The weaknesses of state-of-the-art methods are summarized as follows:

- Difficulty in detecting curvilinear structures with different diameters (thick, median, fine) when presenting together.
- Non-uniform response across various structures due to the lack of dealing with changes in illumination.
- Loss of the curved tree connectivity because of the suppression of junctions and rounded structures.
- Susceptibility to noise, particularly in regions with uniform intensity.

This paper introduces a new method for the detection of curve-like structures, which overcomes the mentioned deficiencies of the state-of-the-art algorithms. The proposed method incorporates the local properties of line-like features derived from the local phase vector-based gaussian derivative quadrature (GDQ) filter, with the coherence and linear anisotropy information of curvilinear patterns acquired from hessian tensor. Due to the fact of relating structural information to a given local orientation and to deal with the presence of various structure diameters together, a set of GDQ filters are used at multiple directions to estimate the local phase vectors. Further, the local phase based-GDQ filter has an inherent feature invariant to changes in intensity, and can discriminate between lines and edges, which in turn contributes to achieving robust hessian analysis. Then, we present a new measure function-based hessian tensor by quantifying the linearity of local anisotropic intensity and the local coherence of target structure to yield a strong and uniform response for various structure widths. The proposed function automatically measures the response without any threshold parameter controlling the sensitivity of the function and indices, except the scale range determining the size of target structure. To make sure that curvilinear structures with different diameters are captured, the responses of the measure function are maximized over multiple angles to produce a response at each pixel. The proposed method scheme (decomposition-filtering-recombination) preserves network connectivity while retaining thin and small curved patterns. The assessment that is conducted on retinal vessels and leaf veins shows the effectiveness of the proposed method, yielding excellent results which outperform the previous methods.

The reminder of this paper is organized as follows. In Sect. 2, the proposed method is presented, where the GDQ deriving the local phase and anisotropic indices-based the proposed measure function are explained in detail. Section 3 shows the experiment results including the quantitative and qualitative assessment and the comparison with existing methods, whereas the discussion and conclusion are stated in Sect. 4.

2 Methodology

In this section, a novel method for the detection of curvilinear structures in 2D biomedical images is proposed. The laminar flow within biological structures such as vessels and veins lead to flow velocity variations, which in turn cause the changes in illumination and non-uniformity of intensity across various curvilinear structures running at different widths and orientations. To deal with variations in intensity and scales, and suppression of undesired structures such as noise, we firstly introduce an oriented gaussian derivative quadrature (GDQ) filter to estimate local phase vector at various angles. Then, a novel measure function for the detection of curved structures is presented based hessian tensor by quantifying the coherence and linearity indices of local anisotropy information to produce a uniform response across various curved structures while being insensitive to noise regions with uniform intensity.

2.1 Local Phase-Based GDQ Filter

Local phase is a key local feature for determining the structural information (lines, edges) of an image and can be defined in 3D space under the concepts of monogenic signal, which assumes that signal comprises of frequencies with bandlimited [9]. We introduce a gaussian derivative quadrature (GDQ) filter to estimate local phase vector. The phase map includes the location and orientation of the image features compared to amplitude carrying their intensity only, see Fig. 1 (first row), where the contour plotting is shown for the magnitude, real part and imaginary part of local phase of an example image. Also, we use the phase-based GDQ filter rather than other intensity-based filters as it is insensitive to changes in illumination (zero DC component), and can exploit line-like patterns, which can be distinguished independently from other image features. According to [9], the GDQ filter has an advantage of allowing a continuous bandwidth with a maximum bandwidth of (2.59) octaves, having less aliasing than other quadrature filters such as log-Gabor and difference of gaussian which are with extended tail. By utilizing the derivative property of Fourier transform, the GDQ filter is defined in frequency domain as follows:

$$G_d(w) = \begin{cases} n_c w^a \exp\left[-\sigma^2 w^2\right] & \textit{if } w \geq 0; \\ 0, & \textit{otherwise} \end{cases} \quad \textit{where, } a \in \mathbb{R} \textit{ and } a \geq 1 \quad (1)$$

$$n_c = 2\frac{\sqrt{\pi}\sigma^{\left(a+\frac{1}{2}\right)}}{\sqrt{\Gamma\left(a+\frac{1}{2}\right)}} \quad (2)$$

where, n_c represents the normalized constant value, Γ is the gamma function, and the peak tuning frequency is given by $\left(\frac{\sqrt{a}}{\sigma}\right)$. The local phase-based the GDQ filter is estimated as a vector response and is defined as $Q^\theta(x)$ at each point x of the image I as follows:

$$Q^\theta(x) = E^\theta(x) + iO^\theta(x), \quad where\, i = \sqrt{-1} \tag{3}$$

The $E^\theta(x)$ and $O^\theta(x)$ denote the even-symmetric and odd-symmetric components (real and imaginary), respectively, at an orientation θ. In practice, various angles are required to capture curvilinear structures distributed at different diameters and directions and therefore an oriented GDQ filter is introduced to derive a set of local phase vectors, where the curvilinear network is decomposed into several sub-directional structures. The filter response is dominantly real for lines as illustrated in Fig. 1-second row, where E^θ yields maximal response, while O^θ has almost zero response. In contrast, O^θ attains high response at edges whereas E^θ is almost zero (see Fig. 1-third row). Hence, we determine line-like structures from real part only, which is obtained as:

$$E^\theta(x) = real\left(F^{-1}\left(G_d^\theta(w) * F(I(x))\right)\right) \tag{4}$$

where F and F^{-1} represents forward and inverse Fourier transforms, respectively.

2.2 Anisotropic Indices-Based Measure Function

The line-like structures derived from the symmetric even (real) part has an anisotropic intensity. We propose a novel measure function-based hessian tensor not only to consider the degree of anisotropy, but also to produce a uniform and strong response for different structure diameters while being more robust against background noise. The proposed function utilizes the linear anisotropy index, which has been reviewed in [10], and the coherence measure used in [11]. Hessian tensor is performed on a gaussian scale space of the $E^\theta(x)$ at scale s, and is computed as a 2×2 matrix $(H^\theta(x, s))$, where eigenvalues (λ_i) of the tensor are sorted based on their magnitudes. To capture anisotropic behavior, we evaluate the linearity of local anisotropic intensity and the local coherence of target structure. The linearity of anisotropy (L_a) quantifies the variation of eigenvalues across various structures, where the change of anisotropy along the curved structures is measured. The coherence index (Coh) provides the actual amount of local anisotropy information. The L_a and Coh indices are defined as:

$$L_a = \frac{\lambda_2 - \lambda_1}{\lambda_1 + 2\lambda_2}, \quad Coh = (\lambda_2 - \lambda_1)^2 \tag{5}$$

These quantities are large for line-like structures while they tend to zero for isotropic structures (homogenous regions), where the greater local anisotropic intensity, the greater the difference between eigenvalues. Then, the proposed measure function $\left(V_s^\theta\right)$ is introduced as:

$$V_s^\theta = Coh \exp\left(-S_t\left(\frac{\lambda_1}{\lambda_2}L_a\right)^2\right) \tag{6}$$

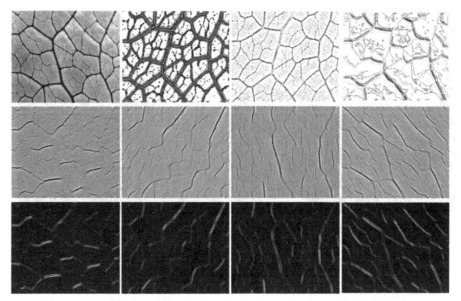

Fig. 1. First row: contour plotting for magnitude, real and imaginary of a leaf image taken from [2], respectively. Second and third rows: real and imaginary parts of local phase at multiple orientations, respectively.

where $S_t = \sqrt{\lambda_1^2 + \lambda_2^2}$ refers to the second order factor of structuredness. To produce invariant response to size and orientation, the final response $F_v(x)$ is obtained by maximizing the measure function $\left(V_s^{\theta}\right)$, at each pixel x, over a range of scales (s), and a set of angles (Θ), as follows:

$$F_v(x) = \{\max_{\theta \in \Theta} \{ \max_{s_{min} \leq s \leq s_{max}} \left(V_s^{\theta}\left[\text{eig } H^{\theta}(x, s)\right]\right)\}\} \tag{7}$$

where the range of scales (s) determines the expected minimal and maximal size of curvilinear structures of interest.

3   Experimental Results

The evaluation of the proposed detection method is conducted on fundus imaging (DRIVE dataset[1] and STARE dataset[2]), leaf image dataset[3], and sub images obtained from mouse retina[4]. Regarding the DRIVE and STARE datasets, the green band is only considered for its better contrast then prepressed using the morphologically approach in [12] for further enhancement. Figure 2 shows the detection results on these datasets,

---

[1] https://www.isi.uu.nl/Research/Databases/DRIVE/download.php.

[2] http://cecas.clemson.edu/~ahoover/stare/.

[3] http://www.imageprocessingplace.com/root_files_V3/image_databases.htm (developed by V. Waghmare).

[4] https://bisque.ece.ucsb.edu/client_service/.

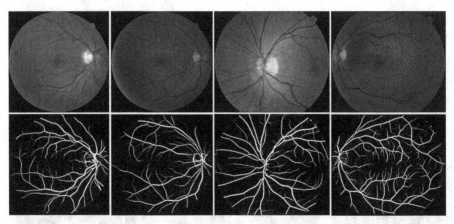

Fig. 2. First row: original images (the first two images from DRIVE while others from STARE). Second row: detection results by our method.

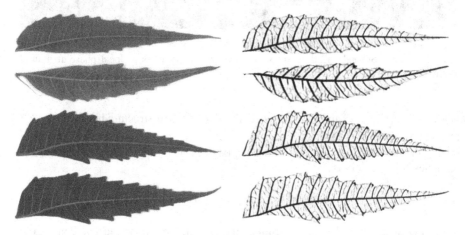

Fig. 3. First column: original leaf images. Second column: detection results by our method.

where the scale range $s = (0.2:0.2:5)$ is set for the DRIVE while $s = (0.5:0.5:4)$ for the STARE, and $\Theta = \{\pi, \pi/18, 2\pi/18, 3\pi/18, \ldots 17\pi/18\}$ is for both datasets. The venation patterns are extracted from leaf images, as illustrated Fig. 3, under consideration $s = (0.5:0.5:3)$, and $\Theta = \{\pi, \pi/12, 2\pi/12, 3\pi/12, \ldots 11\pi/12\}$. The extraction of vessel trees from sub-images of mouse retina is demonstrated in Fig. 4, and is carried out with $\Theta = \{\pi, \pi/12, 2\pi/12, 3\pi/12, \ldots 11\pi/12\}$, and $s = (0.1:0.1:2.2)$ for the first sub-image whereas $s = (0.5:0.5:5)$ for the others.

The performance of the proposed method on the DRIVE and STARE databases is quantitively assessed using three common metrics: accuracy (ACC), specificity (SP) and sensitivity (SE). The obtained results are compared with eleven state-of-the-art algorithms, see Table 1, where local entropy method is used for thresholding the final response of our method. For the DRIVE dataset, the results are calculated using the test

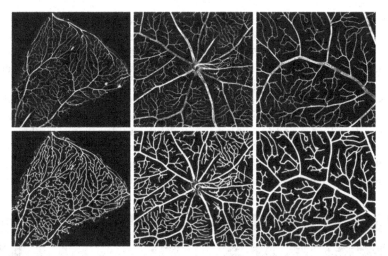

Fig. 4. First row: sub-images from mouse retina. Second row: detection results by our method.

Table 1. Performance comparison of proposed method with benchmark algorithms on DRIVE and STARE databases.

Database	DRIVE			STARE		
Method type	ACC	SP	SE	ACC	SP	SE
Supervised						
Marín [13]	0.945	0.980	0.706	0.952	0.981	0.694
You [14]	0.943	0.975	0.741	0.949	0.975	0.726
Soares [15]	0.944	0.976	0.723	0.948	0.973	0.710
Lázár [16]	0.945	0.972	0.764	0.949	0.975	0.724
Cheng [17]	0.947	0.979	0.725	0.963	0.984	0.781
Unsupervised						
Zhao [18]	0.947	0.978	0.735	0.950	0.976	0.718
Zhao [19]	0.954	0.982	0.742	0.956	0.978	0.780
Lam [20]	0.947	–	–	0.956	–	–
Nguyen [21]	0.940	–	–	0.932	–	–
Fraz [22]	0.943	0.976	0.715	0.944	0.968	0.731
Zhang [23]	0.938	0.972	0.712	0.948	0.975	0.717
Proposed method	**0.956**	**0.976**	**0.747**	**0.954**	**0.970**	**0.760**

set (20 images) with ground truth A (1st_manual). For the STARE database (20 images), the manual segmentations by the first observer are used as ground truth. The results of the algorithm [15] on both fundus imaging databases are obtained from the paper [18], while the rest are taken from published results in their original studies. As illustrated in Table 1, the proposed detection method performs best on the DRIVE against the supervised and unsupervised algorithms by producing the highest accuracy over other methods, higher sensitivity except the method in [16] and competitive specificity to others. On the STARE, the proposed method achieves a higher accuracy except the methods [17, 19] and [20], and a higher sensitivity excluding the methods [17] and [19].

The qualitative comparison is also carried out using randomly chosen retinal and leaf images (see Figs. 5 and 6) with seven state-of-the-art detection algorithms: phase congruency tensor-vesselness (PCT-V) [2], phase congruency tensor-neurite (PCT-N) [2], multiscale bowler-hat transform (MBT) [4], isotropic nonlinear filtering (ISNF) [5], mathematical morphology with curvature evaluation (MCE) [6], vesselness detector

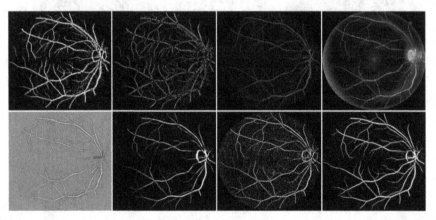

Fig. 5. Detection results of multiple algorithms on a retinal image. First row: PCT-V, PCT-N, MBT, and ISNF, respectively. Second row: MCE, VD, ND, and our method, respectively.

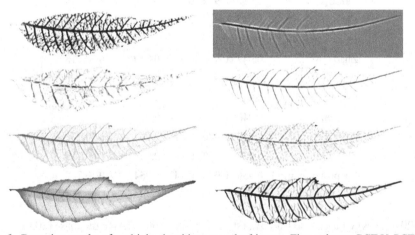

Fig. 6. Detection results of multiple algorithms on a leaf image. First column: PCT-V, PCT-N, MBT, and ISNF, respectively. Second column: MCE, VD, ND, and our method, respectively.

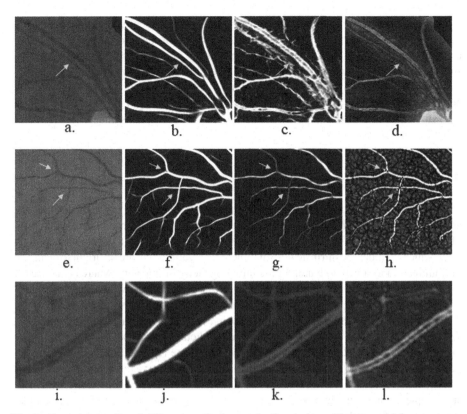

Fig. 7. Detection results of different methods on selective regions. First row: (a) low contrast region, (b) our method, (c) PCT-V, (d) ISNF. Second row: (e) region with bifurcation/crossing, (f) our method, (g) VD, (h) ND. Third row: (i) a vessel with central light reflection, (j) our method, (k) MBT, (l) ISNF.

(VD) [7], and neurite detector (ND) [8]. To have fair comparison, their parameters are optimized to produce best results. As shown in Fig. 5, the ISNF and VD can only detect vessels with large diameter, missing small ones, and produce non-uniform response, while the PCT-V is susceptible to noise, especially in regions with uniform intensity and to light reflection running at the center of wide-diameter pattern. The methods MBT and MCE enhance background regions, which causes a challenge into distinguishing vessel tree, whereas the ND and PCT-N clearly produce false detection in junctions and crossings, losing the connectivity of the tree. In Fig. 6, the proposed method produces superior response on the detection of venation patterns, outperforming the other algorithms which fail into extracting vein network from leaf images. The methods PCT-V, MBT and ND enhance the background artifacts and produces noisy response. The methods ISNF, VD and ND miss the veins with fine diameters and produces discontinuous venation tree. The PCT-N and MCE yield very weak response to extract venation patterns as the PCT-N is sensitive to low intensity veins while the MCE is susceptible the intensity inhomogeneity, enhancing background regions with venation tree.

Further, we closely examined the response of the proposed method, in comparison with other algorithms, in weak contrast, junctions and crossings, and central light reflex, as exhibited in Fig. 7. Based on the obtained results, the proposed method shows its superior to: (I) detect low contrast pattern in inhomogeneous region as compared to the methods PCT-V and ISNF; (II) produce stronger response at bifurcations than VD and ND methods; (III) be more robust when presenting the central light than MBT and ISNF; (IV) attain a more uniform response across various structure widths than other methods.

4 Discussion and Conclusion

The detection of curvilinear patterns is still challenging for existing methods due to variations in intensity, highly curved structures, inhomogeneous background, and variability of pattern diameters. In this paper, we exploit the key advantage of the local phase-GDQ filter, which includes the intrinsic information of line features that are invariant to changes in illumination as a way to handling the inhomogeneity of intensity, whereas the use of the GDQ filter at several orientations deals with various structure widths and directions. Also, the separation between lines and edges by the GDQ increases the robustness of the detection and avoids the duplication problem associated with fusing edges and lines, where the combination of features produces unpredictable response. The success of the measure function-based hessian tensor benefits from the region information of curved patterns, which include local anisotropy intensity, by quantifying the linearity and coherence quantities. These quantities facilitate a robust detection as they are more sensitive to anisotropic intensity, which makes the proposed function sustainable to low-magnitude eigenvalues, whereas being insensitive to noise regions with constant intensity (isotropy regions).

Moreover, the directional decomposition by the GDQ to curved patterns enables more reliable hessian analysis results, where the eigensystem calculation is facilitated because each directional image includes patterns within the same orientation. The final response is produced by maximizing the function outputs over multiple directions to obtain a uniform and strong response across different widths and intensities. Regarding the connectivity of the structure network, the proposed method scheme (decomposition-filtering-recombination) avoids the junction suppression while retaining narrow and low contrast structures.

In conclusion, we introduce a new unsupervised method for the detection of the curvilinear structures in 2D biomedical images. The proposed method utilizes the local phase map estimated by the gaussian derivative quadrature filter to identify line features, and quantifies the geometric information of target structures, by anisotropic linearity and coherence indices, based hessian tensor to propose a novel measure function. The proposed framework has been validated for the extraction of retinal vessel tree and venation pattern, and its results have been compared with state-of-the-art benchmark methods to demonstrate its excellent performance. Our detection method is not only able to detect curvilinear tree with a response standing out from the background more conspicuously, but also can suppress noise, retain small and low contrast structures, produce more uniform response, be more robust in the presence of central light reflection and preserve the connectivity of curved pattern. The focus of the future work is to extend the proposed framework to be applied to the vessel extraction from 3D biomedical images and the crack detection in pavement images.

References

1. Vicas, C., Nedevschi, S.: Detecting curvilinear features using structure tensors. IEEE Trans. Image Process. **24**, 3874–3887 (2015). https://doi.org/10.1109/tip.2015.2447451
2. Obara, B., Fricker, M., Gavaghan, D., Grau, V.: Contrast-independent curvilinear structure detection in biomedical images. IEEE Trans. Image Process. **21**, 2572–2581 (2012). https://doi.org/10.1109/tip.2012.2185938
3. Larese, M., Namías, R., Craviotto, R., Arango, M., Gallo, C., Granitto, P.: Automatic classification of legumes using leaf vein image features. Pattern Recogn. **47**, 158–168 (2014). https://doi.org/10.1016/j.patcog.2013.06.012
4. Sazak, Ç., Nelson, C., Obara, B.: The multiscale bowler-hat transform for blood vessel enhancement in retinal images. Pattern Recogn. **88**, 739–750 (2019). https://doi.org/10.1016/j.patcog.2018.10.011
5. Liu, L., Zhang, D., You, J.: Detecting wide lines using isotropic nonlinear filtering. IEEE Trans. Image Process. **16**, 1584–1595 (2007). https://doi.org/10.1109/tip.2007.894288
6. Zana, F., Klein, J.: Segmentation of vessel-like patterns using mathematical morphology and curvature evaluation. IEEE Trans. Image Process. **10**, 1010–1019 (2001). https://doi.org/10.1109/83.931095
7. Frangi, A.F., Niessen, W.J., Vincken, K.L., Viergever, M.A.: Multiscale vessel enhancement filtering. In: Wells, W.M., Colchester, A., Delp, S. (eds.) MICCAI 1998. LNCS, vol. 1496, pp. 130–137. Springer, Heidelberg (1998). https://doi.org/10.1007/BFb0056195
8. Meijering, E., Jacob, M., Sarria, J., Steiner, P., Hirling, H., Unser, M.: Design and validation of a tool for neurite tracing and analysis in fluorescence microscopy images. Cytometry **58A**, 167–176 (2004). https://doi.org/10.1002/cyto.a.20022
9. Boukerroui, D., Noble, J., Brady, M.: On the choice of band-pass quadrature filters. J. Math. Imaging Vis. **21**, 53–80 (2004). https://doi.org/10.1023/b:jmiv.0000026557.50965.09
10. Peeters, T., Rodrigues, P., Vilanova, A., ter Haar Romeny, B.: Analysis of distance/similarity measures for diffusion tensor imaging. In: Laidlaw, D., Weickert, J. (eds.) Visualization and Processing of Tensor Fields. MATHVISUAL, pp. 113–136. Springer, Heidelberg (2009). https://doi.org/10.1007/978-3-540-88378-4_6
11. Weickert, J.: Coherence-enhancing diffusion filtering. Int. J. Comput. Vis. **31**, 111–127 (1999). https://doi.org/10.1023/a:1008009714131
12. Challoob, M., Gao, Y.: Retinal vessel segmentation using matched filter with joint relative entropy. In: Felsberg, M., Heyden, A., Krüger, N. (eds.) CAIP 2017. LNCS, vol. 10424, pp. 228–239. Springer, Cham (2017). https://doi.org/10.1007/978-3-319-64689-3_19
13. Marín, D., Aquino, A., Gegundez-Arias, M., Bravo, J.: A new supervised method for blood vessel segmentation in retinal images by using gray-level and moment invariants-based features. IEEE Trans. Med. Imaging **30**, 146–158 (2011). https://doi.org/10.1109/tmi.2010.2064333
14. You, X., Peng, Q., Yuan, Y., Cheung, Y., Lei, J.: Segmentation of retinal blood vessels using the radial projection and semi-supervised approach. Pattern Recogn. **44**, 2314–2324 (2011). https://doi.org/10.1016/j.patcog.2011.01.007
15. Soares, J., Leandro, J., Cesar, R., Jelinek, H., Cree, M.: Retinal vessel segmentation using the 2-D Gabor wavelet and supervised classification. IEEE Trans. Med. Imaging **25**, 1214–1222 (2006). https://doi.org/10.1109/tmi.2006.879967

16. Lázár, I., Hajdu, A.: Segmentation of retinal vessels by means of directional response vector similarity and region growing. Comput. Biol. Med. **66**, 209–221 (2015). https://doi.org/10.1016/j.compbiomed.2015.09.008

17. Cheng, E., Du, L., Wu, Y., Zhu, Y., Megalooikonomou, V., Ling, H.: Discriminative vessel segmentation in retinal images by fusing context-aware hybrid features. Mach. Vis. Appl. **25**, 1779–1792 (2014). https://doi.org/10.1007/s00138-014-0638-x

18. Zhao, Y., Wang, X., Wang, X., Shih, F.: Retinal vessels segmentation based on level set and region growing. Pattern Recogn. **47**, 2437–2446 (2014). https://doi.org/10.1016/j.patcog.2014.01.006

19. Zhao, Y., Rada, L., Chen, K., Harding, S., Zheng, Y.: Automated vessel segmentation using infinite perimeter active contour model with hybrid region information with application to retinal images. IEEE Trans. Med. Imaging **34**, 1797–1807 (2015). https://doi.org/10.1109/tmi.2015.2409024

20. Lam, B., Gao, Y., Liew, A.: General retinal vessel segmentation using regularization-based multiconcavity modeling. IEEE Trans. Med. Imaging **29**, 1369–1381 (2010). https://doi.org/10.1109/tmi.2010.2043259

21. Nguyen, U., Bhuiyan, A., Park, L., Ramamohanarao, K.: An effective retinal blood vessel segmentation method using multi-scale line detection. Pattern Recogn. **46**, 703–715 (2013). https://doi.org/10.1016/j.patcog.2012.08.009

22. Fraz, M., et al.: An approach to localize the retinal blood vessels using bit planes and centerline detection. Comput. Methods Programs Biomed. **108**, 600–616 (2012). https://doi.org/10.1016/j.cmpb.2011.08.009

23. Zhang, B., Zhang, L., Zhang, L., Karray, F.: Retinal vessel extraction by matched filter with first-order derivative of Gaussian. Comput. Biol. Med. **40**, 438–445 (2010). https://doi.org/10.1016/j.compbiomed.2010.02.008

Biometrics and Identification

Exposing Presentation Attacks by a Combination of Multi-intrinsic Image Properties, Convolutional Networks and Transfer Learning

Rodrigo Bresan, Carlos Beluzo, and Tiago Carvalho[✉]

Federal Institute of São Paulo, Campinas, SP 13069-901, Brazil
tiagojc@gmail.com

Abstract. Nowadays, adoption of face recognition for biometric authentication systems is widespread, mainly because this is one of the most accessible biometric characteristic. Techniques intended on deceive these kinds of systems by using a forged biometric sample, such as a printed paper or a recorded video of a genuine access, are known as presentation attacks. Presentation Attack Detection is a crucial step for preventing this kind of unauthorized accesses into restricted areas or devices. In this paper, we propose a new method that relies on a combination of the intrinsic properties of the image with deep neural networks to detect presentation attack attempts. Exploring depth, salience and illumination properties, along with a Convolutional Neural Network, proposed method produce robust and discriminant features which are then classified to detect presentation attacks attempts. In a very challenging cross-dataset scenario, proposed method outperform state-of-the-art methods in two of three evaluated datasets.

Keywords: Presentation attack · Spoofing attack · Transfer learning · CNN · Intrinsic image properties

1 Introduction

Biometrics consists in identify a given individual by its physiological traits (e.g., face, iris or fingerprint) or behavioral patterns (e.g., keystroke dynamics, gait) and it have been used on different types of devices for authentication purpose. Attacks to biometric systems are known as presentation or spoofing attacks. It consists in present a synthetic biometric sample, simulating biometric pattern of a valid user, to the system in order to obtain access as a legitimate user.

To fight back presentation attacks, different literature methods have been proposed in the last years. According to Pan *et al.* [10], techniques for Presentation Attack Detection (PAD) can be grouped into four major groups:

© Springer Nature Switzerland AG 2020
J. Blanc-Talon et al. (Eds.): ACIVS 2020, LNCS 12002, pp. 153–165, 2020.
https://doi.org/10.1007/978-3-030-40605-9_14

user behavior modeling, data-driven characterization, user cooperation and hardware-based.

Techniques based on behaviour modeling for PAD consists in models user's behaviors, such as head movements and eye blinking. Data-driven techniques are based on finding artifacts in attempted attacks by exploiting data that came from a standard acquisition sensor. User cooperation based techniques focus on interaction between user and authentication system, such as asking the user to execute some movements. Finally, there are techniques that use extra hardware, such as depth sensors and infrared cameras, to obtain more information about the scenario to finding cues that reveal an attempted attack[1].

Schwartz et al. [16] presented an anti-spoofing method by exploring the use of several visual descriptors for characterizing facial region according its color, texture, and shape properties. To deal with the high dimensionality in final representation vector, the authors proposed to use Partial Least Squares (PLS) classifier, an statistical approach for dimensionality reduction and classification, which was designed to distinguish a genuine biometric sample from a fraudulent one.

Pinto et al. [15] proposed a data-driven method for video PAD based on Fourier analysis in residual noise signature extracted from input videos. Use of well-known texture feature descriptors, such as Local Binary Patterns was also considered in the literature by Maata et al. [9], which focuses on detecting micro-texture patterns that are added into the fake biometric samples during the acquisition process. Approaches based on Differences of Gaussian (DoG) [12,18] and Histogram of Oriented Gradients (HOG) [7,19] were also proposed, but at the cost of final results is affected by illumination conditions and the capture sensor, due to their nature.

Yeh et al. [21] proposed an effective approach against face presentation attacks, based on perceptual image quality assessment, by adopting a Blind Image Quality Evaluatior (BIQE) along with a Effectivate Pixel Similary Deviation (EPSD), to generate new features to use on a multi-scale descriptor, showing it's efficacy when compared to previous works.

In this paper we introduce a new PAD technique which requires no additional hardware components (e.g., depth sensor, infrared sensor). Different intrinsic image properties are estimated and combined with a Convolutional Neural Network (CNN) and applying a transfer learning process we are able to extract robust and discriminative features. These features are then fed into a Extreme Gradient Boosting (XGBoost) classifier and a classification process with two steps is applied in order to classify samples into attack attempt or genuine sample.

Proposed method outperformed many existing literature approaches for face PAD problem, presenting better results in two of three datasets evaluated.

The main contributions of this paper include: (1) proposition of a new method for face PAD, which is based on a combination between

---

[1] Since this paper focus on data-driven techniques, we focused our literature review on this kind of methods.

intrinsic image properties and deep neural networks; (2) evaluation of different intrinsic properties (e.g., saliency, depth and illumination maps) for the PAD problem, which to the best of our knowledge, have never been evaluated in this context; (3) expressive results for both cross and intra dataset protocol in different public datasets; (4) effective application of transfer-learning approach in a PAD context.

2 Proposed Method

The method proposed in this paper can be divided in four main steps as depicted on Fig. 1. First state consists on estimate intrinsic properties from images. Then, we use a ResNet50 to extract bottleneck features which are submitted to the first classification step by an XGBoost classifier. This step calculates probabilities for each video frame to be, or not, part of an attack attempt. Then, these probabilities are used in a final stage, which performs a meta-learning process combining information from illumination, depth, and saliency maps, resulting in a new artifact, referred in this paper as fusion vector. Finally, this fusion vector feed a second XGBoost classifier responsible for the final prediction.

2.1 Intrinsic Images Properties Estimation

In order to extract intrinsic image information from video samples, for each frame, intrinsic image properties are extracted, which generates intermediate level image representations as depicted on Fig. 2.

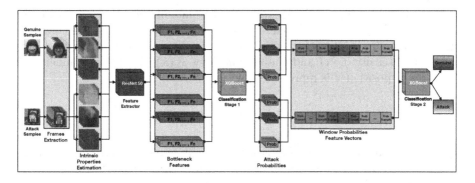

Fig. 1. Overview of the proposed method. Each video sample is split into frames and from each frame, intrisic image properties are calculated. Then, using a ResNet50, proposed method extracts bottleneck features, which are classified by an XGBoost according probability to be an attack. Probabilities of different intrinsic properties are then combined, by using a window of N frames, where N is the small number of frames in a video of evaluated datasets, into a final feature vector which is classified according it average probability of all frames.

Depth Maps. Due to the fact of presentation attacks being frequently reproduced over a flat surface, such as a sheet of paper or a tablet, we believe that
the depth estimation from a given biometric sample can provide relevant information about its authenticity. Our hypothesis is that when presented with a flat
surface, depth map estimated from a sample should differ from a real face.

Proposed method estimates depth maps by using Godard *et al.* [5] method,
which uses stereo images to train a fully convolutional deep neural network
associated with a modified loss function to estimates image depth. This trained
network is then used to estimate depth maps from a single image. As described in
Sect. 2.2, here we also take advantage of transfer learning approach, transferring
weights from the method proposed by Godard *et al.* to our estimator.

Godard *et al.* method's learn a function f which can predict the depth from
a given pixel on a single image. Using an unsupervised learning approach, the
authors propose to reconstruct a given image from another, based on a calibrated
pair of binocular cameras, thus allowing the learning of 3D cues of the original
image. This is performed by finding depth field from the left image, and then
reconstructing the correspondent right image. By using a modified loss function
that outputs the disparity maps, which combines the smoothness, reconstruction
and left-right consistency, the method estimates depth map from a single image.

Illumination Maps. In digital forensics, illumination inconsistencies have been
frequently used to detect image forgeries [1,2]. Inspired by these works, proposed
method also take advantage of illuminant maps to encode illumination information into PAD context. Our hypothesis is that generated illumination maps from
a real face will show differences in its reflection when compared to the generated
illumination map from a face depicted in a flat surface.

To capture illumination information, we calculate illuminant maps from each
frame using the approach proposed by Riess and Angelopoulou [14]. This method
estimates illuminant maps by using the Inverse Intensity-Chromaticity Space
where the intensity $f_c(\mathbf{x})$ and the chromaticity $\chi_c(\mathbf{x})$ of a color channel $c \in
\{R, G, B\}$ at position $\mathbf{x}$ is represented by

$$\chi_c(\mathbf{x}) = m(\mathbf{x}) \frac{1}{\sum_{i \in \{R,G,B\}} f_i(\mathbf{x})} + \gamma_c \ . \tag{1}$$

In Eq. 1, γ_c represents the chromaticity of the illuminant in channel c, whereas
$m(\mathbf{x})$ mainly captures geometric influences, *i.e.* light position, surface orientation
and camera position, and is approximate as described in [17].

Saliency Maps. As in depth and illumination cases, proposed method also
takes advantage of saliency information using the same hypothesis that flat
objects used in PAD will spoil quality in saliency estimation.

Saliency maps are estimated using the method proposed by Zhu *et al.* [24]
which have two major steps: (1) a background modeling using boundary connectivity, which characterizes the spatial layout of image regions with respect to

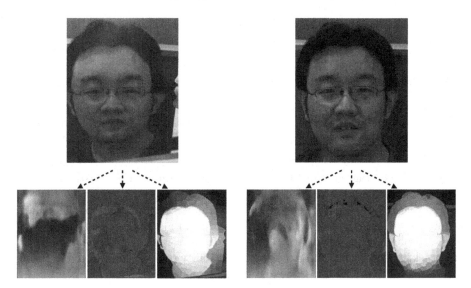

Fig. 2. Intrinsic image properties representation. Comparison between a presentation attack (upper left) with a genuine user (upper right). Below each picture is presented the generated map for depth, illumination, and saliency, respectively.

image boundaries; (2) a principled optimization framework to integrate multiple low-level cues, including proposed background measure. The following equation denotes the method proposed by Zhu *et al.* [24], to generate a saliency map from a single image.

$$BndCon(R) = \frac{|\{p|p \in R, p \in Bnd\}|}{\sqrt{|\{p|p \in R\}|}} \tag{2}$$

where p is a given image patch and *Bnd* is the set of image boundary patches.

2.2 Features Extraction

Once intrinsic image properties maps are estimated, next step of proposed method consists in extract features from each intermediate representation map. To accomplish this task, first we perform an alignment at eye's level on all of our frames and their property maps, followed by a crop on the face region, avoiding background and scene information[2].

Next, proposed method takes advantage of a combination between an well know CNN architecture and the transfer learning process [22]. We choose ResNet50 [6], a robust and effective CNN architecture, associated with ImageNet weights, to extract features from previously generated maps. Removing top layer, ResNet50 works as a feature extractor, which provides feature vectors commonly known as bottleneck features. As the final output of this step, a

---

[2] A classifier which consider scene information could lead to undesirable features and an unfair comparison against literature methods.

feature vector of 2,048 dimensions will be generated, which we will be later on referred to as the bottleneck feature vector.

2.3   Classification

Proposed method uses a two-stages classification pipeline, in which the first classifier is used for frames classification, while the latter one is used for classifying samples (videos) itself.

Stage 1. First stage use an XGBoost [3] classifier, due to its robustness in the task of binary classification when using multiple features. Given a bottleneck feature vector, our classifier returns for each frame, the probability of that frame belong to an attack video, or not. This stage results in 8 probabilities for each frame (probability to be an attack, or not, from frame itself, probability to be an attack, or not, from illuminant map, probability to be an attach, or not, from depth map and probability to be an attack, or not, from salience map).

Stage 2 (Fusion). Given an input video V_P, which already have intrinsic properties estimated, composed by n frames $f_1^P, f_2^P, \cdots, f_n^P$, and where P denotes the intrinsic property extracted from the video ($P \in \{$D, I, S$\}$). In previous stage, we estimated probability for each frame belonging to a class or another, denoted by f_i^P.

Using a fusion-based approach, we combine information from all intrinsic image properties in a way to use all these information together, resulting in a **P**robability **F**eature **V**ector (PFV) defined by

$$PFV = \{p^D, p^I, p^S\} \tag{3}$$

where p^P is given by

$$p^P = f_1^P, f_2^P, \cdots, f_m^P \qquad P \in \{D, I, S\} \tag{4}$$

where m is given by the number of frames into the video with small number of frames in dataset, D, I and S represents depth, illumination and salience maps, respectively.

Finally, PFV vectors are classified using a second XGBoost classifier.

3   Experiments and Results

To evaluate proposed method, different rounds of experiments were performed using three public anti-spoofing datasets, containing samples from genuine accesses and presentation attacks. The adoption of protocols focused in intra-dataset evaluation, where one dataset is tested within the same scenario was performed by following the protocols suggested by datasets' creators. Evaluation of different datasets scenarios, commonly known as inter-dataset or cross-dataset, was also conducted, to assess the performance of proposed method in unknown scenarios. This latter one is the most challenging in the literature, due to the differences in capture conditions that one dataset shows from another one.

Furthermore, it is also paramount to realize that, since we are interested in evaluate the efficiency of each intrinsic property individually, final results reported for depth, illumination, and saliency reflects a majority vote process among all the frames classified on Stage 1.

3.1 Datasets, Metrics, and Setup

To address the efficiency of the proposed method, three publicly available anti-spoofing datasets were selected. The criteria for selection of these datasets among many others available was due to their major adoption in previous works that tackle PAD.

Replay-Attack [4]. Consisting of 1300 video clips from both photo and video attacks from 50 subjects, the Replay-Attack (RA) dataset shows itself as a reliable dataset for the evaluation of the hereby proposed method, once it is presented with different lighting and environmental conditions. In this dataset, three different types of attack are provided: print attacks, mobile attacks, and video attacks. It is separated into three subsets: training set (containing 360 videos); development set (containing 360 videos); testing set (containing 480 videos); and enrollment set (containing 100 videos);

CASIA-FASD [23]. The CASIA-FASD dataset contains a total amount of 600 videos from 50 different subjects, created to provide samples from many of the existent types of presentation attacks. The videos are presented in twelve different scenarios, where each of them is composed by three genuine accesses and three attacks from the same person. Three different resolutions were used to capture (low, normal and high), along with three different types of attack (normal, printed attacks, printed and warped, printed with cut on the eyes region and video-based attacks).

NUAA Photograph Imposter Dataset [18]. The NUAA Photograph Imposter Dataset is composed of 15 subjects, comprising a total of 5,105 valid access images and 7,509 presentation attacks collected through a generic webcam at 20 fps with a resolution of 640 × 480 pixels. The subjects were captured over three sections in different places and lighting conditions. The production of the attack samples was made by shooting a high-resolution photograph with a Canon digital camera.

Metrics. To allow the comparison of the results obtained in this work, we adopt the *Half Total Error Rate* (HTER), which is measured by the mean value between the False Acceptance Rate (FAR), denoted by the rate of attack attempts misclassified as authentic, and the False Rejection Rate (FRR), which is denoted by the rate of authentic samples misclassified as attack. The HTER is measured by

$$\text{HTER} = \frac{\text{FAR} + \text{FRR}}{2} \tag{5}$$

where FAR is the False Acceptance Rate and FRR is the False Rejection Rate.

Experimental Setup. For illumination maps and its segmentation, parameters are the same as the presented in the work of Carvalho *et al.* [2]. For the depth and saliency maps, proposed method uses default parameters as suggested by Godard *et al.* [5] and Zhu *et al.* [24], respectively.

For Stage 1 and Stage 2, classification steps, proposed method uses XGBoost with a *gamma* of 0, a *max_depth* of 6, *gbtree* as booster and a learning rate of 0.3.

Experiments have been conducted by using Python programming language (version 3.6), along with the Keras[3] (version 2.2) and TensorFlow[4] (version 1.8).

3.2   Intra-dataset Evaluation

In intra-dataset evaluation evaluation protocol, we apply the same protocols proposed by each databases' authors, and use HTER metric to measure performance.

As displayed in Table 1, the usage of the fusion outperformed single properties results in Replay Attack, with an HTER value of 3.75%. For CASIA dataset, best results have been achieved using fusion, yielding an HTER result of 9.63%. Finally, results in NUAA dataset using depth maps outperformed all the other features, yielding an HTER of 18.31%.

Table 1. Results (in %) considering the Intra-Dataset protocol for the RA, CASIA and NUAA datasets.

Method	RA HTER	CASIA HTER	NUAA HTER
Raw	6.00	15.74	26.35
Depth	30.25	44.44	**18.31**
Illumination	16.12	16.11	43.65
Saliency	18.37	29.25	31.24
Fusion	**3.75**	**9.63**	26.34

These results present the importance of individual features and increase our hypothesis that different intrinsic properties can be used together to detect attacks. In special, depth maps depicted special representation value in attack detection process.

---

[3] https://keras.io.
[4] https://www.tensorflow.org.

3.3 Cross-Dataset Evaluation

Building a method that is highly adaptable from one face anti-spoofing database to another unknown one has been posed as a major challenge in previous works, and it's an essential ability for real-world applications that rely on face recognition for authentication.

This experiment presents results for the cross-dataset (inter-dataset) evaluation protocol, when one dataset have been used for training while a different one have been used for testing. Table 2 present results when testing method over RA, CASIA and NUAA datasets, respectively.

Table 2. Results (in %) considering the Cross-Dataset Protocol using as test dataset RA (left), CASIA (middle), and NUAA (right).

Train/Test Set	Method	HTER
NUAA/RA	Raw	57.14
	Depth	**49.00**
	Illumination	56.28
	Saliency	62.92
	Fusion	58.64
CASIA/RA	Raw	51.57
	Depth	55.71
	Illumination	**45.21**
	Saliency	48.42
	Fusion	46.71

Train/Test Set	Method	HTER
NUAA/CASIA	Raw	38.33
	Depth	44.81
	Illumination	54.07
	Saliency	48.33
	Fusion	**35.37**
RA/CASIA	Raw	55.55
	Depth	51.11
	Illumination	50.92
	Saliency	**50.74**
	Fusion	59.44

Train/Test Set	Method	HTER
CASIA/NUAA	Raw	38.13
	Depth	**34.11**
	Illumination	50.22
	Saliency	48.37
	Fusion	35.67
RA/NUAA	**Raw**	**51.67**
	Depth	60.35
	Illumination	52.21
	Saliency	58.18
	Fusion	51.88

From presented tables is not difficult to realize that different intrinsic help in different ways for cross-dataset scenario. This fact expose that different kinds of intrinsic properties collaborate differently for each scenario but always aggregating some important information.

Again, better HTERs are achieved when using Depth (training on CASIA dataset and testing on NUAA dataset) and Fusion approaches (training on NUAA dataset and testing on CASIA).

3.4 Comparison with State-of-the-art

Since cross-dataset represents more challenging scenario, this experiment compares achieved results against some state of the art methods. Table 3 summarize best results (HTER) obtained for proposed method compared against some state-of-the-art methods.

When compared against state-of-the-art methods, proposed method outperformed literature in two of three datasets for cross-dataset protocol. Testing on NUAA dataset, proposed method achieved an HTER value of 34.11% when trained on the CASIA dataset, outperforming results obtained in previous works [12,18]. For the CASIA dataset, the best results were attained with the usage of the features fusion, with an HTER of 35.37% when trained on NUAA dataset. The best results for the RA dataset were achieved by the usage of the illumination maps, with an HTER of 45.21%, but outperformed by Yang et al. [20].

Table 3. Comparison among existing approaches for cross-dataset evaluation protocol.

Method	CASIA	RA	NUAA
Yeh *et al.* [21]	39.00	38.10	–
Pinto *et al.* [13]	47.16	49.72	–
Yang *et al.* [20]	42.04	41.36	–
Patel *et al.* [11]	–	**31.60**	–
Tan *et al.* [18]	–	–	45.85
Peixoto *et al.* [12]	–	–	49.85
Raw image	38.33	51.57	38.13
Depth	44.81	49.00	**34.11**
Illumination	50.92	45.21	50.22
Saliency	48.33	48.42	48.37
Fusion	**35.37**	46.71	35.67

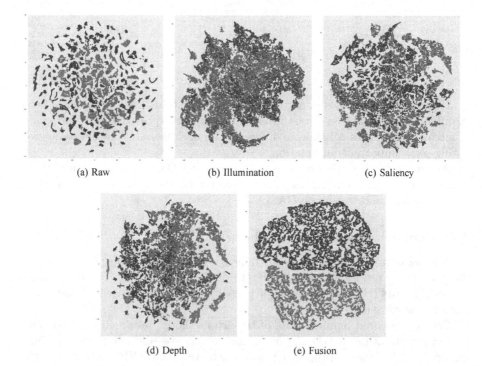

(a) Raw   (b) Illumination   (c) Saliency

(d) Depth   (e) Fusion

Fig. 3. t-SNE features extracted from Replay Attack dataset. Each figure depicts features for an specific intrinsic properties, where blue points represents genuine access samples and red points represent attack samples. Each intrinsic property perform a different degree of separability between samples. Fusion of all of the intrinsic features perform a considerably separability between classes. (Color figure online)

3.5 Intrinsic Properties and Features Analysis

Last experiment performed on proposed method focus on show how each one of intrinsic properties contribute to improve classes separability. This analysis is performed using T-distributed Stochastic Neighbor Embedding (tSNE) [8], which project into a 2D feature space bottleneck features (originally with 2048 dimensions) extracted from each intrinsic property map. Figure 3 depicts feature vectors extracted from Replay Attack dataset. Each figure depicts features for an specific intrinsic properties, where blue points represents genuine access samples and red points represent attack samples. Each intrinsic property perform a different degree of separability between samples. Fusion of all of the intrinsic features perform a considerably separability between classes.

4   Conclusions and Research Directions

In this paper, we have proposed a new method that, by using a two-step classification model, along with intrinsic image properties, such as depth, illumination, and saliency, learn representative features for the task of presentation attack detection. Evaluating the hereby proposed method in three different databases, we reach results outperforming previous works for PAD problem. Findings provided by this paper, such as the efficacy of using image intrinsic properties, can lead to a better understanding on the study and development of new anti-spoofing methods, as well as to provide better insights for development of new datasets. Our results also confirm our hypothesis that by adopting transfer learning techniques along intrinsic image properties, are capable to detect attempts of presentation attacks.

For future works, we intend to investigate other types of intrinsic properties, to better understand the features that may help in the task of distinguishing between an authentic facial biometric sample and a fraudulent one. We also believe that by performing a finetuning step, we could achieve even better results, once that the results attained in this work were achieved by adopting the weights of a pretrained network on data that does not share many similarities with the problem of PAD.

Acknowledgments. We would like to thank São Paulo Research Foundation (FAPESP) (#2017/12631-6), to the National Council for Scientific and Technological Development - CNPq (#423797/2016-6), and to NVIDIA for the donation of a TITAN XP GPU to be used on this research.

References

1. Carvalho, T., Faria, F.A., Pedrini, H., da Silva Torres, R., Rocha, A.: Illuminant-based transformed spaces for image forensics. IEEE Trans. Inf. Forensics Secur. **11**(4), 720–733 (2016). https://doi.org/10.1109/TIFS.2015.2506548

2. de Carvalho, T.J., Riess, C., Angelopoulou, E., Pedrini, H., de Rezende Rocha, A.: Exposing digital image forgeries by illumination color classification. IEEE Trans. Inf. Forensics Secur. **8**(7), 1182–1194 (2013). https://doi.org/10.1109/TIFS.2013. 2265677

3. Chen, T., Guestrin, C.: XGBoost: a scalable tree boosting system. In: Proceedings of the 22nd ACM SIGKDD International Conference on Knowledge Discovery and Data Mining, pp. 785–794. ACM (2016)

4. Chingovska, I., Anjos, A., Marcel, S.: On the effectiveness of local binary patterns in face anti-spoofing. In: 2012 BIOSIG - Proceedings of the International Conference of Biometrics Special Interest Group (BIOSIG), pp. 1–7, September 2012

5. Godard, C., Aodha, O.M., Brostow, G.J.: Unsupervised monocular depth estimation with left-right consistency. In: 2017 IEEE Conference on Computer Vision and Pattern Recognition (CVPR), pp. 6602–6611, July 2017. https://doi.org/10.1109/ CVPR.2017.699

6. He, K., Zhang, X., Ren, S., Sun, J.: Deep residual learning for image recognition. In: Proceedings of the IEEE Conference on Computer Vision and Pattern Recognition, pp. 770–778 (2016)

7. Komulainen, J., Hadid, A., Pietikinen, M.: Context based face anti-spoofing. In: 2013 IEEE Sixth International Conference on Biometrics: Theory, Applications and Systems (BTAS), pp. 1–8, September 2013. https://doi.org/10.1109/BTAS. 2013.6712690

8. van der Maaten, L., Hinton, G.: Visualizing data using t-SNE. J. Mach. Learn. Res. **9**, 2579–2605 (2008)

9. Maatta, J., Hadid, A., Pietikinen, M.: Face spoofing detection from single images using micro-texture analysis. In: 2011 International Joint Conference on Biometrics (IJCB), pp. 1–7, October 2011. https://doi.org/10.1109/IJCB.2011.6117510

10. Pan, G., Wu, Z., Sun, L.: Liveness detection for face recognition. In: Delac, K., Grgic, M., Bartlett, M.S. (eds.) Recent Advances in Face Recognition, chap. 9, pp. 235–252. IntechOpen, Rijeka (2008). https://doi.org/10.5772/6397

11. Patel, K., Han, H., Jain, A.K.: Cross-database face antispoofing with robust feature representation. In: You, Z., et al. (eds.) CCBR 2016. LNCS, vol. 9967, pp. 611–619. Springer, Cham (2016). https://doi.org/10.1007/978-3-319-46654-5_67

12. Peixoto, B., Michelassi, C., Rocha, A.: Face liveness detection under bad illumination conditions. In: 2011 18th IEEE International Conference on Image Processing, pp. 3557–3560, September 2011. https://doi.org/10.1109/ICIP.2011.6116484

13. Pinto, A., et al.: Counteracting presentation attacks in face, fingerprint, and iris recognition. In: Deep Learning in Biometrics, p. 245 (2018)

14. Riess, C., Angelopoulou, E.: Scene illumination as an indicator of image manipulation. In: Böhme, R., Fong, P.W.L., Safavi-Naini, R. (eds.) IH 2010. LNCS, vol. 6387, pp. 66–80. Springer, Heidelberg (2010). https://doi.org/10.1007/978-3-642-16435-4_6

15. da Silva Pinto, A., Pedrini, H., Schwartz, W., Rocha, A.: Video-based face spoofing detection through visual rhythm analysis. In: 2012 25th SIBGRAPI Conference on Graphics, Patterns and Images, pp. 221–228, August 2012. https://doi.org/10. 1109/SIBGRAPI.2012.38

16. Schwartz, W.R., Rocha, A., Pedrini, H.: Face spoofing detection through partial least squares and low-level descriptors. In: 2011 International Joint Conference on Biometrics (IJCB), pp. 1–8, October 2011. https://doi.org/10.1109/IJCB.2011. 6117592

17. Tan, R.T., Ikeuchi, K., Nishino, K.: Color constancy through inverse-intensity chromaticity space. In: Digitally Archiving Cultural Objects, pp. 323–351. Springer, Boston (2008). https://doi.org/10.1007/978-0-387-75807-16

18. Tan, X., Li, Y., Liu, J., Jiang, L.: Face liveness detection from a single image with sparse low rank bilinear discriminative model. In: Daniilidis, K., Maragos, P., Paragios, N. (eds.) ECCV 2010, Part VI. LNCS, vol. 6316, pp. 504–517. Springer, Heidelberg (2010). https://doi.org/10.1007/978-3-642-15567-3_37

19. Yang, J., Lei, Z., Liao, S., Li, S.Z.: Face liveness detection with component dependent descriptor. In: 2013 International Conference on Biometrics (ICB), pp. 1–6, June 2013. https://doi.org/10.1109/ICB.2013.6612955

20. Yang, J., Lei, Z., Li, S.Z.: Learn convolutional neural network for face anti-spoofing. arXiv preprint arXiv:1408.5601 (2014)

21. Yeh, C.H., Chang, H.H.: Face liveness detection based on perceptual image quality assessment features with multi-scale analysis. In: 2018 IEEE Winter Conference on Applications of Computer Vision (WACV), pp. 49–56. IEEE (2018)

22. Yosinski, J., Clune, J., Bengio, Y., Lipson, H.: How transferable are features in deep neural networks? In: Advances in Neural Information Processing Systems, pp. 3320–3328 (2014)

23. Zhang, Z., Yan, J., Liu, S., Lei, Z., Yi, D., Li, S.Z.: A face antispoofing database with diverse attacks. In: 2012 5th IAPR International Conference on Biometrics (ICB), pp. 26–31, March 2012. https://doi.org/10.1109/ICB.2012.6199754

24. Zhu, W., Liang, S., Wei, Y., Sun, J.: Saliency optimization from robust background detection. In: 2014 IEEE Conference on Computer Vision and Pattern Recognition, pp. 2814–2821, June 2014. https://doi.org/10.1109/CVPR.2014.360

Multiview 3D Markerless Human Pose Estimation from OpenPose Skeletons

Maarten Slembrouck[1]([✉])(iD), Hiep Luong[1], Joeri Gerlo[3], Kurt Schütte[2],
Dimitri Van Cauwelaert[1], Dirk De Clercq[3], Benedicte Vanwanseele[2],
Peter Veelaert[1], and Wilfried Philips[1]

[1] TELIN-IPI, Faculty of Engineering and Architecture, Ghent University-imec,
Ghent, Belgium
maarten.slembrouck@ugent.be
[2] Human Movement Biomechanics Research Group, Department of Kinesiology,
KU Leuven, Leuven, Belgium
[3] Department of Movement and Sport Sciences, Ghent University, Ghent, Belgium
https://ipi.ugent.be

Abstract. Despite the fact that marker-based systems for human motion estimation provide very accurate tracking of the human body joints (at mm precision), these systems are often intrusive or even impossible to use depending on the circumstances, e.g. markers cannot be put on an athlete during competition. Instrumenting an athlete with the appropriate number of markers requires a lot of time and these markers may fall off during the analysis, which leads to incomplete data and requires new data capturing sessions and hence a waste of time and effort. Therefore, we present a novel multiview video-based markerless system that uses 2D joint detections per view (from OpenPose) to estimate their corresponding 3D positions while tackling the people association problem in the process to allow the tracking of multiple persons at the same time. Our proposed system can perform the tracking in real-time at 20–25 fps. Our results show a standard deviation between 9.6 and 23.7 mm for the lower body joints based on the raw measurements only. After filtering the data, the standard deviation drops to a range between 6.6 and 21.3 mm. Our proposed solution can be applied to a large number of applications, ranging from sports analysis to virtual classrooms where submillimeter precision is not necessarily required, but where the use of markers is impractical.

Keywords: Markerless human motion · Joint detection · Multiview

1 Introduction

Current experiments in the field of human motion analysis are often analysed with marker-based systems such as Qualisys or Vicon. The biggest drawback of these systems is the time needed to instrument a person with reflective markers. Moreover, during the data capturing process markers may fall off, rendering that

© Springer Nature Switzerland AG 2020
J. Blanc-Talon et al. (Eds.): ACIVS 2020, LNCS 12002, pp. 166–178, 2020.
https://doi.org/10.1007/978-3-030-40605-9_15

recording useless. Beside this, such a system with markers cannot be deployed in numerous applications such as virtual classrooms, athlete analysis during competition and many more.

Recent advances in markerless monocular skeleton detection enable new applications that require semi-accurate tracking of body parts. Such markerless systems provide the solution for the above mentioned drawbacks of marker-based systems. Whereas marker-based systems claim submillimeter accuracy for the markers, markerless systems only obtain an accuracy up to a few centimeters. The reason is that a joint (e.g. an ankle) is not always detected at the anatomically correct position. Depending on the clothing, joints might not even be visible and even humans would have a hard time to locate the exact position of the joints from the videos only.

The changes in planar joint angles are often used for movement analysis, e.g. technical performance in sports or basic clinical gait analysis. For this reason a markerless system has its value despite the fact that it cannot accurately measure rotations along the limbs axis. Markerless systems have been around for a while now. Since the early 2000s, research has been going on to find the location of joints in RGB videos. Most of these approaches relied on shape-from-silhouettes and tried to match a detailed kinematic model. Positional errors were typically larger than 10 cm [6,7,15,16]. Later advances obtained 51 to 100 mm positional errors on the joints [12,13]. More recently, the shift to monocular pose extraction enabled more flexible camera setups [8,9]. However, the reported positional errors of these systems are all typically between 56 and 140 mm. The multiview system that we propose goes one step further. Our system can accurately detect 2D joints, which enables us to obtain positional errors between 24.2 and 49.2 mm.

Apart from obtaining a better accuracy, we also aim at improved robustness. To obtain this goal, we use the existing 2D pose extractor of OpenPose [3] and triangulate the joint positions in 3D. However, we noticed a number of issues, such as self-occlusion, switching limbs and misdetected joints by the 2D pose extractor. We handle all three issues in this paper and present a robust system that can be applied in a wide range of applications due to its flexibility in the number of cameras and the scale in which it can be applied. In the results section, we will show promising results that demonstrate this flexibility.

The outline of the paper is as follows. In Sect. 2 we discuss the pose extraction from monocular video. In Sect. 3 triangulation is explained for at least two image points from different cameras. We explain how we match different persons in the people association step in Sect. 4 and in Sect. 5 the fusion from 2D to 3D at a skeleton level is explained. In Sect. 6 we discuss two use cases where a single person is analysed with 3 and 8 cameras.

2   Real-Time Skeleton Detection

Our goals is to have an accurate robust system that runs in real-time, i.e. at a framerate of at least 20 fps. OpenPose [3,4] is one of the deep learning algorithms that provides real-time skeleton data. OpenPose is currently the only framework to support 25 joint points per person, which makes it very useful to analyse human motion. Our system uses the detected skeletons from OpenPose to reason further about the 3D position of each joint using multiple camera views. Alternative pose extractors were also researched, such as Alpha Pose [10], Cascaded Pyramid Networks [5], Dense Pose [1], PoseFlow [17] and SMPL [2]. However, they only support a limited number of joints to be detected. VNect on the other hand is a close competitor to OpenPose and it also supports for instance the detection of the foot tip, but we were unable to obtain the needed source code to use this framework in our experiments [14]. OpenPose is able to run on an NVidia 1080Ti GPU card at almost 30 fps. We employ two graphics cards in parallel and equally distribute the load over both cards. By stitching multiple images together before feeding it to the OpenPose neural network, we are able to process multiple camera images at a rate of 20–25 fps. More convenient GPUs will run the proposed method between 10 and 15 fps.

In the next section we will discuss how we estimate 3D position from a set of 2D points using least squares triangulation. We need this step for the people association in Sect. 4 as well as the fusion of skeleton point into their 3D positions, taking into account possible misdetections in 2D in Sect. 5.

3   Triangulation

Traditional cameras observe the 3D world by light ray projection on a 2D plane. During this process depth information is lost, which is one of the reasons why it is hard to accurately reconstruct a 3D scene from a single captured image. The use of multiple cameras from multiple viewpoints has proven that 3D reconstruction is possible by using the content of the image and calibration data of the different cameras [11].

The mathematical conversion from a set of 2D points from multiple cameras into a 3D location is often referred to as triangulation. An example of triangulation is shown in Fig. 1. The idea is to estimate the position of point X based on the 2D image positions p_i and fixed known camera calibrations (intrinsic and extrinsic). Due to inaccuracies in the camera calibration and the discretization of the image sensor, the lines will rarely intersect. Therefore, we need to apply an approximated model. In the following paragraphs, we will briefly discuss how we perform triangulation using the minimization of least squares of the distances between the point X and the lines defined by the camera position and the pixel location.

For a single joint, we define two vectors: c_i which represents the camera position of camera C_i and vector a_i which is the vector between the unknown 3D position of the joint x and c_i: $a_i = x - c_i$.

Let the ith line be defined by c_i and a unit vector d_i. Given the principal point (u_0^i, v_0^i) of camera i, we can calculate d_i as follows:

$$\widetilde{d}_{t,i} = R_i^T \begin{pmatrix} m_{ix}(p_{i,u} - u_0^i) \\ m_{iy}(p_{i,v} - v_0^i) \\ f_i \end{pmatrix},$$

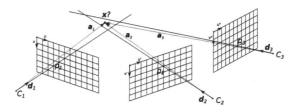

Fig. 1. Triangulation example where we need to estimate the 3D position of point X based on 2D points from multiple cameras. Ideally vectors $\boldsymbol{a}_i$ and $\boldsymbol{d}_i$ would coincide.

$$d_i = \frac{\widetilde{d}_{t,i}}{||\widetilde{d}_{t,i}||},$$

where R_i is the rotation matrix, f_i is the focal length (mm), m_{ix} and m_{iy} represent the horizontal and vertical pixel size on the sensor (mm) and $(p_{i,u}, p_{i,v})$ the projected joint position.

The point to line distance is given by:

$$||\boldsymbol{w}_i|| = \sqrt{\boldsymbol{w}_i \cdot \boldsymbol{w}_i} \quad \text{where} \quad \boldsymbol{w}_i = \boldsymbol{d}_i \times \boldsymbol{a}_i. \tag{1}$$

We now determine the single 3D point $\boldsymbol{x}$ that minimizes the sum of squared point to line distances $\sum_i ||\boldsymbol{w}_i||^2$. This minimum occurs where the gradient is the zero vector ($\boldsymbol{0}$):

$$\nabla\left(\sum_i ||\boldsymbol{w}_i||^2\right) = \boldsymbol{0}.$$

Expanding the gradient,

$$\sum_i \left(2\boldsymbol{d}_i(\boldsymbol{d}_i \cdot \boldsymbol{a}_i) - 2(\boldsymbol{d}_i \cdot \boldsymbol{d}_i)\boldsymbol{a}_i\right) = \boldsymbol{0}.$$

We found that the coordinates of $\boldsymbol{x}$ satisfy a 3×3 linear system,

$$M\boldsymbol{x} = \boldsymbol{b}, \tag{2}$$

where the kth row (a 3-element row vector) of matrix M is defined as

$$M_k = \sum_i \left(d_{ik}\boldsymbol{d}_i - (\boldsymbol{d}_i \cdot \boldsymbol{d}_i)\boldsymbol{e}_k\right)^T$$

with vector $\boldsymbol{e}_k$ the respective unit basis vector, and

$$\boldsymbol{b} = \sum_i \boldsymbol{d}_i(\boldsymbol{c}_i \cdot \boldsymbol{d}_i) - \boldsymbol{c}_i(\boldsymbol{d}_i \cdot \boldsymbol{d}_i).$$

In practice M is almost always not singular. However, in rare circumstances, M could be singular. For example, we could have a system with only two cameras facing each other and a joint close to the line joining the 2 projection centres.

Since this situation is highly unlikely, and can be easily avoided, we exclude it. We use Gaussian elimination to find x in Eq. 2.

Note that self-occlusion of a joint is automatically solved by not taking into account a viewpoint that did not detect the joint. However, at least two cameras need to observe a joint to report a position.

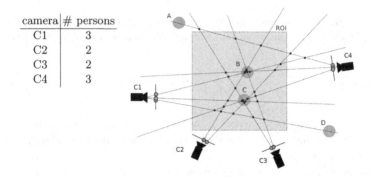

camera	# persons
C1	3
C2	2
C3	2
C4	3

Fig. 2. Clustering of pairwise camera point reconstructions to find different people in the scene. The table indicates the number of detected skeletons in each view. Persons B and C are found, Persons A and D are located outside the region on interest (ROI).

4   People Association

Triangulation of a person's joints relies heavily on the assumptions that the 2D point correspondences in multiple views belong to the same 3D point. However, in case of multiple persons in the area of interest, these correspondences are not so easy to find due to occlusion and possible inaccurate 2D joint positions. As illustrated in Fig. 2.

Let us consider a simple camera network with 4 cameras (Fig. 2). For clarity, we have reduced all skeletons to a single point in the figure. Note that each of these points actually represent 25 joints. The number of possible matches between the different 2D skeletons becomes $(3 + 1)(2 + 1)(2 + 1)(3 + 1) = 144$. We add 1 for each view because none of the skeleton views may correspond to a 3D skeleton in the area of interest. From the 144 combinations, there are also some combinations that cannot result in a skeleton reconstruction since a minimum of two points from different views is required for triangulation.

There is no need to test all 25 individual joints from the same skeleton combinations. To tackle the people association problem, we calculate pairwise correspondences between skeletons from different viewpoints. We assume that joints from a detected skeleton belong to the same person and the spine is correctly detected which is the case in our applications. Therefore, we only consider the spine of a person, defined by the neck and midhip joint (other joints may be chosen for a different application). Valid combinations are those that correspond to a low point to line distance between the triangulated point and the line defined by its 2D detection and the camera position for both the neck and the mid-hip (in the example, two non-parallel lines will always interest. In 3D however, these lines rarely intersect resulting in a reprojection error). In a second phase, the valid matches are clustered based on the 3D distances between the detected spines. 2D skeletons that have been clustered are removed from the search space to avoid that these combinations are again matched with other skeletons later on. Therefore, multiple persons can be calculated from the same set of frames. Additional constraints concerning the region of interest (ROI) may be used to reject persons that are detected in multiple cameras, but are not inside the ROI.

5   Fusion of 2D Joint Projections into 3D Joints

The conversion to a 3D point from multiple 2D joint projections in different views has been discussed in detail above. However, a number of difficulties needs to be addressed, specifically when it comes to pose estimation.

Triangulation supposes that the detected 2D points are accurate. A pose extractor, such as OpenPose, provides a confidence score in range 0.0–1.0 for each of the detected joints. Usually the position of joints is rather accurate. However, when the confidence score of joints is low (e.g. below 0.2), we noticed that we better discard these points in the triangulation process. Another issue with the pose extractor of OpenPose is confusion in a sense of the left and right extremities of a person's body. We especially noticed this problem with the legs. In most cases, the left leg is detected on the left side of the person, but sometimes the left leg is confused with the right leg, or both legs are detected inside the same leg depending on the pose of the person of interest. Both issues demand a suitable solution to avoid discrete changes in the spatio-temporal domain. With a limited number of cameras, it is not always clear what the correct solution should be, especially when multiple cameras suffer from left/right confusion at the same time. We need to be careful to swap the limbs in the correct view and not to pose swapping on correct views. In that case the 3D positions of both legs are correct, with a low point to line distances, but the label left/right might be switched, causing inferior results. Therefore spatio-temporal tracking offers a suitable solution. Figure 3 shows errors that occur frequently in our datasets.

5.1   Handling Limb Ambiguities

After reconstructing a skeleton, we calculate the point to line distances between the reconstructed joint position X_j and the line defined by the camera position C_j and the image location on the image sensor p_{ij} and store them in matrix D which has n rows and m columns where n is the number of cameras and m the number of joints. Only the following are considered in this matrix because they handle switching legs: LHip, LKnee, LAnkle, LHeel, LBigToe, LSmallToe, RHip, RKnee, RAnkle, RHeel, RBigToe and RSmallToe. The same can be done for the arms of a person. We define matrix D as

Fig. 3. Misdetected joints causing difficulties in the 3D matching. OpenPose confused between left (red) and right (green) in two different views captured at the same time and both the left and right leg are detected inside the physical right leg, while the physical left leg remains undetected. (Color figure online)

$$
D = \begin{bmatrix} ||\boldsymbol{w}_{11}||^2 & ||\boldsymbol{w}_{12}||^2 & ||\boldsymbol{w}_{13}||^2 & \cdots & ||\boldsymbol{w}_{1m}^2|| \\ ||\boldsymbol{w}_{21}||^2 & ||\boldsymbol{w}_{22}||^2 & ||\boldsymbol{w}_{23}||^2 & \cdots & ||\boldsymbol{w}_{2m}^2|| \\ \vdots & \vdots & \vdots & \ddots & \vdots \\ ||\boldsymbol{w}_{n1}||^2 & ||\boldsymbol{w}_{n2}||^2 & ||\boldsymbol{w}_{n3}||^2 & \cdots & ||\boldsymbol{w}_{nm}^2|| \end{bmatrix},
$$

where $||\boldsymbol{w}_{ij}||^2$ is the squared point to line distance between the 3D joint position $\boldsymbol{x}_j$ and its detected position on camera j (cfr. Eq. 1). We choose point to line distances over projection errors because the former takes into account the distance between the camera and the detected point.

5.2   Minimizing the Squared Point to Line Distances in Matrix D

Our goal is to remove all limb ambiguities so that we obtain the minimum $\sum_i \sum_j ||\boldsymbol{w}_{ij}||^2$. In order to facilitate real-time processing, we limit the combinations. Some are more likely than others e.g. chances are rather small that the limbs of a skeleton have been switched by more than half of the cameras. Therefore, we constrain the search space to reduce the number of limb reassignments. We are satisfied when all values in D are below a certain threshold T_m (maximum allowed point to line distance). We also use an additional threshold T_u with $T_m < T_u$ to detect and to cope with extreme cases. Both thresholds are not very sensitive. We found that $T_m = 100$ mm and $T_u = 500$ mm correctly fixed the issues with the 2D skeleton detection.

Figure 4 illustrates how the reconstruction is made while coping with switching limbs and misdetected joints. The algorithm starts with the calculation of matrix D. We first check if the maximum value in D is lower than T_u^2. If this is the case, we immediately arrive at the final reconstruction. If this is not the case, we verify if the number of iterations t is smaller than $I_{\max}$. If that is the case, we decide not to investigate any longer to maintain the real-time processing time of the algorithm. Depending on the application, we choose to leave out the joints from the final reconstruction to make sure that no faulty positions are returned.

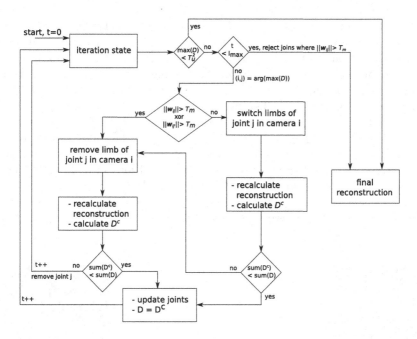

Fig. 4. Flowchart of the proposed algorithm. Joint j' represents the opposite joint j.

At each iteration the following steps are taken: when only one of the two legs has a point to line distance higher than T_m^2, we are almost certain that both legs are detected inside the same leg, meaning one correct location and one incorrect location. Therefore, we decide to test the point to line distances for the reconstruction without the positions of this leg of that particular camera. If both x_{ij} and $x_{ij'}$ are higher than T_m^2, we switch the limbs first. If that does not lead to a better reconstruction, we ignore the limb as well. In all cases, the reconstruction is only accepted, when the sum of the errors of the new candidate joints D^C decreases: $\text{sum}(D^C) < \text{sum}(D)$.

6   Experiments

Two experiments were conducted on two different locations which both were recorded with vision-based cameras and infra-red cameras. The marker-based camera systems (Qualisys and Vicon) have a theoretical submillimeter precision for the marker positions. However, we should keep in mind that due to marker/soft tissue movement it is unlikely that submillimeter precision is reached for the calculation of joint center positions.

The first dataset was recorded at the Sports Science Laboratory Jacques Rogge (SSL-JR) at Ghent University. The vision-based system consisted out of seven 4.5 MP cameras (Manta G-046C, AVT, Stadtroda, Germany). The person of interest ran in a straight line, always in the same direction at different speeds ranging from 2.1 to 5.1 m/s (Fig. 5). The camera images are captured synchronously by two computers at 67 Hz. The running length that can be captured is around 11 m. The infra-red based motion capture system consisted out of ten 1.3 MP cameras (Oqus3+, Qualisys AB, Göteborg, Sweden) operating at a frame rate of 250 Hz. The cameras were fixed to the lab walls, uniformly distributed to measure 4 m of the running length, with a distance to the center of the volume ranging from 3.5 to 7 m. In total 88 Passive IR-reflective 12 mm-sized spherical markers were attached to the subject body and used for full body modeling in Visual 3D software (C-Motion Inc., Germantown, USA). Joint center coordinates of the ankles, knees and hips were exported for comparison. The wand calibration of the setup showed a standard deviation on measured distances of 0.4 mm.

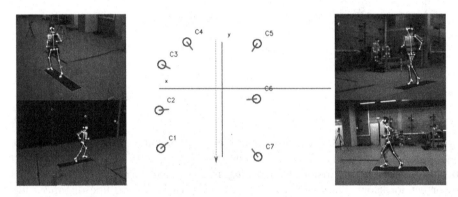

Fig. 5. Camera setup in the Sport Science Lab Jacques Rogge in Ghent on the left (SSL-JR dataset). The red arrow indicated the running direction. On the right we show two camera images with detected 3D skeleton (white) on top. The yellow and cyan lines represent the people association. (Color figure online)

The second dataset was recorded in Leuven, Belgium. Only three 4.5 MP cameras (Manta G-046C, AVT, Stadtroda, Germany) were used operating at a frame rate of 50 Hz. The cameras were located closer to the person of interest in comparison with the first dataset because the measuring volume was only $3 \times 3 \times 2$ m. The person in this dataset is executing stationary movements such as squats and clocks (Fig. 6). A fixed ten camera Vicon system (Vicon MX T20, VICON Motion Systems Ltd., Oxford, UK) supplemented with three additional portable Vicon Vero cameras (Vicon Vero v1.3, VICON Motion Systems Ltd., Oxford, UK) were used. All cameras were sampled at 100 Hz and utilizing a measurement error of 1 mm. In addition, ground reaction forces were collected using two AMTI OR 6 Series force plates sampled at 1000 Hz (Optima, Advanced Mechanical Technology, Inc., Watertown, USA). These force plates were used to determine initial ground contact during the side cut maneuver and check the execution of the clock for the Vicon system. A single researcher placed 39 retro-reflective markers on the participant using palpation to identify the correct attachment site. Markers were placed as shown in Fig. 6 on the trunk, pelvis and both legs and feet, in order to collect kinematic data for the trunk, hip, knee and ankle.

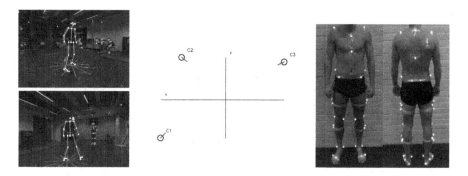

Fig. 6. 3-camera setup in Leuven positioned approximately 2.5 m from the person.

First, we take a look at the graphical representation of the individual results, after which we will evaluate results averaged over all sequences and the spread that can be found in these measurements. We filtered the raw measurements in the spatio-temporal domain from the frame by frame results with a Hanning window of length 7. Such operation slightly improves the results. Figure 7 shows a typical graph produced by the proposed system. We see that the proposed system follows the marker-based positions rather accurately.

Table 1 shows the accuracy averaged per dataset, while Fig. 8 shows the distribution of these numbers. We may conclude that spatio-temporal filtering improves the results by decreasing the standard deviation and average positional error between 1 and 3 mm. For the second dataset we notice an offset in positional errors for a number of joints. The limited number of cameras is most likely the cause of this. Also the cameras in this setup are not entirely evenly distributed around the person of interest. However, the experiments show that even with a limited number of cameras, the proposed method performs well. The offsets between the marker-based and proposed system, bares little significance because the standard deviation is small we reliably detects position changes of the different joints even in case of self-occlusion.

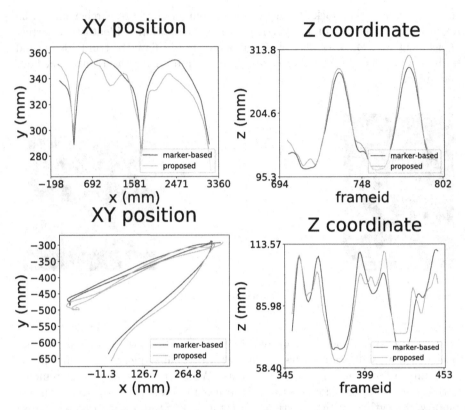

Fig. 7. Typical result of the positions of an ankle joint (top row: SSL-JR dataset, bottom row IPLAY-Leuven dataset). Note the different scales in each of the graphs.

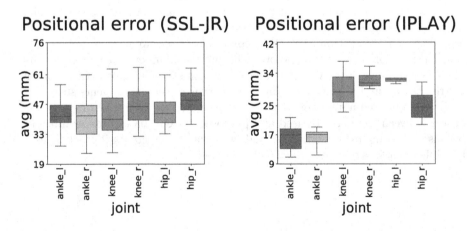

Fig. 8. Average positional error between the marker-based positions and the estimated 3D position for all 33 sequences (SSL-JR) and 9 sequences (IPLAY).

Table 1. Positional errors and standard deviation averaged over 33 sequences (SSL-JR) and 9 sequences (IPLAY-Leuven dataset). All measurements are in mm.

	joint	Unfiltered		Filtered	
		avg	stddev	avg	stddev
S	ankle_l	42.1	23.7	40.6	21.3
S	ankle_r	41.2	21.7	38.8	19.4
L	knee_l	42.2	20.8	40.4	19.1
J	knee_r	46.3	19.9	44.8	18.3
R	hip_l	44.4	18.8	41.6	15.5
	hip_r	50.7	15.2	49.2	12.5
I	ankle_l	19.4	11.9	18.7	8.2
P	ankle_r	16.7	9.6	15.9	6.6
L	knee_l	30.3	10.6	29.8	7.9
A	knee_r	32.6	12.4	32.1	10.0
Y	hip_l	34.2	15.7	33.5	10.9
	hip_r	26.1	14.7	25.1	11.0

7   Conclusion

In this paper we presented a fast and reliable way to convert 2D OpenPose skeleton detections from multiple camera views into 3D skeletons. Our proposed method copes with misdetected joints and switching limbs to extract reliable 3D tracking data for 25 joints of the human body. During our experiments we found that the positional error for the lower limbs are between 15.9 and 49.2 mm and the standard deviation between 6.6 and 21.3 mm. We compared our system to marker-based systems, which claim submillimeter accuracy. The reported accuracy is not as precise as marker-based systems, but much more flexible and can be used in applications which are satisfied with an accuracy of a few millimeter such as entertainment applications, macro body analysis, virtual classrooms...

Acknowledgement. This work was supported by the imec.ICON IPLAY project and a joint cooperation between the Faculty of Engineering and the Faculty of Medicine and Health Sciences at Ghent University. We would like to acknowledge Maxim Steinmeyer and Maarten Van Dyck for their assistance with participant recruitment, data collection and labeling of the IPLAY-Leuven dataset.

References

1. Alp Güler, R., Neverova, N., Kokkinos, I.: DensePose: dense human pose estimation in the wild. In: CVPR, pp. 7297–7306 (2018)

2. Bogo, F., Kanazawa, A., Lassner, C., Gehler, P., Romero, J., Black, M.J.: Keep it SMPL: automatic estimation of 3D human pose and shape from a single image. In: Leibe, B., Matas, J., Sebe, N., Welling, M. (eds.) ECCV 2016. LNCS, vol. 9909, pp. 561–578. Springer, Cham (2016). https://doi.org/10.1007/978-3-319-46454-1_34

3. Cao, Z., Hidalgo, G., Simon, T., Wei, S.E., Sheikh, Y.: OpenPose: realtime multi-person 2D pose estimation using Part Affinity Fields. arXiv (2018)

4. Cao, Z., Simon, T., Wei, S.E., Sheikh, Y.: Realtime multi-person 2D pose estimation using part affinity fields. In: CVPR, pp. 7291–7299 (2017)

5. Chen, Y., Wang, Z., Peng, Y., Zhang, Z., Yu, G., Sun, J.: Cascaded pyramid network for multi-person pose estimation. In: CVPR, June 2018

6. Cheung, G.K.M., Baker, S., Hodgins, J., Kanade, T.: Markerless human motion transfer. In: 3DPVT, pp. 373–378, September 2004

7. Cheung, K., Baker, S., Kanade, T.: Shape-from-silhouette of articulated objects and its use for human body kinematics estimation and motion capture. In: CVPR. IEEE (2003)

8. Du, Y., et al.: Marker-less 3D human motion capture with monocular image sequence and height-maps. In: Leibe, B., Matas, J., Sebe, N., Welling, M. (eds.) ECCV 2016. LNCS, vol. 9908, pp. 20–36. Springer, Cham (2016). https://doi.org/10.1007/978-3-319-46493-0_2

9. Elhayek, A., Kovalenko, O., Murthy, P., Malik, J., Stricker, D.: Fully automatic multi-person human motion capture for VR applications. In: Bourdot, P., Cobb, S., Interrante, V., kato, H., Stricker, D. (eds.) EuroVR 2018. LNCS, vol. 11162, pp. 28–47. Springer, Cham (2018). https://doi.org/10.1007/978-3-030-01790-3_3

10. Fang, H.S., Xie, S., Tai, Y.W., Lu, C.: RMPE: regional multi-person pose estimation. In: ICCV (2017)

11. Hartley, R.I., Sturm, P.: Triangulation. In: Hlaváč, V., Šára, R. (eds.) CAIP 1995. LNCS, vol. 970, pp. 190–197. Springer, Heidelberg (1995). https://doi.org/10.1007/3-540-60268-2_296

12. Hofmann, M., Gavrila, D.M.: Multi-view 3D human pose estimation combining single-frame recovery, temporal integration and model adaptation. In: CVPR, pp. 2214–2221, June 2009

13. Huo, F., Hendriks, E., Paclik, P., Oomes, A.H.: Markerless human motion capture and pose recognition. In: 2009 10th Workshop on Image Analysis for Multimedia Interactive Services, pp. 13–16. IEEE (2009)

14. Mehta, D., et al.: VNect: Real-time 3D human pose estimation with a single RGB camera. ACM Trans. Graph. **36**(4), 44 (2017)

15. Rosenhahn, B., Kersting, U.G., Smith, A.W., Gurney, J.K., Brox, T., Klette, R.: A system for marker-less human motion estimation. In: Kropatsch, W.G., Sablatnig, R., Hanbury, A. (eds.) DAGM 2005. LNCS, vol. 3663, pp. 230–237. Springer, Heidelberg (2005). https://doi.org/10.1007/11550518_29

16. Saboune, J., Charpillet, F.: Markerless human motion capture for gait analysis. arXiv (2005)

17. Xiu, Y., Li, J., Wang, H., Fang, Y., Lu, C.: Pose flow: efficient online pose tracking. In: BMVC (2018)

Clip-Level Feature Aggregation: A Key Factor for Video-Based Person Re-identification

Chengjin Lyu[✉], Patrick Heyer-Wollenberg, Ljiljana Platisa, Bart Goossens, Peter Veelaert, and Wilfried Philips

TELIN-IPI, Ghent University - imec, 9000 Ghent, Belgium
chengjin.lyu@ugent.be

Abstract. In the task of video-based person re-identification, features of persons in the query and gallery sets are compared to search the best match. Generally, most existing methods aggregate the frame-level features together using a temporal method to generate the clip-level features, instead of the sequence-level representations. In this paper, we propose a new method that aggregates the clip-level features to obtain the sequence-level representations of persons, which consists of two parts, i.e., Average Aggregation Strategy (AAS) and Raw Feature Utilization (RFU). AAS makes use of all frames in a video sequence to generate a better representation of a person, while RFU investigates how batch normalization operation influences feature representations in person re-identification. The experimental results demonstrate that our method can boost the performance of existing models for better accuracy. In particular, we achieve 87.7% rank-1 and 82.3% mAP on MARS dataset without any post-processing procedure, which outperforms the existing state-of-the-art.

Keywords: Person re-identification · Convolutional neural network · Feature aggregation

1 Introduction

Person re-identification refers to identifying a person of interest across different images/video sequences, by comparing the tracks of persons from multiple cameras with non-overlapping areas. Recently, it has drawn increasing attention thanks to its potential applications in intelligent video surveillance, such as person tracking [7] and search [28]. It is still a very challenging task due to occlusion, background clutter, as well as the intensive changes in lighting, pose and viewpoint.

Technically speaking, person re-identification tasks could be divided into two sub-tasks: image-based and video-based re-identification. Single image based re-identification methods have achieved impressive results. However, image-based results may be seriously affected by the quality of images, especially when there is significant occlusion which may even lead to false detection on a single frame.

© Springer Nature Switzerland AG 2020
J. Blanc-Talon et al. (Eds.): ACIVS 2020, LNCS 12002, pp. 179–191, 2020.
https://doi.org/10.1007/978-3-030-40605-9_16

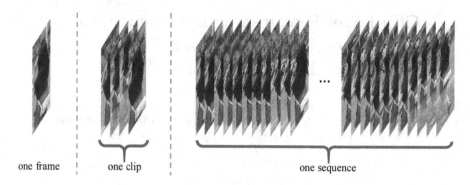

one frame   one clip   one sequence

Fig. 1. Examples of a frame, a clip and a sequence for the same person. A clip is defined as a few frames of fixed length. A sequence contains all the available frames in a track.

Different from image-based methods, video-based re-identification approaches use multiple frames from a complete video sequence to extract the features of a person. It is natural and practical that a sequence of images can provide richer information compared to a single image [33]. The appearances of a person from multiple frames are complementary to each other, which could be used to extract more robust features. Moreover, if a person is moving in a given sequence, spatiotemporal information that describes a person's motion can be summarized from a sequence.

The main idea of video-based person re-identification approaches is to design a spatial-temporal model to extract discriminative representations for video sequences. In general, a typical video sequence lasts a few or even dozens of seconds in person re-identification, which leads to dozens or even hundreds of frames in a sequence. However, due to the physical limitation of GPU's memory, it is almost not realistic to feed all the frames of a sequence into the spatial-temporal model and train the model in a batch with enough sequences. Furthermore, considering the trade-off between the training cost and the information contained within each subsequence (named as clip in this paper), clips of fixed length (usually 4, 6, or a little more) are usually utilized for further feature extraction, which also makes the model suitable for sequences of variable lengths [29]. Examples of a frame, a clip and a sequence for the same person are shown in Fig. 1. Thus, most existing video-based person re-identification approaches focus on the designs of spatial-temporal models for clip-level feature extraction. However, the effect of clip-level feature aggregation has not been studied well.

To explore this issue, we propose a simple yet powerful method to aggregate clip-level features. It is parameter-free, and could be used to boost the performance of existing video-based re-identification methods. Our main contributions in this work are as follows:

1. Propose an Average Aggregation Strategy (AAS) to aggregate clip-level features, where clip-level feature vectors from the same sequence are averaged to represent the whole sequence and then cosine similarities of the averaged vectors are computed for the inference stage.

2. Investigate the influence of the final batch normalization layer for video-based person re-identification, and demonstrate Raw Feature Utilization (RFU) for both training and inference stages.
3. Perform experimental evaluation on MARS dataset. The results show that our proposed method outperforms the existing state-of-the-art without any post-processing procedure.

2   Related Work

In this section, we review related researches including image-based person re-identification, video-based person re-identification, and feature aggregation used in person re-identification.

Image-based person re-identification was first proposed by Gheissari et al. [10] in 2006. In that paper, a combination of color and salient edgel histograms was utilized to perform visual matching on a 44-person dataset. According to the methods of feature generation, the research progress of image-based person re-identification can generally be divided into hand-craft and deep-learning based methods. Hand-craft features such as texture histograms [11], SIFT descriptor [31], and color histograms [5] are widely used. In recent years, with the great success of convolutional neural network (CNN), deep-learning based methods have made distinguished achievements on image-based person re-identification tasks. A siamese model was first chosen to train a CNN for person re-identification with pairs of images [16]. Cheng et al. [3] use triplets of images to train a CNN which can learn both global and local features. Varior et al. [24] introduced long short-term memory (LSTM) module into siamese network to extract spatial relationships. Sun et al. [23] proposed a part-based convolutional baseline (PCB) network which extracts a feature vector consisting of several part-level features. Luo et al. [19] collected and evaluated some effective training tricks in image-based person re-identification.

Video-based person re-identification appeared in the year of 2010 [1,8], where it was first named as multi-shot person re-identification. At first, best match between sets of image-level features was explored. Karaman et at. [14] introduced a conditional random field (CRF) to person re-identification via building a neighborhood topology based on spatial and temporal similarity. Cho et al. [4] proposed an approach to estimate target poses and perform multi-pose model generation and matching. In the second stage, hand-craft features were designed for spatial-temporal information extraction. In [26], spatiotemporal features were first introduced to video-based person re-identification. In [18], body-action units are extracted and then fed to Fisher vectors for final feature generation. More recently, thanks to the emergence of large-scale datasets (e.g., iLIDS-VID [27] and MARS [32]), deep-learned methods have improved the performance of video-based re-identification rapidly. McLaughlin et al. [20] employed a recurrent neural network (RNN) to summarize features from CNNs. Liu et al. [17] built a network which can accumulate motion context from adjacent frames by with the help of RNN. To better extract spatial-temporal information from the videos, attention mechanisms are introduced to this community.

Zhou et al. [35] trained a network to pick out the most discriminative frames with the help of a temporal attention model. Song et al. [21] employed a landmark detector and fully convolutional network to generate region-based quality and representation. Li et al. [15] proposed a multiple attention framework to discover a diverse set of distinctive body parts. Fu et al. [9] introduced a spatial-temporal attention (STA) approach to generate robust clip-level feature representation for video-based person re-identification. Su et al. [22] proposed a k-reciprocal harmonious attention network (KHAN), where spatial and channel attentions are fused as spatial attention and k-reciprocal attention is calculated for temporal attention.

Feature aggregation in person re-identification. In the first works of video-based person re-identification [1,8], multi-match strategies were utilized which take a sequence of images as multiple features for matching. Generally, these multi-match strategies may lead to high computational cost and poor scalability. Therefore, current video-based person re-identification methods usually employ an aggregation step to generate one single feature vector for a video sequence. This step could be max or average pooling [17,20], learned by LSTM [29] or reinforcement learning (RL) [30], for frame-level features. These methods have reached impressive results. What is more, for a long video sequence, the clip-level feature extraction methods could output multiple features to represent the same person. The recent work of [2] introduced a way of competitive snippet-similarity aggregation, which is a variant of multi-match strategy based on clip-level features. However, the aggregation of these clip-level features in video-based person re-identification has not been studied well. In this paper, we propose a roust method for clip-level feature aggregation and assess its performance with the aim to ensure efficient use of the information contained in a video sequence.

3 Methodology

Given an existing model which can be used to extract clip-level spatial-temporal features, where and how to aggregate these features are both essential questions for feature aggregation. In this section, a simple but effective aggregation method which boosts the performance of clip-level features for video-based person re-identification is presented.

3.1 Baseline Model

In this paper, we adopt the STA (Spatial-Temporal Attention) [9] model as our baseline model (see Fig. 2). ResNet50 [12] is chosen as backbone network, which is also the backbone network for many video-based re-identification works. It is worth noting that our method is not restricted to the STA model, and could also work with other video-based person re-identification methods with similar structure. The calculation progress of spatial-temporal attention in the STA model is listed as follows.

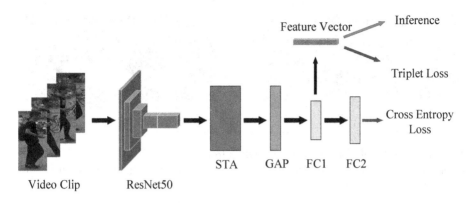

Fig. 2. Pipeline of the baseline model. An input video clip is generated via random sampling. The output is a feature vector that represents the input person.

Suppose that the input clip of length N is represented as $\{I_n\}_{\{n=1:N\}}$, where I_n is a whole-body image of a person. The backbone network generates a set of $H \times W$ feature maps $\{f_n\}_{\{n=1:N\}}$, with $D = 2048$ channels. The attention map a_n is calculated as:

$$a_n(h,w) = \frac{\|\sum_{d=1}^{D} f_n(h,w,d)^2\|_2}{\sum_{h=1}^{H}\sum_{w=1}^{W}\|\sum_{d=1}^{D} f_n(h,w,d)^2\|_2}, \tag{1}$$

where H, W are the height and width of the feature maps, respectively. Then, motivated by the success of part-level features [23], divide these feature and attention maps into M blocks horizontally:

$$\begin{cases} f_n = [f_{n,1}, ..., f_{n,m}, ..., f_{n,M}] \\ a_n = [a_{n,1}, ..., a_{n,m}, ..., a_{n,M}]. \end{cases} \tag{2}$$

After that, the attention score for the mth horizontal region on the nth frame is computed as:

$$s_{n,m} = \sum_{i,j} \|a_{n,m}(i,j)\|_1, \tag{3}$$

where $\{(i,j)\}$ covers the mth horizontal region on the nth frame. Then, an $N \times M$ attention matrix $S = [s_{n,m}]_{N \times M}$ is obtained, which stores the attention scores for different horizontal regions on different frames. The final spatial-temporal attention scores can be calculated as:

$$S(n,m) = \frac{s_{n,m}}{\sum_n \|s_{n,m}\|_1}. \tag{4}$$

Finally, element-wise multiplication of the attention score $S(n,m)$ and horizontal region $f_{n,m}$ is employed to generate the final feature maps.

To obtain the feature vector that represents the input video clip, a global average pooling (GAP) followed by a fully connected (FC) layer is used. For the training progress, the combination of batch-hard triplet loss [13] and cross entropy loss is adopted.

(a)   (b)

Fig. 3. Examples of random sampled images from a sequence. (a) Ideal case that all sampled images are of good quality, (b) Worst case that all sampled images are badly occluded.

3.2 Average Aggregation Strategy

In the baseline model, every video sequence is reduced to N frames via random sampling for both the training and inference stages. Thus, there is no feature aggregation operation on the clip-level features at the stage of inference, for there is only one clip for a sequence. However, representation for a whole sequence using only one N-frame clip is somewhat risky. Sometimes, a clip generated by random sampling could suffer from the low-quality samples, which might lead to false match at the stage of inference. Examples of random sampled images can be found in Fig. 3.

To reduce the uncertainty of a single clip, we use multiple random sampled clips to generate the final feature representation for the whole sequence. The sampled clips from the same sequence are not overlapped with each other. Let v_i denote an arbitrary output feature clip-level vector of the baseline model, the average of all clip-level feature vectors in the same sequence is:

$$\bar{v} = \frac{1}{C} \sum_{i=1}^{C} v_i, \tag{5}$$

where C is the amount of sampled clips, and $\bar{v}$ is the obtained sequence-level feature vector.

Motivated by the successful work of face verification [25], at the stage of inference, the cosine similarity is used to calculated the similarity of two feature vectors, so as to find the best match. Different from Euclidean distance, the cosine similarity only counts on the angle between two vectors instead of their magnitudes in the high-dimensional feature space. The influence of AAS is discussed in Sect. 4.

3.3 Raw Feature Utilization

A batch normalization (BN) layer is widely used in various deep neural networks to speed up the training progress and also improve the performance of networks. In our re-implementation of STA model, we also include a BN layer (see Fig. 4).

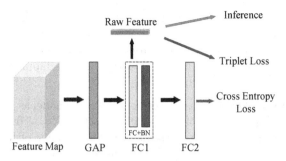

Fig. 4. Illustration of the proposed raw feature utilization. The feature vector before the batch normalization layer is defined as raw feature. This raw feature is utilized for further feature aggregation and the computation of triplet loss for training.

However, in the task of video-based person re-identification, the output of a feature extraction model are feature vectors and then perform future matching based on these feature vectors, which means the direct output of the deep models are feature vectors instead of labels of persons. This is slightly different from the classic computer vision tasks (e.g., image classification and segmentation), where the labels are exactly the outputs of deep neural networks.

Nevertheless, for a person re-identification task, the batch normalization operation of a final feature vector might have harmful effects, since the original distribution of data get changed after batch normalization [19]. In this paper, this change of distribution could influence the performance of subsequent feature aggregation for inference and the computation of the triplet loss function for training. In order to keep the advantage of the final batch normalization layer and also reduce its side effects, we define a way of utilization of raw features in video-based person re-identification for both training and inference stages.

Let v_i^r denote an arbitrary feature vector which is a clip-level representation of a person. It comes after the first fully connect (FC1 in Fig. 4) layer directly, without the final BN operation, which means that v_i^r keeps the original distribution in the feature space. Thus, we name v_i^r as a raw feature. In this paper, we explore the utilization of raw features in both inference and training stages (as shown in Fig. 4). For the inference stage, the final sequence-level feature vector is calculated based on $\{v_i^r\}$ using our average aggregation strategy in Sect. 3.2. For the training stage, raw features are used to compute the triplet loss. The experimental analysis of raw feature utilization is demonstrated in Sect. 4.

4 Experiments

4.1 Experimental Setup

Dataset. MARS dataset [32] is the largest video-based person re-identification dataset, which contains 20,715 video sequences. Every sequence has more than 59 frames on average. Among all these sequences, there are 3,248 distractors, which are distributing samples from false detection and tracking, increasing the difficulty of re-identification significantly. The 1,261 identities are captured by six non-overlapping cameras in this dataset, and each identity shows under at least two cameras. The dataset is split into two non-overlapping train and test sets, containing 625 and 636 identities, respectively.

Implementation Details. We re-implement the STA model as a baseline in our experiments. The input length of a video clip is set to $N = 4$ frames, extracted by random sampling of the original sequence. The number of horizontal parts of a feature map is set to $M = 4$. The input size of an image is 256×128. ResNet50 [12] pretrained on ImageNet [6] is used as the backbone network. The last spatial down-sampling operation of ResNet50 is removed by setting the stride of the last residual block to 1. Thus, the spatial resolution of feature maps is 16×8, with 2048 channels. The channel number of FC1 layers is 512, which leads to a 512-dimensional feature vector. For the training phase, following the common practices [20], random cropping and random horizontal flipping are used to augment the training data. We train this network on one single NVIDIA RTX 2080Ti GPU. Due to the limitation of its memory, we randomly sample 8 identities and 4 clips for each identity to organize a mini-batch with size of 32 for training, which is half of the size 64 in original paper. This might decrease the performance of baseline model a little, while the effectiveness of our proposed method could still be evaluated. For triplet loss, the margin parameter is set to 0.3 as recommended. All the other hyperparameters remain the same with [9]. Note that the baseline model in this paper is the STA model without extra feature fusion and inter-frame regularization strategies, so as to keep the baseline network vanilla.

Evaluation Protocol. In our experiments, we employ Cumulative Matching Characteristics (CMC) curve and mean Average Precision (mAP) as standard evaluation metrics. For CMC curve, we report its value at rank-1, rank-5 and rank-20, respectively.

4.2   Ablation Study

In this paper, we conduct two analytic experiments, namely ablation study, for average aggregation strategy (AAS) and raw feature utilization (RFU). This ablation study could evaluate the effectiveness of these two components in our proposed method. The experimental results are shown in Tables 1 and 2, respectively.

Analysis of AAS. For average aggregation strategy, we carry out experiments to investigate the influence of sampled clip amount C and cosine similarity of final feature vector. The experimental results in Table 1 verify the effectiveness of our average aggregation strategy. In the first column of Table 1, *All* means that all the sampled clips cover the whole video sequence. It shows that the utilization of multiple clips improve the performance of model significantly, e.g., rank-1 accuracy from 83.5% to 85.7%, mAP from 77.5% to 80.0%. It is intuitive

Table 1. Results with different average aggregation strategies.

Number of clips	Cosine similarity	rank-1	rank-5	rank-20	mAP
1	No	83.5	93.8	96.8	77.5
4	No	85.3	94.8	97.1	79.6
All	No	85.7	94.8	97.2	80.0
4	Yes	85.9	94.6	96.9	80.3
All	Yes	86.5	94.7	96.9	80.9

that using all the possible clips to compute an average of all the clip-level features, could generate a more robust feature representation of the whole sequence. Moreover, compared to Euclidean distance (marked as *No* in the second column of Table 1), using cosine similarity to find the best match based on these averaged final features also improve both rank-1 and mAP, leading to 86.5% rank-1 and 80.9% mAP. We believe the reason behind this improvement is that the elements' magnitudes might be not as robust as the vector's angle in a high-dimensional space. Overall, this proposed average aggregation strategy improves the rank-1 by 3.0% and mAP by 3.4%.

Table 2. Ablation study of raw feature utilization.

Inference	Training	rank-1	rank-5	rank-20	mAP
No	No	86.5	94.7	98.2	80.9
Yes	No	87.2	95.8	97.1	82.3
Yes	Yes	87.7	96.4	98.3	82.3

Effectiveness of RFU. The experimental results of Table 1 are based on the feature vectors after the batch normalization, instead of the raw features. Here, we analyze the effect of raw feature utilization for video-based person re-identification. In our experiments, we perform inference using raw features (marked as *Yes* in the first column of Table 2), and then introduce these raw features to train a new model (marked as *Yes* in the second column of Table 2). The experimental results in Table 2 indicate that the utilization of raw features contributes to both inference and training stages. By removing the last batch normalization, the features with original distribution could not only help improve the feature aggregation of inference state but also training using triplet loss. Finally, it improves rank-1 by 1.2% and mAP by 1.4%.

4.3   Comparison with State-of-the-Art

We compare our method to the state-of-the-art methods on the large-scale MARS dataset, including SeeForest [35], ReRank [34], RQEN [21], Diversity [15],

Table 3. Performance comparison on MARS dataset.

Method	rank-1	rank-5	rank-20	mAP
SeeForest [35]	70.6	90.0	97.6	50.7
ReRank [34]	73.9	–	–	68.5
RQEN [21]	77.8	88.8	94.3	71.1
Diversity [15]	82.3	–	–	65.8
RL [30]	83.1	91.3	–	69.9
KHAN [22]	85.7	94.3	97.2	77.8
Snippet-Sim [2]	86.3	94.7	98.2	76.1
STA [9]	86.3	95.7	98.1	80.8
STA + ReRank [9]	87.2	96.2	98.6	**87.7**
Ours	**87.7**	**96.4**	**98.3**	82.3
Ours + ReRank	**88.6**	**96.3**	**98.8**	87.4

RL [30], KHAN [22], Snippet-Sim [2] and STA [9]. Here, we report the result of STA, which is the full version with feature fusion strategy and inter-frame regularization. Table 3 shows our method is superior to the existing state-of-the-art, e.g., reaching rank-1 accuracy 87.7% and mAP 82.3%. In addition, we also report the results after re-ranking, where our method achieves 88.6% on rank-1 accuracy.

5   Conclusion

In this paper, a simple but efficient clip-level feature aggregation method for video-based person re-identification is proposed, which contains Average Aggregation Strategy (AAS) and Raw Feature Utilization (RFU). This parameter-free method can be applied to existing video-based person re-identification models which extract clip-level features, without extra layers or post-processing procedure. An ablation study is conducted to verify the effectiveness of AAS and RFU. Experimental results on MARS dataset demonstrate the improved accuracy of our method. For the future work, an more intelligent way of selecting distinctive frames than random sampling from a long video sequence should be investigated. What is more, the combination of re-identification and tracking tasks is an interesting topic, which could lead to a better cross-camera tracking application.

Acknowledgments. The research leading to these results has received funding from the European Union's Horizon 2020 research and innovation programme under the Marie Skłodowska-Curie grant agreement No. 765866 - ACHIEVE.

References

1. Bazzani, L., Cristani, M., Perina, A., Farenzena, M., Murino, V.: Multiple-shot person re-identification by HPE signature. In: Proceedings of the IEEE International Conference on Pattern Recognition, pp. 1413–1416. IEEE (2010)
2. Chen, D., Li, H., Xiao, T., Yi, S., Wang, X.: Video person re-identification with competitive snippet-similarity aggregation and co-attentive snippet embedding. In: Proceedings of the IEEE Conference on Computer Vision and Pattern Recognition, pp. 1169–1178 (2018)
3. Cheng, D., Gong, Y., Zhou, S., Wang, J., Zheng, N.: Person re-identification by multi-channel parts-based CNN with improved triplet loss function. In: Proceedings of the IEEE Conference on Computer Vision and Pattern Recognition, pp. 1335–1344 (2016)
4. Cho, Y.J., Yoon, K.J.: Improving person re-identification via pose-aware multi-shot matching. In: Proceedings of the IEEE Conference on Computer Vision and Pattern Recognition, pp. 1354–1362 (2016)
5. Das, A., Chakraborty, A., Roy-Chowdhury, A.K.: Consistent re-identification in a camera network. In: Fleet, D., Pajdla, T., Schiele, B., Tuytelaars, T. (eds.) ECCV 2014. LNCS, vol. 8690, pp. 330–345. Springer, Cham (2014). https://doi.org/10.1007/978-3-319-10605-2_22
6. Deng, J., Dong, W., Socher, R., Li, L.J., Li, K., Fei-Fei, L.: ImageNet: a large-scale hierarchical image database. In: Proceedings of the IEEE Conference on Computer Vision and Pattern Recognition, pp. 248–255. IEEE (2009)
7. Dimitrievski, M., Veelaert, P., Philips, W.: Behavioral pedestrian tracking using a camera and lidar sensors on a moving vehicle. Sensors $19(2)$, 391 (2019)
8. Farenzena, M., Bazzani, L., Perina, A., Murino, V., Cristani, M.: Person re-identification by symmetry-driven accumulation of local features. In: Proceedings of the IEEE Conference on Computer Vision and Pattern Recognition, pp. 2360–2367. IEEE (2010)
9. Fu, Y., Wang, X., Wei, Y., Huang, T.: STA: spatial-temporal attention for large-scale video-based person re-identification. In: Proceedings of the Association for the Advancement of Artificial Intelligence (2019)
10. Gheissari, N., Sebastian, T.B., Hartley, R.: Person reidentification using spatiotemporal appearance. In: Proceedings of the IEEE Conference on Computer Vision and Pattern Recognition, vol. 2, pp. 1528–1535. IEEE (2006)
11. Gray, D., Tao, H.: Viewpoint invariant pedestrian recognition with an ensemble of localized features. In: Forsyth, D., Torr, P., Zisserman, A. (eds.) ECCV 2008. LNCS, vol. 5302, pp. 262–275. Springer, Heidelberg (2008). https://doi.org/10.1007/978-3-540-88682-2_21
12. He, K., Zhang, X., Ren, S., Sun, J.: Deep residual learning for image recognition. In: Proceedings of the IEEE Conference on Computer Vision and Pattern Recognition, pp. 770–778 (2016)
13. Hermans, A., Beyer, L., Leibe, B.: In defense of the triplet loss for person re-identification. arXiv preprint arXiv:1703.07737 (2017)
14. Karaman, S., Bagdanov, A.D.: Identity inference: generalizing person re-identification scenarios. In: Fusiello, A., Murino, V., Cucchiara, R. (eds.) ECCV 2012. LNCS, vol. 7583, pp. 443–452. Springer, Heidelberg (2012). https://doi.org/10.1007/978-3-642-33863-2_44
15. Li, S., Bak, S., Carr, P., Wang, X.: Diversity regularized spatiotemporal attention for video-based person re-identification. In: Proceedings of the IEEE Conference on Computer Vision and Pattern Recognition, pp. 369–378 (2018)

16. Li, W., Zhao, R., Xiao, T., Wang, X.: DeepReID: deep filter pairing neural network for person re-identification. In: Proceedings of the IEEE Conference on Computer Vision and Pattern Recognition, pp. 152–159 (2014)
17. Liu, H., et al.: Video-based person re-identification with accumulative motion context. IEEE Trans. Circuits Syst. Video Technol. **28**(10), 2788–2802 (2017)
18. Liu, K., Ma, B., Zhang, W., Huang, R.: A spatio-temporal appearance representation for viceo-based pedestrian re-identification. In: Proceedings of the IEEE International Conference on Computer Vision, pp. 3810–3818 (2015)
19. Luo, H., Gu, Y., Liao, X., Lai, S., Jiang, W.: Bag of tricks and a strong baseline for deep person re-identification. In: Proceedings of the IEEE Conference on Computer Vision and Pattern Recognition Workshops (2019)
20. McLaughlin, N., Martinez del Rincon, J., Miller, P.: Recurrent convolutional network for video-based person re-identification. In: Proceedings of the IEEE Conference on Computer Vision and Pattern Recognition, pp. 1325–1334 (2016)
21. Song, G., Leng, B., Liu, Y., Hetang, C., Cai, S.: Region-based quality estimation network for large-scale person re-identification. In: Thirty-Second AAAI Conference on Artificial Intelligence (2018)
22. Su, X., et al.: k-reciprocal harmonious attention network for video-based person re-identification. IEEE Access **7**, 22457–22470 (2019)
23. Sun, Y., Zheng, L., Yang, Y., Tian, Q., Wang, S.: Beyond part models: person retrieval with refined part pooling (and a strong convolutional baseline). In: Ferrari, V., Hebert, M., Sminchisescu, C., Weiss, Y. (eds.) ECCV 2018. LNCS, vol. 11208, pp. 501–518. Springer, Cham (2018). https://doi.org/10.1007/978-3-030-01225-0_30
24. Varior, R.R., Shuai, B., Lu, J., Xu, D., Wang, G.: A siamese long short-term memory architecture for human re-identification. In: Leibe, B., Matas, J., Sebe, N., Welling, M. (eds.) ECCV 2016. LNCS, vol. 9911, pp. 135–153. Springer, Cham (2016). https://doi.org/10.1007/978-3-319-46478-7_9
25. Wang, F., Xiang, X., Cheng, J., Yuille, A.L.: NormFace: l_2 hypersphere embedding for face verification. In: Proceedings of the 25th ACM International Conference on Multimedia, pp. 1041–1049. ACM (2017)
26. Wang, T., Gong, S., Zhu, X., Wang, S.: Person re-identification by video ranking. In: Fleet, D., Pajdla, T., Schiele, B., Tuytelaars, T. (eds.) ECCV 2014. LNCS, vol. 8692, pp. 688–703. Springer, Cham (2014). https://doi.org/10.1007/978-3-319-10593-2_45
27. Wang, T., Gong, S., Zhu, X., Wang, S.: Person re-identification by discriminative selection in video ranking. IEEE Trans. Pattern Anal. Mach. Intell. **38**(12), 2501–2514 (2016)
28. Xiao, T., Li, S., Wang, B., Lin, L., Wang, X.: Joint detection and identification feature learning for person search. In: Proceedings of the IEEE Conference on Computer Vision and Pattern Recognition, pp. 3415–3424 (2017)
29. Yan, Y., Ni, B., Song, Z., Ma, C., Yan, Y., Yang, X.: Person re-identification via recurrent feature aggregation. In: Leibe, B., Matas, J., Sebe, N., Welling, M. (eds.) ECCV 2016. LNCS, vol. 9910, pp. 701–716. Springer, Cham (2016). https://doi.org/10.1007/978-3-319-46466-4_42
30. Zhang, W., He, X., Lu, W., Qiao, H., Li, Y.: Feature aggregation with reinforcement learning for video-based person re-identification. IEEE Trans. Neural Netw. Learn. Syst. (2019). https://doi.org/10.1109/tnnls.2019.2899588
31. Zhao, R., Ouyang, W., Wang, X.: Person re-identification by salience matching. In: Proceedings of the IEEE International Conference on Computer Vision, pp. 2528–2535 (2013)

32. Zheng, L., et al.: MARS: a video benchmark for large-scale person re-identification. In: Leibe, B., Matas, J., Sebe, N., Welling, M. (eds.) ECCV 2016. LNCS, vol. 9910, pp. 868–884. Springer, Cham (2016). https://doi.org/10.1007/978-3-319-46466-4_52

33. Zheng, L., Yang, Y., Hauptmann, A.G.: Person re-identification: Past, present and future. arXiv preprint arXiv:1610.02984 (2016)

34. Zhong, Z., Zheng, L., Cao, D., Li, S.: Re-ranking person re-identification with k-reciprocal encoding. In: Proceedings of the IEEE Conference on Computer Vision and Pattern Recognition, pp. 1318–1327 (2017)

35. Zhou, Z., Huang, Y., Wang, W., Wang, L., Tan, T.: See the forest for the trees: Joint spatial and temporal recurrent neural networks for video-based person re-identification. In: Proceedings of the IEEE Conference on Computer Vision and Pattern Recognition, pp. 4747–4756 (2017)

Towards Approximating Personality Cues Through Simple Daily Activities

Francesco Gibellini, Sebastiaan Higler, Jan Lucas, Migena Luli,
Morris Stallmann, Dario Dotti, and Stylianos Asteriadis[✉]

Department of Data Science and Knowledge Engineering, Maastricht University,
Maastricht, The Netherlands
stelios.asteriadis@maastrichtuniversity.nl

Abstract. The goal of this work is to investigate the potential of making use of simple activity and motion patterns in a smart environment for approximating personality cues via machine learning techniques. Towards this goal, we present a novel framework for personality recognition, inspired by both Computer Vision and Psychology. Results show a correlation between several behavioral features and personality traits, as well as insights of which type of everyday tasks induce stronger personality display. We experiment with the use of Support Vector Machines, Random Forests and Gaussian Process classification achieving promising predictive ability, related to personality traits. The obtained results show consistency to a good degree, opening the path for applications in psychology, game industry, ambient assisted living, and other fields.

Keywords: Personality recognition · Behavior analysis · Machine learning · Personality traits

1   Introduction

Behavior Understanding through the application of Machine Learning techniques has been an active field of research for a long time thanks to its many applications including video surveillance [9], health care [1] and human-computer interaction [15]. Moreover, personality has been successfully linked to behaviour, for example in [4,13,17]. However, so far, the focus has been placed mostly on investigating specific situations and it remains unclear whether a link to personality psychology can be established in an unconstrained scenario. *By unconstrained we mean that the data that is to be related to personality, is recorded in contexts that allow people to move in and interact with their environment freely.* Furthermore, the environment is not restricted to meet any requirements (other than it is equipped with certain sensors).

Personality describes the characteristics of an individual's mind through quantifiable terms. A common method in the literature, to estimate personality traits is the Big-5 questionnaire [19]. Based on self-annotations, a score for each of the following traits is calculated: *Extraversion, Agreeableness, Conscientiousness, Neuroticism* and *Openness to experience*. Personality computing aims at

© Springer Nature Switzerland AG 2020
J. Blanc-Talon et al. (Eds.): ACIVS 2020, LNCS 12002, pp. 192–204, 2020.
https://doi.org/10.1007/978-3-030-40605-9_17

recognizing those traits using statistical or/and machine learning techniques [21]. Lately, the field has received an increasing amount of interest, mainly thanks to the wide availability of personal information collected through new technologies, such as social media and smart devices. There is a significant body of research in the area of personality synthesis for individualized interfaces (e.g. chatbots), catering for a plethora of needs. All computing domains concerned with personality consider the following three problems: personality recognition, perception and synthesis [21]. The focus of this research is strictly related to the first of these problems: Personality recognition and, in particular, its role in everyday activities, not necessarily within social contexts as has been the focus in literature so far.

How challenging that can be is shown by studies like [5] and [17], that try to predict all personality traits in constrained scenarios. The achieved accuracies are between 70 and 75%. Works similar to [7], and [8] make use of smart phone usage data and achieve slightly higher scores. All those works show that in psychology and personality computing, there is ongoing research in people's behavior and how their personality is externalized through patterns in social or non-social settings. However, most of these works focus on specific application areas (e.g. interactions with smart phones or Skype interviews), not supporting the proposition of a generic framework using as basis, ordinary activities. Recently, authors in [10], showed promising results for personality recognition using posture sequences in an unconstrained indoor context dataset. In their approach, self-assessed personality scores are clustered in higher personality types namely Resilient, Undercontrolled and Overcontrolled [6], and used as new ground-truth labels for their Long short-term memory (LSTM) framework. These personality types encoded combinations of the standard Big-5 traits, but little work was done to investigate the role of each trait using the non-social context dataset [10]. Hence, in this work, we propose a new framework to predict the level of each of the Big-5 personality traits on a per task basis, investigating the relation between the personality display and the type of activity performed. Since personality can be considered as a subjective attribute, the proposed research calls for an interdisciplinary study between Psychology and Computer Vision. Motion features that have been scientifically grounded in the personality psychology community are extracted from the data provided by [10], and correlation analysis is performed to find the most robust features in a non-social context. Finally, our personality recognition experiments and results show additional insights regarding the human body posture as well as motion, and the Big-5 personality traits.

Therefore, the novelty of this work lies on two factors: (1) We investigate which behavioral features are inherently correlated with personality traits during indoor daily activities, providing new insight regarding the relation between non verbal behaviors and personality traits. (2) A new framework for personality recognition is evaluated on a public unconstrained dataset related to a smart home environment [10], providing a reliable system for an application in ambient assisted living settings. The applicability of this framework can be of wide interest, especially in domains such as integrated healthcare, psychology, psychiatry.

Especially, given that the context of our research is coming from daily activities, in settings where people are not interacting with other people, recognizing characteristics related to personality traits can assist in removing biases in human expressivity consciously driven by interactions with others.

2   Related Work

The majority of the studies in Personality Psychology focus on observing social situations in order to find behavioural patterns that can be linked to personality. These patterns have been successfully extracted from textual data [14], social media profiles [21], as well as from video data [20]. In this study, we focus only on the research that relates behavior analysis and personality traits using video data.

In the pioneering work in [3], body positioning during conversation is used as a predictive factor for dominant/submissive personality trait prediction. Body position that emphasizes personal size, with an upright posture is found to be a predictive factor of dominance, while postures minimizing body size are more related to submissive personalities. Similarly, by using posture features (among others), the authors in [5] are able to build classifiers for predicting whether a subject was above or below the median for the Big-5 personality traits. In the findings proposed in the social psychology research of [12], the authors describe the sum of amplitudes as a good predictor for extraversion, agreeableness and emotional stability. The same study shows that movement pattern regularity can help in classifying agreeableness, openness, emotional stability and extraversion. Finally, the authors suggest that movement speed can give a good indication of emotional stability.

Even though the [5] and [12] propose insights that are relevant for this research, they are constrained to very specific scenarios. On the other hand, our research tries to generalize those findings, providing a model that is applicable also in unconstrained scenarios like smart homes.

Linking Human Behavior Understanding to Personality Computing and Personality Psychology in an *unconstrained* scenario is a relatively new approach. In [18], a dataset is proposed with the goal of finding good candidates for job interviews, predicting personality traits using face and audio analysis from unconstrained web videos. For smart home applications, authors in [10] recorded a dataset performing simple everyday tasks in a nonsocial context such as a home environment. Participants' personality traits are clustered in three major types: Resilient, Overcontrolled and Undercontrolled, and used as labels for the recognition framework. On the other hand, the proposed work is centered around the investigation of the role of each of the Big-5 personality traits during everyday tasks.

3   Dataset

This work is based on the dataset provided by [10], the *Personality in a nonsocial context* dataset contains data involving 46 individuals performing simple everyday tasks in an indoor environment (Fig. 1). All tasks were designed based on Activity of Daily Living (ADL) Datasets and problem solving psychological tests, and participants' personality scores were collected using the self-assessment questionnaire in [19].

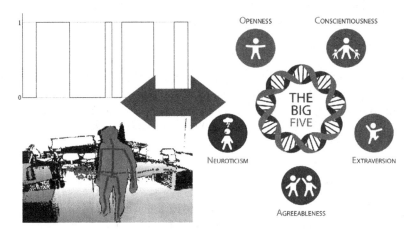

Fig. 1. Output of a binary sensor (top left), view of a camera tracking a human figure in 3D space (bottom left) and the 5 traits as described by the Big-5 personality traits (right)

3.1 Tasks

Since our personality recognition experiments are strictly correlated with the activity performed in each task, for the sake of a better explanation of our results, we re-describe briefly the goal of each task: The subjects were put in different everyday situations, with varying problem solving levels, designed to encourage expressiveness of their personality. In particular:

Task 1: The subjects were asked to find the keys of two cabinets and open them after their search to simulate an everyday simple action.

Task 2: The subjects had to look for a hidden object that could be placed anywhere in the room to mimic an everyday situation, during which the subjects need to concentrate on the given task.

Task 3: They were asked to find a second item but this time, the item did not exist. This simulates an everyday situation that confuses the subjects and, potentially, causes stress.

Task 4: The fourth task was to remember the drawer contents for both of the cabinets. The subjects were allowed to open the drawers as many times and as long as they wished to. This was done to simulate another everyday situation, in which the subjects are encouraged to express movement patterns.

Task 5: Next, the subjects were asked to write down answers to questions about the drawer content. For answering questions about the drawer content, they were also allowed to open the drawers as many times as they wanted. This is considered an everyday task requiring effort.

Task 6: The subjects made a cup of tea to capture behaviour externalization in relaxed situations, especially since the previous task was a challenging one.

4   Method

In this paper, we propose a full framework for personality recognition. First, inspired by previous Personality Computing and Psychology research, behavioral features are extracted. Then, correlation analysis is performed to select the features with inherent relation to personality traits and, finally, machine learning algorithms are tested for the task of personality recognition.

4.1   General Pre-processing

Following the procedure of [10], the door activation sensors are used to separate different tasks from each other, the provided body joint coordinates are given in three dimensions, while all features are computed on a per task basis. Lastly, the personality questionnaire results were normalized per subject to make the highest scored trait be one and lowest be zero.

4.2   Posture

A plethora of studies linked posture with personality traits, [5,10,11,22]. Following [22], we define posture as the Euclidean distances between body joint positions. Further, authors in [3] suggest to use only frames, in which a person is "standing still". Therefore, we define "Standing still" as sequences of frames, in which the person's body joint positions are not moving more than 3% in any of the coordinates compared to a base-frame. In order to be considered, each sequence was chosen to consist of at least 60 frames. Unfortunately, due to camera occlusions, only eleven upper body joints could be considered such that there is a total of $\binom{11}{2} = 55$ joint position differences. The actual features are the means over all detected sequences such that there is a 55-dimensional featry partiure vector $x_{p,t}^{(1)} \in \mathbb{R}^{55} (p = 1, \ldots 42, \ t = 1, \ldots, 6)^1$ describing a person's posture for every task.

4.3   Movement Sum of Amplitudes

Authors in [12] relate amplitude features to agreeableness, extraversion and emotional stability. We describe this feature mainly by Euclidean distances between

---

[1] Note that at the time of this study, the dataset contained data for 42 participants.

coordinates. In particular, given all the joints displacement, we select as reference joint the one with the lowest mean displacement. Then, we compute the distance between the reference joint (i.e. spine-middle) and the other moving joints (i.e. both hands, wrists and the head), after they have been normalized per sequence, in order to consider relative body size changes due to perspective effects. This representation allows to see how far the moving joint goes from the still one. The distances of all combinations for a still and a moving joint are calculated throughout all frames per subject. Finally, the sub-features used to represent this high level feature are the minimum, maximum, mean of the movements, while the frequency of a movement being wider than a threshold was also considered. Here, the threshold is set to the maximum distance minus the mean distance between the joints for every subject individually. So, a feature vector $x_{p,t}^{(2)} \in \mathbb{R}^{24}(p = 1, \ldots 42, t = 1, \ldots, 6)$ (see Footnote 1) for every participant/task results, p indicates the number of participant and t the task.

4.4   Movement Pattern Regularity

In [12], a relationship between agreeableness, openness, emotional stability, extraversion, and movement pattern regularities is discovered. We investigate a subject's movement pattern by using three binary magnetic sensors positioned on pieces of furniture related to the task in hand, each time. As the magnetic switches are placed on the cabinet drawers, when the drawer opens, the sensor sends a positive signal, whereas, when the drawer closes, the sensor sends a negative signal. In this way, the period of time in which the drawer is activated in every task can be computed looking at the sensor signals. In particular, the mean, minimum, maximum and total activation time and number of activations for each of the sensors were added to the feature sets. For task 1 (looking for keys), task 5 (answering questions) and task 6 (making tea), there were too few overall sensor activations and, thus, for those tasks, binary sensor features are not included into the final feature set such that there is a third feature vector $x_{p,t}^{(3)} \in \mathbb{R}^{15}(p = 1, \ldots 42, t = 2, 3, 4)$ (see Footnote 1).

4.5   Movement Speed

Movement speed is captured by calculating the distances from the position of a reference joint (the head) in 3D space over time. The Euclidean measure is used to determine the distance a joint has moved in an interval of 30 frames. The maximum, minimum, median and mean values are then calculated over the resulting speeds and a fourth feature vector $x_{p,t}^{(4)} \in \mathbb{R}^4(p = 1, \ldots 42, t = 1, \ldots, 6)$ (see Footnote 1) is added to the feature set.

The overall feature representation discussed in the previous subsections results into six high dimensional feature spaces (one per task) with a total of 98 feature-variables for tasks 2, 3 and 4 and 83 for the remaining tasks for each of the 42 observations.

5   Feature Selection

Per construction, not all features are suitable for predicting all personality traits from all tasks. Therefore, we first investigate the correlations between self-annotated personality traits and the feature representation described above, in order to select features for the recognition task. Further, this provides an insight of the system's reliability and a first bridging between data and personality characteristics.

5.1   Correlation

For each of the Big-5 traits, Pearson's correlation coefficient is calculated. Table 1 shows the top correlated features per task per trait (absolute values). It can

Table 1. Top correlated features per task per trait

Task	Trait	Top feature	Pearson's R (absolute)
1	Extraversion	Mean amplitude spine, left hand	0.4071
	Agreeableness	Amplitude max. neck, right hand	0.3597
	Conscientiousness	Posture right shoulder, spine	0.4947
	Neuroticism	Amplitude min. spine, left hand	0.3065
	Openness to experience	Posture spine, right wrist	0.3019
2	Extraversion	Posture left wrist, left hand	0.3255
	Agreeableness	Amplitude frequency spine, head	0.2416
	Conscientiousness	Posture right shoulder, head	0.3009
	Neuroticism	Speed mean	0.3316
	Openness to experience	Amplitude max. neck, head	0.4129
3	Extraversion	Amplitude frequency spine, head	0.3059
	Agreeableness	Amplitude min. neck, head	0.2897
	Conscientiousness	Amplitude max. spine, left wrist	0.3869
	Neuroticism	Speed mean	0.4524
	Openness to experience	Amplitude min. neck, head	0.4389
4	Extraversion	Amplitude min. spine, head	0.3826
	Agreeableness	Amplitude frequency neck, left hand	0.3566
	Conscientiousness	Amplitude min. spine, left wrist	0.3184
	Neuroticism	Posture right wrist, right hand	0.5142
	Openness to experience	Amplitude frequency neck, left wrist	0.5426
5	Extraversion	Posture right wrist, right hand	0.2127
	Agreeableness	Speed mean	0.2727
	Conscientiousness	Amplitude frequency spine, hand left	0.3976
	Neuroticism	Speed mean	0.4686
	Openness to experience	Amplitude frequency neck, left hand	0.3376
6	Extraversion	Posture right shoulder, head	0.3206
	Agreeableness	Amplitude min. spine, head	0.4142
	Conscientiousness	Posture shoulder, neck	0.3553
	Neuroticism	Posture head, right elbow	0.4703
	Openness to experience	Amplitude mean neck head	0.3562

be noted that, for every personality trait, there is at least one feature with a correlation coefficient >0.4. The correlation values of features-traits tuples are mostly dependent on the analyzed task, showing a relation between the task requirements, body motion and personality. For example, a subject's score in conscientiousness has the highest correlation with observing the subject's posture while performing task 1 (finding two keys and opening drawers). Openness to experience shows the highest correlation to amplitude features while performing task 3 (looking for a non-existing object) and task 4 (try to remember drawer contents). Neuroticism is linked to posture during performance of task 4 and task 6 (making tea) as well as to speed features while performing task 3 and task 5 (questionnaires).

Previous research was able to correlate movement speed with neuroticism (see Sect. 2) and, therefore, our observation is consistent with state-of-the-art in Psychology. Moreover, amplitude features have been linked to a subject's extraversion, agreeableness and neuroticism by works in Psychology (Sect. 2) which is mostly consistent with our observations. In fact, the highest correlations for extraversion and agreeableness were achieved by amplitude features (task 1: finding keys and task 6: making tea respectively). This may indicate that extrovert and agreeable subjects move their body with wider and smoother movements compared to the others.

Finally, only features that were significantly correlated (e.g. higher than 0.3) with personality traits were used for personality prediction.

6   Personality Recognition

After discovering feature correlations per trait per task, a question that arises is what is the predictive power of apparent behavioral features regarding personality recognition. Our goal, thus, is to infer personality scores for each one of the traits from the features on a per task basis.

For evaluating the predictive power of motion features with regards to personality traits, each subject's Big-5 score was translated into a label indicating whether the subject's score is below or above the median for that trait, similar to the approach in [5]. Thus, for each trait, the recognition task is formulated as a binary problem (i.e. predicting whether a subject scored above or below the median for every trait, given a set of motion features). Note that, for each task-trait combination, one classifier was trained such that there is a total of $6 * 5 = 30$ trained models in the end.

6.1   Machine Learning Models

One of the goals of this paper is to infer which types of tasks are more suitable to enhance personality display. Given our experiment settings on a per task basis, training and test data is limited for any Neural Network algorithm. Therefore, for this experiment, three models from the scikit-learn package [16] are trained and evaluated, namely Support Vector Machines, Gaussian Process Classifiers and Random Forests.

Support Vector Machines (SVM) and Gaussian Process Classifier (GPC) are trained with an RBF kernel to account for possible non-linearities in classifying data-points. Random Forest (RF) is an ensemble method that builds multiple decision-trees and uses a combination of each decision-tree's prediction to make a final prediction. In our RF model we use the Gini criterion for measuring the quality of a split in a tree, there need to be at least three samples at a node to apply a split, and there is no restriction in the minimum number of points in a leaf.

The GPC model makes use of latent function f, placing a GP prior on it, then squashing it through a link function to obtain the probabilistic classification. Its purpose is to allow a convenient formulation of the model. f is integrated out during prediction. The model implements the logistic link function, for which the integral is approximated in the binary case. The GP prior mean is assumed to be 0.

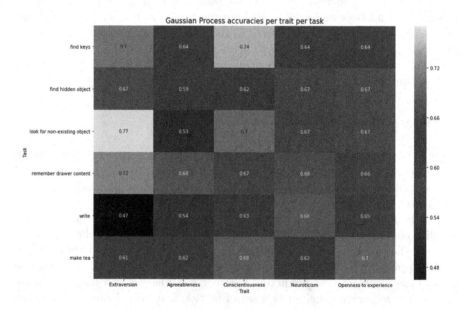

Fig. 2. Accuracy achieved by Gaussian Process classifiers per trait (x-axis) per task (y-axis)

6.2   Results

Models are evaluated in a 10-fold cross-validation fashion where, for each fold, we use 90% of the data for training and 10% for testing. The GPC-model with an RBF-kernel shows the best overall performance with a mean accuracy of 0.6510

(Fig. 2), compared to SVM (0.6025) and Random Forest models (0.5738). For all three models, extraversion from task 3 (looking for non-existing objects) obtained the highest accuracy, showing a high correlation between the selected feature and the task design. In fact, given the level of problem solving required for this task, extroversion is expressed more clearly than other traits. Moreover, for all three models, neuroticism is recognized above the mean accuracy in tasks 2 (finding object), 3 (looking for a nonexisting objects), 4 (remember drawer contents), 5 (answering questions). This implicates that, to recognize the neuroticism trait, our feature selection is robust to the choice of classifiers during tasks that require the completion of a goal under pressure. Conscientiousness trait is best predicted in task 1 by all models, indicating that, during tasks where the problem solving level is lower, conscientious subjects scan the room in more calm and clear way than in a task where the pressure of completing the goal is higher. Finally, for predicting agreeableness and open to experience traits, all models obtained different results, indicating that these traits are hard to recognize with this dataset, in which the recordings were made in a non-social context.

Additionally, in Table 2, we summarize the features used for each task. The highest prediction accuracy is obtained in task 3 (looking for non-existing objects, about 0.77–0.8 for every classifier) predicting the extraversion trait, indicating that the amplitude and posture features correlated with this trait correspond with real motion expressed by extroverted subjects.

7   Discussion

The classification experiments produce interesting insights into the relation between the way humans perform tasks and their personality. People externalize their personality more recognizably when they are either under pressure, because they cannot complete a task (looking for a non-existing object), or very concentrated (remembering drawer contents). Moreover, intuitively, a person's conscientiousness is externalized when performing a task with lower problem solving level (finding keys) and the prediction experiments support that hypothesis. Furthermore, we found a high correlation between posture and conscientiousness in task 1 (find keys and open drawers). However, similar results in task 2 (looking for an object) and task 3 (looking for non-existing object) were expected, but not observed. One reason might be that the engineered features do not capture conscientiousness well enough in all the searching tasks, or subjects became more stressed during tasks with higher problem solving level, and therefore, they failed to show conscientious behavior.

Previous research was able to correlate movement speed with neuroticism (see Sect. 2) and, therefore, our correlation findings are consistent with state-of-the-art in psychology. Moreover, amplitude features have been linked to a subject's extraversion, agreeableness and emotional stability by research in psychology which is mostly consistent with our observations. The highest correlations for extraversion and agreeableness were achieved by amplitude features (task 1: finding keys and task 6: making tea respectively). However, neuroticism is more correlated with speed (task 3: looking for a non-existing item and

Table 2. Best Models, their tasks and features that are used per trait

Trait	Best task	Features
Extraversion	Look for non-existing object	Amplitude: freq. spine-head, minimum spine-left hand, mean spine-left hand; Posture: left shoulder-right shoulder, left shoulder-neck, head-left shoulder
Agreeableness	Making tea	Speed: maximum; Amplitude: mean neck-right hand, min. spine-head, min. spine-right hand, freq. neck-hand; Posture: wright hand-wright wrist
Conscientiousness	Look for non-existing object	Amplitude: min. neck-left wrist, min. spine-head, mean spine-head, max. spine-left wrist, freq. spine-right hand; Posture: right wrist-left elbow
Neuroticism	Remember drawer contents	Posture: right hand-right wrist, right hand-right elbow, right hand-left elbow, Amplitude: min. spine-head, freq. spine-left wrist
Openness to experience	Make tea	Amplitude: freq. spine-head, freq. neck-head, mean neck-head, min. neck-right hand, min. spine-left hand; Posture: right shoulder-right elbow

task 5: answering questions) and posture features (task 4: remembering drawer contents and task 6: making tea). These findings are consistent with research in psychology since [3] links posture to submissive behaviour and [11] suggests "a strong relationship" between submissive behaviour and neuroticism. Thus, the direction of this research can be considered as a promising field for further research towards creating a system to automatically estimate personality traits in smart environments.

8   Conclusion and Future Work

To answer the main research question "How can personality be predicted with data obtained in unconstrained scenarios?", an end-to-end personality recognition framework was presented. Firstly, based on literature review, features that can be robustly engineered and are likely to be linked to personality are selected (Sect. 2). Differing from other works in human behaviour understanding, feature extraction is not constrained to certain postures or specific situations. Subjects could move freely in the environment and could follow any strategy to complete the tasks (Sect. 5).

Our findings suggest that the employed features, gathered through simple every-day tasks, have correlations with self-annotated personality scores

(Sect. 5.1). The correlation results are promising as they link skeleton data obtained in unconstrained scenarios with personality and are consistent with Psychology literature (Sect. 5.1). Furthermore, the personality recognition experiments on a per task basis, lead to novel results compared to [10], and an important conclusion can be drawn: Different tasks make persons externalize different personality traits (Sect. 7). Future work could concentrate on further improvement of the ML-models as well as more refined fusion methods, like the Dempster-Shafer fusion method proposed in [2], could be applied to produce a better final output making use of multiple classifier results.

References

1. Atchison, K.A., Dubin, L.F.: Understanding health behavior and perceptions. Dent. Clin. Am. **47**(1), 21–39 (2003)
2. Bagheri, M.A., Hu, G., Gao, Q., Escalera, S.: A framework of multi-classifier fusion for human action recognition. In: 2014 22nd International Conference on Pattern Recognition, pp. 1260–1265. IEEE (2014)
3. Ball, G., Breese, J.: Relating personality and behavior: posture and gestures. In: Paiva, A. (ed.) IWAI 1999. LNCS (LNAI), vol. 1814, pp. 196–203. Springer, Heidelberg (2000). https://doi.org/10.1007/10720296_14
4. Batrinca, L., Lepri, B., Mana, N., Pianesi, F.: Multimodal recognition of personality traits in human-computer collaborative tasks. In: Proceedings of the 14th ACM International Conference on Multimodal Interaction, pp. 39–46. ACM (2012)
5. Batrinca, L.M., Mana, N., Lepri, B., Pianesi, F., Sebe, N.: Please, tell me about yourself: automatic personality assessment using short self-presentations. In: Proceedings of the 13th International Conference on Multimodal Interfaces, pp. 255–262. ACM (2011)
6. Block, J., Block, J.H.: The role of ego-control and ego-resiliency in the organization of behavior. In: Development of Cognition, Affect, and Social Relations, pp. 49–112. Psychology Press (2014)
7. Chittaranjan, G., Blom, J., Gatica-Perez, D.: Who's who with big-five: analyzing and classifying personality traits with smartphones. In: 2011 15th Annual International Symposium on Wearable Computers, pp. 29–36. IEEE (2011)
8. Chittaranjan, G., Blom, J., Gatica-Perez, D.: Mining large-scale smartphone data for personality studies. Pers. Ubiquitous Comput. **17**(3), 433–450 (2013)
9. Dotti, D., Popa, M., Asteriadis, S.: Unsupervised discovery of normal and abnormal activity patterns in indoor and outdoor environments. In: VISIGRAPP (5: VISAPP), pp. 210–217 (2017)
10. Dotti, D., Popa, M., Asteriadis, S.: Behavior and personality analysis in a nonsocial context dataset. In: Proceedings of the IEEE Conference on Computer Vision and Pattern Recognition Workshops, pp. 2354–2362 (2018)
11. Gilbert, P., Allan, S.: Assertiveness, submissive behaviour and social comparison. Br. J. Clin. Psychol. **33**(3), 295–306 (1994)
12. Koppensteiner, M.: Motion cues that make an impression: predicting perceived personality by minimal motion information. J. Exp. Soc. Psychol. **49**(6), 1137–1143 (2013)
13. Lepri, B., Subramanian, R., Kalimeri, K., Staiano, J., Pianesi, F., Sebe, N.: Connecting meeting behavior with extraversion - a systematic study. IEEE Trans. Affect. Comput. **3**, 443–455 (2012)

14. Majumder, N., Poria, S., Gelbukh, A., Cambria, E.: Deep learning-based document modeling for personality detection from text. IEEE Intell. Syst. **32**(2), 74–79 (2017)
15. Pantic, M., et al.: Social signal processing: the research agenda. In: Moeslund, T., Hilton, A., Krüger, V., Sigal, L. (eds.) Visual Analysis of Humans, 511–538. Springer, London (2011). https://doi.org/10.1007/978-0-85729-997-0_26
16. Pedregosa, F., et al.: Scikit-learn: machine learning in Python. J. Mach. Learn. Res. **12**, 2825–2830 (2011)
17. Pianesi, F., Mana, N., Cappelletti, A., Lepri, B., Zancanaro, M.: Multimodal recognition of personality traits in social interactions. In: Proceedings of the 10th International Conference on Multimodal Interfaces, pp. 53–60. ACM (2008)
18. Ponce-López, V., et al.: ChaLearn LAP 2016: first round challenge on first impressions - dataset and results. In: Hua, G., Jégou, H. (eds.) ECCV 2016. LNCS, vol. 9915, pp. 400–418. Springer, Cham (2016). https://doi.org/10.1007/978-3-319-49409-8_32
19. Rammstedt, B., John, O.P.: Measuring personality in one minute or less: a 10-item short version of the big five inventory in English and German. J. Res. Pers. **41**(1), 203–212 (2007)
20. Ricci, E., Varadarajan, J., Subramanian, R., Rota Bulo, S., Ahuja, N., Lanz, O.: Uncovering interactions and interactors: joint estimation of head, body orientation and f-formations from surveillance videos. In: Proceedings of the IEEE International Conference on Computer Vision, pp. 4660–4668 (2015)
21. Vinciarelli, A., Mohammadi, G.: A survey of personality computing. IEEE Trans. Affect. Comput. **5**(3), 273–291 (2014)
22. Yang, X., Tian, Y.L.: EigenJoints-based action recognition using Naive-bayes-nearest-neighbor. In: 2012 IEEE Computer Society Conference on Computer Vision and Pattern Recognition Workshops, pp. 14–19. IEEE (2012)

Person Identification by Walking Gesture Using Skeleton Sequences

Chu-Chien Wei[✉], Li-Huang Tsai, Hsin-Ping Chou, and Shih-Chieh Chang

National Tsing Hua University, Hsinchu 30013, Taiwan, R.O.C.
zz0201zz@gmail.com, lihuangtsai@gapp.nthu.edu.tw,
alan.durant.chou@gmail.com, scchang@cs.nthu.edu.tw

Abstract. When coping with person identification problem, previous approaches either directly take raw RGB as inputs or use more sophisticated devices to capture other information. However, most of the approaches are sensitive to the changes of environment and different clothing, little variation may lead to failure identification. Recent research shows that "gait" (i.e., a person's manner of walking) is a unique trait of a human being. Motivated by this, we propose a novel method to identify people by their gaits. In order to figure out the characteristic of individual gait, we are interested in utilizing skeletal information, which is more robust to the diversification of environment and appearance. To effectively utilize skeletal data, we analyze the spatial relationship of joints and transform the 3D skeleton coordinates into relative distances and angles between joints, and then we use a bidirectional long short-term memory neural network to explore the temporal information of the skeleton sequences. Results show that our proposed method can outperform previous methods on BIWI and IAS-Lab datasets by gaining 10.33% accuracy improvement on average.

Keywords: Person identification · Bidirectional LSTM · Attention · Skeleton information

1 Introduction

Person identification is currently an active research field in computer vision and machine learning techniques. As the importance of security and safety of people is continuously growing day by day in society, person identification has attracted great concern in automatic surveillance systems. For example, we can automatically identify whether a person is allowed to enter a private area with a person identification technique Lavi et al. [3]. Besides security concerns, the identification system can also be widely applied in tasks such as human tracking for intelligent robots.

Recent research Prakash et al. [13] has investigated the attributes of human gait and found that human gait provide a way of locomotion by combined

---

C.-C. Wei and L.-H. Tsai—The authors contribute equally to this paper.

© Springer Nature Switzerland AG 2020
J. Blanc-Talon et al. (Eds.): ACIVS 2020, LNCS 12002, pp. 205–214, 2020.
https://doi.org/10.1007/978-3-030-40605-9_18

efforts of the brain, nerves and muscles. With a comprehensive gait analysis, they demonstrate that "gait", which means the way a person walks, is varying from person to person. Motivated by this observation, our goal is to identify people through the analysis of their walking postures.

To achieve this goal, the first challenge is to design a model that is not only capable of distinguishing different people but also powerful for intra-class variability. The second challenge is that person identification inherently contains a large number of classes but with very few training examples in each class.

Previous methods that directly take visual input are likely to encounter many biometric challenges, e.g., pose variation, illumination variation, occlusion, facial expression. Önsen et al. [15]. Because the appearance of a person may change over time due to different environment. Different appearances of a person will cause a significant decrease in person identification performance. And individual features are usually insufficient to match people due to inherent intra-class differences correctly.

To alleviate the impact caused by different clothes and illumination, some identification methods use people's silhouettes as input data. However, for silhouette-based methods, the changes in camera's perspectives and accessories such as hats, bags, etc. can still result in different silhouettes.

Concerning the weak points of RGB, there are methods considering people silhouettes and advanced 3D imaging devices such as Microsoft Kinect to dealing with it. There is a growing interest in making use of depth images and skeleton sequences as input data. With respect to RGB-based methods, depth-based method can take advantage of 4D input, and thus resist the shortcomings of RGB-based methods. In both RGB-based and depth-based methods, input data are often be preprocessed before training due to the high dimensionality of images and the limitation of the computational ability, though the preprocessing procedure could be time-consuming and may cause loss of information. On the contrary, using skeletal data only requires little preprocessing. The main advantage of using skeleton-based methods is that a small number of joint positions can effectively represent human motion. In addition, there is little research on person identification utilizing human representation based on 3D skeleton information.

To address the issues we mentioned above, we present a novel method for person identification based on skeleton information in 3D coordinates and soft-biometrics. Skeleton sequences provide trajectories of human joints. The advantages of using skeletal data are the robustness of the changes in clothing and illumination and the invariance to the perspective of cameras. Our approach uses a richer representation to take as much spatio-temporal information as possible from the skeleton. This representation is a combination of several soft-biometrics cues extracted from skeletal data.

To obtain the spatial relationship between joints, we extract features by computing the relative positions of the joints and reference joints for each frame of skeleton. Previous methods usually use long short-term memory (LSTM) to capture the dynamics of joints. However, traditional LSTM can only get information from history. Hence, we use bidirectional LSTM to get more information not only from history but also from the future. Our main contributions are as follows:

(i) We propose a novel approach to extract the spatial relationship between each joint of skeletal data and meanwhile overcome the intra-class variance by decoupling the person identification process from factors of environment, appearances, etc.

(ii) The proposed method achieves the state-of-the-art on IAS-Lab dataset in person identification task.

2   Related Works

Most person identification research uses RGB images as input data and formulate the identification problem as a similarity task. They calculate the similarity between input images such as a siamese network. Our main differences from existing methods are: (i) We leverage only the modality of skeletal data. That is, we identify people only by their dynamic motion rather than information from appearance. (ii) We use a sequence of a skeleton as input and determine the identity of the person without metric learning.

According to the data used, person identification methods can be divided into several categories:

RGB-based methods: Most of the existing works [19, 21] belong to this category. [19] proposed an effective human part-aligned representation for handling the body part misalignment problem in person identification. And [21] made use of GANs for person re-identification. Those approaches aim to design discriminant features from appearance characteristics to identify people. However, RGB-based variant suffers from clothes change, pose change, different location, and illuminating conditions.

Silhouette-based methods: This approach transforms RGB images into a human silhouette and thus avoid suffering from color information. To extract the features, [9, 18] use geodesic distance between parts of their body and soft-biometrics for person identification. However, the silhouette of a person may be changed by different camera's view and accessories such as bags.

Depth-based methods: As the gradually spread of depth-sensing devices, the research for computer vision was revolutionized, and this technique has also been used to solve person identification problems [6, 8]. There are two categories of this method. The first category integrates appearance and depth information together [4, 7]. The second category is based on geometric features. In the previous approach [11], a person's posture is reconstructed and placed into a standard pose before identify, such that point clouds coming from different frames can be merged to compose a model. Nevertheless, if a person is carrying something or wearing additional accessories such as hats, those extra things may be regarded as a part of his body.

Skeleton-based methods: According to [2], the skeletal information can be extracted in real-time from depth data. Hence skeletal information can be collected rapidly. There are several works on computer vision, particular in the field of action recognition using skeletal information to overcome the drawbacks of previous methods [16,17,20]. But it is not well investigated in the field of person re-identification. All the related works take advantage of skeletal information that can represent people's locomotion and is robust to the changes of appearance and environment.

In the next section, we detail the feature extraction process from skeletal data and then describe the model we adopted.

3   Proposed Method

In this section, we describe the detail of our method. First, we explain the method for extracting features from skeleton sequences. Then we give a introduction to *Long Short-Term Memory* (LSTM); its extension model, *Bidirectional LSTM*; a technique to boost recurrent neural network performance, *Attention* mechanism.

3.1   Feature Extraction from Skeleton

To enhance the robustness of the model, we have to against changes in appearance and illumination. Our proposed method ignores image information and focuses on 3D information of skeleton sequences. We use skeleton sequences to identify humans according to their gaits. To extract spatial features, we transform each frame of the original skeletal data to the distance and angle between joints. For temporal features, we partition the skeleton sequences into several instances and use bidirectional LSTM to capture the dynamics.

As for spatial features, we first separate 20 joints into 5 groups, each group containing 4 joints. Figure 1 shows the schematic of skeletal data. Joints with the same color belong to the same group, that is, joint number 1–4 belong to group 1, 5–8 belong to group 2, ..., and so on. Next, we select a reference joint located at the center part of the body, which seldom moves during walking for each group. In the remaining part of this section, we will use the joint numbers, as illustrated in Fig. 1 to represent these joints.

Then, we calculate the distance d_i between the reference joint and other joints in the same group, i is joint's number. Consequently, we get a distance matrix $D = [d_1, d_2, ..., d_{15}]$, which contains 15 numbers representing the distance of each joint and its group reference joint. Next, to extract angles information, we set a reference vector for reference joint in group 2 through group 5. Then, we compute vectors between the reference joint and other joints in the same group. Finally, we calculate the angles a_i for the reference vector and other vectors in the same group, and hence we get an angle matrix $A = [a_1, a_2, ..., a_{12}]$. The reference joint and vector of each group shows below:

- Group 1: Reference joint name, *Spine.*
  - Reference joint: 2

- Group 2: Reference joint name, *Left Shoulder.*
  - Reference joint: 5
  - Reference vector: (5, 3)
- Group 3: Reference joint name, *Right Shoulder*
  - Reference joint: 9
  - Reference vector: (9, 3)
- Group 4: Reference joint name, *Left Hip*
  - Reference joint: 13
  - Reference vector: (13, 1)
- Group 5: Reference joint name, *Right Hip*
  - Reference joint: 17
  - Reference vector: (17, 1)

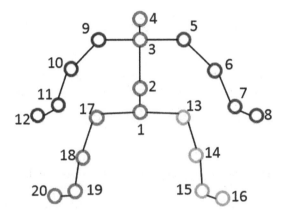

Fig. 1. Schematic diagram of skeleton and joints number. (Color figure online)

We also extract additional features and get a feature matrix $B = [b_1, b_2, ..., b_{15}]$. The additional features are as follow:

- Distance between knees, ankles, foots, elbows, and wrists
- Distance between knee and elbow for each side, and distance between knee and hand for each side.
- Angles of knees, thigh, elbows, and arm

Consequently, we have a feature matrix $F = [D, A, B]$ for each frame and the matrix F contains 42 features.

3.2   Bidirectional Recurrent Attention Model

Since standard RNNs process sequences in time series, they tend to ignore future information. If we can utilize future messages just like how we do in past information, it will be very beneficial for many sequence annotation tasks. To overcome the limitations of conventional RNN. Schuster et al. [14] proposed *bidirectional recurrent neural network* (BRNN). The basic idea of BRNN is that each training sequence has two RNN with different directions, forward and backward, and both are connected to an output layer, so this structure can completely leverage past and future context information. Therefore, we adopt bidirectional long short-term memory (BLSTM) for person identification in order to learn people's walking posture forward and backward.

Attention Mechanism can take all the input into account and learning relative importance between the input and output elements. We add this mechanism into our BLSTM model to enhance the ability to extract information. Equation (1) shows the attention we use [10].

$$a^t = \sum_l^L softmax(d_t W h) \qquad (1)$$

4   Experiments

In this section, we first describe our experiment settings, then present the experimental results from the proposed method and compared the result of other works.

4.1   Datasets

Our goal is to identify people based on their gaits. Most the datasets are catered to human activity analysis and action recognition. Among publicly available identification benchmarks, we use BIWI [12] and IAS-Lab [12] datasets, which are widely used for long-term people identification from RGB-D cameras. These datasets include synchronized RGB images, depth images, persons' segmentation maps, and skeletal data. We will introduce these two datasets in the following section.

As for BIWI, it contains 50 training and 56 testing sequences of 50 people. Figure 2 shows the RGB images, depth images, visualized skeletal data, and segmentation maps of training and testing set.

- Training set: 50 people perform a certain amount of action in front of the Kinect camera, such as rotating around the vertical axis, moving heads, and walking towards the camera.
- Testing set: 28 out of 50 people presenting in the training set were also recorded in two testing videos. These test sequences were collected on different days and at different locations relative to the training set, so most subjects have different clothing. And there are a Still sequence and a Walking sequence of each person in testing set.
  - Still: Each person is still or slightly moving in place.
  - Walking: Each person performs some walks frontally diagonally with respect to the Kinect.

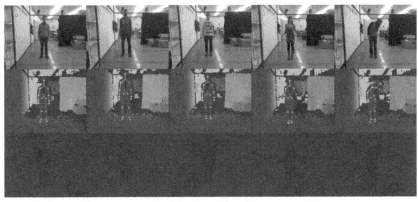

(a) Training set

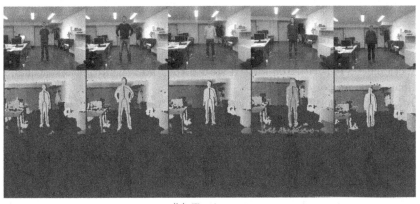

(b) Testing set

Fig. 2. Samples of RGB images, depth images, skeleton and user map data for five people from BIWI dataset [1].

As for IAS-Lab, it contains 11 training and 22 testing sequences of 11 different people. For every subject, we recorded three sequences shown in Fig. 3.

- Training set: Each person rotates on himself and performs some walks.
- Testing set: It contains 2 testing sets, *TestingA* and *TestingB*. Each testing set contains 11 people and every person takes the same action as in the training set.
  - TestingA: This set is acquired with people wearing different clothes from the training set.
  - TestingB: This set is collected in a different room and with different illumination, but people wear the same clothes as in the training set.

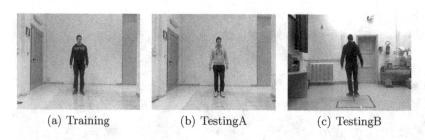

(a) Training   (b) TestingA   (c) TestingB

Fig. 3. Sample of RGB images from IAS-Lab dataset.

4.2 Experimental Setup

To preprocess the input of LSTM, we use a sliding window with stride size 1 to partition the skeleton sequences into several instances and take one instance as a sample of input data. An input data in frame i, denoted as x_i, includes 42 features consisting of distances and angles. Because it takes about 1 second per footstep, and the frame rate is 10, so the timesteps is set to be 30. We use a bidirectional LSTM with 100 hidden states, then add the attention mechanism to enhance model performance.

To select the best hyperparameters and features for input, we use 10-fold cross-validation to evaluate the performance of each model. After data preprocessing, there are 6739 training sequences, 7118 and 7910 testing sequences of 11 people in IAS-Lab dataset. For BIWI dataset, there are 15716 training sequences for 50 people, 3969 and 3525 testing sequences for 28 people.

4.3 Compared Model

We compare the performance of our method with the method using BIWI and IAS-Lab dataset. We briefly describe the method as follow:

- Grassmann manifold [5]: This method uses 3D skeleton sequences to exploit the dynamic motion by projecting skeletal data onto Grassmann manifold. Then align the sequences by dynamic time warping. Finally, use nearest neighbor method for person identification.

4.4 Result Summary

Table 1 shows the testing accuracy (%) on IAS-Lab dataset.

Table 2 shows the testing accuracy (%) of BIWI dataset, and there are 50 people in training set and 28 people in the testing set. The results on both testing set outperform Grassmann manifold.

Compare to the results with Grassmann manifold, we carry out a better performance on all testing set. On IAS-Lab dataset, we improve 12% and 10.3% for TestingA and TestingB. On BIWI dataset, we improve 8.44% and 10.59% for Walking and Still testing set.

Table 1. Results of IAS-Lab dataset.

Accuracy on IAS-Lab dataset		
Methods	TestingA	TestingB
Grassmann manifold	45.5%	63.6%
Our method + LSTM	49.7%	67.7%
Our method + BLSTM	50.6%	70.4%
Our method + BLSTM + Attention	**57.5%**	**73.9%**

Table 2. Results of BIWI dataset.

Accuracy on BIWI dataset		
Methods	Walking	Still
Grassmann manifold	10.71%	28.57%
Our method + LSTM	14.3%	30.73%
Our method + BLSTM	15.82%	32.45%
Our method + BLSTM + Attention	**19.15%**	**39.16%**

5   Conclusion

In this paper, we propose a novel person identification method to analyze the walking posture of humans by utilizing skeletal data. Because of the effective usage of skeletal data and unique attribute of human gait, we can decouple the person identification process from factors of environment and appearances. Thus, our method is more robust to the changes of appearances than previous methods and achieves the state-of-the-art on IAS-Lab dataset.

References

1. BIWI dataset kernel description. http://robotics.dei.unipd.it/reid/index.php/8-dataset/2-overview-biwi
2. Shotton, J., et al.: Real-time human pose recognition in parts from single depth images. In: CVPR 2011, pp. 1297–1304 (2011)
3. Lavi, B., Serj, M.F., Ullah, I.: Survey on deep learning techniques for person re-identification task. Pattern Recognit. (2018)
4. Baltieri, D., Vezzani, R., Cucchiara, R., Utasi, A., Benedek, C., Szirányi, T.: Multiview people surveillance using 3D information. In: 2011 IEEE International Conference on Computer Vision Workshops (ICCV Workshops), pp. 1817–1824 (2011)
5. Elaoud, A., Barhoumi, W., Drira, H., Zagrouba, E.: Analysis of skeletal shape trajectories for person re-identification. In: Blanc-Talon, J., Penne, R., Philips, W., Popescu, D., Scheunders, P. (eds.) ACIVS 2017. LNCS, vol. 10617, pp. 138–149. Springer, Cham (2017). https://doi.org/10.1007/978-3-319-70353-4_12. https://hal.archives-ouvertes.fr/hal-01703945
6. Haque, A., Alahi, A., Li, F.F.: Recurrent attention models for depth-based person identification, November 2016

7. Oliver, J., Albiol, A., Albiol, A.: 3D descriptor for people re-identification. In: Proceedings of the 21st International Conference on Pattern Recognition (ICPR 2012), pp. 1395–1398 (2012)
8. Karianakis, N., Liu, Z., Chen, Y., Soatto, S.: Reinforced temporal attention and split-rate transfer for depth-based person re-identification. In: Ferrari, V., Hebert, M., Sminchisescu, C., Weiss, Y. (eds.) ECCV 2018. LNCS, vol. 11209, pp. 737–756. Springer, Cham (2018). https://doi.org/10.1007/978-3-030-01228-1_44
9. Wang, L., Tan, T., Ning, H., Weiming, H.: Silhouette analysis-based gait recognition for human identification. IEEE Trans. Pattern Anal. Mach. Intell. **25**(12), 1505–1518 (2003). https://doi.org/10.1109/TPAMI.2003.1251144
10. Luong, T., Pham, H., Manning, C.D.: Effective approaches to attention-based neural machine translation. In: Proceedings of the 2015 Conference on Empirical Methods in Natural Language Processing, Lisbon, Portugal, pp. 1412–1421. Association for Computational Linguistics, September 2015. https://doi.org/10.18653/v1/D15-1166, https://www.aclweb.org/anthology/D15-1166
11. Munaro, M., Basso, A., Fossati, A., Gool, L.V., Menegatti, E.: 3D reconstruction of freely moving persons for re-identification with a depth sensor. In: 2014 IEEE International Conference on Robotics and Automation (ICRA), pp. 4512–4519 (2014)
12. Munaro, M., Basso, A., Fossati, A., Van Gool, L., Menegatti, E.: 3D reconstruction of freely moving persons for re-identification with a depth sensor. In: 2014 IEEE International Conference on Robotics and Automation (ICRA), pp. 4512–4519, May 2014. https://doi.org/10.1109/ICRA.2014.6907518
13. Prakash, C., Kumar, R., Mittal, N.: Recent developments in human gait research: parameters, approaches, applications, machine learning techniques, datasets and challenges. Artif. Intell. Rev. **49**(1), 1–40 (2018). https://doi.org/10.1007/s10462-016-9514-6
14. Schuster, M., Paliwal, K.: Bidirectional recurrent neural networks. IEEE Trans. Signal Process. **45**(11), 2673–2681 (1997). https://doi.org/10.1109/78.650093
15. Önsen Toygar, E.A., Afaneh, A.: Person identification using multimodal biometrics under different challenges. Pattern Recognit. (2017)
16. Yan, S., Xiong, Y., Lin, D.: Spatial temporal graph convolutional networks for skeleton-based action recognition. In: AAAI 2018, pp. 1297–1304 (2018)
17. Du, Y., Wang, W., Wang, L.: Hierarchical recurrent neural network for skeleton based action recognition. In: 2015 IEEE Conference on Computer Vision and Pattern Recognition (CVPR), pp. 1110–1118 (2015)
18. Zeng, W., Wang, C., Yang, F.: Silhouette-based gait recognition via deterministic learning. Pattern Recognit. **47**(11), 3568–3584 (2014)
19. Zhao, L., Li, X., Zhuang, Y., Wang, J.: Deeply-learned part-aligned representations for person re-identification. In: Proceedings of the IEEE International Conference on Computer Vision, pp. 3219–3228 (2017)
20. Zheng, W., Li, L., Zhang, Z., Huang, Y., Wang, L.: Relational network for skeleton-based action recognition. In: International Conference on Multimedia and Expo (ICME) 2019 (2018)
21. Zheng, Z., Zheng, L., Yang, Y.: Unlabeled samples generated by GAN improve the person re-identification baseline in vitro. In: Proceedings of the IEEE International Conference on Computer Vision, pp. 3754–3762 (2017)

Verifying Kinship from RGB-D Face Data

Felipe Crispim$^{(\boxtimes)}$, Tiago Vieira, and Bruno Lima

Institute of Computing, Federal University of Alagoas, Maceio, AL, Brazil
fcc@ic.ufal.br

Abstract. We present a kinship verification (KV) approach based on Deep Learning applied to RGB-D facial data. To work around the lack of an adequate 3D face database with kinship annotations, we provide an online platform where participants upload videos containing faces of theirs and of their relatives. These videos are captured with ordinary smartphone cameras. We process them to reconstruct recorded faces in tridimensional space, generating a normalized dataset which we call Kin3D. We also combine depth information from the normalized 3D reconstructions with 2D images, composing a set of RGBD data. Following approaches from related works, images are organized into four categories according to their respective type of kinship. For the classification, we use a Convolutional Neural Network (CNN) and a Support Vector Machine (SVM) for comparison. The CNN was tested both on a widely used 2D Kinship Verification database (KinFaceW-I and II) and on our Kin3D for comparison with related works. Results indicate that adding depth information improves the model's performance, increasing the classification accuracy up to 90%. To the extent of our knowledge, this is the first database containing depth information for Kinship Verification. We provide a baseline performance to stimulate further evaluations from the research community.

Keywords: Kinship Verification · Face biometrics · Structure from motion · 3D reconstruction

1 Introduction

In their daily lives, people receive information and interact in a three-dimensional world. Despite the high complexity and cost of its shapes, nowadays, researchers in computer vision and similar fields are increasingly improving techniques and extracting the benefits of this additional dimension. Even applications to unlock smartphones by 2D facial recognition can be deceived (commonly known as face recognition spoofing mechanisms). Using 3D information improves system's robustness as it is less dependent on environment interference such as illumination. Indeed, Apple has added active infrared illumination for face recognition so users can unlock their mobile devices using its FaceID technology. Regardless, a son was able to unlock his mother's phone due to their facial resemblance.

© Springer Nature Switzerland AG 2020
J. Blanc-Talon et al. (Eds.): ACIVS 2020, LNCS 12002, pp. 215–226, 2020.
https://doi.org/10.1007/978-3-030-40605-9_19

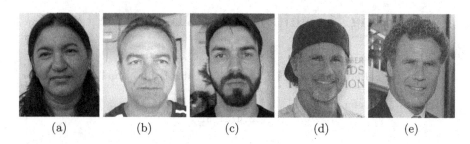

(a)   (b)   (c)   (d)   (e)

Fig. 1. Challenges related to facial Kinship Verification. People in (a) and (b) are twins (sister and brother, respectively). Individuals in images (b) and (c) are father and son, respectively. On the other hand, unrelated people can present similar faces as can be seen in the resemblance between actors Chad Smith (d) and Will Ferrell (e).

Kinship Verification (KV) from face images is a challenging task (*cf.* Fig. 1) due to many difficulties. It must handle similarities between non-relatives (which might reduce inter-class similarity distances) while dealing with relatives with different appearances (which could increase within-class distance). There are also variations in gender, age, ethnicity, and half-sibling categories. As a consequence, verifying kinship from faces is a broadly open research topic in computer vision and biometrics.

To the extent of our knowledge, works associated with KV are limited to two-dimensional information, either through photos or videos. There is, however, a degradation of these methodologies due to variations in lighting, expression, pose and registration [16]. In this work, we present a Kinship Verification (KV) approach by considering not only 2D images, but also depth information from faces.

In order to overcome the lack of a suitable face image database containing both depth information and kinship annotations, we provide an online platform where individuals can upload videos from their faces and their kin. From the face videos, we compute the 3D reconstruction and obtain the depth for each individual's face. RGBD images from each pair of faces are then fed into two classifiers: a Support Vector Machine, to provide a baseline, and a Convolutional Neural Network (CNN). The goal is to analyze whether depth information can contribute to the Kinship Verification problem. Moreover, we test our CNN classifier on widely used kinship database benchmarks, namely, KinFaceW-I and KinFaceW-II. When only 2D face images are considered, results are consistent with current state-of-the-art. When depth information is taken into account, results experiment an improvement of 5 (five) percentage points in accuracy.

Our contribution is two-fold. Firstly, we collected the first database with 3D information and kinship annotations. Secondly, we tested both a traditional and a contemporaneous classification techniques to provide performance accuracies to serve as baseline for future approaches. Additionally, we verify that using 3D facial information improves the model's performance, suggesting that this is an interesting path to pursue in future applications.

This paper is organized as follows. Section 2 presents an overview of works associated with Kinship Verification (KV) published so far. We provide information on methods of collecting, pre-processing, normalizing and classifying face images in Sect. 3. Results are reported in Sect. 4 and we draw final remarks and suggest future work in Sect. 5.

2   Related Work

Face verification and recognition have been extensively tackled by the scientific community and there are many commercial applications available [2,8]. Currently, new problems are increasingly more popular, such as age estimation [23] and expression recognition [12]. Among emerging areas of facial analysis, Kinship Verification (KV) is a recent topic in biometrics, provided that the first work addressing this problem was published by Fang et al. in 2010 [7]. Interest has grown due to possible applications in: (i) Forensic science, such as mitigation of human trafficking, disappearance of children and refugee crises; and (ii) Automatic annotation and reduction of the search space in large databases. Research interest also comes from the fact that KV is a challenging and open problem.

Indeed, faces carry many information about a person and they have been quite used in tasks of recognition. Somanath and Kambhamettu [20] explain how faces can verify blood-relations. As the faces develop, their features become sharper and more alike to their relatives. Parts of faces such as the eyes, nose, ears, cheekbones, and jaw, can supply helpful attributes as relative position, shape, size and color to kinship verification.

As Dibeklioglu explains [6], there are basically two types of kinship analysis: verification and recognition. The former consists of binary identification, which means identifying whether two people are related or not. The latter differentiates the type of kinship between two people (siblings or father-son).

Conventional KV methods use engineered (hand-crafted) features to be extracted from faces for further combination and classification such as Local Binary Patterns (LBP), Gabor features and others [3,22]. We highlight that, to the extent of our knowledge, no work presented thus far has used depth images to tackle the KV problem. As opposed to using hand-crafted features, Deep Learning (DL) techniques for computer vision such as Convolutional Neural Networks (CNNs) are widely used sources for obtaining additional, hierarchical representations [9].

Robinson et al. presented in 2018 the largest kinship database with 1000 family trees [15]. They used a semi-supervised labelling process to improve a pre-annotated clustering. Although the amount of images is large and, therefore, well suited for Deep Learning approaches, they provide 2D face images only, with no depth information whatsoever.

Ozkan et al. have proposed the use of Adversarial Generative Networks to synthesize faces of children by analyzing their parents [14]. One possible application is the increment in database size by generating new samples. The approach is applied to 2D images only.

3   Methodology

In this section we describe the process of collecting, preprocessing and classifying our dataset entitled Kin3D (Sect. 3.1). Then we explain the classification methods we tested, namely CNN and SVM, as well as the metrics we compute to assess model's performances.

3.1   Kin3D Database Collection

Our goal to build our Kin3D dataset is threefold: (1) to prepare, as much as possible, a reduced noise facial dataset with depth information; (2) to include labeled relatives in a database; (3) to classify kinship using known machine learning techniques as deep neural network. As a result, we have reconstructed pairs of relatives and use both RGB and depth images (D) to conduct our validation experiment.

In order to construct the Kin3D dataset, we faced two challenges. Firstly, face scanners are neither cheap nor much portable to be transported into people's homes as an attempt to scan relatives. Besides, *smartphones* are showing increasingly higher spatial and exposure resolutions, which results in 3D reconstructions with greater quality. Secondly, people might not be willing to attend an event to have their faces scanned if there is no type of compensation.

In order to facilitate collecting videos from participants, the task of recording their faces was executed with their own smartphone cameras, in uncontrolled environments. Besides, university students who willingly engaged in this activity had extra scores taken into account.

To enable this research, an online form was created in order to collect annotated data from users. Through the form, one can send a video of his face and another one for each of his relatives. Each video must comprise the lower, median and upper regions of the face, with three different slopes as shown in first stage (upper-left) of the pipeline in Fig. 2. Thus, details that would be occluded on a frontal static image are also captured. The only suggestions were smooth camera movement and a reasonably bright recording environment.

3.2   Structure-from-Motion Pipeline and 3D Reconstruction

From each video, a reduced number of frames is extracted so that they cover all movement. Structure-from-Motion technique takes these frames as input to reconstruct the face as a 3D mesh.

Prior to the 3D reconstruction, we perform a 2D normalization using frontal face features such as the eyes. Frames that present face as frontal as possible are used to compose our RGB dataset, being aligned and cropped into a image of 64×64 pixels as follows:

- Face region of interest and its landmarks detection in the RGB image;
- From eyes' landmarks the image is translated, rotated and scaled so that the eyes lie on a horizontal line;

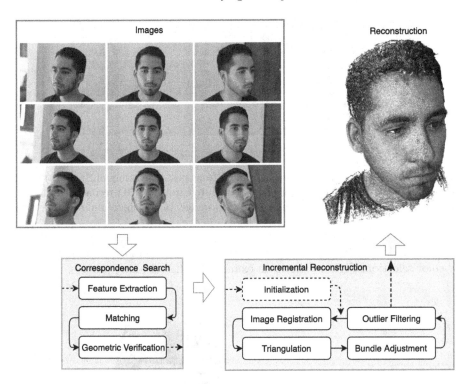

Fig. 2. Structure-from-Motion (SFM) pipeline. A set of images (upper-left) serves as input for feature detection and matching (lower, green box). After geometric verification, the 3D reconstruction happens (blue box) resulting in a set of (x, y, z) and (R, G, B) values, as shown in the upper-right corner of the figure. (Color figure online)

– The matrix used as affine transformations in RGB images is also used to normalize the normal and depth images, as shown in Fig. 3.

The main process of 3D reconstruction is illustrated in Fig. 2. The input is a set of overlapping facial images of the same person. That process of reconstructing is divided into three parts [17,18]:

1. Feature detection and extraction;
2. Feature matching and geometric verification;
3. Structure and motion reconstruction.

In the reconstruction process, pixels from each subsequent frame are matched while the vertices of the 3D mesh are generated. After the mesh is complete, a depth and a normal map are extracted from frontal view, as shown in Fig. 3.

At the end of this process, most of the resulting point cloud show a reasonable amount of noise due to the uncontrolled environment: illumination, camera quality and distance between the recorded person and the camera. Using the software MeshLab [5], we perform a cleaning process by removing the farthest noise just

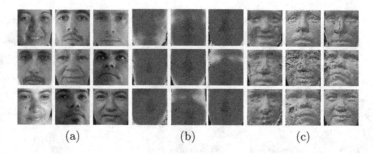

(a)   (b)   (c)

Fig. 3. Examples of picture channels obtained from post processing of SFM. (a) Normalized 2D face images. (b) Depth maps computed from the 3D reconstruction. (c) Map of normal vectors represented as one RGB image. (Color figure online)

selecting and excluding it, and for the merged noise from face we remove them by color selection. Figure 4 shows an example of that cleaning process.

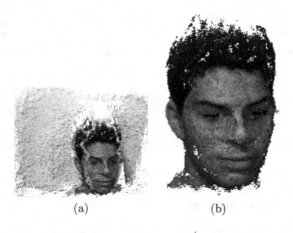

(a)   (b)

Fig. 4. Point cloud with and without noise. (Color figure online)

Graphic Processing Units (GPUs) are designed for heavy workload and throughput. Their parallelism helps to speed up many operations in which they would last hours or days to be finished on some Central Process Units (CPUs). We have used a GPU to generate our point clouds and depth images from about 70 facial images extracted from a video.

3.3   Classification Models

In order to evaluate our classification models consistently, we performed the following steps:

1. We reproduced the results from [24] to assess the applicability of CNN to classify the widely used 2D kinship databases KinFaceW-I and KinFaceW-II. This first evaluation was useful since the number of samples is very small which may cause the model to overfit. We mitigated this by using data augmentation.
2. Since our database is novel, we test a well known classification method – Support Vector Machine – onto features obtained from 2D face images only. To this end, we follow the procedure proposed recently [21].
3. Finally, we apply the known CNN topology for performing KV. To this end, we apply the same model developed in the Stage 1 to classify our database (RGB and RGBD) to evaluate whether depth improves the accuracy or not.

The basic structure of Convolutional Neural Network (CNN) used in this project consists of several concatenated (in parallel) neural nets. The simplest network contains two convolutional layers which are connected to two max-pooling layers, then followed by two fully-connected layers and ending with a soft-max layer, as shown in Table 1. Input images are of size $39 \times 39 \times 6$ (39 wide, 39 high, 6 color channels).

Table 1. Topology of the Convolutional Neural Network before concatenation.

Conv1	Pool1	Con2	Pool2	FC
conv-32	max-2	conv-32	max-2	FC1-128
				FC2-128

The convolutional layers basically are parameterized by: the number of maps, the size of the maps and filter sizes. Our first convolutional layer receives 6 feature maps of size 39×39 and after, with $5 \times 5 \times 6$ filters, it generates 32 maps. These filters slide, or convolve, around the input image with a stride of 1. Since the image size is small, it's used a padding to ensure that the output has the same length as the original input. To initialize weights of the convolutional layers, we use a normal distribution with zero mean and a standard deviation of 0.05. A Rectifier Linear Unit (ReLU) function is used as activations.

In general, pooling layers are based in operations such as max and average. We have chosen max pooling layers because they show better accuracy. These layers have filters of size 2×2 applied to a stride of 2, downsampling the input image along both width and height. Therefore, every operation takes a max among 4 values.

Just like the humans who look at specific parts of the face to recognize who the person is, in our task we decompose the relative's RGB faces into ten individual parts, following the approach in [24]. Next we input each one to a neural network since it has been known that CNN can learn better than in a holistic way. After the convolutional, max-pooling and fully connected layers, we concatenate the ten networks. The same is also done with the depth images.

In this approach with RGB+Depth, there are twenty nets. The concatenation of the rgb and depth networks are also concatenated. Linked to this layer there is an output layer that is added to complete our network. This layer has one neuron per class in the classification task, totaling 2 neurons to relatives and non-relatives. A softmax activation function is applied.

We compile our model using the optimizer Adam [10], which is an improved version of the stochastic gradient descent algorithm that incorporates time and learning rate adaptive. Furthermore, binary crossentropy as the loss function was applied since our targets are in categorical format.

Given all the possible kinship classes, a one-versus-one (OvO) strategy was adopted. Thus, our CNN was trained to validate the relation between one father (or mother) and his son (or daughter). In addition to this strategy being faster, it is also more appropriate for smaller sets. We generate arbitrary sinthetic negative examples by combining people from different families.

To train and validate our deep convolutional neural network we used Keras and TensorFlow [1,4] on NVIDIA's CUDA programming framework [13].

4   Results

4.1   Database Collection

As a result of people's participation, we collected smartphone videos from 120 individuals. When organized by categories, the number of pairs are described in Table 2. Both positive and negative pairs have the same size. The negative pairs are generated from the swapping between positive pairs. We recognize that Kinship Verification is a very challenging problem and that a much larger dataset is recommended for a more representative approach. Regardless, we evaluate the CNN with care, keeping track of training and validation accuracies to prevent overfitting. We acknowledge that, in order to best represent a broad problem such as kinship, collecting a much larger dataset is advisable. That being said, we are still receiving videos through our form to be rebuilt and analyzed. Moreover, it is important to highlight that we use data augmentation, using rotations and vertical flip, multiplying the number of samples threefold.

Table 2. Number of kin pairs for each kinship category.

Category	Number of pairs	Number of samples after data augmentation
Father-Son (FS)	14	42
Father-Daughter (FD)	9	27
Mother-Son (MS)	27	81
Mother-Daughter (MD)	11	33

4.2   Testing a First Topology

KinFaceW [11] is divided into KinFaceW-I and KinFaceW-II and they are ones of the most known datasets used for Kinship evaluation. Using the concatenated CNN that we adapted from [24], we apply it through the KinFaceW. The main changes in our model was that we used only two convolutional layers and added one fully-connected layer. In relation to weight initialization, Zhang et al. used a Gaussian distribution with zero mean ($\mu = 0$) and a standard deviation of $\sigma = 0.01$. But we noticed that the standard deviation with that value did not allowed our concatenated CNN to learn. Thus, we initialized the weights with standard deviation $\sigma = 0.5$. Accuracies were approximate to those obtained in [24], as can be seen in Table 3.

Table 3. Verification accuracy (%) on KinFace dataset.

Methods	KinFaceW-I				Methods	KinFaceW-II			
	FS	FD	MS	MD		FS	FD	MS	MD
CNN-Points [24]	71.8	76.1	78.0	84.1	CNN-Points [24]	89.4	81.9	89.9	92.4
Our method	69.2	77.8	72.6	83.8	Our method	86.2	83.3	84.1	88.8

4.3   Our Dataset

Firstly, a Support Vector Machine (SVM) was developed to supply a baseline. SVMs are well suited for classification of complex tasks but with not-so-big datasets. The SVM is applied to 128-dimensional embedded face vectors as described by [19].

Two types of CNN, representative feature learning methods, were trained. The first one learned to verify kinship from only RGB images. The second one learned to verify kinship from RGB plus depth images to assess the contribution of depth information. We evaluate our dataset to the four categories in kinship research community: father-son (FS), father-daughter (FD), mother-son (MS) and mother-daughter (MD). Results are shown in Table 4 and Fig. 5.

Table 4. Verification accuracy (%) on Kin3D dataset.

Methods	Kin3D			
	FS	FD	MS	MD
SVM	69.9	56.6	66.6	73.8
CNN (RGB)	71.5	75.0	86.4	90.5
CNN (RGB-D)	76.4	76.9	88.4	94.3

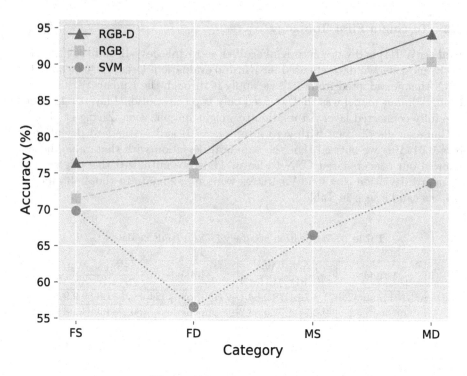

Fig. 5. Classification accuracies.

5 Conclusion

In this paper, we presented a novel use of 3D data for kinship verification (KV). Given the technological advancements in smartphone cameras, computer hardware and image processing software, the addition of 3D information to computer vision and machine learning tasks seems to be the best way to make such tasks more feasible. Our dataset, Kin3D, in addition to information provided about kin relationships also provides information such as age and ethnicity that can be used for studies related to age, synthesis of children faces based on parents, among others.

Overall, the experiments have shown that depth information contributes to the model's performance. We acknowledge that, in order to better represent a broad problem such as kinship, collecting a much larger dataset is advisable. Nevertheless, we hope this database can provide an initial contribution to the research community focused on Kinship Verification (KV) and interested in investigating further the use of 3D information to tackle this task.

Further work consists in constantly increment the database size and investigate whether each face's Point Cloud (PC) could provide further information to the models. Some questions could be tackled, such as; (i) how can neural nets performances be compared to traditional curvatures analyses; (ii) can we use generative adversarial networks to synthesize childrens faces by analyzing their parents. Studies are being conducted and results will be eventually reported.

Acknowledgments. The authors would like to acknowledge the Msc's grant provided by FAPEAL (state's research support foundation) and Institute of Computing's students who participated in the research activities.

References

1. Abadi, M., et al.: TensorFlow: a system for large-scale machine learning. In: Proceedings of the 12th USENIX Conference on Operating Systems Design and Implementation, OSDI 2016, pp. 265–283. USENIX Association, Berkeley (2016). http://dl.acm.org/citation.cfm?id=3026877.3026899
2. Belhumeur, P.N., Hespanha, J.P., Kriegman, D.J.: Eigenfaces vs. fisherfaces: recognition using class specific linear projection. IEEE Trans. Pattern Anal. Mach. Intell. **19**(7), 711–720 (1997). https://doi.org/10.1109/34.598228
3. Bottino, A., Islam, I.U., Vieira, T.F.: A multi-perspective holistic approach to kinship verification in the wild. In: 2015 11th IEEE International Conference and Workshops on Automatic Face and Gesture Recognition (FG), Ljubljana, pp. 1–6. IEEE, May 2015. https://doi.org/10.1109/FG.2015.7284834, http://ieeexplore.ieee.org/lpdocs/epic03/wrapper.htm?arnumber=7284834
4. Chollet, F., et al.: Keras (2015). https://github.com/fchollet/keras
5. Cignoni, P., Callieri, M., Corsini, M., Dellepiane, M., Ganovelli, F., Ranzuglia, G.: MeshLab: an open-source mesh processing tool. In: Scarano, V., Chiara, R.D., Erra, U. (eds.) Eurographics Italian Chapter Conference. The Eurographics Association (2008). https://doi.org/10.2312/LocalChapterEvents/ItalChap/ItalianChapConf2008/129-136
6. Dibeklioglu, H.: Visual transformation aided contrastive learning for video-based kinship verification. In: The IEEE International Conference on Computer Vision (ICCV), October 2017
7. Fang, R., Tang, K.D., Snavely, N., Chen, T.: Towards computational models of kinship verification. In: 2010 17th IEEE International Conference on Image Processing (ICIP), pp. 1577–1580. IEEE, September 2010. https://doi.org/10.1109/ICIP.2010.5652590, http://ieeexplore.ieee.org/lpdocs/epic03/wrapper.htm?arnumber=5652590chenlab.ece.cornell.edu/projects/KinshipVerification/
8. Freire, A., Lee, K.: Face recognition in 4- to 7-year-olds: processing of configural, featural, and paraphernalia information. J. Exp. Child Psychol. **80**(4), 347–371 (2001). https://doi.org/10.1006/jecp.2001.2639. http://www.sciencedirect.com/science/article/pii/S0022096501926396
9. Goodfellow, I., Bengio, Y., Courville, A.: Deep Learning. MIT Press, Cambridge (2016). http://www.deeplearningbook.org
10. Kingma, D.P., Ba, J.: Adam: A method for stochastic optimization. https://www.arxiv-vanity.com/papers/1412.6980/
11. Lu, J., Zhou, X., Tan, Y.P., Shang, Y., Zhou, J.: Neighborhood repulsed metric learning for kinship verification. IEEE Trans. Pattern Anal. Mach. Intell. **36**(2), 331–345 (2014). https://doi.org/10.1109/TPAMI.2013.134
12. Nigam, S., Singh, R., Misra, A.K.: Efficient facial expression recognition using histogram of oriented gradients in wavelet domain. Multimed. Tools Appl. **77** (2018). https://doi.org/10.1007/s11042-018-6040-3

13. NVIDIA Corporation: CUDA Programming Guide 9.0. NVIDIA Corporation (2018)
14. Ozkan, S., Orkan, A.: KinshipGAN: synthesizing of kinship faces from family photos by regularizing a deep face network. In: 2018 25th IEEE International Conference on Image Processing (ICIP), pp. 2142–2146 (2018). https://doi.org/10.1109/ICIP.2018.8451305
15. Robinson, J.P., Shao, M., Wu, Y., Liu, H., Gillis, T., Fu, Y.: Visual kinship recognition of families in the wild. IEEE Trans. Pattern Anal. Mach. Intell., 1 (2018). https://doi.org/10.1109/TPAMI.2018.2826549
16. Savran, A., Gur, R., Verma, R.: Automatic detection of emotion valence on faces using consumer depth cameras. In: 2013 IEEE International Conference on Computer Vision Workshops, pp. 75–82, December 2013. https://doi.org/10.1109/ICCVW.2013.17
17. Schönberger, J.L., Frahm, J.M.: Structure-from-motion revisited. In: Conference on Computer Vision and Pattern Recognition (CVPR) (2016)
18. Schönberger, J.L., Zheng, E., Frahm, J.-M., Pollefeys, M.: Pixelwise view selection for unstructured multi-view stereo. In: Leibe, B., Matas, J., Sebe, N., Welling, M. (eds.) ECCV 2016. LNCS, vol. 9907, pp. 501–518. Springer, Cham (2016). https://doi.org/10.1007/978-3-319-46487-9_31
19. Schroff, F., Kalenichenko, D., Philbin, J.: FaceNet: a unified embedding for face recognition and clustering. In: Proceedings of the IEEE Conference on Computer Vision and Pattern Recognition, pp. 815–823 (2015)
20. Somanath, G., Kambhamettu, C.: Can faces verify blood-relations? In: 2012 IEEE Fifth International Conference on Biometrics: Theory, Applications and Systems (BTAS), pp. 105–112, September 2012. https://doi.org/10.1109/BTAS.2012.6374564
21. Thilaga, P.J., Khan, B.A., Jones, A.A., Kumar, N.K.: Modern face recognition with deep learning. In: 2018 Second International Conference on Inventive Communication and Computational Technologies (ICICCT), pp. 1947–1951, April 2018
22. Vieira, T.F., Bottino, A., Laurentini, A., De Simone, M.: Detecting siblings in image pairs. Vis. Comput. 30(12), 1333–1345 (2014). https://doi.org/10.1007/s00371-013-0884-3
23. Xing, J., Li, K., Hu, W., Yuan, C., Ling, H.: Diagnosing deep learning models for high accuracy age estimation from a single image. Pattern Recognit. 66, 106–116 (2017)
24. Zhang, K., Huang, Y., Song, C., Wu, H., Wang, L.: Kinship verification with deep convolutional neural networks. In: Xie, X., Jones, M.W., Tam, G.K.L. (eds.) Proceedings of the British Machine Vision Conference (BMVC), pp. 148.1–148.12. BMVA Press, September 2015. https://doi.org/10.5244/C.29.148

VA-StarGAN: Continuous Affect Generation

Dimitrios Kollias[✉] and Stefanos Zafeiriou[✉]

Imperial College London, London, UK
{dimitrios.kollias15,s.zafeiriou}@imperial.ac.uk

Abstract. Recent advances in Generative Adversarial Networks have shown impressive results for the task of facial affect synthesis. The most successful architecture is StarGAN, which is effective, but can only generate a discrete number of expressions. However, dimensional emotion representations, usually valence (indicating how positive or negative an emotional state is) and arousal (measuring the power of the emotion activation), are more appropriate to represent subtle emotions appearing in everyday human computer interactions. In this paper, we adapt Star-GAN for continuous emotion synthesis and propose VA-StarGAN; we use a correlation-based loss instead of the usual MSE; we adapt the discriminator network to account for continuous output; we exploit and utilize the in-the-wild Aff-Wild and AffectNet databases; we propose a trick for generating the target domain when training the generator. Qualitative experiments illustrate the generation of realistic images, whilst comparison with state-of-the-art approaches shows the superiority of our method. Quantitative experiments (in which the synthesized images are used for data augmentation in training Deep Neural Networks) further validate our development.

Keywords: Facial affect synthesis · VA-StarGAN · StarGAN · Valence · Arousal · Continuous generation · Aff-Wild · AffectNet

1 Introduction

Facial image editing and manipulation is used in numerous applications, such as the movie industry, image post-production and dubbing, fashion, e-commerce, generation of faces for facial expression recognition and lipreading. The problem of facial editing is mainly an image-to-image translation one. The approach mainly used for this problem has been to build person-specific models [2,23] or to perform manual editing. However, in recent years, Generative Adversarial Networks (GANs) have been introduced and thrived in a variety of application areas, thus GAN-based approaches have been also used for facial image editing.

The first GAN-based solution to image editing was the pix2pix [6] that used a conditional GAN (cGAN) to learn a mapping from an input to an output image. Pre-aligned image pairs were needed during training, which is a drawback of this

© Springer Nature Switzerland AG 2020
J. Blanc-Talon et al. (Eds.): ACIVS 2020, LNCS 12002, pp. 227–238, 2020.
https://doi.org/10.1007/978-3-030-40605-9_20

method. Next, CycleGAN [28] was introduced, for cases where labels between two domains are available. The key idea has been to build upon the power of the pix2pix architecture, but also to allow to point the model at two discrete, unpaired collections of images. This was achieved through the idea of a full translation cycle (cycle-consistency loss) that determined how good the entire translation system was, thus improving the two respective generators at the same time. Nevertheless, this method is not scalable, especially for image translations on multiple domains, since two pairs of generators and discriminators are needed for each domain translation.

StarGAN [3] was suggested to solve this problem, as it adopts a unified approach in which a single generator is trained to map an input image to one of multiple target domains, selected by the user. StarGAN fuses the target domain attributes with the given image, by concatenating them channel-wise. The Star-GAN can be used to train a model for facial expression transfer using discrete expressions. StarGAN, apart from synthesizing expressions, is able to change other facial attributes, such as age, hair, color or gender. Despite its generality, StarGAN can only change a particular aspect of a face among a discrete number of attributes, defined by the annotation granularity of the dataset. For instance, for the facial expression synthesis task, StarGAN is trained on RaFD [18], which has only 8 binary labels for facial expressions. Additionally RaFD is very small in terms of size (around 4,800 images) and it is a lab-controlled and posed expression database. Therefore, StarGANs' capability to generate affect has not been tested on in-the-wild facial analysis, particularly for emotion recognition. Additionally, in the results presented in the original StarGAN paper, the input domain is fixed to the neutral expression; StarGAN is not tested when the other six expressions constitute the input. What is more, StarGAN generates discrete emotions; it cannot generate continuous affective states.

Next, GANimation [21] was proposed. It translates a facial image according to the activation of certain facial Action Units (AUs) and their intensities. In addition, an attention based generator was introduced to promote the robustness of the model for distracting backgrounds and illuminations. Even though AU is a quite comprehensive model for describing facial motion, detecting AUs is still an open problem both in controlled [24], as well as in unconstrained recording conditions [5,7]. In particular, in unconstrained conditions, the accuracy in detecting certain AUs is not high enough yet [4]. One of the reasons is the lack of annotated data and the resulting large cost of annotation which has to be performed by highly trained experts. In additional, AUs cannot describe all possible lip motion patterns produced during speech. Hence, the GANimation model cannot be used in a straightforward manner for transferring speech. Finally, AUs do not cover the whole spectrum of affect; this is something that dimensional emotion representations can provide.

Although, the most frequently used emotion representation is the categorical one through the seven basic categories, the dimensional emotion representation [22,25] is more appropriate to represent subtle, i.e., not only extreme, emotions appearing in everyday human computer interactions [17]. The categorical model also has many limitations: a user can have more feelings than the seven basic ones; people have a different understanding of emotions, depending on their social and cultural background; there is poor resolution in characterizing emotionally ambiguous examples (e.g., happily surprised), or in discriminating between fear, disgust and sad. To the contrary, this is not the case with the dimensional model [13,15], in which each affective state is represented. In this paper, we focus on the problem of generating [8,9] continuous affective states. According to a well-known and widespread dimensional model, all emotions can be discriminated by their position in the resulting coordinate system, the 2D Valence-Arousal (VA) Space. The advantage of this model in comparison to the categorical one is that it can lead to a very accurate assessment of the actual emotional state; valence and arousal are the emotion-underlying dimensions and can be used to distinguish between different internal emotional states. Figure 1 shows the 2-D VA-Space, with valence ranging from very positive to very negative and arousal ranging from very active to very passive.

Fig. 1. The 2-D VA Space

Let us now discuss the contributions of the paper. At first, we extend Star-GAN and propose VA-StarGAN for continuous dimensional emotion generation according to valence and arousal. We suggest to use a domain regression loss, based on the Concordance Correlation Coefficient (CCC) instead of the usual Mean Squared Error (MSE) that is adopted for regression. Additionally, we reformulate the discriminator network so as to include on top a regression layer that estimates the valence and arousal continuous values. We train VA-StarGAN using in-the-wild databases and not lab-controlled ones. To circumvent the need for pairs of training images of the same person under different expressions, we use the large Aff-Wild video database and the AffectNet database. We examine two different settings: (i) first train VA-StarGAN on Aff-Wild and then fine-tune it on AffectNet and (ii) train VA-StarGAN on the union of Aff-Wild and Affect-Net. Finally, during training in these settings, we use a trick, in which we make sure that each batch contains an equal number of images, with annotations from the 4 quadrants of the 2D VA-Space.

2   VA-StarGAN: The Proposed Framework

2.1   Problem Statement and Notations

Our training images are colour RGB ones, denoted as $I_{y_l} \in \mathbb{R}^{H \times W \times 3}$, with each one having a domain label $y_l \in \mathbb{R}^2$ which is a vector containing the valence and arousal annotated values. Valence and arousal range in $[-1, 1]$. VA-StarGAN translates the input image I_{y_l} into an output image $I_{y'_l}$ given the valence and arousal target y'_l.

2.2   VA-StarGAN: Network Architectures

Our framework is based on Generative Adversarial Networks that contain two sub-networks: a generator G and a discriminator D.

Generator. The generator architecture consists of three convolutional down-sampling layers (each followed by instance normalization and relu) followed by six layers of residual modules, followed by two transpose convolutional up-sampling layers (each followed by instance normalization and relu) and finally ends with a convolutional layer with tanh. The instance normalization layers are used to improve training stability. The input to the generator G is a channel-wise concatenation of the source image and the target domain labels. In more detail, given the input image $I_{y_l} \in \mathbb{R}^{H \times W \times 3}$ and the vector $y'_l \in \mathbb{R}^2$ which is the target domain, valence and arousal, the input of the generator is the concatenation $(I_{y_l}, y'_l) \in \mathbb{R}^{H \times W \times 5}$. The output of the generator is a facial image $I_{y'_l} = G(I_{y_l} | y'_l) \in \mathbb{R}^{H \times W \times 3}$. The generator is trained to realistically transform the facial affect in the input image to the desired affect in the output image. Note that, during training, the generator is used twice: first to map the input image to the output one $I_{y_l} \rightarrow G(I_{y_l} | y'_l) = I_{y'_l}$, and then to render it back $I_{y'_l} \rightarrow G(I_{y'_l} | y_l) = I'_{y_l}$ (cycle consistency). Ideally I'_{y_l} and I_{y_l} should be equal, i.e., should be the same image.

Discriminator. The discriminator architecture consists of seven convolutional down-sampling layers (in each: kernel size is 4×4, stride is 2 and padding is 1, leaky relu follows) and then there are two branches: (i) in the first (top) one, a convolutional layer follows (outputting 1 channel, with kernel size 3×3, stride 1 and padding 1) and (ii) in the second (bottom) one, another convolutional layer follows (outputting 2 channels, with kernel size $H/64 \times W/64$, stride 1 and zero padding). The input to the discriminator D is an image which is either real (I_r) or fake, meaning that it is produced by the generator ($I_f = G(I_{y_l} | y'_l)$). The output of the discriminator is: (i) the prediction whether the given image I is real or fake and (ii) the valence and arousal estimate. The top branch of the discriminator produces a matrix $Y_I \in \mathbb{R}^{H/64 \times W/64}$ where I is the input image and $Y_I[i, j]$ represents the probability of the overlapping patch ij to be real. The final prediction (real or fake) is then the average value of predictions of every patch area. The bottom branch is the regression head (with linear activation function) that gives the final valence and arousal estimates. All in all, the discriminator evaluates the generated images in terms of their photorealism and the desired affect fulfillment. As a result, the generator is encouraged to produce realistic images whose depicted affect is similar to the target labels (Fig. 2).

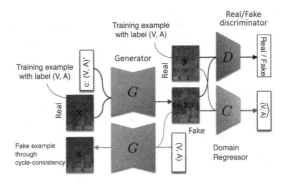

Fig. 2. Overview of VA-StarGAN during training

2.3   Training Trick

In the original StarGAN, during training, random training samples are selected and target domain labels are created based on the permutation of the original labels of the selected training samples. In our case, we decided to split the training samples into 4 categories according to their label value; these 4 categories correspond to the 4 quadrants of the 2D VA-Space. In other words, samples annotated with positive arousal and valence belonged to the first category, samples annotated with positive arousal and negative valence belonged to the second category, samples annotated with negative arousal and valence belonged to the third category and samples annotated with positive valence and negative arousal belonged to the fourth category. During training, samples were randomly selected from all 4 categories. We used this trick because if we did not, our generated images with negative arousal were not of very good quality. This was because all valence-arousal annotated databases are imbalanced and do not contain many samples for negative arousal, as the latter is not frequently recorded in everyday interactions (and thus the images and videos collected, do not contain a lot of such samples). Through this trick, we managed to achieve better results.

3   Loss Functions

Generator's Adversarial Loss. We input a real image I_{y_l}, with label y_l and target label y'_l, to the generator G that outputs a fake image $G(I_{y_l}|y'_l)$. We then pass this fake image $G(I_{y_l}|y'_l)$ to the discriminator D that outputs the probability of the image to be real or fake, $D_{rf}(G(I_{y_l}|y_l'))$. Then, the adversarial loss is defined as follows:

$$\mathcal{L}_{G_{adv}} = -\mathbb{E}_{I_{y_l}, y_l'}[D_{rf}(G(I_{y_l}|y_l'))] \tag{1}$$

Generator's Affect (Regression) Loss. While reducing the adversarial loss, the generator G must also reduce the error produced by the valence and arousal regression head on top of the discriminator D. In this way, G not only learns to render realistic samples but also learns to satisfy the target facial affect encoded by the target label y_l'. This loss is the valence and arousal regression loss with fake images used to learn the regression head on top of D. We feed a real image I_{y_l} with a target label y_l' to G, who outputs a fake image $G(I_{y_l}|y_l')$, which is then passed to D that outputs the valence and arousal estimates $D_v(G(I_{y_l}|y_l'))$ and $D_a(G(I_{y_l}|y_l'))$ respectively. This loss is based on the Concordance Correlation Coefficient (CCC) and is defined as follows:

$$\mathcal{L}_{G_{affect}} = 2 - (CCC(D_v(G(I_{y_l}|y_l')), y_l') + CCC(D_a(G(I_{y_l}|y_l')), y_l')), \tag{2}$$

where CCC takes as input all valence (arousal) predictions and labels and outputs the correlation value; it takes values in $[-1, 1]$ and is defined as:

$$CCC(x, y) = \frac{2s_{xy}}{s_x^2 + s_y^2 + (\bar{x} - \bar{y})^2}, \tag{3}$$

where s_x and s_y are the variances of the valence/arousal labels and predicted values respectively, $\bar{x}$ and $\bar{y}$ are the corresponding mean values and s_{xy} is the covariance value.

Generator's Reconstruction (Cycle-Consistency) Loss. The adversarial loss alone is not sufficient to produce good images, as it leaves the model underconstrained. It enforces that the generated output be of the appropriate affect, but it does not enforce that the input and output are recognizably the same. In other words it does not guarantee that the face in both the input and output images of the generator correspond to the same person. The cycle consistency loss addresses this issue. It relies on the expectation that if you convert an image to the other domain and back again, by successively feeding it through the generator, you should get back something similar to what you put in.

Thus by using a cycle consistency loss we force the generator to maintain the identity of each individual by penalizing the difference between the original image I_{y_l} and its reconstruction $G(G(I_{y_l}|y_l')|y_l)$. To produce realistic images it is critical for the generator to model both low and high frequencies. Our discriminator network already enforces high-frequency correctness by restricting our attention to the structure in local image patches. To also capture low-frequencies, we adopt a pixel-wise ℓ-1 norm as a reconstruction loss for the generator:

$$\mathcal{L}_{G_{recon}} = \mathbb{E}_{I_{y_l}, y_l', y_l}[\|G(G(I_{y_l}|y_l')|y_l) - I_{y_l}\|_1] \tag{4}$$

Discriminator's Adversarial Loss. We input a real image I_{y_l}, with label y_l and target label y_l', to the generator G that outputs a fake image $G(I_{y_l}|y_l')$. We then pass: (i) this fake image $G(I_{y_l}|y_l')$ to the discriminator D that outputs the probability of the image to be real or fake, $D_{rf}(G(I_{y_l}|y_l'))$, and (ii) the real image I_{y_l} to the discriminator D that outputs the probability of the image to be real or fake, $D_{rf}(I_{y_l})$. Then, the adversarial loss is defined as follows:

$$\mathcal{L}_{D_{adv}} = -\mathbb{E}_{I_{y_l}}[D_{rf}(I_{y_l})] + \mathbb{E}_{I_{y_l},y_l'}[D_{rf}(G(I_{y_l}|y_l'))] \tag{5}$$

Discriminator's Wasserstein GAN Loss. To avoid local saturation of the Discriminator's Adversarial Loss, which will lead to vanishing gradients in the discriminator, we additionally use the Wasserstein GAN (WGAN) [1] objective with gradient penalty. In WGAP, the authors proposed to add a gradient penalty for the discriminator network, which is computed as the norm of the gradients with respect to the input to the discriminator. This loss helps stabilizing the training process and generating higher quality images. It is defined as follows:

$$\mathcal{L}_{WGAN} = \mathbb{E}_I[(\|\nabla_I D(I)\|_2 - 1)^2] \tag{6}$$

where I is sampled uniformly along a straight line between a pair of real and fake images.

Discriminator's Affect (Regression) Loss. This loss is the valence and arousal regression loss of real images used to learn the regression head on top of the discriminator. We feed a real image I_{y_l} with label y_l to D that outputs the valence and arousal estimates $D_v(I_{y_l})$ and $D_a(I_{y_l})$ respectively. This loss is based on the CCC and is defined as follows:

$$\mathcal{L}_{D_{affect}} = 2 - (CCC(D_v(I_{y_l}), y_l) + CCC(D_a(I_{y_l}), y_l)), \tag{7}$$

Total Losses. The loss functions used to optimize G and D are written as:

$$\begin{aligned}
\mathcal{L}_D &= \lambda_{adv} * (\mathcal{L}_{D_{adv}} + \mathcal{L}_{WGAN}) + \lambda_{affect} * \mathcal{L}_{D_{affect}}, \\
\mathcal{L}_G &= \lambda_{adv} * \mathcal{L}_{G_{adv}} + \lambda_{affect} * \mathcal{L}_{G_{affect}} + \lambda_{recon} * \mathcal{L}_{G_{recon}},
\end{aligned} \tag{8}$$

where λ_{adv}, λ_{affect} and λ_{recon} are the hyper-parameters that control the relative importance of every loss term.

4   Details

Databases. The *Aff-Wild* [11,27] has been the first large scale captured in-the-wild database that has been annotated for valence and arousal (within the range $[-1,1]$). It served as benchmark for the Aff-Wild Challenge organized in CVPR 2017. It consists of 298 Youtube videos and displays reactions of 200 subjects. The total number of frames in this database is around 1.25M. Recently this database has been extended [12,14,16] to contain basic expression and action unit annotations. The *AffectNet database* [19] contains around 1M facial images downloaded from the Internet. About 400 K of the retrieved images were manually annotated for the 7 discrete emotions (plus contempt) and the valence and arousal (within the range $[-1, 1]$).

Pre-processing. We used the SSH detector [20] based on the ResNet and trained on the WiderFace dataset [26] to extract face bounding boxes and five facial landmarks from all images. These landmarks were used next for face alignment. All cropped and aligned images were then resized to $128 \times 128 \times 3$ pixel resolution and their intensity values were normalized to the range $[-1, 1]$.

Implementation and Settings. We train VA-StarGAN on either the union of Aff-Wild and AffectNet or first on Aff-Wild and then fine-tune it on AffectNet. In all experiments, batch size was 128. The weight coefficients $\lambda_{adv}, \lambda_{affect}, \lambda_{recon}$ for the loss terms in Eq. 8 were set to 1, 5, 10, respectively. Training takes around a day on a Quadro GV100 Volta GPU. The Tensorflow platform was used.

Baseline Networks. Since there is no prior work that performs image-to-image translation on a given face image conditioned on target valence and arousal, we adapt the state-of-the-art GANimation framework (that performs continuous generation) to fit our case. We established a GANimation baseline that tries to straightforwardly adapt the GANimation principles.

5   Experimental Results

This section provides a thorough evaluation of our VA-StarGAN. We first present some qualitative results that demonstrate our model's ability to deal with in-the-wild images and also demonstrate its capability to generate a wide range of different facial affect: (i) we visually compare the generated images by our approach to those of GANimation; (ii) we evaluate our training trick and (iii) we compare VA-StarGAN trained with CCC or MSE as affect regression losses. Then, we present quantitative results: a data augmentation strategy which uses the synthesized data produced by our approach (or GANimation), as additional data to train DNNs, for valence-arousal prediction. Let us mention that the two different settings described before resulted in similar results.

5.1   Qualitative Results

At first, we trained VA-StarGAN with CCC or MSE loss as Affect (Regression) Loss and the baseline -GANimation- with the databases described in Sect. 4. Figure 3 shows: (i) the real images that were fed to the generator of VA-StarGAN, (ii) the generated images when the Affect (Regression) Loss was based on the CCC and when it was based on the MSE, (iii) the generated images by GANimation and (iv) the labels of the generated images (first value corresponds to valence and second one to arousal). It can be observed that VA-StarGAN produces much more realistic images than GANimation, which, in most cases, show artifacts and in some cases certain levels of blurriness. In general, errors in the attention mechanism occur when the input contains extreme expressions. The attention mechanism does not seem to sufficiently weight the color transformation, causing transparencies. Additionally, when using the MSE in the Affect (Regression) Loss, generated images are more blurred and contain some artifacts, compared to the generated images when the Affect (Regression) Loss is based on CCC.

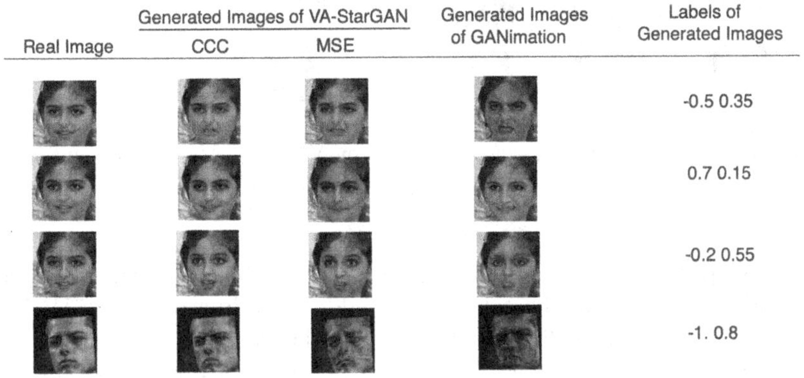

Real Image	Generated Images of VA-StarGAN		Generated Images of GANimation	Labels of Generated Images
	CCC	MSE		
				-0.5 0.35
				0.7 0.15
				-0.2 0.55
				-1. 0.8

Fig. 3. Visual comparison of generated images by VA-StarGAN (with different losses) and GANimation.

Next, we trained VA-StarGAN, with and without using the training trick described in Sect. 2.3, and the baseline GANimation. Figure 4 shows: (i) the real images that were fed to the generator of VA-StarGAN, (ii) the generated images when VA-StarGAN was trained with and without the training trick described in Sect. 2.3, (iii) the images generated by GANimation and (iv) the labels of the generated images (first value corresponds to valence and second one to arousal). Again, it can be observed that images generated by VA-StarGAN were better than the ones generated by GANimation that suffered from artifacts. It can also be seen that the training trick that we adopted, helped VA-StarGAN to generate photorealistic images and avoid generation of serious artifacts.

5.2 Quantitative Results

In the following, we also utilize the networks presented in previous section. In more detail, we utilize: (i) VA-StarGAN with the training trick (we denote this VA-StarGAN), (ii) VA-StarGAN without the training trick (we denote this VA-StarGAN-no trick), (iii) VA-StarGAN trained using MSE as the Affect (Regression) Loss (we denote this VA-StarGAN-MSE) and (iv) GANimation. We feed real images from Aff-Wild and AffectNet to these networks and generate fake

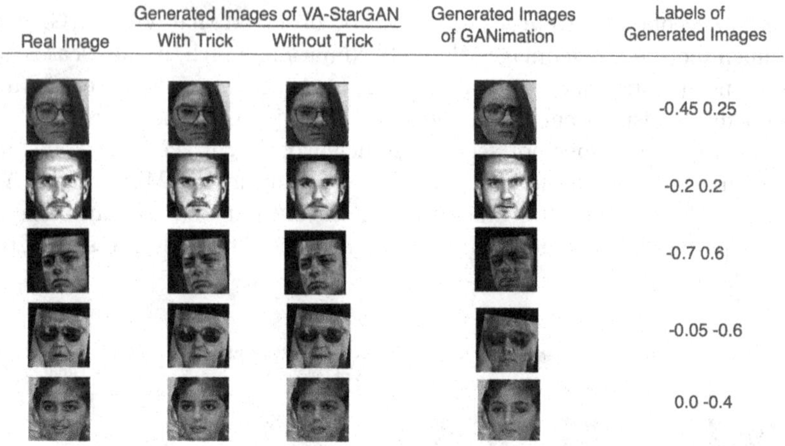

| Real Image | Generated Images of VA-StarGAN With Trick   Without Trick | Generated Images of GANimation | Labels of Generated Images |

Fig. 4. Visual comparison of generated images by VA-StarGAN (when using and not using the training trick defined in Sect. 2.3) and GANimation

images. We perform a data augmentation strategy which uses the synthesized data produced by each of the networks, as additional data to train a DNN (we select VGGFACE), for both valence-arousal prediction.

Table 1 shows the performance in terms of CCC (as a valence-arousal tuple) of the VGGFACE network trained on either AffectNet or Aff-Wild, augmented with the generated images from the above described four networks. One can see that generated images from VA-StarGAN helped VGGFACE to obtain the best performance in both Aff-Wild and AffectNet databases. Table 2 compares the performance of the VGGFACE - trained on AffectNet or Aff-Wild, augmented

Table 1. CCC evaluation of VGGFACE trained on AffectNet or Aff-Wild, augmented with generated images from the four approaches

Databases	VA-StarGAN	VA-StarGAN-no trick	VA-StarGAN-MSE	GANimation
	CCC: (V-A)	CCC: (V-A)	CCC: (V-A)	CCC: (V-A)
AffectNet	**(0.61–0.48)**	(0.59–0.42)	(0.56–0.38)	(0.53–0.43)
Aff-Wild	**(0.55–0.40)**	(0.53–0.36)	(0.52–0.33)	(0.52–0.34)

Table 2. CCC evaluation of state-of-the-art methods vs VGGFACE trained on Affect-Net or Aff-Wild, augmented with generated images from VA-StarGAN

	AffectNet	Aff-Wild
	CCC: (V-A)	CCC: (V-A)
VGGFACE also trained with generated images by VA-StarGAN	**(0.61–0.48)**	**(0.55–0.40)**
AlexNet [19]	(0.60–0.34)	–
VGGFACE [10,11]	–	(0.51–0.33)

with the generated images from VA-StarGAN - to that of two state-of-the-art methods. It can be seen that the described network outperformed both state-of-the-art methods by a large margin (on average 7.5% and 5.5%).

6 Conclusions

In this paper we presented our approach for facial affect synthesis in terms of valence and arousal. We developed VA-StarGAN for continuous emotion synthesis. A qualitative and quantitative analysis was performed; it validated VA-StarGAN's ability to generate photorealistic images.

References

1. Arjovsky, M., Chintala, S., Bottou, L.: Wasserstein GAN. arXiv preprint arXiv:1701.07875 (2017)
2. Blanz, V., Vetter, T., et al.: A morphable model for the synthesis of 3D faces. In: SIGGRAPH, vol. 99, pp. 187–194 (1999)
3. Choi, Y., Choi, M., Kim, M., Ha, J.W., Kim, S., Choo, J.: StarGAN: unified generative adversarial networks for multi-domain image-to-image translation. In: Proceedings of the IEEE Conference on Computer Vision and Pattern Recognition, pp. 8789–8797 (2018)
4. Dapogny, A., Bailly, K., Dubuisson, S.: Confidence-weighted local expression predictions for occlusion handling in expression recognition and action unit detection. Int. J. Comput. Vis. **126**(2–4), 255–271 (2018)
5. Han, S., et al.: Optimizing filter size in convolutional neural networks for facial action unit recognition. In: Proceedings of the IEEE Conference on Computer Vision and Pattern Recognition, pp. 5070–5078 (2018)
6. Isola, P., Zhu, J.Y., Zhou, T., Efros, A.A.: Image-to-image translation with conditional adversarial networks. In: Proceedings of the IEEE Conference on Computer Vision and Pattern Recognition, pp. 1125–1134 (2017)
7. Jiang, B., Valstar, M.F., Pantic, M.: Action unit detection using sparse appearance descriptors in space-time video volumes. In: Face and Gesture 2011, pp. 314–321. IEEE (2011)
8. Kollias, D., Cheng, S., Pantic, M., Zafeiriou, S.: Photorealistic facial synthesis in the dimensional affect space. In: Proceedings of the European Conference on Computer Vision (ECCV), p. 0 (2018)
9. Kollias, D., Cheng, S., Ververas, E., Kotsia, I., Zafeiriou, S.: Generating faces for affect analysis. arXiv preprint arXiv:1811.05027 (2018)
10. Kollias, D., Nicolaou, M.A., Kotsia, I., Zhao, G., Zafeiriou, S.: Recognition of affect in the wild using deep neural networks. In: Proceedings of the IEEE Conference on Computer Vision and Pattern Recognition Workshops, pp. 26–33 (2017)
11. Kollias, D., et al.: Deep affect prediction in-the-wild: Aff-wild database and challenge, deep architectures, and beyond. Int. J. Comput. Vis. **127**(6–7), 907–929 (2019)
12. Kollias, D., Zafeiriou, S.: Aff-wild2: Extending the Aff-wild database for affect recognition. arXiv preprint arXiv:1811.07770 (2018)
13. Kollias, D., Zafeiriou, S.: A multi-component CNN-RNN approach for dimensional emotion recognition in-the-wild. arXiv preprint arXiv:1805.01452 (2018)

14. Kollias, D., Zafeiriou, S.: A multi-task learning & generation framework: valence-arousal, action units & primary expressions. arXiv preprint arXiv:1811.07771 (2018)
15. Kollias, D., Zafeiriou, S.: Exploiting multi-CNN features in CNN-RNN based dimensional emotion recognition on the omg in-the-wild dataset. arXiv preprint arXiv:1910.01417 (2019)
16. Kollias, D., Zafeiriou, S.: Expression, affect, action unit recognition: Aff-wild2, multi-task learning and arcface. arXiv preprint arXiv:1910.04855 (2019)
17. Kollias, D., Marandianos, G., Raouzaiou, A., Stafylopatis, A.G.: Interweaving deep learning and semantic techniques for emotion analysis in human-machine interaction. In: 2015 10th International Workshop on Semantic and Social Media Adaptation and Personalization (SMAP), pp. 1–6. IEEE (2015)
18. Langner, O., Dotsch, R., Bijlstra, G., Wigboldus, D.H., Hawk, S.T., Van Knippenberg, A.: Presentation and validation of the radboud faces database. Cogn. Emot. **24**(8), 1377–1388 (2010)
19. Mollahosseini, A., Hasani, B., Mahoor, M.H.: Affectnet: a database for facial expression, valence, and arousal computing in the wild. arXiv preprint arXiv:1708.03985 (2017)
20. Najibi, M., Samangouei, P., Chellappa, R., Davis, L.S.: SSH: single stage headless face detector. In: Proceedings of the IEEE International Conference on Computer Vision, pp. 4875–4884 (2017)
21. Pumarola, A., Agudo, A., Martinez, A.M., Sanfeliu, A., Moreno-Noguer, F.: Ganimation: anatomically-aware facial animation from a single image. In: Proceedings of the European Conference on Computer Vision (ECCV), pp. 818–833 (2018)
22. Russell, J.A.: Evidence of convergent validity on the dimensions of affect. J. Pers. Soc. Psychol. **36**(10), 1152 (1978)
23. Thies, J., Zollhofer, M., Stamminger, M., Theobalt, C., Nießner, M.: Face2face: real-time face capture and reenactment of RGB videos. In: Proceedings of the IEEE Conference on Computer Vision and Pattern Recognition, pp. 2387–2395 (2016)
24. Valstar, M., Pantic, M.: Fully automatic facial action unit detection and temporal analysis. In: 2006 Conference on Computer Vision and Pattern Recognition Workshop (CVPRW 2006), pp. 149–149. IEEE (2006)
25. Whissel, C.: The dictionary of affect in language, emotion: theory, research and experience. In: Plutchik, R., Kellerman, H. (eds.) The Measurement of Emotions, vol. 4, Academic Press, New York (1989)
26. Yang, S., Luo, P., Loy, C.C., Tang, X.: Wider face: a face detection benchmark. In: Proceedings of the IEEE Conference on Computer Vision and Pattern Recognition, pp. 5525–5533 (2016)
27. Zafeiriou, S., Kollias, D., Nicolaou, M.A., Papaioannou, A., Zhao, G., Kotsia, I.: Aff-wild: valence and arousal in-the-wild challenge. In: Proceedings of the IEEE Conference on Computer Vision and Pattern Recognition Workshops, pp. 34–41 (2017)
28. Zhu, J.Y., Park, T., Isola, P., Efros, A.A.: Unpaired image-to-image translation using cycle-consistent adversarial networks. In: Proceedings of the IEEE International Conference on Computer Vision, pp. 2223–2232 (2017)

Fast Iris Segmentation Algorithm for Visible Wavelength Images Based on Multi-color Space

Shaaban Sahmoud[1(✉)] and Hala N. Fathee[2]

[1] Fatih Sultan Mehmet Vakif University, Istanbul, Turkey
shaaban.sahm@gmail.com
[2] Mosul University, Mosul, Iraq
hala.fathee@uomosul.edu.iq

Abstract. Iris recognition for eye images acquired in visible wavelength is receiving increasing attention. In visible wavelength environments, there are many factors that may cover or affect the iris region which makes the iris segmentation step more difficult and challenging. In this paper, we propose a novel and fast segmentation algorithm to deal with eye images acquired in visible wavelength environments by considering the color information form multiple color spaces. The various existing color spaces such as RGB, YCbCr, and HSV are analyzed and an appropriate set of color models is selected for the segmentation process. To accurately localize the iris region, a set of convenient techniques are applied to detect and remove the non-iris regions such as pupil, specular reflection, eyelids, and eyelashes. Our experimental results and comparative analysis using the UBIRIS v2 database demonstrate the efficiency of our approach in terms of segmentation accuracy and execution time.

Keywords: Iris recognition · Unconstrained environments · Iris segmentation · Unconstrained iris segmentation · Visible wavelength iris segmentation · Multi-color segmentation approach

1 Introduction

Iris recognition systems have been considered as one of the most robust, accurate, and fast biometric identification systems. This importance is due to many reasons such as the stability of iris code over human life, the high uniqueness even for identical twins, the high randomness of iris texture, and the ability of template extraction without touching or contacting [9,11]. The iris is a thin, circular structure in our human eye that contains high number of distinguishing features such as furrows, freckles, and ridges, and it is responsible for controlling the pupil size. Since the development of the first iris biometric system, it becomes a popular identification method and it has been used in many applications such as border control stations, security checkpoints, and secret gates that needs verification [12,25].

© Springer Nature Switzerland AG 2020
J. Blanc-Talon et al. (Eds.): ACIVS 2020, LNCS 12002, pp. 239–250, 2020.
https://doi.org/10.1007/978-3-030-40605-9_21

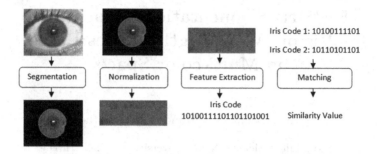

Fig. 1. The steps of normal iris recognition system.

A classical iris recognition system usually includes four main steps: segmentation, normalization, feature extraction, and matching [7]. Figure 1 shows the necessary steps regarding the iris recognition system. The segmentation step includes accurately localizing the interior and exterior boundaries of the iris region, and then non-iris regions that may appear because of eyelashes and eyelids are determined and removed [11]. The iris segmentation is the most significant and difficult step in iris recognition system since all remaining steps depends on its output. If any error happens in this step, the iris code may significantly degrade as a result which causes false identification or verification in the matching step [8]. To more simplify the segmented iris region for easy feature extraction, in the second step called normalization, the circular shape of the iris is converted to rectangular shape as shown in Fig. 1. The conversion process is normally accomplished by transforming from Cartesian coordinates to normalized polar coordinates. In the third step called feature extraction, the unique iris code is extracted from the pixels of the iris region using a convenient filter or method. The resulted iris code from feature extraction step is an array of binary numbers with a fixed size depending on the selected feature extraction method. The iris codes are compared with each other in the last step by utilizing a comparing metric such as Hamming Distance (HD) which is used in most of iris recognition systems.

One of most challenging problems of iris recognition algorithms is the deal with iris images picked in unconstrained environments and in visible wavelength [16]. In unconstrained environments, the iris image is taken by a normal camera rather than using special Near Infrared Light (NIR) camera as in the traditional iris biometric systems. Using normal camera adds more challenging conditions to the iris region like blurring, lighting variations, and occlusions by eyelids or eyelashes. Furthermore, in unconstrained environments, the iris image can be picked using different distances between human eyes and the camera which generates iris with different size regions [22]. Figure 2 shows four samples from constrained and unconstrained iris images. In this paper, we focus on dealing with segmentation step in unconstrained environments by utilizing various existing color space models such as RGB, YCbCr, and HSV. While most of current iris segmentation methods use only one color space such as RGB [17]

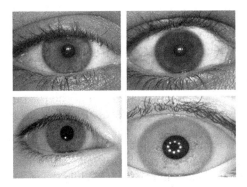

Fig. 2. Sample images for unconstrained environments (first row) from UBIRIS database [18] and constrained environments (second row) from CASIA database [15].

and YCbCr [21], our main motivation of this paper is to utilize a set of color components from different color models to enhance and fasten the segmentation process. To the best of our knowledge, this is the first paper that uses different components from more than one color space. Our experimental results and comparative analysis on iris images from the UBIRIS v2 database demonstrate the efficiency of our multi-color spaces approach in terms of both the segmentation accuracy and the execution time.

The rest of this paper is organized as follow: Sect. 2 reviews the current important research work on unconstrained iris segmentation methods. In Sect. 3, we explain the steps of our proposed multi-color space segmentation algorithm. Our experimental results and performance discussions are given in Sect. 4. Section 5 concludes the paper and gives our future research directions.

2   Related Work

Recently, the research work that deal with iris segmentation in unconstrained environments has significantly increased. By revising the literature, we find that there are different algorithms were proposed to handle this problem using different approaches. In this section, we review some of the common research work in this domain.

In [23], Tan et al. proposed an approach to both increase the consistency of iris code bits and enhance the quality of the weight map of the proposed system using a Zernike moment based encoding algorithm. The proposed approach penalizes the unstable or fragile bits and rewarding the consistent bits. Rapaka et al. [20] proposed a new segmentation mechanism by merging the multi-level Otsu algorithm with an improved swarm optimization algorithm as a pre-processing stage. Moreover, a modified active contours method is employed to accurately localize the iris borders even in the cases of non-circular irises. In [1], the authors proposed another segmentation method based on active contours where they utilized fusion techniques to expand and shrink the active contour for accurate iris region localizing. In 2012, Radman et al. [19] employed the Gabor filter to obtain the pupil centre of the eye, then the inner and outer circle of iris are determined by applying the Daugman's integro differential operator, and finally they by delete the non-iris regions using a method called the live-wire. Wan et al. [24] presented a circular technique to segment non-ideal iris images by using anisotropic diffusion. Both of the iris boarders are localized by using Laplace Pyramid (LP) that detects all reflections and reveals exterior borders.

Proenca [17] proposed a new method for ocular recognition that used two disparate stages. The first stage analyzes the iris image texture and takes advantage of visible-light multi-component information. While the second stage parameterizes the eyelids form and detects a surrounding area that used to detect the eyelids, eyelash and skin regions. Sahmoud and Abuhaiba [21] proposed a new segmentation method that used the K-means algorithm in a pre-processing stage to determine the candidate region of iris, after that a fast circular Hough transform is applied to detect the iris and pupil borders, and then a set of new proposed techniques are applied to remove the non-iris regions. In another research [6], the authors proposed a modified and fast Hough transform with new strategy to localize iris edges by using multi-arcs and multi-lines. Yuting et al. [27] presented a new technique for unconstrained iris segmentation by using convolution network with dilated convolutions. The deep learning and convolutional networks are used in many other research papers to handle the iris segmentation problem as in [3–5,13,26].

3   Proposed Multi-color Space Approach for Iris Segmentation

The proposed segmentation method in this paper is designed to achieve accurate segmentation of the iris region with low execution time. In literature, many researchers depend on the circular Hough transform in order to localize the iris and pupil where both of them are treated as circles, where in unconstrained environments the iris may has a non-circular shape as in the off-axis iris images. As a result, segmentation algorithms spend additional time in removing the non-iris regions that resulted from using circular Hough transform which more slow the segmentation process. In our proposed segmentation approach, we used a fast color controlling/thresholding technique to isolate the iris region based on multiple color components from three different color spaces/models. After that the pupil, eyelashes, and eyelids are removed using different techniques as will described in following steps.

3.1   Extracting the Iris Color Regions

In the proposed segmentation algorithm the color models are used to segment human iris color regions from the color image of human eyes. The iris color space is characterized by the selection of appropriate multiple color components from different models. We use three color models YCbCr, RGB, and HSV, which are widely used in image processing applications, to develop the proposed iris color space. By determining the limits of each color component, our algorithm becomes able to accurately extract the iris color pixels/regions and excluding the non-iris pixels/regions. The limits of each color component are estimated experimentally, where 540 iris images for 108 human eyes are used from the UBIRIS v2 database [18]. The iris images are first segmented manually and all types of noise are isolated. After that the images are converted to other color models where the lower and upper thresholds/limits of the pixel intensity values for all components are computed using the following equations:

$$\mu_i = \frac{1}{(m \times n)} \sum_{x=1}^{m} \sum_{y=1}^{n} I(x,y), \quad \bar{\mu} = \frac{1}{k} \sum_{1}^{k} \mu_i, \quad \alpha = \sqrt{\frac{\sum_{1}^{k}(\mu_i - \bar{\mu}^2)}{k}} \quad (1)$$

$$LCC = \bar{\mu} - 1.5\alpha, \quad HCC = \bar{\mu} + 1.5\alpha \quad (2)$$

Table 1. Threshold values for the color components used in our iris color space

Color component	LCC	HCC
R (RGB)	60	155
G (RGB)	25	150
Y (YCbCr)	52	139
Cr (YCbCr)	121	156

where μ_i is the average of the intensities of the ith sample image $I(x,y)$ for the considered color component (i.e. R, G, B, Cb, Cr and Y). m and n denote the size of image, and $\bar{\mu}$ and α represent the mean and standard deviation of the intensities for a color component using k sample iris images, respectively. LCC and HCC denote the lower and upper limits for the considered color components, respectively. A pixel is classified to be an iris pixel if its color component value falls within the LCC and HCC thresholds. Our preliminary experimental results show that using 1.5α is the optimal value since using larger values increases the probability of classifying non-iris pixels as iris pixels, and using small values increases the probability of incorrectly excluding iris pixels. Figure 3 shows the pixel distribution plot of each component of the three considered color spaces. Based on our preliminary experiments, we found that the optimal set of color components to accurately segment the iris pixels is R component of RGB, G component of RGB, Y component of YCbCr, and Cr component of YCbCr. From Fig. 3, we note that some color components such as B of RGB, H of HSV, and S of HSV are not appropriate for segmenting iris regions since the iris pixels are distributed over large space on their domains.

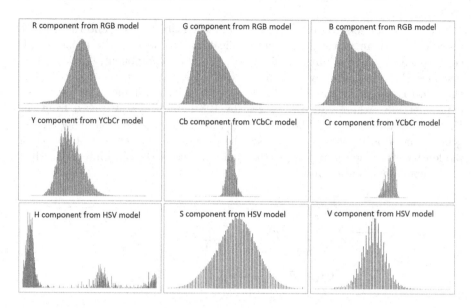

Fig. 3. Distribution of iris color spaces for nine color components from three color spaces using 540 iris images from UBIRIS v2 database.

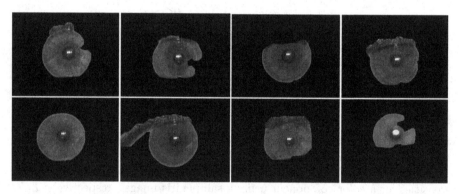

Fig. 4. Results of the proposed multi-color space iris segmentation using 8 iris images from UBIRIS v2 database.

The experimental values of lower and upper thresholds for each color component that used in this paper are given in Table 1. For each pixel in the input eye image, it is considered as an iris pixel if it achieves the conditions of the four color components as given in the following:

$$Iris(x,y) = \begin{cases} 1 & if(LCC_R \leq R(x,y) \leq HCC_R) \ and \ (LCC_G \leq G(x,y) \leq HCC_G) \\ & and \ (LCC_Y \leq Y(x,y) \leq HCC_Y) \ and \ (LCC_{Cr} \leq Cr(x,y) \leq HCC_{Cr}) \\ 0 & otherwise \end{cases}$$

The results of the proposed multi-color space iris segmentation are shown in Fig. 4. As we can see, the proposed multi-color space segmentation algorithm leads to accurate segmentation for iris regions, and successfully isolates the sclera regions and specular reflections resulting from various lighting conditions. On the other hand, there are still some non-iris regions need to be isolated which are pupil regions, eyelashes noise regions, and eyelids noise regions. Therefore, we conduct another noise removing step to completely clean the iris region and remove all non-iris region as will be discussed in the next subsection. Furthermore, one of the most important advantages of our proposed segmentation algorithm is its very fast execution process, since we are using a thresholding mechanism that significantly decreases the computation complexity and reduces the search area for the iris features in next steps.

3.2   Removing Noise

Isolating the Pupil Region. In unconstrained environments, pupil removal becomes a difficult task since its color is close to iris color and the intensity contrast between the pupil and the iris is very low, especially for irises with dark colors. As a result, our mult-color thresholding mechanism can not isolate the pupil region. Moreover, the pupil is usually covered by specular reflections due to outdoor conditions or the used image capture devices. Therefore, we perform an additional step to isolate the pupil region using the following steps:

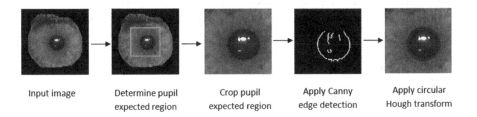

| Input image | Determine pupil expected region | Crop pupil expected region | Apply Canny edge detection | Apply circular Hough transform |

Fig. 5. Steps of pupil isolating process.

1. The resulted iris image obtained from previous step is enhanced by applying some morphological operations to fill the holes and delete the small noise points.
2. Restrict the search area for pupil by determining a small area as the pupil expected region. This is done by computing the center (denoted by c) of the biggest block resulted from previous thresholding step, and then the square with the same center c and diagonal length equals $2r_{avg}$ is considered as the pupil expected region, where r_{avg} is the average radius of irises in the UBIRIS v2 database.
3. Adjust the pupil expected region by mapping its intensity to new values for more focusing on dark intensities.

4. Apply the enhanced Canny edge detection [2] to obtain the edge map of the reduced image.
5. Apply the circular Hough transform [14] to find the pupil borders, assuming that it has a circular shape.

Figure 5 shows the resulted image from each step using a sample iris image from UBIRIS v2. Note that applying the circular Hough transform on only a small square image rather than searching all the eye image significantly reduces the execution time and avoids miss-classification errors that may occur by other edge points in the image.

Isolating the Eyelashes and Eyelids Regions. The upper eyelashes and eyelids are usually the biggest cause of iris occlusions. Furthermore, the isolation of eyelashes and eyelids is considered as one of the most important and difficult tasks in iris segmentation algorithms especially when dealing with unconstrained environments. In this study, we utilize the method presented in [21] to remove the non-iris regions resulted from occlusions of eyelashes and eyelids. This method is efficient and very fast when dealing with images captured in visible wavelength. The steps of this method can be summarized as follow:-

1. Isolate two small rectangle images from the two sides of the localized iris circle.
2. Apply horizontal Canny edge detection on the two isolated images, and enhance the resulted binary image by morphological operations.
3. Determine the noisiest point in each side and draw an arc between them to isolate the upper eyelid. The radius and the center of the arc are determined exactly as explained in [21].

4   Experimental Results and Discussion

In this section, we investigate the segmentation performance of our proposed multi-color space iris segmentation algorithm. To evaluate the performance of our algorithm, it is implemented using MATLAB software, and it is applied on iris images from the UBIRIS v2 database [18]. The UBIRIS v2 is public and one of the most common databases used in testing the performance of unconstrained iris segmentation algorithms. In this study, we used two separated sets of iris images from the UBIRIS v2, the first set is used in pre-processing step to estimate the LCC and HCC values for different color components as described in Sect. 3, and the second set is used for testing. The testing set contains 600 iris images collected from 120 different eyes. Examples of correct segmented iris images using the proposed algorithm in different iris occlusion situations are presented in Fig. 6. As shown in this figure, the proposed algorithm accurately detects the region of iris and the iris boundaries after the first step, even in the existence of specular reflections and noisy situations. The specular reflections are implicitly removed from the resulted iris region without any extra effort (see second row of

Fig. 6). Moreover, the pupil is localized and removed by searching on the reduced iris area as explained in previous section, which prevents false detections resulted from noisy edge points in the eye image. The upper eyelashes regions are also deleted from iris region perfectly without affecting the iris valid pixels.

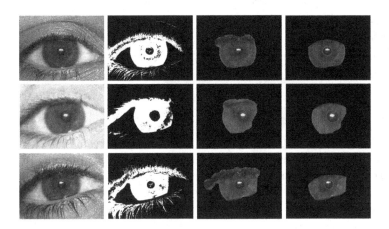

Fig. 6. Samples of correct segmented iris images using the proposed algorithm.

The proposed algorithm is compared with a set of well-known and state-of-the-art iris segmentation algorithms proposed by Daugman [7], Wildes [25], Libor Masek [10], and Sahmoud et al. [21]. We implemented Sahmoud et al. algorithm and used the public code of Daugman and Libor Masek algorithms that are available on Internet. The results of Wildes algorithm are taken from previous research that used the same database [19]. Table 2 shows a comparison between the proposed segmentation algorithm and the other four selected segmentation algorithms. The segmentation of each iris is considered to be correct if two conditions are satisfied. First, the iris and the pupil boundaries should be localized correctly. Second, the upper and the lower eyelids' regions should be correctly detected and removed from the iris region.

The results of Table 2 show that the proposed segmentation algorithm outperforms Daugman and Libor Masek algorithms, and it is competitive with Wildes and Sahmoud et al. algorithms. On the other hand, the proposed algorithm is the fastest one since it uses a fast thresholding mechanism and avoids the heavy and high computational techniques such as line and circular Hough transform as in other algorithms. The presented results confirm the efficiency of the proposed algorithm when dealing with unconstrained iris images, and emphasizes the effectiveness of the proposed multi-color components method that used in the first step of our algorithm. It is important to mention that our multi-color components method can be used independently in other research work as a pre-processing step to quickly determine the expected iris region. From our results, we noted that the proposed algorithm fails in segmenting irises with very dark

Table 2. Performance comparison using segmentation accuracy and average execution time in seconds.

Segmentation algorithm	Segmentation accuracy (%)	Average time of execution (s)
Libor Masek [10]	95.6	3.51
Daugman [7]	95.22	3.23
Wildes [25]	97.27	–
Sahmoud et al. [21]	97.5	2.14
Proposed	97.15	0.94

(as in dark brown irises) and very light (as in light blue irises) colors since their pixels usually fall outside the computed LCC and HCC values. To handle this issue, in one of our future works, we are planning to add an intelligent mechanism to our segmentation algorithm to be able to handle all color types of human irises.

5   Conclusion

In this paper, we proposed a novel and fast segmentation algorithm to deal with eye images acquired in visible wavelength and unconstrained environments. The proposed algorithm utilizes the color information of multiple color spaces. In the first step, the RGB, YCbCr, and HSV color spaces are analyzed based on their importance regarding the iris pixels, and an appropriate set of color components is selected for the segmentation process. To accurately localize the iris region, a set of convenient techniques are applied to detect and remove the non-iris regions such as pupil, eyelids, and eyelashes. Our experimental results and comparative analysis using the UBIRIS v2 database and four reference segmentation algorithms demonstrate the efficiency of our approach in terms of segmentation accuracy and low execution time.

References

1. Abdullah, M.A., Dlay, S.S., Woo, W.L., Chambers, J.A.: Robust iris segmentation method based on a new active contour force with a noncircular normalization. IEEE Trans. Syst. Man Cybern. Syst. **47**(12), 3128–3141 (2016)
2. Bao, P., Zhang, L., Wu, X.: Canny edge detection enhancement by scale multiplication. IEEE Trans. Pattern Anal. Mach. Intell. **27**(9), 1485–1490 (2005)
3. Bazrafkan, S., Thavalengal, S., Corcoran, P.: An end to end deep neural network for iris segmentation in unconstrained scenarios. Neural Netw. **106**, 79–95 (2018)
4. Bezerra, C.S., et al.: Robust iris segmentation based on fully convolutional networks and generative adversarial networks. In: 2018 31st SIBGRAPI Conference on Graphics, Patterns and Images (SIBGRAPI), pp. 281–288. IEEE (2018)

5. Chen, Y., Wang, W., Zeng, Z., Wang, Y.: An adaptive CNNs technology for robust iris segmentation. IEEE Access **7**, 64517–64532 (2019)
6. Chen, Y., Adjouadi, M., Barreto, A., Rishe, N., Andrian, J.: A computational efficient iris extraction approach in unconstrained environments. In: 2009 IEEE 3rd International Conference on Biometrics: Theory, Applications, and Systems, pp. 1–7. IEEE (2009)
7. Daugman, J.: How iris recognition works. In: The Essential Guide to Image Processing, pp. 715–739. Elsevier (2009)
8. Daugman, J.G.: High confidence visual recognition of persons by a test of statistical independence. IEEE Trans. Pattern Anal. Mach. Intell. **15**(11), 1148–1161 (1993)
9. Jain, A.K., Flynn, P., Ross, A.A.: Handbook of Biometrics. Springer, US (2007). https://doi.org/10.1007/978-0-387-71041-9
10. Masek, L., et al.: Recognition of human iris patterns for biometric identification. Ph.D. thesis, Master's thesis, University of Western Australia (2003)
11. Ng, R.Y.F., Tay, Y.H., Mok, K.M.: A review of iris recognition algorithms. In: 2008 International Symposium on Information Technology, vol. 2, pp. 1–7. IEEE (2008)
12. O'Gorman, L.: Comparing passwords, tokens, and biometrics for user authentication. Proc. IEEE **91**(12), 2021–2040 (2003)
13. Osorio-Roig, D., Rathgeb, C., Gomez-Barrero, M., Morales-González, A., Garea-Llano, E., Busch, C.: Visible wavelength iris segmentation: a multi-class approach using fully convolutional neuronal networks. In: 2018 International Conference of the Biometrics Special Interest Group (BIOSIG), pp. 1–5. IEEE (2018)
14. Pedersen, S.J.K.: Circular hough transform. Aalborg University, Vision, Graphics, and Interactive Systems, vol. 123, no. 6 (2007)
15. Phillips, P.J., Bowyer, K.W., Flynn, P.J.: Comments on the casia version 1.0 iris data set. IEEE Transactions on Pattern Analysis and Machine Intelligence 29(10), 1869–1870 (2007)
16. Proenca, H.: Iris recognition: on the segmentation of degraded images acquired in the visible wavelength. IEEE Trans. Pattern Anal. Mach. Intell. **32**(8), 1502–1516 (2009)
17. Proença, H.: Ocular biometrics by score-level fusion of disparate experts. IEEE Trans. Image Process. **23**(12), 5082–5093 (2014)
18. Proenca, H., Filipe, S., Santos, R., Oliveira, J., Alexandre, L.A.: The UBIRIS. v2: a database of visible wavelength iris images captured on-the-move and at-a-distance. IEEE Trans. Pattern Anal. Mach. Intell. **32**(8), 1529–1535 (2009)
19. Radman, A., Jumari, K., Zainal, N.: Fast and reliable iris segmentation algorithm. IET Image Process. **7**(1), 42–49 (2013)
20. Rapaka, S., Kumar, P.R.: Efficient approach for non-ideal iris segmentation using improved particle swarm optimisation-based multilevel thresholding and geodesic active contours. IET Image Process. **12**(10), 1721–1729 (2018)
21. Sahmoud, S.A., Abuhaiba, I.S.: Efficient iris segmentation method in unconstrained environments. Pattern Recogn. **46**(12), 3174–3185 (2013)
22. Sahmoud, S.A.I.: Enhancing Iris Recognition (2011)
23. Tan, C.W., Kumar, A.: Accurate iris recognition at a distance using stabilized iris encoding and zernike moments phase features. IEEE Trans. Image Process. **23**(9), 3962–3974 (2014)

24. Wan, H.L., Li, Z.C., Qiao, J.P., Li, B.S.: Non-ideal iris segmentation using anisotropic diffusion. IET Image Process. **7**(2), 111–120 (2013)
25. Wildes, R.P.: Iris recognition: an emerging biometric technology. Proc. IEEE **85**(9), 1348–1363 (1997)
26. Xu, Y., Chuang, T.C., Lai, S.H.: Deep neural networks for accurate iris recognition. In: 2017 4th IAPR Asian Conference on Pattern Recognition (ACPR), pp. 664–669. IEEE (2017)
27. Yang, Y., Shen, P., Chen, C.: A robust iris segmentation using fully convolutional network with dilated convolutions. In: 2018 IEEE International Symposium on Multimedia (ISM), pp. 9–16. IEEE (2018)

A Local Flow Phase Stretch Transform for Robust Retinal Vessel Detection

Mohsin Challoob[(⊠)] and Yongsheng Gao

School of Engineering, Griffith University, Brisbane, Australia
mohsin.challoob@griffithuni.edu.au,
yongsheng.gao@griffith.edu.au

Abstract. This paper presents a new method for reliably detecting retinal vessel tree using a local flow phase stretch transform (LF-PST). A local flow evaluator is proposed to increase the local contrast and the coherence of the local orientation of vessel tree. This is achieved by incorporating information about the local structure and direction of vessels, which is estimated by introducing a second curvature moment evaluation matrix (SCMEM). The SCMEM evaluates vessel patterns as only features having linearly coherent curvature. We present an oriented phase stretch transform to capture retinal vessels running at various diameters and directions. The proposed method exploits the phase angle of the transform, which includes structural features of lines and curved patterns. The LF-PST produces several phase maps, in which the vessel structure is characterized along various directions. To produce an orientation invariant response, all phases are linearly combined. The proposed method is tested on the publicly available DRIVE and IOSTAR databases with different imaging modalities and achieves encouraging segmentation results outperforming the state-of-the-art benchmark methods.

Keywords: Retinal vessel · Vessel detection · Local flow phase stretch transform

1 Introduction

The detection and examination of retinal vessel tree plays an essential role in clinical applications for the diagnosis of many eye diseases such as diabetic retinopathy [1, 2]. The manual annotation of vessel network by graders is a laborious and time-consuming process due to the complexity of vessel structure. Automatic vessel segmentation can considerably decrease workload, enabling high throughput process in clinics. However, the automatic segmentation of vessel tree continues to be a challenging task because of many factors including low intensity contrast between vessel and background, variation of vessel width, uneven background illumination, and the presence of noise and pathological regions [3, 4]. Many segmentation algorithms have been introduced for the detection of retinal vessel tree [3–14]. Nevertheless, several challenges and limitations remain unsolved, as follows. First, the retinal vessel tree possesses a wide range of widths, where many vessels, particularly thin ones, are with a weak and unstable local intensity contrast and be overwhelmed in inhomogeneous background. If more attention

© Springer Nature Switzerland AG 2020
J. Blanc-Talon et al. (Eds.): ACIVS 2020, LNCS 12002, pp. 251–261, 2020.
https://doi.org/10.1007/978-3-030-40605-9_22

is paid on the detection of the low contrast thin vessels, the wide vessels may be over-enhanced, and some false positive vessels may be generated. In contrast, when more emphasis is placed on wide vessels, poor contrast vessels are more likely to be lost. Secondly, besides vessel tree, a retinal image includes optic disc and may have pathological regions. Many of existing segmentation methods suffer from the false-positive detections due to the presence of such regions. Further, a very limited number of methods has been developed for detecting vessel tree in scanning laser ophthalmoscopy (SLO) image. The majority of the existing segmentation algorithms only work for detecting retinal vessels in fundus image but fails to perform when applied on SLO images. Detecting retinal vessels in a SLO image encounters multiple difficulties such as strong light reflex, high curvature changes, more background artefacts, and many dark regions because of low light exposure during the image acquisition, making the segmentation process more challenging.

In this paper, a novel segmentation method for retinal vessel detection using a local flow phase stretch transform (LF-PST) is presented. A local flow evaluator is introduced to perform as an enhancing filter for increasing the local contrast and the coherence of the local orientation of retinal vessel tree. This is obtained by guiding the partial differential derivatives of gaussian smoothed vessel map with the orientation estimated by a second curvature moment evaluation matrix. The phase stretch transform (PST) is adapted to be an orientation-sensitive method and is employed for the detection of retinal vessels. We utilize the phase angle of the oriented PST for its functionality of being a detector for features like curved structures, and its brightness level equalization property. The LF-PST detects retinal vessels over multiple orientations to capture different vessel widths, producing several directional vessel maps. The final vessel tree is computed by fusing all directional phases of LF-PSTs to obtain a rotationally invariant response. The proposed method is assessed on different imaging modalities. Encouraging experimental results demonstrate the ability of the proposed method to work on different imaging modalities (fundus and SLO). This validates the performance stability of the proposed method against the state-of-the-art algorithms, which tend to be degraded when applied on SLO images.

The rest of the paper is organized as follows. The proposed method is introduced and explained in detail in Sect. 2. In Sect. 3, experimental results are conducted, and compared with other methods, while the discussion and conclusion are presented in Sect. 4.

2   Proposed Method

In this paper, a new method for reliably detecting retinal vessel tree is introduced using a local flow phase stretch transform (LF-PST). First, the green channel of a retinal image to be analyzed is used and converted to grey scale image because of its better vessel-background intensity contrast. Then, the morphologically enhanced approach with disk-shaped structuring element [10] is applied to pre-process the grey scale image. The LF-PST is explained in detail as follows.

2.1 Local Flow Evaluator

A local flow evaluator (LFE) is introduced to provide vessel tree with reliable information about local shape and direction of vessels, which improves the local contrast of vessel structure and the coherence of its local orientation. To achieve this, the partial differential derivatives (PDDs) of gaussian smoothed vessels is guided with the orientation estimated by a proposed second curvature moment evaluation matrix (SCMEM). The SCMEM utilizes the second moment matrix (SMM) used in studies [15, 16] as a local descriptor for flow-like structures, and the curvature evaluation introduced in [17] for measuring vessel-like structure by a gaussian profile whose curvature differs along the crest line. Fusing the curvature evaluation with the SMM can lead to stabilizing the directional behavior of the LFE by making the proposed SCMEM immune to noise while more sensitive to changes in orientation, producing robust information about vessel topology and its orientation. To implement our LFE, let $I(i, j)$ be an input image of finite size $M \times N$, where $0 \leq i \leq M - 1$ and $0 \leq j \leq N - 1$, the proposed SCMEM is calculated as follows:

$$JC_\rho(\nabla I) = K_\rho * \left(\nabla I \nabla I^T\right) = \begin{bmatrix} K_\rho * \left(\frac{\partial I}{\partial x}\right)^2 & K_\rho * \frac{\partial I}{\partial x}\frac{\partial I}{\partial y} \\ K_\rho * \frac{\partial I}{\partial x}\frac{\partial I}{\partial y} & K_\rho * \left(\frac{\partial I}{\partial y}\right)^2 \end{bmatrix} \tag{1}$$

where JC_ρ denotes the SCMEM and K_ρ is the curvature evaluation represented by Laplacian of gaussian function with standard deviation $\rho = 5$, where the sign of the curvature is approximated by the sign of the Laplacian. The JC_ρ includes two orthonormal eigenvectors. The first vector $v_1 || \nabla I$ points to the direction of the gradient, whereas the second vector $v_2 || \nabla I^T$ refers to the direction of the level lines. In this paper, the second vector is normalized and only considered for estimating the local orientation of vessels. The second directional derivative V_{ww}, which is steered with orientation determined by JC_ρ, is calculated as:

$$V_{ww} = c^2 v_{xx} + cs v_{xy} + s^2 v_{yy} \tag{2}$$

where c and s are the first and second elements of the normalized eigenvector respectively, and v_{xx}, v_{xy} and v_{yy} are the second order derivatives of gaussian smoothed vessels with standard deviation $\sigma = 2.5$. The LFE output (Enh_{LF}) is defined as:

$$Enh_{LF}(i, j) = I(i, j) + H_c(V_{ww}(i, j)) * G(i, j) \tag{3}$$

where $G = |\nabla I|$, and $H_c()$ is a continuous ramp function defined as $H_c(V_{ww}) = -(1 + \tanh(V_{ww} - \varepsilon))$. The $\tanh()$ is the hyperbolic tangent function and the parameter ε controls the sensitivity of vessel detection. Larger values of ε make the vessel detection less sensitive, while smaller values increase the detection sensitivity (in our experiments it is set to 0.5).

2.2 Local Flow Phase Stretch Transform

Phase stretch transform (PST) is a physics-inspired approach emulating the propagation of light through a diffractive channel with warped dispersive characteristic, which was originally introduced as a photonic time stretch method in [18, 19]. It has recently been used for the enhancement of impaired images in [20] with an orientation-insensitive kernel as shown in Fig. 1. In this paper, the PST kernel is adapted to be an orientation-sensitive as illustrated in Fig. 2. The LF-PST applies a 2-D nonlinear phase kernel to an input data in frequency domain. The amount of the phase assigned to the input relies on frequency, where higher amount of phase is applied to higher frequency features of the input image. According to [20], we exploit two unique properties from the PST for the vessel detection. The first is that the output of phase angle is corresponding to the even-order derivatives of input data with weighting factors. The key is the even order image derivatives encode shape information about local structures, detecting different image features. Also, the weighting factor is related to the assigned frequency, which can be selected to emphasize various structures of interest. The other property is that the output phase is inversely related to the brightness level of the input image. This crucial characteristic makes the transform more effective for the detection of structures in dark regions by equalizing input image brightness. The output of the LF-PST, represented by $LFPST(i, j, \theta_d)$, is its phase in spatial domain and is defined as:

$$LFPST(i, j, \theta_d) = \angle \langle IFFT2\{PK(p, q, \theta_d).FFT2\{Enh_{LF}(i, j)\}\}\rangle \quad (4)$$

where $\angle.$ denotes the phase angle operator, $FFT2$ and $IFFT2$ are two-dimensional fast Fourier transform and its inverse, respectively, and $PK(p, q, \theta_d)$ is the phase kernel with 2-D frequency variables (p and q) and an orientation (θ_d). The $PK(p, q, \theta_d)$ is computed using a nonlinear frequency dependent phase $\varphi(p, q, \theta_d)$ as:

$$PK(p, q, \theta_d) = e^{j\varphi(p,q,\theta_d)} \quad (5)$$

$$\varphi(p, q, \theta_d) = \varphi_{polar}(r, \theta_p) = F(r)O(\theta_p) \quad (6)$$

$$F(r) = S_t . \frac{W.r.tan^{-1}(W.r) - (1/2).ln(1 + (W.r)^2)}{W.r_{max}.tan^{-1}(W.r_{max}) - (1/2).ln(1 + (W.r_{max})^2)} \quad (7)$$

$$O(\theta_p) = cos^2(\theta_T).H_s\left(\frac{\theta_T}{\pi} + \frac{1}{2}\right).H_s\left(\frac{1}{2} - \frac{\theta_T}{\pi}\right) \quad (8)$$

$$\theta_T = \left| tan^{-1}\left(\frac{sin\left(tan^{-1}\left(\frac{q}{p}\right)\right).cos(\theta_d) - cos\left(tan^{-1}\left(\frac{q}{p}\right)\right).sin(\theta_d)}{cos\left(tan^{-1}\left(\frac{q}{p}\right)\right). cos(\theta_d) + sin\left(tan^{-1}\left(\frac{q}{p}\right)\right).sin(\theta_d)}\right) \right| \quad (9)$$

where $F(r)$ and $O(\theta_p)$ are frequency and angular components of the phase kernel, respectively, $r = \sqrt{p^2 + q^2}$, and r_{max} is the maximum frequency. The $tan^{-1}()$, $sin()$, and $cos()$ are the inverse tangent, sine and cosine functions, respectively, whereas $ln()$ and H_s are the natural logarithm and heaviside step functions, respectively, and the parameters S_t and W are the strength and wrap of the phase profile. These parameters (S_t and W) control the amount of the phase applied to each frequency at the input image, and they are set to be unity, which have no impact on obtained results, making the amount of phase applied the main parameter in this paper. This in turn reduces the complexity in determining the LF-PST parameters. The angular component $O(\theta_p)$ is preforming as

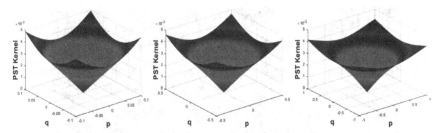

Fig. 1. The gradient of orientation-insensitive PST kernels with various frequencies (0.1, 0.5, 1, respectively).

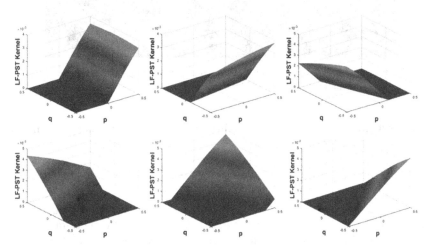

Fig. 2. The gradient of orientation-sensitive LF-PST kernels with various angles: first row (π, $\pi/6$, $2\pi/6$) and second row ($3\pi/6$, $4\pi/6$, $5\pi/6$), respectively.

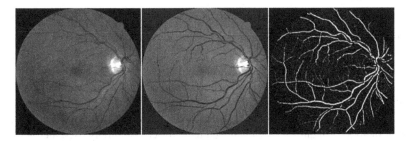

Fig. 3. Original image, LFE response, and final response of LF-PST, respectively.

a filter to determine the orientation for where the kernel of the LF-PST is applied, and this filter is designed by the rotation matrix and cosine function, which is with a fixed bandwidth of $\pi/2$ radius.

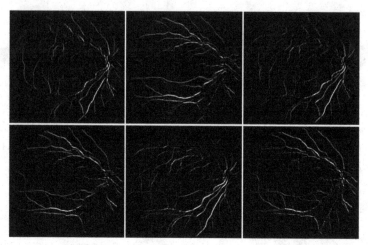

Fig. 4. Output phases of LF-PSTs with various orientations: first row with angles (π, $\pi/6$, $2\pi/6$) and second row with angles ($3\pi/6$, $4\pi/6$, $5\pi/6$), respectively.

The final vessel tree (F_{vess}), which is illustrated in Fig. 3, is produced by combining all phases of the LF-PSTs linearly as:

$$F_{vess}(i, j) = \sum_{\theta_d \in \Theta} LFPST(i, j, \theta_d) \tag{10}$$

where Θ denotes a set of angles with consideration: $\Theta = \{\pi, \pi/6, 2\pi/6, 3\pi/6, 4\pi/6, 5\pi/6\}$. Figure 4 demonstrates a set of directional vessel tree generated with Θ. As a postprocessing step, the local entropy thresholding approach is employed to threshold the vessel tree response (F_{vess}), where the co-occurrence matrix is used to compute the local entropy for obtaining an optimal threshold value.

3   Results

The performance of proposed method is evaluated on two publicly available DRIVE and IOSTAR databases. The DRIVE contains fundus images, and its test set (20 images) with ground truth A (1st_manual) is used for our experiments. The IOSTAR database consists of 30 images acquired with a scanning laser ophthalmoscopy (SLO) technology. The proposed method is quantitatively measured using three commonly used metrics: accuracy (ACC), specificity (SP) and sensitivity (SE). The proposed method is performed on a PC with Intel Core i5, 3.40 GHz, 8.00 GB RAM using MATLAB R2018b, consuming in average 3.9 s on a DRIVE image and 9.3 s on an IOSTAR image. The amount of the applied phase is experimentally examined as shown in Fig. 5 by using four frequency values (0.1, 0.5, 1, 1.5) on both databases. When a small frequency is used, more vessel pixels are detected, but detection results are more sensitive to noise and non-vessel interfaces. Also, the wide vessels, especially the ones with central light reflex, are falsely detected as two small vessels. In contrast, when a large frequency is assigned, more vessels are left undetected, close vessels are merged together, and wide vessels become dilated. Based on these experiments, the amount of phase applied to the DRIVE images is set to 0.5, while for the SLO images it is set to 1. Figure 6 illustrates the segmentation results obtained by the proposed algorithm.

The performance of the proposed method is compared with eleven state-of-the-art algorithms using the average values of the used metrics (see Table 1), including supervised methods [6, 7] and unsupervised ones [3–5, 8, 9, 11–14]. To have fair comparison, the proposed method is assessed against the methods that used the same ground truth as our method, so the results of paper [3] on the DRIVE are excluded from Table 1 due to the use of the second observer results as ground truth. The results of the methods [11–14] on the IOSTAR database are taken from the published results in [3] and the results of methods [6] and [11] on the DRIVE are from [9] and [21], respectively. The rest are obtained from the published results in their original papers. As illustrated in Table 1, the proposed method achieves encouraging results. For the IOSTAR database, the proposed method obtains the highest average accuracy and specificity against other methods. Also, it produces higher sensitivity than the methods in [11, 13]. For the DRIVE database, the proposed method produces higher accuracy than other algorithms except [5, 14], higher sensitivity than the methods of [5–7, 11, 13], and higher specificity than the methods of [4, 12, 13]. The proposed method is also qualitatively evaluated with two benchmark methods [17, 22] on an example image from the IOSTAR database (see Fig. 7). The benchmark methods enhance background region, making identifying vessel tree difficult, showing their weaknesses in the detection of retinal vessel tree in the presence of high curvature changes, more background artefacts, and large dark regions.

4 Discussion and Conclusion

In general, it is challenging to design a vessel detection method to be feasible across different imaging modalities. Also, a very few algorithms have been introduced in literature to extract retinal vessel tree from SLO images in comparison with those ones presented for fundus images, where the SLO image is generally associated with problems of strong light reflex, high curvature changes, and many dark regions. According to the achieved results, the proposed method demonstrates its reliability to detect different vessel width classes (wide, median and fine) and its stability to work on two dissimilar imaging modalities (fundus and SLO).

Also, the proposed method shows an effective performance over previously reported methods for segmenting SLO images by producing the highest accuracy and specificity. The segmentation results of many of previous reported methods are degraded when applied to the SLO image, as demonstrated in Table 1 and Fig. 7. The integration of the curvature evaluation with the SMM results in stabilizing the directional behavior of the SCMEM, making the LFE immune to noise while more sensitive to changes in orientation particularly in the regions with high curvature changes. This is achieved by providing reliable information about local shape and direction of vessels, which increases the local contrast of vessel structure and the coherence of its local orientation. This in turn leads to preserving more vessels especially the ones with low intensity contrast while reducing the detection of false positives. Also, since retinal vessels can present in

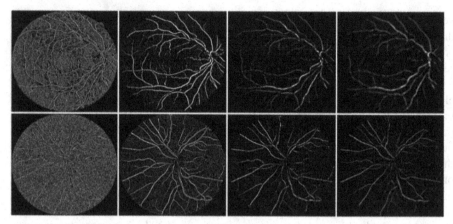

Fig. 5. The LF-PST responses. First row: four frequencies (0.1, 0.5,1,1.5) applied on a DRIVE image, respectively. Second row: the same frequencies applied on a SLO image, respectively.

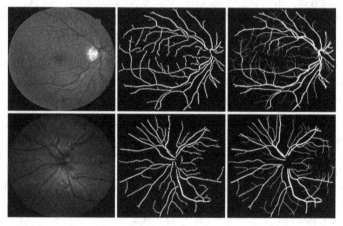

Fig. 6. First column: original images from DRIVE and IOSTAR datasets, respectively. Second column: segmentation results of the proposed method. Third column: ground truth

any orientation and be with various widths and lengths, the use of a set of the directional LF-PSTs whose outputs can be fusing linearly to cover the entire range of possible angles is necessary. This contributes to the detection of vessels with low contrast at different widths. The crucial properties of the PST of being a local shape detector and brightness level equalization makes the proposed method robust to detect vessels in the presence of high curvature changes, dim areas and inhomogeneous intensity. The LF-PST kernel includes a built-in logarithmic function, which equalizes the brightness in the input image by giving a higher gain for lower brightness and vice versa. The use of this function in the LF-PST improves the vessel detection results particularly in the SLO image, which has many dark regions. Also, the proposed method offers an advantage of the controllability over detection results via the amount of the applied phase in the

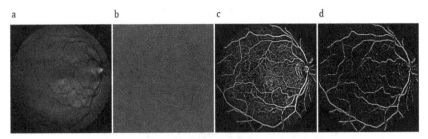

Fig. 7. Comparison of the proposed method against benchmark methods: (a) a randomly chosen IOSTAR image, (b) Zana [17], (c) Nguyen [22], (d) our method.

Table 1. Performance comparison of the proposed method against benchmark methods.

Database	IOSTAR			DRIVE		
Method	ACC	SP	SE	ACC	SP	SE
Zhao [3]	0.948	0.967	0.772	–	–	–
Zhang [4]	0.951	0.974	0.754	0.947	0.972	0.774
Sazak [5]	–	–	–	0.959	0.981	0.718
Soares [6]	–	–	–	0.944	0.976	0.723
Marín [7]	–	–	–	0.945	0.980	0.706
Lam [8]	–	–	–	0.947	–	–
Zhao [9]	–	–	–	0.947	0.978	0.735
Frangi [11]	0.920	0.931	0.704	0.933	0.975	0.646
Azzopardi [12]	0.941	0.967	0.761	0.944	0.970	0.765
Bankhead [13]	0.911	0.921	0.726	0.937	0.971	0.702
Zhao [14]	0.928	0.935	0.757	0.953	0.978	0.744
Proposed Method	**0.957**	**0.975**	**0.750**	**0.952**	**0.974**	**0.732**

LF-PST kernel with an angle, providing the proposed method with more flexibility to reduce false detection while retaining vessels, and to deal with the variability of vessel width. While we have shown the effectiveness of the proposed method on the detection of retinal vessels in two different imaging modalities, the LF-PST is also feasible for the detection of crack tree in pavement images, see Fig. 8, where encouraging results are achieved on pavement images suffering from the problems of noise, shadow, low contrast and occlusion.

In conclusion, this paper presents a new unsupervised method using the proposed local flow phase stretch transform (LF-PST) for the reliable detection of retinal vessel tree. The properties of the LF-PST of being an orientation-sensitive and a detector for line-like structures with its brightness equalization make its performance robust for the detection of retinal vessel tree in different imaging modalities, producing successful segmentation results, which outperforms the state-of-the-art benchmark algorithms.

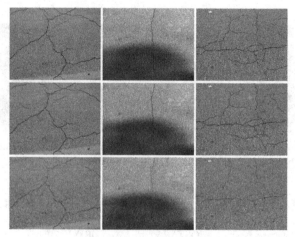

Fig. 8. First row: original pavement images taken from database in [23]. Second row: detection results of our method overlaid on original images with blue color. Third row: ground truths overlaid with red color.

However, the main drawback of the proposed method is that it still produces the false detection around optic disc area. The directions for future work will include: (1) the development of 3D- LF-PST as there is an increase in the use of 3D vessel detection in medical clinics; (2) the extension of the proposed method to detect other curvilinear structures such as vein patterns in leaf images, and roads in aerial images.

References

1. Bibiloni, P., González-Hidalgo, M., Massanet, S.: A survey on curvilinear object segmentation in multiple applications. Pattern Recogn. **60**, 949–970 (2016). https://doi.org/10.1016/j.patcog.2016.07.023
2. Annunziata, R., Garzelli, A., Ballerini, L., Mecocci, A., Trucco, E.: Leveraging multiscale Hessian-based enhancement with a novel exudate inpainting technique for retinal vessel segmentation. IEEE J. Biomed. Health Inf. **20**, 1129–1138 (2016). https://doi.org/10.1109/jbhi.2015.2440091
3. Zhao, Y., Zheng, Y., Liu, Y., Zhao, Y., Luo, L., Yang, S., Na, T., Wang, Y., Liu, J.: Automatic 2-D/3-D vessel enhancement in multiple modality images using a weighted symmetry filter. IEEE Trans. Med. Imaging **37**, 438–450 (2018). https://doi.org/10.1109/TMI.2017.2756073
4. Zhang, J., Dashtbozorg, B., Bekkers, E., Pluim, J., Duits, R., ter Haar Romeny, B.: Robust retinal vessel segmentation via locally adaptive derivative frames in orientation scores. IEEE Trans. Med. Imaging **35**, 2631–2644 (2016). https://doi.org/10.1109/TMI.2016.2587062
5. Sazak, Ç., Nelson, C., Obara, B.: The multiscale bowler-hat transform for blood vessel enhancement in retinal images. Pattern Recogn. **88**, 739–750 (2019). https://doi.org/10.1016/j.patcog.2018.10.011
6. Soares, J., Leandro, J., Cesar, R., Jelinek, H., Cree, M.: Retinal vessel segmentation using the 2-D Gabor wavelet and supervised classification. IEEE Trans. Med. Imaging **25**, 1214–1222 (2006). https://doi.org/10.1109/tmi.2006.879967

7. Marín, D., Aquino, A., Gegundez-Arias, M., Bravo, J.: A new supervised method for blood vessel segmentation in retinal images by using gray-level and moment invariants-based features. IEEE Trans. Med. Imaging **30**, 146–158 (2011). https://doi.org/10.1109/tmi.2010.2064333

8. Lam, B., Gao, Y., Liew, A.: General retinal vessel segmentation using regularization-based multiconcavity modeling. IEEE Trans. Med. Imaging **29**, 1369–1381 (2010). https://doi.org/10.1109/tmi.2010.2043259

9. Zhao, Y., Wang, X., Wang, X., Shih, F.: Retinal vessels segmentation based on level set and region growing. Pattern Recogn. **47**, 2437–2446 (2014). https://doi.org/10.1016/j.patcog.2014.01.006

10. Challoob, M., Gao, Y.: Retinal vessel segmentation using matched filter with joint relative entropy. In: International Conference on Computer Analysis of Images and Patterns, pp. 228–239. Springer (2017). https://doi.org/10.1007/978-3-319-64689-3_19

11. Frangi, A., Niessen, W., Vincken, K., Viergever, M.: Multiscale vessel enhancement filtering. In: International Conference on Medical Image Computing and Computer-Assisted Intervention, pp. 130–137. Springer (1998). https://doi.org/10.1007/bfb0056195

12. Azzopardi, G., Strisciuglio, N., Vento, M., Petkov, N.: Trainable COSFIRE filters for vessel delineation with application to retinal images. Med. Image Anal. **19**, 46–57 (2015). https://doi.org/10.1016/j.media.2014.08.002

13. Bankhead, P., Scholfield, C., McGeown, J., Curtis, T.: Fast retinal vessel detection and measurement using wavelets and edge location refinement. PLoS ONE **7**, e32435 (2012). https://doi.org/10.1371/journal.pone.0032435

14. Zhao, Y., Liu, Y., Wu, X., Harding, S., Zheng, Y.: Retinal vessel segmentation: an efficient graph cut approach with retinex and local phase. PLoS ONE **10**, e0122332 (2015). https://doi.org/10.1371/journal.pone.0122332

15. Vicas, C., Nedevschi, S.: Detecting curvilinear features using structure tensors. IEEE Trans. Image Process. **24**, 3874–3887 (2015). https://doi.org/10.1109/tip.2015.2447451

16. Weickert, J.: Coherence-enhancing shock filters. In: Joint Pattern Recognition Symposium, pp. 1–8. Springer (2003). https://doi.org/10.1007/978-3-540-45243-0_1

17. Zana, F., Klein, J.: Segmentation of vessel-like patterns using mathematical morphology and curvature evaluation. IEEE Trans. Image Process. **10**, 1010–1019 (2001). https://doi.org/10.1109/83.931095

18. Bhushan, A., Coppinger, F., Jalali, B.: Time-stretched analogue-to-digital conversion. Electron. Lett. **34**, 839–841 (1998). https://doi.org/10.1049/el:19980629

19. Han, Y., Jalali, B.: Photonic time-stretched analog-to-digital converter: fundamental concepts and practical considerations. J. Lightwave Technol. **21**, 3085–3103 (2003). https://doi.org/10.1109/jlt.2003.821731

20. Suthar, M., Asghari, H., Jalali, B.: Feature enhancement in visually impaired images. IEEE Access. **6**, 1407–1415 (2018). https://doi.org/10.1109/access.2017.2779107

21. Cheng, E., Du, L., Wu, Y., Zhu, Y., Megalooikonomou, V., Ling, H.: Discriminative vessel segmentation in retinal images by fusing context-aware hybrid features. Mach. Vis. Appl. **25**, 1779–1792 (2014). https://doi.org/10.1007/s00138-014-0638-x

22. Nguyen, U., Bhuiyan, A., Park, L., Ramamohanarao, K.: An effective retinal blood vessel segmentation method using multi-scale line detection. Pattern Recogn. **46**, 703–715 (2013). https://doi.org/10.1016/j.patcog.2012.08.009

23. Zou, Q., Cao, Y., Li, Q., Mao, Q., Wang, S.: CrackTree: automatic crack detection from pavement images. Pattern Recogn. Lett. **33**, 227–238 (2012). https://doi.org/10.1016/j.patrec.2011.11.004

Evaluation of Unconditioned Deep Generative Synthesis of Retinal Images

Sinan Kaplan[1], Lasse Lensu[1]([envelope]) [iD], Lauri Laaksonen[1], and Hannu Uusitalo[2] [iD]

[1] Computer Vision and Pattern Recognition Laboratory, Lappeenranta-Lahti University of Technology LUT, P.O. Box 20, 53850 Lappeenranta, Finland
`lasse.lensu@lut.fi`
[2] Department of Ophthalmology, Faculty of Health and Biotechnology, Tampere University and Tays Eye Center, Tampere, Finland

Abstract. Retinal images have been increasingly important in clinical diagnostics of several eye and systemic diseases. To help the medical doctors in this work, automatic and semi-automatic diagnosis methods can be used to increase the efficiency of diagnostic and follow-up processes, as well as enable wider disease screening programs. However, the training of advanced machine learning methods for improved retinal image analysis typically requires large and representative retinal image data sets. Even when large data sets of retinal images are available, the occurrence of different medical conditions is unbalanced in them. Hence, there is a need to enrich the existing data sets by data augmentation and introducing noise that is essential to build robust and reliable machine learning models. One way to overcome these shortcomings relies on generative models for synthesizing images. To study the limits of retinal image synthesis, this paper focuses on the deep generative models including a generative adversarial network and a variational autoencoder to synthesize images from noise without conditioning on any information regarding to the retina. The models are trained with the Kaggle EyePACS retinal image set, and for quantifying the image quality in a no-reference manner, the generated images are compared with the retinal images of the DiaRetDB1 database using common similarity metrics.

Keywords: Deep generative model · Generative adversarial network · Variational autoencoder · Retinal image

1 Introduction

The unique structure of the eye makes it possible to noninvasively image the retina, a part of our central nervous system. This opportunity and the known relation of retinal changes to several eye, neuronal and systemic diseases makes it attractive for the researchers to study the images for screening and diagnosing diseases like diabetic retinopathy, age-related macular degeneration glaucoma, cerebrovascular and cardiovascular diseases, multiple sclerosis and Alzheimer's

© Springer Nature Switzerland AG 2020
J. Blanc-Talon et al. (Eds.): ACIVS 2020, LNCS 12002, pp. 262–273, 2020.
https://doi.org/10.1007/978-3-030-40605-9_23

disease [21]. The diagnosis of abnormal medical conditions is mainly based on the visual evaluation of the signs typical for the disease. The abnormalities are traditionally analyzed by experts giving their statements based on their education and expertise. This work is time consuming and often subjected to variations between the judgments of individual experts. In addition, quantitative changes in these images are usually more difficult to determine and hard to grade accurately. Therefore, automatic image processing methods are a well-motivated option to help an expert's work and for enabling more accurate quantification of the changes, follow-ups of progressive diseases and high-throughput screening programs [26].

The relevant information about the aforementioned diseases can be gathered from red-green-blue (RGB) images of the retina [21]. While the development in the field of retinal image analysis improves with the help of advances in technology, the demand for retinal image data increases. To study and detect possible abnormalities connected with eye and systemic diseases, the availability of retinal data is crucial to utilize machine learning methods. Although there are publicly available retinal image data sets provided by research institutes and hospitals, these data sets suffer from unbalanced occurrences of abnormalities (related to medical conditions or symptoms visible on the images) and inaccurate annotations.

As a remedy for the lacking retinal image data sets, there is considerable potential in image generation for further development of retinal image analysis methods. While natural medical data is subject to regulations related to the acquisition, storing and anonymization of the medical data, synthetic data avoids the ethical concerns and can be shared and published without privacy issues. In addition, the generated data can be used to improve the performance of baseline supervised methods for lesion detection and image segmentation. This can be achieved by generating synthetic images for a particular context by introducing noise to the machine learning methods in order to build robust models. This way it is be possible to increase the balanced heterogeneity and size of the data sets.

An important approach in terms of generating retinal images is to apply generative models. In this paper, deep generative models are studied and employed on the Kaggle EyePACS retinal image data set to generate synthetic retinal images. The aim is (1) to apply deep generative models including a generative adversarial network (GAN) and a variational autoencoder (VAE) for generating synthetic retinal images without conditioning on extra information and (2) to utilize a similarity-based statistical image quality assessment method for quantitative evaluation of the generated retinal images. The work is based on the master's thesis project of the first author [19].

2   Retinal Image Generation and Deep Generative Models

Related studies for retinal image generation include can be divided into traditional and deep learning based approaches. In [12], the aim was to reconstruct "only" the vessel tree. In [2,14], the focus was on reconstructing the retina by preserving the major features of the retina: the optic disk, vessel tree, and fovea. The former approach produced unrealistic vascular network structures, thus, further development in the latter succeeded to reconstruct more realistic retinal images while preserving the vascular structure.

One of the recent and most interesting developments in deep learning research [9,23] is related to deep generative models. GANs [16] and VAEs [22] are such generative models. Related to retinal image generation, [1] presents a two-stage GAN for first generating label images for the retinal features and then the actual images. In [4], the focus is on AMD and generation of images as a proxy between humans and the automatic methods. The previous studies in this context focused on generating retinal image by conditioning on vessel trees [1,4,7]. In this paper, however, GANs and VAEs are compared to study the limits of deep generative models for synthesizing retinal images without conditioning on any extra information.

2.1 Generative Adversarial Network

In 2014, Goodfellow et al. [16] proposed an adversarial network framework as an alternative generative model estimation process for deep generative networks, called GAN. The main principle of a GAN is that there are two neural networks called the generator (G) and the discriminator (D) that compete to maximize their gains. The main goal for both networks is to improve their capabilities to generate and discriminate the data. In this context, G draws samples from a random noise distribution and D discriminates whether the samples are drawn from G or from real data (the training data).

2.2 Variational Autoencoder

Different from the GANs, there is a probabilistic graphical deep generative model proposed by [22] or Rezende et al. [29] (combination of deep neural networks and Bayesian inference models) to generate data samples from a latent space representation. This generative directed graphical model is called a VAE.

In a VAE, the latent space is used to draw samples from a probability density function (PDF) and these samples are used to generate new data. The latent space primarily provides the information about the underlying hidden structure inside the data [18].

A VAE consists of two primary blocks that are the *Encoder* and the *Decoder*. These building blocks are formed from a multilayer perceptron (MLP) to learn the parameters of the latent space.

2.3 Retinal Image Quality Assessment

There is ambiguity regarding how to quantitatively evaluate the quality of data generated by deep generative models [30]. This is mostly because the reference for each image does not exist, which makes the evaluation correspond with reduced-reference or no-reference image quality evaluation. The studies conducted with GANs and VAEs have suggested that the quality of the generated images can be evaluated by performing classification, segmentation and feature extraction methods within the problem domain [3,5,16,22,27,31,32]. Particularly, Inception score, Fréchet inception distance (FID) and sliced Wasserstein distance (SWD) have been commonly used as quantitative metrics of the generated images. In this paper, we applied FID and SWD to measure the quality of the generated retinal images. FID computes the Wasserstein-2 distance between the distributions of values in the embedding layer for the real and fake samples by using a trained classifier. In this paper, we used Inception net [3] trained with the samples from the Kaggle EyePACS set. SWD is a measure of the cost of transforming one distribution to another one using a given cost function [3].

In addition to FID and SWD, we propose a statistical parameters based image quality assessment method. The method is originated from the study in [13] by taking into account the structure of the generated retinal images. The parameters are used to measure the similarity between the generated images and a high-quality retinal image set. For this purpose, DiaRetDB1 [20] is considered as a benchmark. In the scope of this paper, by considering preceding research on retinal images [8], *mean, variance, entropy, skewness* and *kurtosis* are used to extract the statistical features from the retinal images for the quality assessment.

3   Experiments and Results

3.1   Data Sets

We employed two different data sets for conducting the experiments. The Kaggle EyePACS retinal data set [11] was used for model training, whereas DiaRetDB1 [20] was used for the quality assessment of the generated retinal images.The Kaggle EyePACS set consists of 35,126 images with varying resolutions. DiaRetDB1 set contains 89 high resolution images - W $\times$ H (1500×1152). Examples from both sets are shown in Fig. 1.

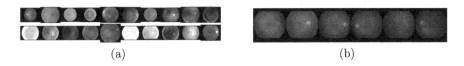

(a)   (b)

Fig. 1. Example images: the Kaggle EyePACS data set used for training; DiaRetDB1 data set used for quality evaluation.

3.2   Architectural Details of Deep Generative Models

This section covers the proposed architectural designs of the GAN and the VAE for the experiments. The implementation of the methods, the associated hyper-parameters for each model and the extra figures are accessible on Github[1].

Architecture of the Generative Adversarial Network. By having [10,15, 24,28,28] as a guideline, we propose a GAN architecture seen in Figs. 2 and 3. The detailed description of the architecture is given in [19].

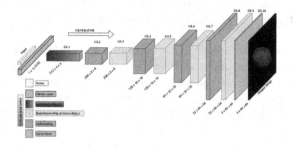

Fig. 2. Architecture of the proposed generator network which is a part of the generative adversarial network, based on convolutional neural network units with their parameters including the number of layers and their dimensions. GL stands for generative layer.

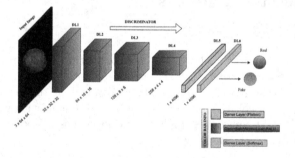

Fig. 3. Architecture of the proposed discriminator network which is a part of the generative adversarial network, based on convolutional neural network units with their parameters including the number of layers and their dimensions. DL stands for discriminator layer.

Architecture of the Variational Autoencoder. The VAE architecture is rather simple and it is based on a MLPs as seen in Fig. 4. By taking [22,29] as the basis, both the encoder and the decoder blocks are designed to have only one hidden layer with 512 neurons in total. The encoder block takes an image ($3 \times 64 \times 64$) from the training set and encodes it into the latent space

---

[1] The source code and the materials: https://github.com/kaplansinan/MasterThesis.

by learning the mean μ and variance σ^2. Similarly, the decoder samples a noise vector z from the normal distribution with the learned mean μ and variance σ^2 and reconstructs it as an image with the dimensions of $(3 \times 64 \times 64)$.

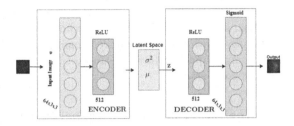

Fig. 4. The proposed variational autoencoder architecture based on multilayer perceptrons which are composed by a single hidden layer both in the encoder and decoder.

3.3 Retinal Image Generation with the Generative Adversarial Network

Examples of the generated retinal images from the proposed GAN are demonstrated in Fig. 5(a). While qualitatively analyzing the generated images, one can see that the network is able to capture the global structure of the retina, including the *shape, optic disk, fovea, and macula.* The generated images share same diversity in terms of color as in the Kaggle EyePACS set (refer to Fig. 1). However, detailed structures of the retina such as vessel trees are not generated clearly.

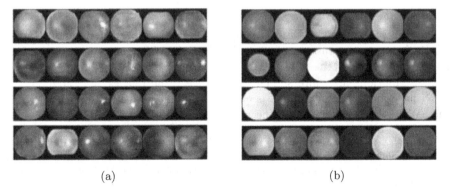

(a)   (b)

Fig. 5. Randomly chosen examples of the generated retinal images: (a) generative adversarial network, (b) variational autoencoder.

3.4   Retinal Image Generation with the Variational Autoencoder

Examples of the generated retinal images with the VAE can be seen in Fig. 5(b). As in the case of the GAN, the VAE also generates the retinal images by preserving the global structure of the retina that contains an optic disk, fovea, macula and the overall shape of the retina. Yet, the VAE is also unable to generate the vessel trees as a local structure of the retina. Furthermore, the VAE mostly generates the retinal images with the dominant red channel. In addition to that, the generated retinal images appear blurry because of the mean squared error measuring the pixel-to-pixel distance.

3.5   Quantitative Analysis of Generated Retinal Images

In order to assess the quality of the generated retinal images quantitatively, we follow the approach presented in Sect. 2.3. For the quality assessment, we present both FID and SWD metrics.

Table 1 shows that the visual quality of the generated images by the VAE is higher than the images generated by the GAN. As it is noted in the studies [3,31], there is a negative correlation between the FID and the visual quality of the generated images.

Table 1. Fréchet inception distance assessment between the Kaggle EyePACS set and the generated retinal images both from the generative adversarial network (GAN) and variational autoencoder (VAE), and DiaRetDB1

	GAN	VAE	DiaRetDB1
EyePACS	161.202	125.00	130.45

Table 2 presented the similarity of distributions represented by the SWD metric between the subsets of The Kaggle EyePACS images and the retinal images generated by the GAN and VAE. To note that SWD is lower when the two distributions are more similar. In this case, the images generated by the GAN have more diversity which is closer to the one of the real data set. Thus, the SWD is lower compared to the images generated by the VAE.

Table 2. Sliced Wasserstein distance between the Kaggle EyePACS subsets and the images generated by the generative adversarial network (GAN) and variational autoencoder (VAE).

	Subset1	Subset2	Subset3	Subset4	Subset5
GAN	0.045	0.048	0.043	0.053	0.053
VAE	0.13	0.127	0.17	0.15	0.134

In terms of the evaluation based on the statistical features, the mean, variance, skewness, kurtosis, and entropy were determined for the benchmark DiaRetDB1, the retinal images generated via the GAN and VAE, and five randomly selected subsets of the Kaggle EyePACS set. Each set contains 89 retinal images in total as the DiaRetDB1 set consists of 89 images. Once the features were extracted, the next step was to compute the cosine similarity between the features of (1) DiaRetDB1 and the retinal images generated by the GAN, (2) DiaRetDB1 and the retinal images generated by the VAE, and (3) DiaRetDB1 and the subset of retinal images from the Kaggle EyePACS set. The analysis of each statistical feature of the retinal images generated by the GAN and VAE is given in Fig. 6. As can be seen in the figure, the similarity between the retinal images generated by the VAE and DiaRetDB1 is higher than the similarity between the retinal images generated by the GAN and DiaRetDB1 in terms of the mean, variance, skewness and kurtosis. This can be explained by the varying colors of the retinal images generated by the GAN, which is opposite of the retinal images in DiaRetDB1 as it contains mostly retinal images with the dominant red channel. However, in the case of entropy values, the results are quite different from the aforementioned cases as seen in Fig. 6(i–j). It can be explained by the fact that the pixel values in each individual image of each set have low entropy because of the uniformity in the colors. This causes the entropy values between each data set to have similar values, thereby high similarity between entropy values.

In addition to the figures, related figures for the other five subsets from the The Kaggle EyePACS can be accessed together with the histogram analysis of each statistical feature of each set on Github[2]. The subsets are chosen to cross validate variations within The Kaggle EyePACS data set.

4 Discussion

Availability of synthetic retinal images is crucial for further development and validation of retinal image analysis methods. To generate retinal images from existing retinal image sources, methods based on deep learning including deep generative models offer potential options.

As an application of the deep generative models, a GAN and a VAE were chosen to generate synthetic retinal images. During the training process of both the GAN and VAE, we explored the following outcomes for the retinal image generation: (1) The overall structure of the retina is successfully generated by applying the GAN and VAE. However, neither one of them is able to model the vessel tree structure clearly. (2) The GAN generates the retinal images with distinctive colors as in the training set whereas the VAE captures only the dominant red channel. This is because of the constraints in the VAE in which the data is generated from the normal distributions of retinal images, the GAN generates sharp retinal images, whereas the VAE generates blurry images because

---

[2] The source code and the materials: https://github.com/kaplansinan/MasterThesis.

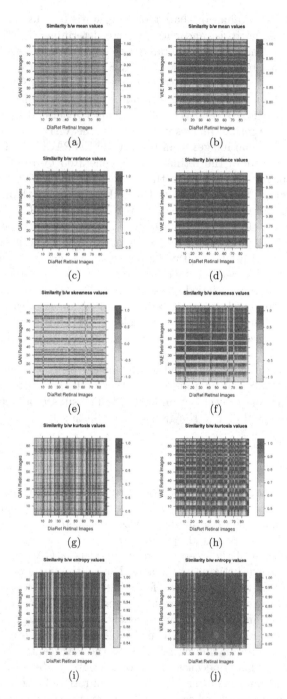

Fig. 6. Comparison of similarity measurements b/w the GAN and DiaRetDB1 and VAE: (a-b) Mean; (c-d) Variance; (e-f) Skewness; (g-h) Kurtosis and (i-j) Entropy.

of the mean squared error used to compute the pixel-to-pixel distance between the generated and the actual retinal images.

As the GAN and VAE generate retinal images with a global structure, the image quality evaluation was based on the global statistical features including the mean, variance, kurtosis, skewness, and entropy. The quantitative analysis of the generated images by the proposed similarity based quality assessment method reveals the following outcome: the similarity between the generated retinal images by the VAE and the benchmark data set is higher than the generated retinal images by the GAN and randomly chosen subsets from the Kaggle EyePACS set. This is because of the fact that the VAE often generates retinal images with the dominant red channel.

By considering the quality of the generated retinal images from the application perspective, using these images can be considered as introducing noise to a supervised learning algorithm for lesion detection and image segmentation. Another possible scenario where these images can be applied is that they can be used for generating vessel tree structures. To do so, one can use these generated images as an additional conditioning input information.

The recent studies [6,17,25] have shown the potential of combining GANs and VAEs together to synthesize images. Therefore, the retinal image synthesis can be studied to see whether the combination of a GAN and a VAE is capable of generating the vascular tree structure in an unconditioned way. Also, GANs and VAEs can be applied to retinal image generation tasks by conditioning the models on features of retinal images specific for the subsequent use.

5   Conclusion

This paper investigates the utilization of deep generative models for the retinal image generation in an unconditioned manner. In the experimental analysis, we showed that the overall architecture of the retina can be synthesized by applying unconditioned deep generative models. However, the studied deep generative models were unable to generate detailed structures of the retina such as the vascular tree. The proposed quality assessment method revealed that the overall similarity between the retinal images generated by the VAE and the benchmark set (DiaRetDB1) was higher than the similarity between the retinal images generated by the GAN and the benchmark set (DiaRetDB1). The generated images reveal that without conditioning on any additional data, the studied deep generative models are only able to synthesize retinal images with the dominant visible parts in the retina like the fovea and the optic disk. This shows the lack of generalization capability of these models and their limited learning process in which they tend to learn only the dominant features in the data.

References

1. Andreini, P., Bonechi, S., Bianchini, M., Mecocci, A., Scarselli, F., Sodi, A.: A two stage GAN for high resolution retinal image generation and segmentation. Technical report, University of Siena, July 2019. https://arxiv.org/abs/1907.12296v1

2. Bonaldi, L., Menti, E., Ballerini, L., Ruggeri, A., Trucco, E.: Automatic generation of synthetic retinal fundus images: vascular network. Procedia Comput. Sci. **90**, 54–60 (2016)
3. Borji, A.: Pros and cons of GAN evaluation measures. arXiv preprint arXiv:1802.03446 (2018)
4. Burlina, P.M., Joshi, N., Pacheco, K.D., Liu, T.Y.A., Bressler, N.M.: Assessment of deep generative models for high-resolution synthetic retinal image generation of age-related macular degeneration. JAMA Ophthalmol. **137**(3), 258–264 (2019). https://doi.org/10.1001/jamaophthalmol.2018.6156, https://jamanetwork. com/journals/jamaophthalmology/fullarticle/2720489
5. Burt, P., Adelson, E.: The laplacian pyramid as a compact image code. IEEE Trans. Commun. **31**(4), 532–540 (1983)
6. Choi, E., Biswal, S., Malin, B., Duke, J., Stewart, W.F., Sun, J.: Generating multi-label discrete electronic health records using generative adversarial networks. arXiv preprint arXiv:1703.06490 (2017)
7. Costa, P., et al.: Towards adversarial retinal image synthesis. arXiv preprint arXiv:1701.08974 (2017)
8. Davis, H., Russell, S., Barriga, E., Abramoff, M., Soliz, P.: Vision-based, real-time retinal image quality assessment. In: 22nd IEEE International Symposium on Computer-Based Medical Systems, CBMS 2009, pp. 1–6. IEEE (2009)
9. Deng, L., Yu, D., et al.: Deep learning: methods and applications. Found. Trends® Sig. Process. **7**(3–4), 197–387 (2014)
10. Denton, E.L., Chintala, S., Fergus, R., et al.: Deep generative image models using a laplacian pyramid of adversarial networks. In: Advances in Neural Information Processing Systems, pp. 1486–1494 (2015)
11. Diabetic retinopathy detection — kaggle, May 2017. https://www.kaggle.com/c/ diabetic-retinopathy-detection
12. Fang, L., et al.: Fast acquisition and reconstruction of optical coherence tomography images via sparse representation. IEEE Trans. Med. Imaging **32**(11), 2034–2049 (2013)
13. Fasih, M.: Retinal image quality assessment using supervised classification. Ph.D. thesis, École Polytechnique de Montréal (2014)
14. Fiorini, S., Ballerini, L., Trucco, E., Ruggeri, A.: Automatic generation of synthetic retinal fundus images. In: Eurographics Italian Chapter Conference, pp. 41–44 (2014)
15. Gauthier, J.: Conditional generative adversarial nets for convolutional face generation. Class Project for Stanford CS231N: Convolutional Neural Networks for Visual Recognition, Winter Semester 2014, 5 (2014)
16. Goodfellow, I., et al.: Generative adversarial nets. In: Advances in Neural Information Processing Systems, pp. 2672–2680 (2014)
17. Gorijala, M., Dukkipati, A.: Image generation and editing with variational info generative adversarial networks. arXiv preprint arXiv:1701.04568 (2017)
18. Hoff, P.D., Raftery, A.E., Handcock, M.S.: Latent space approaches to social network analysis. J. Am. Stat. Assoc. **97**(460), 1090–1098 (2002)
19. Kaplan, S.: Deep generative models for synthetic retinal image generation. Master's thesis, Lappeenranta University of Technology (2017). http://urn.fi/URN:NBN:fi-fe201708047855
20. Kauppi, T., et al.: Diaretdb1-standard diabetic retino-pathy database. IMAGERET Optimal Detection and Decision-Support Diagnosis of Diabetic Retinopathy (2007)

21. Keane, P.A., Sadda, S.R.: Retinal imaging in the twenty-first century: state of the art and future directions. Ophthalmology **121**(12), 2489–2500 (2014)
22. Kingma, D.P., Welling, M.: Auto-encoding variational bayes. arXiv preprint arXiv:1312.6114 (2013)
23. LeCun, Y., Bengio, Y., Hinton, G.: Deep learning. Nature **521**(7553), 436–444 (2015)
24. Ledig, C., et al.: Photo-realistic single image super-resolution using a generative adversarial network. arXiv preprint arXiv:1609.04802 (2016)
25. Makhzani, A., Shlens, J., Jaitly, N., Goodfellow, I., Frey, B.: Adversarial autoencoders. arXiv preprint arXiv:1511.05644 (2015)
26. Patton, N., et al.: Retinal image analysis: concepts,applications and potential. Prog. Retin. Eye Res. **25**(1), 99–127 (2006). https://doi.org/10.1016/j.preteyeres.2005.07.001, https://linkinghub.elsevier.com/retrieve/pii/S1350946205000406, 00587
27. Pu, Y., et al.: Variational autoencoder for deep learning of images, labels and captions. In: Advances in Neural Information Processing Systems, pp. 2352–2360 (2016)
28. Radford, A., Metz, L., Chintala, S.: Unsupervised representation learning with deep convolutional generative adversarial networks. arXiv preprint arXiv:1511.06434 (2015)
29. Rezende, D.J., Mohamed, S., Wierstra, D.: Stochastic backpropagation and approximate inference in deep generative models. arXiv preprint arXiv:1401.4082 (2014)
30. Theis, L., Oord, A.V.D., Bethge, M.: A note on the evaluation of generative models. arXiv preprint arXiv:1511.01844 (2015)
31. Vertolli, M.O., Davies, J.: Image quality assessment techniques show improved training and evaluation of autoencoder generative adversarial networks. arXiv preprint arXiv:1708.02237 (2017)
32. Yin, W., Fu, Y., Sigal, L., Xue, X.: Semi-latent GAN: learning to generate and modify facial images from attributes. arXiv preprint arXiv:1704.02166 (2017)

Image Analysis

Dynamic Texture Representation Based on Hierarchical Local Patterns

Thanh Tuan Nguyen[1,2(✉)], Thanh Phuong Nguyen[1], and Frédéric Bouchara[1]

[1] Université de Toulon, Aix Marseille Université, CNRS, LIS, Marseille, France
[2] Faculty of IT, HCMC University of Technology and Education,
HCM City, Vietnam
tuannt@hcmute.edu.vn

Abstract. A novel effective operator, named HIerarchical LOcal Pattern (HILOP), is proposed to efficiently exploit relationships of local neighbors at a pair of adjacent hierarchical regions which are located around a center pixel of a textural image. Instead of being thresholded by the value of the central pixel as usual, the gray-scale of a local neighbor in a hierarchical area is compared to that of all neighbors in the other region. In order to capture shape and motion cues for dynamic texture (DT) representation, HILOP is taken into account investigating hierarchical relationships in plane-images of a DT sequence. The obtained histograms are then concatenated to form a robust descriptor with high performance for DT classification task. Experimental results on various benchmark datasets have validated the interest of our proposal.

Keywords: Dynamic texture · Hierarchical local pattern · Hierarchical encoding · LBP · Video representation

1 Introduction

Efficiently encoding dynamic textures (i.e., textural structures repeated in a temporal domain) is a decisive task of various applications in computer vision, e.g., facial expressions [37,47], tracking objects [11,18,43], fire and smoke detection [6], etc. To this end, many approaches have been proposed for DT representation in which the main problems (e.g., turbulent motions, noise, illumination, etc.) are addressed in order to improve the discrimination power in DT recognition. These approaches are roughly grouped into the following categories. First, *optical-flow-based methods* [10,24,28,30] are mainly based on direction properties of normal flow to effectively capture the turbulent motion characteristics of DTs in sequences. In the meantime, *model-based methods* [7,17,32,44] have utilized Linear Dynamical System (LDS) [38] and its variants in order to deal with the complication of chaotic motions (e.g., turbulent water) and camera moving features (e.g., panning, zooming, and rotations). On the other side, filter bank techniques are exploited by *filter-based methods* to reduce the negative impacts of illumination and noise on video representation [3,37]. Motivated by geometry theory, *geometry-based methods* estimate self-similarity features using fractal analysis to fight against problems of environmental changes,

© Springer Nature Switzerland AG 2020
J. Blanc-Talon et al. (Eds.): ACIVS 2020, LNCS 12002, pp. 277–289, 2020.
https://doi.org/10.1007/978-3-030-40605-9_24

such as Dynamic Fractal Spectrum (DFS) [46], Multi-Fractal Spectrum (MFS) [45], Wavelet-based MFS [15], Spatio-Temporal Lacunarity Spectrum (STLS) [35]. Recently, *learning-based methods* are interested in, particularly deep learning techniques thanks to their high accuracy in classifying DTs. Two trends of those can be listed as follows: (i) exploiting Convolutional Neural Network (CNN) for capturing deep features [1,2,31]; (ii) using kernel sparse coding for learning featured dictionaries for DT description [33,34]. In the meanwhile, *local-feature-based methods* have also achieved promising rates with simple and efficient computation. They have principally addressed Local Binary Pattern (LBP) operator [27] and its variants to analyze videos. Two main techniques of those are mostly prompted for video description as follows: Volume LBP (VLBP) [47] for encoding dynamical features in consideration of spatio-temporal relationships on three consecutive frames; and LBP-TOP [47] for capturing motion and shape clues by using LBP on three orthogonal planes of sequences.

In consideration of gray-level differences between a center pixel and its local neighbors, LBP-based variants have acquired the promising rates on DT classification. However, they have remained several limitations, such as sensitivity to noise, near uniform regions [21,40], and large dimension [36,39,47]. Addressing those obstacles, we introduce in this work a novel and effective operator HILOP to capture local relationships between a pair of hierarchical regions. Accordingly, a center pixel is encoded by comparing the gray value of each neighbor in the first hierarchical region with all of those in the other. HILOP is then involved with analyzing plane-images of a DT sequence to structure spatio-temporal features. The obtained probability distributions are concatenated and normalized to form a descriptor with more discrimination power for DT classification. In short, the major contributions of this work can be listed as follows.

- A novel, efficient local operator HILOP is proposed to capture textural features based on analyzing a pair of hierarchical regions where the local neighbors in a hierarchy are consecutively thresholded by all of those in the other instead of by the center pixel as the existing LBP-based variants.
- Multi-hierarchy HILOP encoding allows to enrich appearance information by addressing more further supporting regions.
- An efficient framework for DT representation is presented in which spatio-temporal features are effectively captured by taking advantage of HILOP.

2   Proposed Method

As mentioned above, using a simple computation, local-feature-based methods have achieved promising results in DT classification. However, limitations of their performances are often caused by problems of sensitivity to noise, illumination, and near uniform regions. In this section, we first take a look of LBP and its variants as well as their uses in DT representation. We then propose a novel, simple operator HILOP in order to investigate local relationships between hierarchical supporting regions. Finally, an efficient framework for DT description is presented to take advantage of the beneficial properties of HILOP for addressing above restrictions of LBP-based variants.

2.1   A Brief Review of LBP and Its Operation for DT Description

In consideration of relationships between a pixel and its surrounding regions, Ojala et al. [27] introduced a simple operator LBP in order to capture local characteristics for still image representation. Appropriately, let $\mathcal{I}$ denote a 2D gray-scale image. A LBP code for a pixel $\mathbf{q} \in \mathcal{I}$ is featured by comparing the gray-level differences between $\mathbf{q}$ and its local neighbors $\{\mathbf{p}_i\}_{i=1}^P$ as

$$\mathrm{LBP}_{P,R}(\mathbf{q}) = \sum_{i=0}^{P-1} f\big(\mathcal{I}(\mathbf{p}_i) - \mathcal{I}(\mathbf{q})\big)2^i \tag{1}$$

in which P means quantity of $\mathbf{q}$'s neighbors that are sampled on a circle of radius R using an interpolated calculation, $\mathcal{I}(.)$ points out the gray-value of a pixel, and function $f(.)$ is defined as follows.

$$f(x) = \begin{cases} 1, x \geq 0 \\ 0, \text{otherwise.} \end{cases} \tag{2}$$

As the result of that, it takes a large dimension (i.e., 2^P distinct values) for describing a still image. Therefore, two conventional mappings should be applied in practice to deal with this restriction: $u2$ with $P(P-1)+3$ bins and $riu2$ with $P+2$ for structuring uniform patterns and rotation-invariant uniform patterns respectively. Furthermore, other mapping techniques can be also remarkable to enhance the encoding power, such as Local Binary Count [49] - a substitution for addressing uniform characteristics, topological mapping $TAP^{\mathcal{A}}$ [19].

Inspired by the simple and efficient properties of operator LBP in still image encoding, several efforts have taken it into account DT representation. First, Zhao et al. [47] structured a voxel by considering its P neighbors along with its two symmetrical voxels and $2P$ corresponding neighbors placed in the previous and posterior frames. All these neighbors and two symmetrical voxels are then thresholded by the concerning voxel to form a VLBP code of $3P + 2$ binary bits. Due to the huge dimension of VLBP (i.e., 2^{3P+2} bins), it is limited in real applications. In order to handle this shortcoming, Zhao et al. [47] considered a voxel and its P neighbors on orthogonal planes of a video to shape LBP-TOP patterns. The obtained probability distributions are then concatenated and normalized to form a descriptor with 3×2^P dimensions. Motivated by the LBP-TOP concept, many methods have been proposed to improve the discrimination power: CVLBC [48] - a combination of CLBC [49] and VLBP; CVLBP [39] - an integration of CLBP [13] and VLBP; CLSP-TOP [21], CSAP-TOP [22], and HLBP [40] - dealing with noise and illumination problems.

2.2   Hierarchical Local Patterns

Let $\Omega_1 = \{\mathbf{p}_i\}_{i=1}^N$ and $\Omega_2 = \{\mathbf{q}_j\}_{j=1}^N$ be two different hierarchies of supporting regions of a pixel $\mathbf{q}$ in a texture image $\mathcal{I}$, so that $\Omega_1 \cap \Omega_2 = \emptyset$. Each neighbor $\mathbf{p}_i$

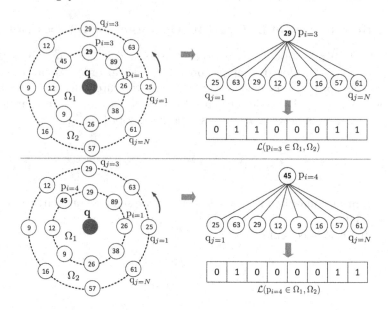

Fig. 1. An instance of structuring at $\mathbf{p}_{i=3}, \mathbf{p}_{i=4} \in \Omega_1$ based on $\{\mathbf{q}_j\}_{j=1}^{N=8}$ of Ω_2.

in hierarchical region Ω_1 is encoded as a binary string of N bits by considering the difference of $\mathbf{p}_i$'s gray value with that of all $\mathbf{q}_j \in \Omega_2$ as

$$\mathcal{L}(\mathbf{p}_i \in \Omega_1, \Omega_2) = \{g(\mathcal{I}(\mathbf{q}_j) - \mathcal{I}(\mathbf{p}_i))\}_{j=1}^N \tag{3}$$

in which $\mathcal{I}(.)$ returns the gray-value of a pixel, $g(.)$ is identical to Eq. (2). Figure 1 graphically shows an instance of this computation using $N = 8$ neighbors for each circle-hierarchical region. Accordingly, two-hierarchical pattern of $\mathbf{q}$ based on a pair of supporting regions (Ω_1, Ω_2) is featured by addressing all neighbors $\mathbf{p}_i$ of supporting region Ω_1 as follows.

$$\Gamma_{\Omega_1, \Omega_2}(\mathbf{q}) = [\mathcal{L}(\mathbf{p}_i \in \Omega_1, \Omega_2)]_{i=1}^N \tag{4}$$

It should be noted that $\Gamma(.)$ is absolutely different from structuring difference-based patterns that are introduced in [16], i.e., RD-LBP and AD-LBP. Specifically, in this work, all $\mathbf{q}_j \in \Omega_2$ are thresholded by each $\mathbf{p}_i \in \Omega_1$ to be able to point out N patterns. In contrast to that, RD-LBP [16] is formed by comparing a pairwise of $(\mathbf{q}_j, \mathbf{p}_j)$ in parallel to achieve only one pattern, while AD-LBP [16] is computed by addressing the differences of pixels in the same regions.

In order to forcefully enrich discriminative information, we address the function $\Gamma(.)$ on multi-region of adjacent hierarchies to capture more useful features in further areas. According to that, let $\mathcal{D} = \{\Omega_1, \Omega_2, ..., \Omega_m\}$ be a set of hierarchical supporting regions of a pixel $\mathbf{q} \in \mathcal{I}$, so that $\Omega_k \bigcap \Omega_{k+1} = \varnothing$. HIerarchical LOcal Pattern (HILOP) for $\mathbf{q}$ is formed as follows.

$$\text{HILOP}_{\mathcal{I}}(\mathbf{q}, \mathcal{D}) = [\Gamma_{\Omega_k, \Omega_{k+1}}(\mathbf{q})]_{k=1}^m \tag{5}$$

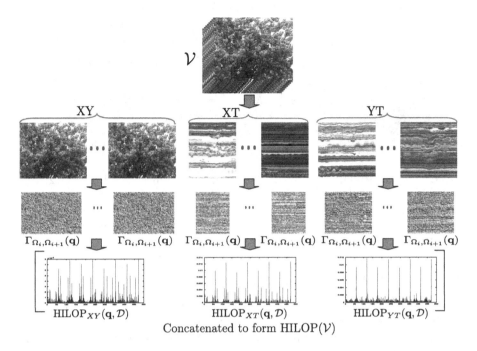

Fig. 2. Our proposed framework of encoding a video $\mathcal{V}$.

2.3 DT Representation Based on HILOP Patterns

In this section, a simple framework for efficiently structuring shape information and motion clues of DTs is proposed by exploiting the advantages of HILOP's properties for video analysis. For an input video $\mathcal{V}$, the proposed framework takes three main stages as follows. Firstly, the video $\mathcal{V}$ is split into plane-images subject to its three orthogonal planes $\{XY, XT, YT\}$ (see Fig. 2 for a graphical illustration). Secondly, the proposed operator HILOP is taken into account analyzing these plane-images to capture hierarchical features based on a set of multi-layer supporting regions $\mathcal{D}$. Finally, the obtained histograms are concatenated and normalized to produce a robust descriptor HILOP($\mathcal{V}$) as

$$\mathrm{HILOP}(\mathcal{V}) = [\mathrm{HILOP}_{XY}(\mathbf{q}, \mathcal{D}), \mathrm{HILOP}_{XT}(\mathbf{q}, \mathcal{D}), \mathrm{HILOP}_{YT}(\mathbf{q}, \mathcal{D})] \quad (6)$$

Our proposed descriptor HILOP($\mathcal{V}$) takes the following beneficial properties in order to improve the discrimination power:

- The HILOP operator structures hierarchical patterns by considering relationships of a pair of regional hierarchies, instead of those between a center pixel and its local neighbors as conducted in LBP-based variants.
- Thanks to taking HILOP into account analyzing plane-images of a DT video, spatio-temporal properties of DTs are efficiently encoded to construct the robust descriptor with high performance in DT classification.

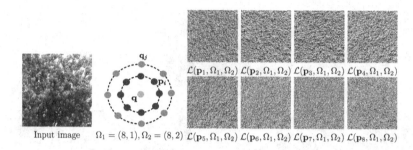

Fig. 3. Several HILOP patterns structured using a hierarchy $\mathcal{D} = \{(8,1),(8,2)\}$.

- Incorporation of hierarchical features captured in multi-supporting hierarchies allows to enrich more forceful discriminative information.

3 Experiments

In order to verify the execution of our descriptor HILOP, we address it for DT classification task on various benchmark datasets, i.e., UCLA [38], DynTex [29], and DynTex++ [12]. A linear multi-class SVM classifier implemented in the library of LIBLINEAR[1] [9] is employed with the default settings. The acquired results are then compared to those of the state-of-the-art approaches.

3.1 Experimental Settings

To be compliant with LBP encoding, a supporting region $\Omega_i \in \mathcal{D}$ should be located by P neighbors which are interpolated on a circle of radius R_i at a center pixel $\mathbf{q}$, i.e., $\Omega_i = (P, R_i)$. According to that, we address $\mathcal{D} = \{(8,1),(8,2),(8,3),(8,4),(8,5)\}$ to investigate further hierarchical local regions (i.e., $N = 8$). For computing a histogram, we use $u2$ mapping for each pattern $\mathcal{L}(.)$ to capture HILOP uniform features. As a result, the final descriptor HILOP has $(|\mathcal{D}| - 1) * 3P(P(P-1)+3)$ dimensions (see Table 2 for specific instances), where $|\mathcal{D}|$ denotes the number of hierarchical regions involved in, i.e., $|\mathcal{D}| = 5$ in this case. Several HILOP samples of this encoding are shown in Fig. 3.

3.2 Datasets and Protocols

In order to verify the performance of our proposal, we detail in this section features of benchmark datasets along with their experimental protocols for DT recognition task. In addition, Table 1 shows their properties in brief for a look.

   UCLA [38] includes 200 DT videos in $110 \times 160 \times 75$ dimension that are categorized into 50 classes with four sequences for each of them (see Fig. 4(a) for some samples). For DT classification issue, a tiny version of $48 \times 48 \times 75$ sequences is usually used and arranged into challenging sub-sets as follows.

---

[1] https://www.csie.ntu.edu.tw/~cjlin/liblinear.

Fig. 4. Samples of DT videos in UCLA (a) and DynTex (b) datasets.

– *50-class:* Using the scheme of 50 original classes along with two following protocols: *leave-one-out* (LOO) [3,41] and *4-fold cross validation* [21,40].
– *9-class* and *8-class:* 200 DT sequences are grouped to form scheme *9-class* with the following labels and corresponding numbers of sequences: "boiling water" (8), "plants" (108), "flowers" (12), "fire" (8), "fountains" (20), "water" (12), "smoke" (4), "sea" (12), and "waterfall" (16). Because of the dominant quantity in "plants" category, it is eliminated to establish scheme *8-class* with more challenges in DT recognition [46]. Following the setting protocol in [12,21], a half of sequences in each categories is randomly selected in order to train a classifying model, and the rest for the testing phase. The final rates on these schemes are reported as the average rates of 20 runtimes.

DynTex [29] consists of more than 650 high-quality DT sequences which are recorded in different conditions of environment (see Fig. 4(b) for some particular instances). Similar to the setting in [2,3,8], rates of DT classification are obtained by using LOO protocol for all the following DynTex variants:

– *DynTex35* is composed by taking out 35 videos from DynTex and splitting them into sub-videos in the following ways: randomly clipping each of them at partition points of X, Y, and T axes but not at the center of them to achieve 8 non-overlapping sub-sequences; 2 more obtained by splitting along its T axis. A the result, 35 categories are correspondingly formed with 10 sub-videos for each [3,40,47].
– *Alpha* consists of three categories of 20 DynTex sequences labeled as follows: "Sea", "Grass", and "Trees".
– *Beta* contains 162 DynTex sequences categorized into 10 groups with various numbers of samples: "sea", "vegetation", "trees", "flags", "calm water", "fountains", "smoke", "escalator", "traffic", and "rotation".
– *Gamma* also includes 10 groups of 264 DynTex sequences with different quantities: "flowers", "sea", "naked trees", "foliage", "escalator", "calm water", "flags", "grass", "traffic", and "fountains".

DynTex++ is constructed by 345 original sequences of DynTex which are pre-processed in order to retain dominant chaotic motions [12]. The outputs are then fixed in dimension of $50 \times 50 \times 50$ and arranged into 36 groups with 100 sub-videos for each, i.e., 3600 DTs in total. Following the setting set in [3,12,24], a half of samples in each group is randomly selected for training phase and the rest for testing. The final rate is calculated by averaging 10 trials.

Table 1. A summary of main properties of DT datasets.

Dataset	Sub-dataset	#Videos	Resolution	#Classes	Protocol
UCLA	50-class	200	$48 \times 48 \times 75$	50	LOO and 4fold
	9-class	200	$48 \times 48 \times 75$	9	50%/50%
	8-class	92	$48 \times 48 \times 75$	8	50%/50%
DynTex	DynTex35	350	Different dimensions	10	LOO
	Alpha	60	$352 \times 288 \times 250$	3	LOO
	Beta	162	$352 \times 288 \times 250$	10	LOO
	Gamma	264	$352 \times 288 \times 250$	10	LOO
DynTex++		3600	$50 \times 50 \times 50$	36	50%/50%

Note: LOO and 4fold are leave-one-out and four cross-fold validation respectively. 50%/50% denotes a protocol of taking randomly 50% samples for training and the remain (50%) for testing.

3.3   Experimental Results

The specific executions of our proposed descriptor HILOP on different DT datasets are presented in Table 2, in which the highest rates are in bold. It can be validated from this table that encoding DT features in consideration of local relationships on hierarchical regions has pointed out a robust descriptor with promising power, as expected in Sects. 2.2 and 2.3. Furthermore, it is also verified that taking into account multi-supporting regions makes the proposed descriptor more discriminative. Specifically, the settings of $|\mathcal{D}| = 4$ and $|\mathcal{D}| = 5$ have reported the best DT recognition rates (see Table 2). Due to the more "stable" performance, $\mathcal{D} = \{(8,1),(8,2),(8,3),(8,4),(8,5)\}$ is addressed for a comparison with state of the art. In general, the performance of our proposal is more efficient than most of methods (see Table 3), except deep learning techniques using a complex framework for learning DT features. Hereafter, we express in detail the effectiveness of HILOP on the particular DT datasets.

UCLA Dataset: Thanks to exploiting hierarchical features, the proposed HILOP descriptor obtains promising results on this scenario. More specifically, its performance using the comparing parameters is at 99.5% for both *50-LOO* and *50-4fold*, the highest rates compared to all existing methods, including deep learning approaches, i.e., PCANet-TOP [2] and DT-CNN [1] (see Table 3). In terms of DT recognition on *9-class* and *8-class* schemes, our descriptor gains a promising rate of 97.8% on *9-class*, but just 96.3% on *8-class*. In comparison with the typical LBP-based approaches, its ability is only better than that of VLBP [47] (96.3%, 91.96%), LBP-TOP [47] (96%, 93.67%), and CVLBP

Table 2. Classification rates (%) on DT benchmark datasets.

Descriptor		UCLA				DynTex				Dyn++
$\mathcal{D} = \{(P, \{R\})\}$	#bins	50-LOO	50-4fold	9-class	8-class	Dyn35	Alpha	Beta	Gamma	
$\{(8, \{1, 2\})\}$	1416	98.00	98.50	96.40	93.15	98.00	**98.33**	88.89	92.42	96.05
$\{(8, \{1, 2, 3\})\}$	2832	98.50	98.50	96.95	95.22	98.57	**98.33**	89.51	92.42	96.19
$\{(8, \{1, 2, 3, 4\})\}$	4248	99.00	**99.50**	97.55	**96.41**	99.43	96.67	90.12	**92.80**	96.06
$\{(8, \{1, 2, 3, 4, 5\})\}$	5664	**99.50**	**99.50**	**97.80**	96.30	**99.71**	96.67	**91.36**	92.05	**96.21**

Note: *50-LOO* and *50-4fold* denote results on *50-class breakdown* using leave-one-out and four cross-fold validation. Dyn35 and Dyn++ are shortened for *DynTex35* sub-set and DynTex++ respectively.

Table 3. Comparison of recognition rates (%) on benchmark DT datasets

Group	Dataset	UCLA				DynTex				Dyn++
	Encoding method	50-LOO	50-4fold	9-class	8-class	Dyn35	Alpha	Beta	Gamma	
A	FDT [24]	98.50	99.00	97.70	99.35	98.86	98.33	93.21	91.67	95.31
	FD-MAP [24]	99.50	99.00	99.35	**99.57**	98.86	98.33	92.59	91.67	95.69
B	AR-LDS [38]	89.90^N	-	-	-	-	-	-	-	-
	KDT-MD [4]	-	97.50	-	-	-	-	-	-	-
	Chaotic vector [44]	-	-	85.10^N	85.00^N	-	-	-	-	-
C	3D-OTF [45]	-	87.10	97.23	99.50	96.70	83.61	73.22	72.53	89.17
	WMFS [15]	-	-	97.11	96.96	-	-	-	-	-
	NLSSA [5]	-	-	-	-	-	-	-	-	92.40
	DFS [46]	-	**100**	97.50	99.20	97.16	85.24	76.93	74.82	91.70
	2D+T [8]	-	-	-	-	-	85.00	67.00	63.00	-
	STLS [35]	-	99.50	97.40	99.50	98.20	89.40	80.80	79.80	94.50
D	MBSIF-TOP [3]	99.50^N	-	-	-	98.61^N	90.00^N	90.70^N	91.30^N	97.12^N
	DNGP [37]	-	-	**99.60**	99.40	-	-	-	-	93.80
E	VLBP [47]	-	89.50^N	96.30^N	91.96^N	81.14^N	-	-	-	94.98^N
	LBP-TOP [47]	-	94.50^N	96.00^N	93.67^N	92.45^N	98.33	88.89	84.85^N	94.05^N
	DDLBP with MJMI [36]	-	-	-	-	-	-	-	-	95.80
	CVLBP [39]	-	93.00^N	96.90^N	95.65^N	85.14^N	-	-	-	-
	HLBP [40]	95.00^N	95.00^N	98.35^N	97.50^N	98.57^N	-	-	-	96.28^N
	CLSP-TOP [21]	99.00^N	99.00^N	98.60^N	97.72^N	98.29^N	95.00^N	91.98^N	91.29^N	95.50^N
	MEWLSP [42]	96.50^N	96.50^N	98.55^N	98.04^N	99.71^N	-	-	-	98.48^N
	WLBPC [41]	-	96.50^N	97.17^N	97.61^N	-	-	-	-	95.01^N
	CVLBC [48]	98.50^N	99.00^N	99.20^N	99.02^N	98.86^N	-	-	-	91.31^N
	CSAP-TOP [22]	**99.50**	99.50	96.80	95.98	**100**	96.67	92.59	90.53	-
	Our HILOP	**99.50**	99.50	97.80	96.30	99.71	96.67	91.36	92.05	96.21
F	DL-PEGASOS [12]	–	97.50	95.60	-	-	-	-	-	63.70
	Orthogonal Tensor DL [34]	-	99.80	98.20	99.50	-	87.80	76.70	74.80	94.70
	Equiangular Kernel DL [33]	-	-	-	-	-	88.80	77.40	75.60	93.40
	st-TCoF [31]	-	-	-	-	-	100*	100*	98.11*	-
	PCANet-TOP [2]	99.50*	-	-	-	-	96.67*	90.74*	89.39*	-
	D3 [14]	-	-	-	-	-	100*	100*	98.11*	-
	DT-CNN-AlexNet [1]	-	99.50*	98.05*	98.48*	-	100*	99.38*	**99.62***	98.18*
	DT-CNN-GoogleNet [1]	-	99.50*	98.35*	99.02*	-	100*	100*	**99.62***	**98.58***

Note:"-" means "not available". Superscript "*" indicates results using deep learning algorithms. "N" indicates rates with 1-NN classifier. *50-LOO* and *50-4fold* denote results on *50-class breakdown* using leave-one-out and four cross-fold validation respectively. Dyn35 and Dyn++ are abbreviated for *DynTex35* sub-set and DynTex++ respectively. Evaluations of VLBP and LBP-TOP operators are referred to the evaluations of implementations in [31,40]. Group A denotes *optical-flow-based approaches*, B: *model-based*, C: *geometry-based*, D: *filter-based*, E: *local-feature-based*, F: *learning-based*.

[39] (96.9%, 95.65%) respectively (see Group E in Table 3). It may be caused by the similarity of turbulent motion properties on regional hierarchies in DT sequences of two schemes. In the meanwhile, other LBP-based variants have better recognition rates, such as CVLBC [48] (99.2%, 99.02%), CLSP-TOP [21] (98.6%, 97.72%), MEWLSP [42] (98.55%, 98.04%), and WLBPC [41] (97.17%, 97.61%) respectively, but most of them have not been validated on the challenging DynTex dataset, except CLSP-TOP. However, in this case, the CLSP-TOP's performance is also not better than ours on DynTex (see Group E in Table 3).

DynTex Dataset: It can be verified from Table 2 that the performance of HILOP descriptor has been steadily increased along with more hierarchical supporting regions taken into account. For DT classification on *DynTex35*, our obtained rate is 99.71%, the highest in comparison with that of all existing methods, except CSAP-TOP [22] with 100% (see Table 3). However, its dimension is up to over double, 13200 bins compared to 5664 of our (see Table 2), as well as not better than our performance on other schemes, e.g., *9-class* and *8-class* of UCLA. MEWLSP [42] also has the same our ability, but not working well on 50 categories of UCLA and not been validated on the challenging sub-sets of DynTex (i.e., *Alpha, Beta, Gamma*). In terms of DT recognition on DynTex variants, our proposed framework achieves rates of 96.67%, 91.36%, 92.05% on *Alpha, Beta*, and *Gamma* respectively. It can be seen from Table 3 that our results are mostly better than most of state of the art, except deep learning approaches, i.e., st-TCoF [31], D3 [14], and DT-CNN [1] in which DT characteristics are captured by utilizing complicated algorithms in many layers of learning process. It should be noted that this shortcoming has restricted taking them into account mobile applications due to the limited resources of mobile devices.

DynTex++ Dataset: It can be observed from Table 2 that our descriptor achieves over 96% on this scheme for all settings of regional hierarchies. With the highest rate of 96.21% for the comparing setting, ours is better than most of shallow methods (see Table 3), except MEWLSP [42], HLBP [40], and MBSIF-TOP [3]. However, as mentioned above, those have neither been verified on DynTex variants (MEWLSP) nor outperformed ours on DynTex (MBSIF-TOP), and on *50-class* of UCLA (MEWLSP, HLBP). Furthermore, learning methods with complex computation also obtain lower performances compared to ours (see Group F in Table 3), except those of using deep learning techniques, i.e., DT-CNN [1] with the best rate of 98.58% involving with GoogleNet architecture.

4   Conclusions

A simple and efficient operator HILOP have been proposed to capture local features in consideration of hierarchical regions around an image pixel. For DT representation, HILOP is involved with video analysis for addressing shape information and motion cues through plane-images of a DT sequence. Concatenation of the obtained probability distributions forms the discriminative descriptor, which has been proved in above experiments for DT classification on different datasets. In the further contexts, it can be exploited HILOP for moment images/volumes [20,25], Gaussian-based outcomes [23,26] to make the description more robust against negative impacts of DT encoding problems.

References

1. Andrearczyk, V., Whelan, P.F.: Convolutional neural network on three orthogonal planes for dynamic texture classification. Pattern Recognit. **76**, 36–49 (2018)
2. Arashloo, S.R., Amirani, M.C., Noroozi, A.: Dynamic texture representation using a deep multi-scale convolutional network. JVCIR **43**, 89–97 (2017)
3. Arashloo, S.R., Kittler, J.: Dynamic texture recognition using multiscale binarized statistical image features. IEEE Trans. Multimed. **16**(8), 2099–2109 (2014)
4. Chan, A.B., Vasconcelos, N.: Classifying video with kernel dynamic textures. In: CVPR, pp. 1–6 (2007)
5. Baktashmotlagh, M., Harandi, M.T., Lovell, B.C., Salzmann, M.: Discriminative non-linear stationary subspace analysis for video classification. IEEE Trans. PAMI **36**(12), 2353–2366 (2014)
6. Barmpoutis, P., Dimitropoulos, K., Grammalidis, N.: Smoke detection using spatio-temporal analysis, motion modeling and dynamic texture recognition. In: EUSIPCO, pp. 1078–1082 (2014)
7. Chan, A.B., Vasconcelos, N.: Modeling, clustering, and segmenting video with mixtures of dynamic textures. IEEE Trans. PAMI **30**(5), 909–926 (2008)
8. Dubois, S., Péteri, R., Ménard, M.: Characterization and recognition of dynamic textures based on the 2D+T curvelet transform. Signal Image Video Process. **9**(4), 819–830 (2015)
9. Fan, R., Chang, K., Hsieh, C., Wang, X., Lin, C.: LIBLINEAR: a library for large linear classification. JMLR **9**, 1871–1874 (2008)
10. Fazekas, S., Chetverikov, D.: Analysis and performance evaluation of optical flow features for dynamic texture recognition. Signal Process. Image Commun. **22**(7–8), 680–691 (2007)
11. Garrigues, M., Manzanera, A., Bernard, T.M.: Video extruder: a semi-dense point tracker for extracting beams of trajectories in real time. J. R. Time IP **11**(4), 785–798 (2016)
12. Ghanem, B., Ahuja, N.: Maximum margin distance learning for dynamic texture recognition. In: Daniilidis, K., Maragos, P., Paragios, N. (eds.) ECCV 2010. LNCS, vol. 6312, pp. 223–236. Springer, Heidelberg (2010). https://doi.org/10.1007/978-3-642-15552-9_17
13. Guo, Z., Zhang, L., Zhang, D.: A completed modeling of local binary pattern operator for texture classification. IEEE Trans. IP **19**(6), 1657–1663 (2010)
14. Hong, S., Ryu, J., Im, W., Yang, H.S.: D3: recognizing dynamic scenes with deep dual descriptor based on key frames and key segments. Neurocomputing **273**, 611–621 (2018)
15. Ji, H., Yang, X., Ling, H., Xu, Y.: Wavelet domain multifractal analysis for static and dynamic texture classification. IEEE Trans. IP **22**(1), 286–299 (2013)
16. Liu, L., Zhao, L., Long, Y., Kuang, G., Fieguth, P.W.: Extended local binary patterns for texture classification. Image Vis. Comput. **30**(2), 86–99 (2012)
17. Mumtaz, A., Coviello, E., Lanckriet, G.R.G., Chan, A.B.: Clustering dynamic textures with the hierarchical EM algorithm for modeling video. IEEE Trans. PAMI **35**(7), 1606–1621 (2013)

18. Nguyen, T.P., Manzanera, A., Garrigues, M., Vu, N.: Spatial motion patterns: action models from semi-dense trajectories. IJPRAI **28**(7), 1460011 (2014)
19. Nguyen, T.P., Manzanera, A., Kropatsch, W.G., N'Guyen, X.S.: Topological attribute patterns for texture recognition. Pattern Recognit. Lett. **80**, 91–97 (2016)
20. Nguyen, T.P., Vu, N., Manzanera, A.: Statistical binary patterns for rotational invariant texture classification. Neurocomputing **173**, 1565–1577 (2016)
21. Nguyen, T.T., Nguyen, T.P., Bouchara, F.: Completed local structure patterns on three orthogonal planes for dynamic texture recognition. In: IPTA, pp. 1–6 (2017)
22. Nguyen, T.T., Nguyen, T.P., Bouchara, F.: Completed statistical adaptive patterns on three orthogonal planes for recognition of dynamic textures and scenes. J. Electron. Imaging **27**(05), 053044 (2018)
23. Nguyen, T.T., Nguyen, T.P., Bouchara, F.: Smooth-invariant gaussian features for dynamic texture recognition. In: ICIP, pp. 4400–4404 (2019)
24. Nguyen, T.T., Nguyen, T.P., Bouchara, F., Nguyen, X.S.: Directional beams of dense trajectories for dynamic texture recognition. In: Blanc-Talon, J., Helbert, D., Philips, W., Popescu, D., Scheunders, P. (eds.) ACIVS 2018. LNCS, vol. 11182, pp. 74–86. Springer, Cham (2018). https://doi.org/10.1007/978-3-030-01449-0_7
25. Nguyen, T.T., Nguyen, T.P., Bouchara, F., Nguyen, X.S.: Momental directional patterns for dynamic texture recognition. CVIU (2020, in press). https://doi.org/10.1016/j.cviu.2019.102882
26. Nguyen, T.T., Nguyen, T.P., Bouchara, F., Vu, N.-S.: Volumes of blurred-invariant gaussians for dynamic texture classification. In: Vento, M., Percannella, G. (eds.) CAIP 2019. LNCS, vol. 11678, pp. 155–167. Springer, Cham (2019). https://doi.org/10.1007/978-3-030-29888-3_13
27. Ojala, T., Pietikäinen, M., Mäenpää, T.: Multiresolution gray-scale and rotation invariant texture classification with local binary patterns. IEEE Trans. PAMI **24**(7), 971–987 (2002)
28. Peh, C., Cheong, L.F.: Synergizing spatial and temporal texture. IEEE Trans. IP **11**(10), 1179–1191 (2002)
29. Péteri, R., Fazekas, S., Huiskes, M.J.: Dyntex: a comprehensive database of dynamic textures. Pattern Recognit. Lett. **31**(12), 1627–1632 (2010)
30. Péteri, R., Chetverikov, D.: Dynamic texture recognition using normal flow and texture regularity. In: Marques, J.S., Pérez de la Blanca, N., Pina, P. (eds.) IbPRIA 2005. LNCS, vol. 3523, pp. 223–230. Springer, Heidelberg (2005). https://doi.org/10.1007/11492542_28
31. Qi, X., Li, C.G., Zhao, G., Hong, X., Pietikainen, M.: Dynamic texture and scene classification by transferring deep image features. Neurocomputing **171**, 1230–1241 (2016)
32. Qiao, Y., Xing, Z.: Dynamic texture classification using multivariate hidden Markov model. IEICE Trans. **101–A**(1), 302–305 (2018)
33. Quan, Y., Bao, C., Ji, H.: Equiangular kernel dictionary learning with applications to dynamic texture analysis. In: CVPR, pp. 308–316 (2016)
34. Quan, Y., Huang, Y., Ji, H.: Dynamic texture recognition via orthogonal tensor dictionary learning. In: ICCV, pp. 73–81 (2015)
35. Quan, Y., Sun, Y., Xu, Y.: Spatiotemporal lacunarity spectrum for dynamic texture classification. CVIU **165**, 85–96 (2017)
36. Ren, J., Jiang, X., Yuan, J., Wang, G.: Optimizing LBP structure for visual recognition using binary quadratic programming. SPL **21**(11), 1346–1350 (2014)
37. Rivera, A.R., Chae, O.: Spatiotemporal directional number transitional graph for dynamic texture recognition. IEEE Trans. PAMI **37**(10), 2146–2152 (2015)
38. Saisan, P., Doretto, G., Wu, Y.N., Soatto, S.: Dynamic texture recognition. In: CVPR, pp. 58–63 (2001)

39. Tiwari, D., Tyagi, V.: Dynamic texture recognition based on completed volume local binary pattern. MSSP **27**(2), 563–575 (2016)
40. Tiwari, D., Tyagi, V.: A novel scheme based on local binary pattern for dynamic texture recognition. CVIU **150**, 58–65 (2016)
41. Tiwari, D., Tyagi, V.: Improved weber's law based local binary pattern for dynamic texture recognition. Multimed. Tools Appl. **76**(5), 6623–6640 (2017)
42. Tiwari, D., Tyagi, V.: Dynamic texture recognition using multiresolution edge-weighted local structure pattern. Comput. Electr. Eng. **62**, 485–498 (2017)
43. Wang, H., Kläser, A., Schmid, C., Liu, C.: Dense trajectories and motion boundary descriptors for action recognition. IJCV **103**(1), 60–79 (2013)
44. Wang, Y., Hu, S.: Chaotic features for dynamic textures recognition. Soft Comput. **20**(5), 1977–1989 (2016)
45. Xu, Y., Huang, S.B., Ji, H., Fermüller, C.: Scale-space texture description on sift-like textons. CVIU **116**(9), 999–1013 (2012)
46. Xu, Y., Quan, Y., Zhang, Z., Ling, H., Ji, H.: Classifying dynamic textures via spatiotemporal fractal analysis. Pattern Recognit. **48**(10), 3239–3248 (2015)
47. Zhao, G., Pietikäinen, M.: Dynamic texture recognition using local binary patterns with an application to facial expressions. IEEE Trans. PAMI **29**(6), 915–928 (2007)
48. Zhao, X., Lin, Y., Heikkilä, J.: Dynamic texture recognition using volume local binary count patterns with an application to 2D face spoofing detection. IEEE Trans. Multimed. **20**(3), 552–566 (2018)
49. Zhao, Y., Huang, D.S., Jia, W.: Completed local binary count for rotation invariant texture classification. IEEE Trans. IP **21**(10), 4492–4497 (2012)

Temporal-Clustering Based Technique for Identifying Thermal Regions in Buildings

Antonio Adán[✉], Juan García, Blanca Quintana, Francisco J. Castilla, and Víctor Pérez

Visual Computing and Robotics Laboratory, Castilla La Mancha University, Ciudad Real, Spain
{antonio.adan,juan.gaguilar,blanca.quintana,fcojavier.castilla, victor.perez}@uclm.es

Abstract. Nowadays, moistures and thermal leaks in buildings are manually detected by an operator, who roughly delimits those critical regions in thermal images. Nevertheless, the use of artificial intelligence (AI) techniques can greatly improve the manual thermal analysis, providing automatically more precise and objective results. This paper presents a temporal-clustering based technique that carries out the segmentation of a set of thermal orthoimages (STO) of a wall, which have been taken at different times. The algorithm has two stages: region labelling and consensus. In order to delimit regions with similar temporal temperature variation, three clustering algorithms are applied on STO, obtaining the respective three labelled images. In the second stage, a consensus algorithm between the labelled images is applied. The method thus delimitates regions with different thermal evolutions over time, each characterized by a temperature consensus vector. The approach has been tested in real scenes by using a 3D thermal scanner. A case study, composed of 48 thermal orthoimages at 30 min-intervals over 24 h, are presented.

Keywords: 3D data processing · 3D thermal models · Thermal data analysis

1 Introduction

Thermal characterization of buildings is a challenging topic that can be framed in the context of energy saving and efficient construction. Usually, an expert performs a thermal assessment of a building by taking local measurements in walls or any other structural component. Hand-sensors, such as contact thermometers are used to collect temperature in specific locations and times. Thermal cameras are also commonly used by engineers with the aim of finding damages or thermal leakages in buildings.

Although human exploration provides quick and on-line results, also entails subjective and, sometime, wrong and unprecise conclusions. The inclusion of new technologies and the use of intelligent algorithms can complement the action of humans and bring new and challenging research lines in the context of thermal analysis of buildings.

J. Blanc-Talon et al. (Eds.): ACIVS 2020, LNCS 12002, pp. 290–301, 2020.
https://doi.org/10.1007/978-3-030-40605-9_25

2   Previous Work

One of the last advances in this matter concerns the creation of 3D thermal models. These models are obtained from thermal point clouds, which are been collected using LiDARs [1], photogrammetric cameras [2] or depth cameras [3], combined with infrared cameras. All these systems provide the temperature of specific regions which can be precisely identified into a three-dimensional space, thus overtaking the functionality of earlier 2D based sensors.

Beyond this technology, in the last two years a few autonomous mobile platforms that yield 3D thermal models have been presented. A representative example of autonomous systems are that of Borrmann et al. [4] and Adan et al. [5]. In [4] the platform automatically creates low resolution thermal mesh models of buildings indoors. The system performs thermal scans but with a limited vertical FOV. In [5] the scanning platform is complete in the sense that the sensorial setup (3D laser scanner+colour camera+thermal camera) can obtain 360 thermal data per scan and, if necessary, rotate vertically in order to cover a higher vertical field of view (FOV) of the scene.

Apart from sensors, the artificial intelligence is now flourishing into the world of automatic thermal analysis. The use of IA techniques is currently focused on identifying specific problems and zones in structural elements and facades, such as bridges [6], leakages [7] and humid zones. Most of the authors process thermal images of the scene at a specific time and obtain visual results that an expert can further use to make conclusions. Some of the most representative works are referenced in the next paragraphs.

In [8], a thermal image is divided in several zones according to several temperature ranges which are beforehand imposed. The surface within each range is considered as an isotherm in the wall. Lopez et al. [9] carry out a segmentation of the image by imposing a unique threshold. The image is divided into low and high temperature groups. The algorithm detects windows and energy loss.

In order to identify and classify the type of windows within walls, Demisse et al. [10] segment a thermal point cloud into plane regions and impose several assumptions. They assume that each segmented region contains windows and that coldest regions belong to a window. According the temperature distribution of the window points this is classified as open, closed or damage. Fernandez-Lorca et al. [11], segment the facades into three regions corresponding to glasses, windows and walls. For each component, they search for locations with minimum and maximum temperature values and directly assign these locations to cold and heat leakage sources.

An interesting method is accomplished by Golparvar-Fard et al. in [12]. The authors obtain a thermal point cloud that is manually split in structural components, each with an associate thermal map. These maps are later compared to the ones obtained from a computational fluid dynamics (CFD) analysis. Finally, they obtain a 3D visualization of the performance deviation of the same areas. In [13] the heat loss and gain are calculated in areas with potential problems.

Some limitations and disadvantages of the current proposals are presented below:

- Most of the segmentation algorithms are based on imposing or calculating a set of thresholds, which is not an efficient method when the scene has more than two segments. It is known that thresholding is an elementary process that might entail risks in the image processing results.
- Many methods impose hard conditions and assumptions, which rest consistency and applicability for real cases.
- Owing to the simplicity of the data processing, the aforementioned techniques do not guarantee the same functionality and results under temporal or environment changes.

Additionally, the aforementioned references process a static image of the scene at a specific time and day. Nevertheless, the results obtained for the same scene at different times by using image processing techniques can be different. Figure 1 illustrates two thermal images of the same wall taken at 10 a.m and 3 p.m. It is clear the disparity of the images.

The idea of this paper lies on processing, not one but many, consecutive thermal images. There exist a few articles that deal with several thermal images at different times and provide interesting results. A representative case is that of Natephra et al. [14, 15]. The research is here focused on characterise the scene by using a visual temporal representation and the thermal comfort level of a room. The authors do not propose a specific algorithm, but a pipeline of connected sub-processes carried out from multiple commercial software tools.

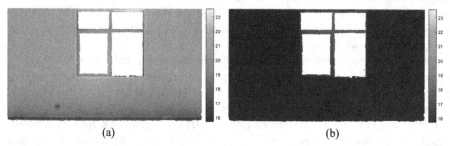

(a)   (b)

Fig. 1. (a) Thermal orthoimages of a wall scanned at 10 a.m and 3 p.m. For a better visualization, the temperature is painted by using a colour code.

3 Objectives and Contributions

The general objective of this paper is to progress in energy analysis methodologies for building indoors with the help of current artificial intelligence techniques applied on thermal data. Specifically, we aim to introduce new temporal-based segmentation algorithms on the thermal images of the structural elements (i.e. walls) of buildings and facilities. The input of the process is a set of thermal orthoimages of a wall, extracted from a 3D thermal model, which are taken at consecutive times during a day and the output is a set of segments identified on the wall with different dynamic thermal evolutions.

The main contributions of this research are as follow:

- Originality. The application of artificial intelligent temporal algorithms on thermal images is a new topic in the field of energy analysis.
- Robustness. The segmentation yielded with local and static based techniques are very sensitive to thermal changes and disturbances in the scene. Nevertheless, our approach is robust to energy changes and external circumstances, such as the momentary weather variations, the sun rays breaking through or the central heating changes.
- Instead of the classic 2D-based techniques, our approach is extended to the third dimension, thus providing a 3D semantic thermal model in which different zones with different thermal properties are precisely located.
- Utility. We believe that this 3D semantic thermal model can be useful for engineers and architects because they can easily explore, identify and delimitate interesting regions from an energy point of view, such as heat leakages or bridges.

4 Methodology

4.1 The 3D Thermal Model of a Building

In our case, 3D thermal information consists of the 3D coordinates of a set of points of the scene together with their associated temperature. The acquisition process is composed of three steps:

The first stage deals with the scanner/camera calibration problem and provides a reduced point cloud corresponding to the FOV of the camera. This calibration process essentially calculates matrix M in the Eq. (1), in which (X_p, Y_p, Z_p) are the coordinates of a point in the scanner coordinate system and (X_f, Y_f) are the coordinates (in pixels) of the corresponding point projected onto the thermal image. Matrix M is calculated by using a large number of known 3D coordinates and their associated pixels in the thermal image. After M is calculated, the temperature of each pixel is associated to its corresponding 3D point.

$$\begin{pmatrix} \lambda X_f \\ \lambda Y_f \\ \lambda \end{pmatrix} = M \begin{pmatrix} X_p \\ Y_p \\ Z_p \\ 1 \end{pmatrix} \tag{1}$$

The second stage copes with the matching of additional thermal images onto the 360-point cloud. Since the FOV of a thermal camera is usually limited, in order to cover a 360-thermal scan, the thermal camera must rotate around the Z-axis of the scanner and take several camera shots, matching the thermal images with their corresponding points.

In the third stage, the former thermal point cloud is subsequently aligned with others, which are obtained from new scanner's positions, eventually generating a complete thermal point cloud model of the scene.

The thermal model can be now segmented into a set of parts, which are the structural elements (SE) of the scanned building indoor. SEs are basically architectural components of the building, such as ceiling, floor, columns and walls. This process basically consists of segmenting the whole point cloud into sets of points that fit to vertical and horizontal planes (see complete information in [16]). The point cloud segmented into "thermal SEs" at a specific time t is denoted as $\Omega(t)$. Each structural element can afterwards be analyzed separately as we explain in the next paragraphs.

Figure 2(a) and (b) illustrates the outputs of the first two aforementioned steps for a simple example. The final 3D thermal model and the segmentation results are shown in Fig. 2(c) and (d).

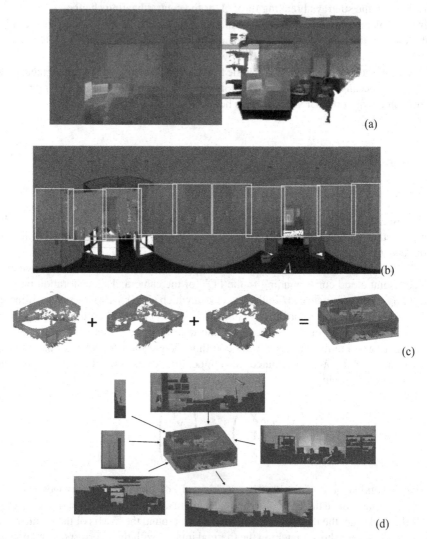

(a)

(b)

(c)

(d)

Fig. 2. (a) Assigning temperature to 3D points. (b) Overlapping several thermal images to a 360-point cloud. (c) Aligning several thermal point clouds. (d) Segmentation of a thermal point cloud into thermal SEs.

4.2   The Temporal 3D Thermal Model

When the scanning session is repeated at another time t', a dual 3D thermal model $\Omega(t')$ is obtained. The new model maintains the former geometry (i.e. the SEs), but changes its temperature. In general, for a set of n sessions, taken at evenly-spaced intervals of time $\{t_1, t_2, \ldots t_n\}$, a temporal 3D thermal model $\Omega = \{\Omega(t_1), \Omega(t_2), \ldots \Omega(t_n)\}$ is generated.

Each SE, at any time t, can be geometrically defined by the coordinates of their vertices and the normal vector. On the other hand, each SE has got its associated points, containing their respective temperatures at times $\{t_1, t_2, \ldots t_n\}$. A more formal explanation follows in the next paragraphs.

Let A be one of the structural elements of the scene (e.g. a wall of a room), defined by four vertices $\{v_1, v_2, v_3, v_4\}$ and the normal vector u. Let S_A be the point cloud associated to A at time t_1. Since A has been sensed from the same scanner positions in different sessions at evenly-spaced intervals of time $\{t_1, t_2, \ldots t_n\}$, S_A can be maintained for each observation of A.

Let T_1 be the set of temperature values assigned to S_A at time t_1. Assuming that a SE is a rectangular shape, A can be represented by an image I_1, which contains the points projected into the plane of normal vector u, together with their corresponding temperatures T_1. In summary, the structural element A is represented now by a grey image (or a coded-colour image) in which a grey level (or a colour code) encodes the temperature.

For n observations of A, we finally create a set of n thermal images $\{I_1, I_1, \ldots I_n\}$. Assuming images $m \times p$ in size, a three-dimensional thermal-data structure D ($m \times p \times n$) is obtained. In order to analyse D, we consider that each sensed point P is described by an n-vector that describes the evolution of the temperature of P along the period of time that it is observed. Typically, the period of time is set on 24 h, and the number of observations (temperature data) are performed at intervals of 20–30 min. Formally, P is characterized by the ($1 \times n$) temperature vector $T(P) = \{T_1(P), T_2(P), \ldots, T_n(P)\}$.

4.3   Processing the Temporal 3D Thermal Model

The objective of the analysis process is to calculate a set of regions of A that have a similar temperature evolution during the observation period. To do this, we have applied three different clustering techniques on the vectors $T(P)$, P extended to all the points sensed in A.

We have selected one of the existing methods per clustering category: partition based-techniques, hard clustering and hierarchical clustering. The chosen techniques are: k-means [17], the mixture of Gaussian distributions (MGD) [18], the Birch method [19]. A density based-technique, specifically the DBSCAN technique [20], has also been tested, yielding erratic results. Eventually this method has been rejected. Each of the three selected techniques has different advantages and limitations.

I.   K-means makes partitions of the data into k distinct clusters based on distance to the centroid of a cluster. It is easy to implement and fast, but the number of clusters must be beforehand imposed, and it is sensitive to outliers. On the other hand, this method could be inefficient under disparity in size and density of the clusters.

II.   Hard clustering is a method that assigns each data point to exactly one cluster. The algorithm first fits a Gaussian mixture model to data. The multivariate normal components of the fitted model can represent clusters.

III. Hierarchical clustering builds a multilevel hierarchy of clusters by creating a cluster tree. The main disadvantage of this method is the low scalability.

As a result of applying a particular clustering algorithm, the wall is split into several segments corresponding to the groups. Since a set of vectors $T(P)$ is associated to each group, a characteristic prototype vector (or signature descriptor) can be assigned per cluster. We denote ${}^i\xi_j$ and ${}^i\Phi_j$ as the region and the prototype vector of the j-th cluster obtained from the i-th algorithm.

The consensus between the segments and prototype vectors coming from algorithms I to III have been modelled through two distance metrics. The first distance d_1 between two algorithms (i) and (k) is formally defined by Eq. (2), in which symbol $\langle . \rangle$ signifies the cardinal of a set.

$$d_{1n}(i, k) = 1 - max_{\arg\{j,m/j=1...h, m=1...h\}} \frac{1}{L*W} \sum_{j,m}^{h} \left\langle {}^i\xi_j \cap {}^k\xi_m \right\rangle \qquad (2)$$

In this equation, the number of clusters is denoted as h. Note that d_1 is normalised to [0, 1]. To define the consensus regions $\xi_j, j = 1 ... h$, we discard distances above threshold μ_1 (in our case $\mu_1 = 0.3$) and then find the best intersection of regions for the rest of valid distances.

Distance d_2 is defined by using the prototype vectors according with Eq. (3). This is a minimum mean square error (MMSE) estimator applied to each pair of prototype vectors from clustering algorithms i and k.

$$d_{2n}(i, k) = \frac{d_2(i, k)}{D} \qquad (3)$$

Where

$$d_2(i, k) = min_{\arg\{j,m / j=1...h, m=1...h\}} \sum_{j,m}^{h} \left\| {}^i\phi_j, {}^k\phi_m \right\| \qquad (4)$$

$$D = \max\{d_2(i, k), \ i, k = 1, ... 3\} \qquad (5)$$

As in the previous case, the pairs with associate distances above a threshold μ_2 are discarded. The consensus prototype vectors $\Phi_j, j = 1 ... h$, are found by averaging the rest of valid associated prototype vectors.

5   Experimental Results

The method proposed has been tested in real indoor environments by using a platform with a 3D laser scanner and a thermal camera [21]. In this section, we first show the results of our method on a representative case study.

Our scanner covers an area of $360° \times 100°$ per scan, taking 47 s and yielding 5 million coordinates. The thermal camera has a resolution of 640×512 pixels, with a

FOV of 45° × 37° at a frequency of 30 Hz. The temperature range in the HighMode is between 233° and 823 °K with a precision of 0.4 °K.

The scenes are walls which are sequentially scanned during 24 h at time intervals of 30 min, generating thermal data matrices D_F and D_M of 640 × 480 × 48. Different external thermal conditions were applied to the scene. The central heat went off randomly during two intervals of three hours and the large windows of the adjacent walls were covered in order not to let sun rays coming in the room. Additionally, the sun's rays light up the wall in the morning during several hours.

In order to illustrate several temperature evolutions, Fig. 3 shows the signature $T(P)$ corresponding to three different points of the wall. It is clear different temperature variation for different zones of the wall.

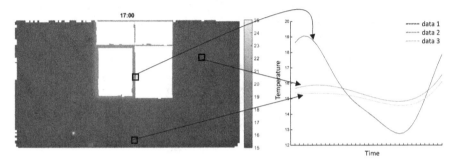

Fig. 3. Characteristic thermal vectors of three points of the wall

The goal of our experiment is to extract a set of regions that have a particular temporal thermal evolution. To do this, we explore the results yielded from the aforementioned three clustering techniques.

After applying the k-means algorithm on the datasets D_F and D_M, we distinguish clearly four zones with different thermal evolution that can be visualised by means of their corresponding signature descriptors. We have labelled these zones as: window-frame, east zone, west zone and borderline zone. From a visual inspection of the clusters, it is clear that the results from GMD are quite similar to the k-means' and that the hierarchical method is also coherent with the earlier segmented regions. Figure 4 presents the regions and the prototype vectors for four clusters. It is evident that the prototype vectors of methods I, II and III are quite close.

Tables 1 and 2 present the distances d_1 and d_2 and make more evident the previous conclusions. Figure 5 shows the final consensus regions and signatures. Black colour is assigned to no-consensus points as well as the lack of data.

From the consensus results, several conclusions can be made. Region 1 corresponds to the left side of the wall and remain a little warmer than the right side (Region 2) during the experiment. This might be due because of the north-south orientation of the wall and in the fact that the external right part would be better sheltered from wind and rain. Region 3 is associated to the window frame located on the wall. It is clear that the

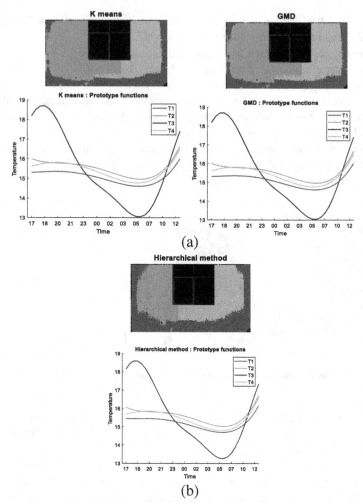

Fig. 4. (a) Regions $^I\xi_j$, $^{II}\xi_j$, $^{III}\xi_j$ in different colours and (b) their corresponding prototype vectors $^I\Phi_j$, $^{II}\Phi_j$, $^{III}\Phi_j$ for four clusters. The temperature of the glass is not sensed by the scanner.

material of the window (aluminium) follows, with a certain thermal inertia, the outdoor temperature. Region 4 is located in the left and right borderlines of the wall, and in the lower part, next to the floor. By far these regions remain with lower temperature and could be thought as potential humid zones or thermal leakages.

Figure 6 illustrates more examples in which our temporal-clustering based technique identifies separate thermal regions.

Table 1. Values of the normalised distance d_1

d1	I	II	III
I	0	0.21	0.38
II	0.21	0	0.37
III	0.38	0.37	0

Table 2. Values of the normalised distance d_2

d1	I	II	III
I	0	0.004	1
II	0.004	0	0.97
IV	1	0.97	0

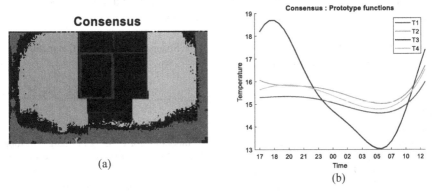

Fig. 5. The consensus clustering (a) Consensus regions ξ_j and (b) consensus prototype vectors Φ_j.

Fig. 6. Thermal labeling in four walls using temporal clustering.

6 Conclusions

This paper presents an original approach to characterize the thermal evolution of indoor walls of buildings over time. The method processes walls' thermal orthoimages that have been extracted from 3D thermal models of the building at different times. Four clustering techniques have been applied over this temporal dataset, each obtaining different zones located within the walls.

In order to make the method more consistent, the results obtained from three clustering algorithms are put in common by using two distance metrics. As a result of this, the consensus regions are precisely identified on the wall and characterized by prototype thermal vectors. The thermal temporal evolution can thus be discussed, yielding interesting and no-evident conclusion.

This work means a primary research result in the field of automatic thermal dynamic characterization of buildings, which will be developed in the future. Many improvements and challenging goals will be considered in a short term. Among them, we aim to extend this investigation to the whole building structure and identify conflictive zones. Our final objective is to generate complete thermal-labeled 3D models of buildings.

Acknowledgment. This work has been supported by the Spanish Economy and Competitiveness Ministry [DPI2016-76380-R project].

References

1. Wang, C., Cho, Y.K., Gai, M.: As-Is 3D thermal modeling for existing building envelopes using a hybrid LIDAR system. J. Comput. Civ. Eng. **27**(6), 645–656 (2013). https://doi.org/10.1061/(ASCE)CP.1943-5487.0000273
2. Ham, Y., Golparvar-Fard, M.: An automated vision-based method for rapid 3D energy performance modeling of existing buildings using thermal and digital imagery. Adv. Eng. Inform. **27**(3), 395–409 (2013). https://doi.org/10.1016/j.aei.2013.03.005
3. Rangel, J., et al.: 3D thermal imaging: fusion of thermography and depth cameras. In: Conference on Quantitative InfraRed Thermography (2014)
4. Borrmann, D., et al.: A mobile robot based system for fully automated thermal 3D mapping. Adv. Eng. Inform. **28**(4), 425–440 (2014). https://doi.org/10.1016/j.aei.2014.06.002
5. Adán, A., Prieto, S.A., Quintana, B., Prado, T., García, J.: An autonomous thermal scanning system with which to obtain 3D thermal models of buildings. In: Mutis, I., Hartmann, T. (eds.) Advances in Informatics and Computing in Civil and Construction Engineering, pp. 489–496. Springer, Cham (2019). https://doi.org/10.1007/978-3-030-00220-6_58
6. Garrido, I., Lagüela, S., Arias, P., Balado, J.: Thermal-based analysis for the automatic detection and characterization of thermal bridges in buildings. Energy Build. **158**, 1358–1367 (2018). https://doi.org/10.1016/j.enbuild.2017.11.031
7. Hoegner, L., Stilla, U.: Building facade object detection from terrestrial thermal infrared image sequences combining different views. In: ISPRS Annals of Photogrammetry, Remote Sensing and Spatial Information Sciences, vol. II-3/W4, pp. 55–62, March 2015. https://doi.org/10.5194/isprsannals-ii-3-w4-55-2015
8. González-Aguilera, D., Rodriguez-Gonzalvez, P., Armesto, J., Lagüela, S.: Novel approach to 3D thermography and energy efficiency evaluation. Energy Build. **54**, 436–443 (2012). https://doi.org/10.1016/j.enbuild.2012.07.023
9. López-Fernández, L., Lagüela, S., González-Aguilera, D., Lorenzo, H.: Thermographic and mobile indoor mapping for the computation of energy losses in buildings. Indoor Built Environ. **26**(6), 771–784 (2017). https://doi.org/10.1177/1420326X16638912

10. Demisse, G.G., Borrmann, D., Nuchter, A., Nüchter, A., Nuchter, A., Nüchter, A.: Interpreting thermal 3D models of indoor environments for energy efficiency. J. Intell. Rob. Syst. Theor. Appl. **77**(1), 55–72 (2015). https://doi.org/10.1007/s10846-014-0099-5

11. Fernández-Llorca, D., Lorente, A.G., Fernández, C., Daza, I.G., Sotelo, M.A.: Automatic thermal leakage detection in building facades using laser and thermal images. In: Moreno-Díaz, R., Pichler, F., Quesada-Arencibia, A. (eds.) EUROCAST 2013. LNCS, vol. 8112, pp. 71–78. Springer, Heidelberg (2013). https://doi.org/10.1007/978-3-642-53862-9_10

12. Golparvar-Fard, M., Ham, Y.: Automated diagnostics and visualization of potential energy performance problems in existing buildings using energy performance augmented reality models. J. Comput. Civ. Eng. **28**(1), 17–29 (2014). https://doi.org/10.1061/(ASCE)CP.1943-5487.0000311

13. Ham, Y., Golparvar-Fard, M.: Automated cost analysis of energy loss in existing buildings through. In: ISARC 2013 - 30th International Symposium on Automation and Robotics in Construction and Mining, Held in Conjunction with the 23rd World Mining Congress, pp. 1065–1073 (2013)

14. Natephra, W., Motamedi, A., Yabuki, N., Fukuda, T.: Integrating 4D thermal information with BIM for building envelope thermal performance analysis and thermal comfort evaluation in naturally ventilated environments. Build. Environ. **124**, 194–208 (2017). https://doi.org/10.1016/j.buildenv.2017.08.004

15. Natephra, W., Motamedi, A., Yabuki, N., Fukuda, T., Michikawa, T.: Building envelope thermal performance analysis using BIM-based 4D thermal information visualization. In: Conference: 16th International Conference on Computing in Civil and Building Engineering (ICCCBE2016) (2016)

16. Adán, A., Huber, D.: Reconstruction of wall surfaces under occlusion and clutter in 3D indoor environments, Robotics Institute, Carnegie Mellon University, Pittsburgh, PA CMU-RI-TR-10-12 (2010)

17. Hartigan, J.A., Wong, M.A.: Algorithm AS 136: a K-means clustering algorithm. Appl. Stat. **28**(1), 100 (1979). https://doi.org/10.2307/2346830

18. McLachlan, G.J., Peel, D.: Finite Mixture Models. Wiley, New York (2000)

19. Zhang, T., Ramakrishnan, R., Livny, M.: BIRCH: an efficient data clustering method for very large databases. In: SIGMOD 1996 Proceedings of the 1996 ACM SIGMOD International Conference on Management of Data, pp. 103–114 (1996)

20. Ester, M., Kriegel, H.-P., Sander, J., Xu, X.: A density-based algorithm for discovering clusters in large spatial databases with noise. In: Proceedings of the Second International Conference on Knowledge Discovery and Data Mining, pp. 226–231 (1996)

21. Quintana, B., Prieto, S.A., Adán, A., Vázquez, A.S.: Semantic scan planning for indoor structural elements of buildings. Adv. Eng. Inform. (2016). https://doi.org/10.1016/j.aei.2016.08.003

Distance Weighted Loss for Forest Trail Detection Using Semantic Line

Shyam Prasad Adhikari[1] and Hyongsuk Kim[1,2][✉]

[1] Division of Electronics and Information Engineering, Chonbuk National University,
Jeonju 54896, Republic of Korea
[2] Core Research Institute of Intelligent Robots, Chonbuk National University,
Jeonju 54896, Republic of Korea
hskim@jbnu.ac.kr
https://home.jbnu.ac.kr/robotv/index.htm

Abstract. Unlike structured urban roads, forest trails do not have defined shape or appearance and have ambiguous boundaries making them challenging to be detected. In this work we propose to train a deep convolutional encoder-decoder network with a novel distance weighted loss function for end to end learning of unstructured forest trail. The forest trail is annotated with "semantic line" representing the trail, and a L1 distance map is derived from the binarized ground truth. We propose to use the distance map to weigh the loss function to guide the focus of the network on the forest trail. The proposed loss function penalizes low activations around the ground truth and high activations in areas further away from the trail. The proposed loss function is compared against other commonly used loss functions by evaluating the performance on the publicly available IDSIA forest trail dataset. The proposed method leads to higher trail detection accuracy with 2.52%, 4.69% and 8.18% improvement in mean intersection over union (mIoU) over mean squared error, Jaccard loss and cross-entropy, respectively.

Keywords: Semantic lines · Forest trail detection · Distance transform · Distance weighted loss

1 Introduction

Navigation in highly unstructured environments like trails in forests or mountains is extremely challenging for autonomous robots due to the numerous variations present in natural environment, and the absence of structured pathways or distinct lane markings. A robot capable of autonomously navigating off-road environments would become invaluable aid in search-and-rescue missions, wilderness monitoring and mapping etc.

The resurgence of neural networks in the form of "deep" neural networks (DNNs) [1] has improved the performance in various high-level computer vision tasks [2–5]. DNNs have also been successfully applied for road and lane detection

© Springer Nature Switzerland AG 2020
J. Blanc-Talon et al. (Eds.): ACIVS 2020, LNCS 12002, pp. 302–311, 2020.
https://doi.org/10.1007/978-3-030-40605-9_26

in highways, and structured urban settings [6–8]. The task of trail detection is related to the task of road (lane) detection however the two are fundamentally different. Unlike structure urban road, natural environment has infinite variations and the absence of structured pathways or distinct lane markings makes the problem of trail detection more challenging.

Off-road trail detection has been primarily approached as a segmentation problem [9–11]using classical computer vision techniques. Neural networks(NNs) have also been used for autonomous navigation in unstructured natural environments [12–15]. Hadsell et al. [12] used a self-supervised NN in conjunction with a stereo module to classify the terrain in front of the robot as ground or obstacle. Guisti et al. [13] and Smolyanskiy et al. [14] used DNN as a supervised classifier to output the heading direction of a trail compared to the viewing direction of a quadrotor. Both of these works predict the instantaneous heading direction of the trail. However, the trail itself is not detected and localized. Adhikari et al. [15] used a segment-then-detect approach for trail detection. A patch based DNN was used to segment the trail and dynamic programming was used as a post-processing step to detect and localize the trail line. However, due the ambiguity between trail and non-trail patches, the accuracy of the patch based segmentation is low which affects the accuracy of the detected trail.

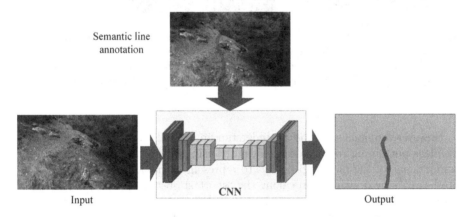

Fig. 1. The proposed method of learning semantic lines for forest trail detection using convolutional neural networks.

In this work we propose an end-to-end learning system for forest trail detection using semantic lines, as shown in Fig. 1. Unlike the segment-then-detect approach, semantic line is used to annotate the trail and a fully convolutional neural network is used for learning the semantic lines. Moreover, we propose a novel distance weighted loss function for training the fully convolutional neural network. The proposed loss function focuses the attention of the network on the trail by penalizing low activations around the ground truth and high activations in areas further away from the trail. The proposed loss function is shown to produce more accurate forest trails compared to other commonly used loss functions.

2   Semantic Line for Trail Detection

One of the factors enabling the rapid adoption and increased performance of deep neural networks is the availability of huge amount of data for training. However, for supervised training of DNNs this data has to be annotated manually with ground truth. It is expensive and time consuming to prepare large scale ground truth annotations. Manual annotation is particularly time consuming for semantic segmentation where per-pixel annotation is required. For tasks like forest trail detection, the cluttered background with amorphous "stuffs", and ambiguous boundaries without lane markings makes it difficult even for humans to annotate correctly.

Fig. 2. Annotating forest trail with bounding box is impractical and dense annotation for trail segmentation is difficult as the trail does not have defined shape or appearance, and at times merges seamlessly with the surrounding environment. However, it is simple and meaningful to annotate with a semantic line representing the concept of a forest trail.

However, humans are good at understanding visual scenes and communicating ideas pertaining to the scene efficiently using graphical representation. For example, one of the meaningful ways to represent a "navigable" path on a wide forest-trail is by drawing a line along the "shortest obstacle-free" path, as shown in Fig. 2. The trail in the image is not marked, have no defined shape or appearance, and the boundaries are ambiguous as the trail merges seamlessly with the surrounding environment. Unlike the per-pixel annotations required for semantic segmentation, the trail is annotated by drawing few pixel thick lines across the length of the visible trail. Such a simple line, henceforth referred as semantic line, can also encode higher-level concepts about obstacle avoidance and path planning required for navigating forest trails. This graphical representation of the trail line is intuitive and can be annotated easily even for complex natural environments.

Structured road lanes have also been annotated with lines [17] for lane detection using deep learning and can be considered as semantic lines. However, there is a fundament difference between the semantic lines representing road lanes and natural trails. Urban road lanes are clearly guided by distinct lane markings present on the road and are not subjective. However, no such guiding marks are available for natural trail and the annotated trail is often subjective [15] making them more challenging. In this work we use the IDSIA forest trail dataset [16] to demonstrate the effectiveness of the proposed approach.

3   Distance Weighted Loss Function

The commonly used loss function for learning DNNs with 2D outputs like semantic segmentation are the categorical cross-entropy loss (CCE), mean squared error (MSE) and Jaccard coefficient (JC). To mitigate the effect of class-imbalance in the training data, these losses are generally weighted using frequency based techniques. The weights assigned to the different classes are inversely proportional to their frequency of occurrence in the training set. Hence, the dominant classes are assigned lower weights compared to rarely occurring classes so as to give equal importance to all the classes while training. Though effective for mitigating the class imbalance problem, the high weights assigned to rare labels produce noisy gradients and may lead to unstable optimization.

The semantic line annotation used in this study results in highly imbalanced classes with a strong bias towards the background class. To mitigate the effect of the unbalanced data, we propose a novel loss function given by,

$$L_{DW} = \frac{1}{N \times M} \sum_{j=1}^{N} \sum_{i=1}^{M} \{(1 - g_{ij}) \cdot d(i,j) \cdot p_{ij} - g_{ij} \cdot d_{\max} \cdot \log(p_{ij})\} \qquad (1)$$

where $g_{ij} \in \{0, 1\}$ is the ground truth label where background $= 0$ and trail $= 1$, $M \times N$ is the spatial dimension of the target, $p_{ij} \in [0, 1]$ is the result of sigmoid non-linearity applied to the output layer and $d(i,j)$ is the value of the distance map at location (i, j) and $d_{\max}$ is the maximum value in the map. Instead of the frequency based weighting, the proposed loss function, Eq. (1), uses a distance map based weighting. The distance map is generated by computing the distance transform on the ground truth annotation with the semantic line as foreground. The pixels far away from the semantic line are weighted more compared to those in the proximity of the line.

The first term in Eq. (1) has effect only on the background pixels, whereas the second term affects only the foreground pixels. In the first term we multiply p_{ij} with $d(i,j)$ to penalize high activations in areas of the background. The loss term effectively penalizes trail pixels which should not be there. The faraway background pixels are penalized more than the nearby background pixels. The second term is the log-likelihood of the foreground pixel. This term penalizes low activations and encourages high activations near the ground truth. This term is weighted by $d_{\max}$ to bring both parts the loss function to similar scales.

4   Learning Semantic Lines Using Encoder-Decoder Network

Convolutional encode-decoder networks have been used successfully for various computer vision tasks that require a 2D output [3–5]. The encoder-decoder architecture is adopted in this study for learning of forest trails using semantic lines, as shown in Fig. 1. A U-Net [18] like architecture is used in this study and the details of the network are given in Table 1.

Table 1. The architecture of the encoder decoder network for detecting forest trail using semantic lines.

Layers	Output size	Input layers	Filters
C1	$192 \times 320 \times 3$	Input image	3×3 conv
VGG_C1	$96 \times 160 \times 32$	C1	3×3 VGG_conv, max pool
VGG_C2	$48 \times 80 \times 64$	VGG_C1	3×3 VGG_conv_block, max pool
VGG_C3	$24 \times 40 \times 96$	VGG_C2	3×3 VGG_conv_block, max pool
VGG_C4	$12 \times 20 \times 128$	VGG_C3	3×3 VGG_conv_block, max pool
VGG_C5	$12 \times 20 \times 12$	VGG_C4	3×3 VGG_conv_block
ResNet_D4	$24 \times 40 \times 12$ (U4)	VGG_C5	2×2 conv2d_transpose
	$24 \times 40 \times 12$(R4)	U4	3×3 ResNet_dconv_block
	$24 \times 40 \times 140$	R4, VGG_C4	Concatenation, dropout (0.8)
ResNet_D3	$48 \times 80 \times 140$ (U3)	ResNet_D4	2×2 conv2d_transpose
	$48 \times 80 \times 12$(R3)	U3	3×3 ResNet_dconv_block
	$48 \times 80 \times 108$	R3, VGG_C3	Concatenation, dropout (0.8)
ResNet_D2	$96 \times 160 \times 108$ (U2)	ResNet_D3	2×2 conv2d_transpose
	$96 \times 160 \times 12$(R2)	U2	3×3 ResNet_dconv_block
	$96 \times 160 \times 76$	R2, VGG_C2	Concatenation, dropout (0.8)
ResNet_D1	$192 \times 320 \times 76$ (U1)	ResNet_D2	2×2 conv2d_transpose
	$192 \times 320 \times 12$(R1)	U1	3×3 ResNet_dconv_block
	$192 \times 320 \times 44$	R1, VGG_C1	Concatenation, dropout (0.8)
Output Layer	$192 \times 320 \times 1$	ResNet_D1	1×1 conv, sigmoid

The encoder consists of VGG [19] style blocks where three consecutive 3×3 convolutions are followed by 2×2 max pooling to successively reduce the spatial resolution of the feature maps. In the VGG block, the feature maps after each convolution are batch normalized and non-linearly transformed using the ReLU activation as shown in Fig. 3(a). The decoder follows the encoder in U-Net like fashion. Each block in the decoder consists of up-sampling using transposed convolution followed by a ResNet [20] like de-convolution (transposed convolution) block, as shown in Fig. 3(b). The output is then concatenated with the corresponding feature maps (before max-pooling) from the encoder. The final layer is a convolutional layer that maps the input to a single channel output. Sigmoid non-linearity, Eq. (2), is then applied to output a saliency map indicating the confidence of each pixel belonging to the trail line. The network is then trained

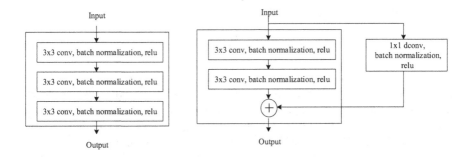

Fig. 3. a. VGG style block used in the encoder, b. ResNet-style block used in the decoder of the network.

from scratch using the proposed loss function of Eq. (1).

$$\tilde{y}_{ij} = p(x_{ij}) = \frac{1}{(1 + \exp(-x_{ij}))} \tag{2}$$

5 Experiments and Results

We evaluate the proposed method on a subset of the IDSIA forest trail dataset. The IDSIA forest trail dataset contains images of natural forest trail captured by three head-mounted cameras oriented in different directions. Image level annotations for instantaneous heading direction of the trail are implicit for this dataset. However, our intended target is not to find the instantaneous heading direction, but to detect the local segment of the trail visible in the image. Only the images captured from the front facing camera of folders 001 and 002 were used as the forest trail was visible only from the front facing camera and the images of folders 001 and 002 were captured under identical settings. A total of 3835 images were extracted and the trail was annotated with semantic lines and the corresponding distance maps were computed. 2860 images were used for training, 472 for validation and 503 images were set aside for testing. The training and test images were taken from different sections of the trail without overlap. The high resolution images were down sampled to 232×360 pixels to reduce computation and memory requirements. The training data was augmented with vertical mirrors, and random crops of size 192×320 were generated during runtime.

The network was trained from scratch with the network parameters initialized using Xavier initialization. We train the network with a batch size of 12 using Adam optimizer with initial learning rate of $1e^{-}4$ for a total of 40 epochs. The learning rate was reduced to half after 20 epochs. At the end of each epoch the model was evaluated on the validation set using Eq. (1), and the model with the lowest validation error was selected as the final model for test. The experiments were carried out using Tensorflow on a machine with NVIDIA Titan X GPU.

The mean intersection over union (mIoU) given by

$$IoU = \frac{G \cap P}{G \cup P} \tag{3}$$

is chosen as the evaluation metric on the test set, G is the ground truth mask and P is the binary prediction mask obtained after thresholding ($\tau = 0.5$) the output of the network. The performance of the proposed loss is compared to the commonly used CCE, MSE and JC. The target labels are one-hot encoded to compute CCE, MSE and JC, hence the number of channels at the output is two for these losses. Softmax was applied to these feature maps for scaling the output. To reduce the effect of class-imbalance in the training data, weights based on the inverse of class frequency is used for CCE and MSE. We report the performance by training the network five times with different random seeds. The final result is computed as the average of the prediction on the original and its mirror image. The mIoU performance of the different losses is presented in Table 2. From Table 2 we see that the proposed method leads to more accurate trails than CCE, JC and MSE. Unlike the other loss functions where the final test accuracy varies widely with the initialization condition, the proposed loss function leads to consistent results irrespective of the initialization condition.

Table 2. Performance of the proposed distance weighted loss function on the test set.

S.N	Loss	mIoU % (test set)
1	CCE	64.2 ± 1.55
2	MSE	67.7 ± 1.43
3	JC	66.3 ± 1.75
4	Distance weighted (proposed)	69.5 ± 0.22

In Fig. 4, we also plot the empirical cumulative distribution of the mIoU for a set of models trained under similar settings using the different loss functions. From Fig. 4 we can see that under a weak evaluation setting (the trail is said to be positively detected if mIoU > 0.3) CCE performs better than MSE and JC. But, under a strong evaluation setting (trail is said to be positively detected if mIoU > 0.5) the performance of CCE is way below that of MSE and JC. However, the proposed loss function shows superior performance compared to all the other losses under both the weak and strong evaluation settings.

Some qualitative results of end-to-end trail detection using the proposed method are presented in Fig. 5. Some examples where the mIoU between the ground truth and the predicted trail is greater than 60% is shown in Fig. 5(a) to (e). Some failure cases of the proposed method where the mIoU is less than 30% is presented in Fig. 5(f) to (j). We can observe that the network outputs thicker trail lines then the ground truth. This is in part due to low distance based penalty assigned to background pixels in the vicinity of the ground truth

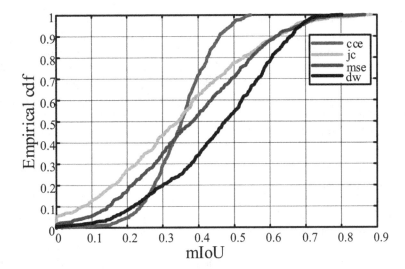

Fig. 4. Empirical cumulative distribution function of mIoU on the test set obtained by model trained using different loss functions. (cce: cross entropy, jc: Jaccard coefficient, mse: mean squared error, dw: distance weighted (proposed) loss).

and in part due to the architecture of the decoder. Simple post-processing step like erosion can be used to refine the output, however no postprocessing is done in the experiments.

In the absence of clear guiding landmarks or boundary, the process of annotating trails with semantic lines is subjective. Though we did not employ multiple

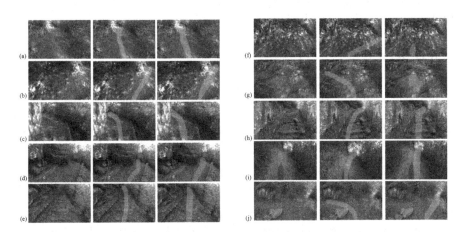

Fig. 5. Some qualitative results of trail detection using the proposed method. The test image, image with ground truth overlapped and the prediction overlapped are presented in the first, second and third column, respectively. (a–e) Results where the mIoU > 0.6, (f–j) failure cases where the mIoU < 0.3.

annotators to estimate the effect of human subjectivity, from Fig. 5(h), (i) and (j) it can be observed that human subjectivity can have significant effect on the accuracy of the trained model. Though the trained model outputs valid "navigable" trails for these images, their mIoU with the ground truth is less than 30% rendering them as poor detections. While preparing the training data, the trails were annotated with a single line. Even the trails bifurcating into multiple paths were annotated with a line representing the dominant path. From Fig. 5(j), it can be observed that such trail bifurcations have to be handled in an appropriate manner.

6   Conclusion

A novel distance weighted loss function for end to end learning of unstructured forest trail using convolutional encoder-decoder network with semantic lines was proposed. Unlike the pixel wise labelling used in semantic segmentation, the forest trail was annotated with "semantic line". A distance map derived from the ground truth was then used to weigh the loss function to guide the focus of the network on the forest trail. The proposed loss function produced more accurate and consistent trail predictions compared to other commonly used loss functions like the categorical cross-entropy, mean squared error and dice coefficient on the IDSIA forest trail dataset. Future work will include the development of techniques to output accurate and thinner trails and evaluation the proposed method on other applications and datasets.

Acknowledgments. This work was supported in part by the Korea Research Fellowship Program through the National Research Foundation (NRF) of Korea funded by the Ministry of Science and ICT (NRF-2015H1D3A1062316), and the Basic Science Research Program through the NRF funded by the Ministry of Education (NRF-2019R1A6A1A09031717 and NRF-2019R1A2C1011297).

References

1. Krizhevsky, A., Sutskever I., Hinton G.E.: ImageNet classification with deep convolutional neural networks. In: Advances in Neural Information Processing Systems, pp. 1097–1105 (2012)
2. Ren, S., He, K., Girshick, R., Sun, J.: Faster R-CNN: towards real-time object detection with region proposal networks. In: Advances in Neural Information Processing Systems, pp. 91–99 (2015)
3. Badrinarayanan, V., Kendall, A., Cipolla, R.: SegNet: a deep convolutional encoder-decoder architecture for scene segmentation. IEEE Trans. Pattern Anal. Mach. Intell. **39**(12), 2481–2495 (2017)
4. Mayer, N., et al.: A large dataset to train convolutional networks for disparity, optical flow, and scene flow estimation. In: Proceedings of the IEEE Conference on Computer Vision and Pattern Recognition, pp. 4040–4048 (2016)
5. Yang, J., Price, B., Cohen, S., Lee, H., Yang, M.H.: Object contour detection with a fully convolutional encoder-decoder network. In: Proceedings of the IEEE Conference on Computer Vision and Pattern Recognition, pp. 193–202 (2016)

6. Huval, B., et al.: An empirical evaluation of deep learning on highway driving. arXiv:1504.01716 (2015)

7. Chen, C., Seff, A., Kornhauser, A, Xiao, J.: DeepDriving: learning affordance for direct perception in autonomous driving. In: Proceedings of the IEEE International Conference on Computer Vision, pp. 2722–2730 (2015)

8. Bojarski, M., et al.: End to end learning for self-driving cars. arXiv:1604.07316 (2016)

9. Rasmussen, C., Scot, D.: Shape-guided superpixel grouping for trail detection and tracking. In: Proceedings of the IEEE/RSJ International Conference on Intelligent Robots and Systems, pp. 4092–4097 (2008)

10. Rasmussen, C., Lu, Y., Kocamaz, M.: Appearance contrast for fast, robust trail-following. In: Proceedings of the IEEE/RSJ International Conference on Intelligent Robots and Systems, pp. 3505–3512 (2009)

11. Santana, P., et al.: Tracking natural trails with swarm-based visual saliency. J. Field Robot. **30**(1), 64–86 (2013)

12. Hadsell, R., et al.: Learning long-range vision for autonomous off-road driving. J. Field Robot. **26**(2), 120–144 (2009)

13. Giusti, A., et al.: A machine learning approach to visual perception of forest trails for mobile robots. IEEE Robot. Autom. Lett. **1**(2), 661–667 (2016)

14. Smolyanskiy, N., Kamenev, A., Smith, J., Birchfield, S.: Toward low-flying autonomous MAV trail navigation using deep neural networks for environmental awareness. arXiv:1705.02550 (2017)

15. Adhikari, S.P., et al.: Accurate natural trail detection using a combination of a deep neural network and dynamic programming. Sensors **18**(1), 178 (2018)

16. Homepage: A Machine Learning Approach to Visual Perception of Forest Trails for Mobile Robots. http://people.idsia.ch/~guzzi/DataSet.html. Accessed 8 Aug 2019

17. Neven, D., De Brabandere, B., Georgoulis, S., Proesmans, M., Van Gool, L.: Towards end-to-end lane detection: an instance segmentation approach. In: 2018 IEEE Intelligent Vehicles Symposium (IV), pp. 286–291 (2018)

18. Ronneberger, O., Fischer, P., Brox, T.: U-Net: convolutional networks for biomedical image segmentation. In: Navab, N., Hornegger, J., Wells, W.M., Frangi, A.F. (eds.) MICCAI 2015. LNCS, vol. 9351, pp. 234–241. Springer, Cham (2015). https://doi.org/10.1007/978-3-319-24574-4_28

19. Simonyan, K., Zisserman, A.: Very deep convolutional networks for large-scale image recognition. In: International Conference Learning Representation (2015)

20. He, K., Zhang, X., Ren, S., Sun, J.: Deep residual learning for image recognition. In: Proceedings of the IEEE Conference on Computer Vision and Pattern Recognition, pp. 770–778 (2016)

Localization of Map Changes
by Exploiting SLAM Residuals

Zoltan Rozsa[1,2]([:envelope:]) [iD], Marcell Golarits[1] [iD], and Tamas Sziranyi[1,2] [iD]

[1] Machine Perception Research Laboratory of Institute for Computer Science
and Control (SZTAKI), Kende u. 13-17, Budapest 1111, Hungary
sziranyi.tamas@sztaki.mta.hu
[2] Faculty of Transportation Engineering and Vehicle Engineering,
Budapest University of Technology and Economics (BME KJK),
Műegyetem rkp. 3, Budapest 1111, Hungary
zoltan.rozsa@logisztika.bme.hu

Abstract. Simultaneous Localization and Mapping is widespread in both robotics and autonomous driving. This paper proposes a novel method to identify changes in maps constructed by SLAM algorithms without feature-to-feature comparison. We use ICP-like algorithms to match frames and pose graph optimization to solve the SLAM problem. Finally, we analyze the residuals to localize possible alterations of the map. The concept was tested with 2D LIDAR SLAM problems in simulated and real-life cases.

Keywords: SLAM · Change detection · Autonomous driving

1 Introduction

Simultaneous Localization and Mapping (SLAM) is the basis of the navigation of today's mobile robots and possibly of the future's autonomous vehicles. Mobile robots traverse a path more than once, this offers the possibility of loop closing. Especially true this statement for automated guided vehicles (AGVs) and the connected vehicles of the future. One autonomous car may travel on a road just once, but others will navigate on this road continuously and can utilize the information acquired earlier. From time-to-time, changing of the environment is inevitable. However, one of the biggest problems of the present SLAM algorithms is the robustness against moving objects (short term) and environment change (long term). Using robust features (and RANSAC like algorithms [19]) to estimate the relative transformation can partly eliminate this problem. However, the detected features can be on moving objects as well and this would cause incorrect relative motion estimation. In general, motion estimation is enhanced by increasing the number of matched features, but it will certainly result in decreased performance if there is some change in the scenes where the match is done. Trying to find these changes to eliminate these problems, update the map, or construct the correct 4D reconstruction [20] can be an exhausting search. We propose a solution to automatically detect these changes.

© Springer Nature Switzerland AG 2020
J. Blanc-Talon et al. (Eds.): ACIVS 2020, LNCS 12002, pp. 312–324, 2020.
https://doi.org/10.1007/978-3-030-40605-9_27

1.1   Contributions

The paper contributes to the following:

- A new methodology is proposed to localize possible changes of SLAM maps instead of brute force search.
- The method does not require any additional step, just the basic SLAM need to be executed and error change need to be analyzed.
- Numerous 2D graph SLAM tests have been tested and evaluated to generate the proof of concept.

1.2   Outline of the Paper

The paper is organized as follows: Sect. 2 surveys the literature about the related topics. Section 3 briefly explains the theoretical backgrounds and Sect. 4 describes the proposed method and the concept of change localization in detail. Sections 5 shows our test results. Finally, Sect. 6 draws some conclusions.

2   Related Works

Today, real-time scan matching and SLAM algorithms [7] available in industrial and market products based on only 2D laser scanner data. However, SLAM can be realized with many sensor types part of the autonomous driving kit. One of the most effective algorithms is the ORB-SLAM which is applicable for mono, stereo, and RGB-D cameras as well [12], but there are particular SLAM algorithms for other 2.5D depth sensors like multi-layer LIDARs [3]. SLAM algorithms can be categorized in many ways besides the sensors for which is applicable. We can distinguish feature-based like the ORB-SLAM semi-direct [5] and direct SLAMs [13], minimizing reprojection or photometric error. There can be 2D and 3D methods based on the motion assumption. In the case of 2D the optimization problem we differentiate filtering-based (e.g. extended Kalman filter - EKF) SLAMs and graph SLAM. In this paper, we deal with direct matching (in case of relative motion and loop closure estimation as well) and graph optimization based SLAMs, like [10]. The motion can be arbitrary, but it will be assumed to be two-dimensional because of ground vehicles and simplicity.

2.1   2D Pose Graph SLAM

This sub-problem alone and the related optimization is quite complex and has been widely researched. There are different approaches to simplify the solution of this nonlinear least squares optimization problem. One possibility can be the linear approximation [4], others try to separate the problem to linear and nonlinear parts [8] (if the orientation is known, position estimation is linear). Besides, dimension reduction is a research direction too. Wang et al. proved a few theories for graph optimization problems in the case of spherical covariance matrices [23]. They proved that six-dimensional least square optimization problem of the trivial pose-graph (3 nodes, 3 edges) can be reduced to the optimization of one variable and for another trivial problem (two anchor nodes) they provided solutions in closed form. Later, they showed that one step point feature SLAM can be also reduced to a one variable optimization [22]. Also, the general point feature SLAM of $3m + 2n$ variables to an m variables optimization problem (m is the number of poses and n is the number of features). It can be explained by its separable structure and with the fact that features can be considered as poses in the graph. These are the basis of our research.

2.2 SLAM and Change Detection

Change detection is important for many reasons as surveillance, statical monitoring of buildings, traffic forecasting or path planning of vehicles. It can be realized with different sensors like mobile-laser scanners [24] or cameras [15]. It can be done with different data structures like point-cloud generated with depth sensors [21] or mono cameras [15] by Structure from Motion (SfM) or on 2D image pairs with conventional methods [16] or deep networks [1]. The common point and disadvantage of these processes are that we have to do the comparison from frame to frame with a previous image or a submap.

Object graphs can be applied for visual place recognition [14] or solution of the whole SLAM problem [11]. Some semantic SLAM algorithms are capable of detecting objects which are inconsistent with previous measurements [17]. However, they require high-level interpretation and limited to objects constructing the pose graph. Other methods just ignore the inconsistent objects. Making robust SLAM algorithms in dynamic environments and analyzing their behavior is an actual topic [25], a survey about them can be found in [18], but these researches deal with moving objects instead of long term changes.

3 The Problem Formulation

In the following, a brief introduction will be given to the problem of 2D graph SLAMs. A detailed explanation can be found in [6]. We assume to have a heterogeneous graph with just robot poses or known data association (feature identification is correctly done by the front-end). The quantity and effect of false loop closures (perceptual aliasing [9]) is negligible, because of the vehicle, robot (e.g. patrolling ones) making rounds, we have a good position estimation.

$X = (p_1, \cdots, p_n)^T$ is a vector of position parameters, where $p_i = (x_i, y_i, \theta_i)$ describes the pose of node i. Let $z_{i,j}$ and $\Omega_{i,j}$ be the mean and the information matrix of a measurement between the node i and the node j. In case of a given conguration of the nodes p_i and p_j, $\tilde{z}_{i,j}(p_i, p_j)$ is the prediction of a measurement representing the relative transformation between two nodes. The function of computing the difference between the expected observation and the real observation is:

$$e(p_i, p_j, z_{i,j}) = z_{i,j} - \tilde{z}_{i,j}(p_i, p_j) \tag{1}$$

The error has assumed to have normal distribution with zero mean. The goal of a maximum likelihood approach is to find the conguration of the nodes p^* that minimizes the negative log-likelihood $F(p)$ of all observations:

$$F(p) = \sum_{i,j} e_{i,j}^T \Omega_{i,j} e_{i,j} \tag{2}$$

thus, we aim to solve the following equation:

$$p^* = argmin_p F(p) \tag{3}$$

4   Proposed Method

In the following, we assume in our change localization that the relative transformation estimation between two nodes is done by some frame matching algorithm [7] using all the points of the frames. This means relatively small change affects on the estimated transformation too. Usually, these algorithms produce some kind of score of the match, but these strongly depend on the frames we match (the scene and position influences). They cannot be directly compared, used for direct change localization (an illustrative example is shown in Fig. 1). However, the environment changes will produce an extra error in the pose estimation, which will appear in the residual error of the pose graph optimization.

4.1   Assumptions

We will investigate in the following the case when a new loop closure edge is detected and added to the graph. First, we will make some theoretical assumptions and practical simplifications:

- Loop closures mainly have local effects. This assumption is necessary to identify the neighborhood of the loop closing edge as the location of the change. In the investigated cases: we circle on a small graph (eg. AGVs on material handling system); or traveling and loop closing on a 'detachable' sub-graph of a large graph (road network).
- The uncertainty of positions will not increase with the new loop closure edge. Let $F_n(p)$ and $F_{n+1}(p)$ the minimized log-likelihood after the nth and $(n+1)^{th}$ graph optimization and so loop closure (n is much higher than number of rounds). F-s are assumed to be close to the optimal solution in the current configuration and also to the global optimum in case of high n and round numbers. Then $\det(\Omega^n) > \det(\Omega^{n+1})$. It is true that the uncertainty of each new position estimates is increasing, but it will be decreasing with every loop closure edges. This holds either in case of uncertainty derived from the sensor model (as [2] proved) or using identity information matrix as frequently applied in SLAM optimization back-ends.
- Approximately identical variance σ_e for all $e_{i,j}$. Because of the locally investigated environment, we assume uniformly distributed measurements and consequently loop closures.

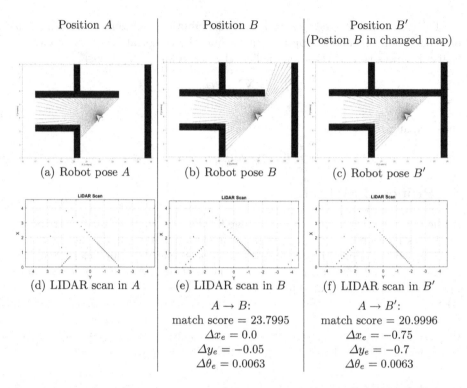

Fig. 1. Example about changing environment can cause wrong relative motion estimation and the uncertainty of match score (bad relative motion estimation with a higher score). Ground truth relative motion: $\Delta x_{gt} = -\frac{\sqrt{2}}{2}, \Delta y_{gt} = -\frac{\sqrt{2}}{2}, \Delta \theta_{gt} = 0.0$

- Residual error is supposed to have a Gaussian distribution. Let $\Omega_{i,j}$ be the identity matrix. After n (high enough) loop closure $F(p) = \sum e_{i,j}^T e_{i,j}$, and the sum of error squares can be approximated as a normal distribution (instead of χ^2) with the expected value:

$$\mu_{\Sigma e^2} = n\sigma_e^2 \tag{4}$$

and variance square:

$$\sigma_{\Sigma e^2}^2 = n2\sigma_e^4 \tag{5}$$

where σ_e is the variance of $e_{i,j}$ and the index Σe^2 refer to the sum of error squares. The equations above can be deduced from the central limit theorem. This simple form is true for independent and identically distributed variables as we assumed earlier. If that is not the case, $\sigma_{e_{i,j}}$-s significantly differ, σ_e^2 and σ_e^4 still can be substituted with one number, the mean of $\sigma_{e_{i,j}}^2$ and $\sigma_{e_{i,j}}^4$. Despite the fact the problem is not linear and p^* node configuration is continuously changing, we found the distribution above to be a good approximation for $F(p^*)$ as the proportion $\frac{\text{number of loop closure edges}}{\text{number of not loop closure graph edges}} \approx 1$ (or higher), as it can be, and was in the cases we investigated.

Using the assumption above, the expected value of the squared error of the next loop closure is approximated as $\mu_{e^2} = \sigma_e^2$ and its variance $\sigma_{e^2} = \sqrt{2}\sigma_e^2$. The threshold for the new error term to decide whether it fits the current approximation of the distribution or not (it indicates a change in the map with 3σ rule certainty):

$$th = \sigma_e^2 + 3\sqrt{2}\sigma_e^2 \tag{6}$$

It is one of the main ideas of the proposed change localization. It is important to determine a number n where we can start our inspection (previously we are far from the optimum, the residuals are unstable and most importantly the distribution assumption does not hold). It depends on the scale and number of the laps (it has to be minimum 2), the unknown variance of the errors (determined by the environment). We monitor the change of the specific function $f(p^*) = \frac{F(p^*)}{n}$ in order to decide this n. When the value of this function numeric derivative approaches 0 (becoming smaller than $3 \cdot 10^{-6}$ in our tests) we consider n large enough. Then, we assume that the difference of residual terms will not frequently change and the distribution of the squares can be approximated as a Gaussian one. The reason for that $\frac{d}{dn}\frac{n\sigma_e^2}{n} = \frac{d}{dn}\sigma_e^2 = 0$, the variance of the errors assumed to be approximately identical, so it does not depend on the number of loop closures.

4.2   Processing Steps

The steps of the proposed method:

1. Compute the earlier defined specific function $f(n)_{|p^*} = \frac{F(n)_{|p^*}}{n}$ and its derivative $\frac{d}{dn}f(n)_{|p^*}$ to examine limit.
2. Define the threshold for change detection at n (number of loop closures) $\frac{d}{dn}[G * f(n)_{|p^*}]$ approaches 0. It will be a higher n value than $\frac{d}{dn}f(n)_{|p^*}$ would result, so we will be closer to the optimum solution of the sub-graph.
3. We have to examine peaks in $\frac{d}{dn}[G * F(n)_{|p^*}]$. Gaussian smoothing is proposed for avoiding single extreme values indicating temporarily local minimums in the graph optimization. Maximums indicating changes should be present in more than one loop closures (depending on the distance between them), because a change in the map can be visible from more than one viewpoint. Also, near extreme values should be checked, because periodic maximums after the change with decreasing values are reasonable (explained later). However, a new maximum with a higher value can mean a new change.
4. In case of an extreme value (possible change candidate) the original value $\frac{d}{dn}F(n)_{|p^*}$ must be compared to the threshold value defined as $\sigma_e^2(1 + 3\sqrt{2})$ (Eq. 6) to decide whether it is salient value (possible change) or not.

4.3   Illustration of the Process

In the following, we would like to introduce the change localization process through a test representing typical data. The parameters of the test case are the following. Map: no. 2 (illustrated in Fig. 2), noise in robot position: 0.5 m, number of changes in the map: 5, rounds without change - with change: 10–5. Detailed explanation about the parameters and tests can be found in Sect. 5.

Figure 3 shows that in this test, in the case of a unchanged environment the residual error approximately linearly increasing with the number of loop closure edges. This fact supports our earlier assumption that the expected value of the matching error in a local environment is approximately the same in the case of approximately equidistant measurements from different positions. The first change can be detected, when there is a loop closing edge that includes a node where the change is perceivable. This will result in an error that does not fit into its previous distribution. This higher error introduced by the map change will be appearing as a step in the error term. However, when the map change will not be visible by the sensor, the approximately linear residual increase will set again, because the consequent frame can be well-matched in the changed environment too. Only relative motion estimation between the old and the new map (loop closure) will result in outstanding error terms. In every round, when the new (displaced or disappeared) objects of the map will be perceivable, a new high error term is added to the current residual values. However, this should decrease every time as we make the rounds. Now in the new state of the map, the edges representing this new state will dominate (also because over time loop closures will be linked to the changed map). Finally, the varying of the residual error should set back to the linear increasing state, but all the error term resulted by the change will not be eliminated. This offset represents the contradictory edges in the graph. The edges and nodes (now outliers) can be filtered out and the map should be updated.

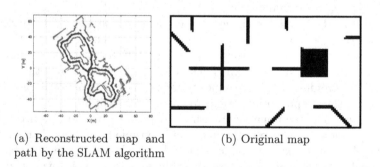

(a) Reconstructed map and path by the SLAM algorithm

(b) Original map

Fig. 2. Example map no. 2

We illustrated the specific function $f(p^*)$ in a basic case Fig. 4 where 0 m noise added and 0 object changed in the environment (e.g. a tram circles on a fixed path) for comparison to the examined case with changes Fig. 3. In this case, a decreasing tendency can be observed in the measured range.

5   Test Results

Our concept and method for change localization have been evaluated in many test cases in a virtual environment and in real-life as well. In this section, the circumstances of the tests and the achieved results are presented.

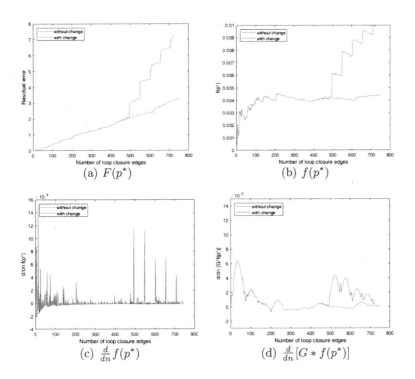

Fig. 3. The characteristic indicator functions of the test case in unchanged (blue) and changed (orange) environment (Color figure online)

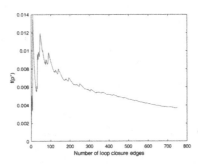

Fig. 4. $f(p^*)$ in the basic case (0 m noise and 0 object changed in the env.)

5.1 Quantitative Evaluation

In our synthetic tests, we used MATLAB environment and its robotics tool-box for the generation of test cases, simulate the robot environment perception and for the LIDAR SLAM [7]. First, we created some maps manually called *baseMaps*, then we placed in these *baseMaps* objects with random shape and random position. We used two maps with these random objects in each simulation to test our theorem. One of the maps was used as long term unchanged environment. To approximate real-world experiments, even more, we ran our tests with noise added to the robot path. We used three different *baseMaps*; three different amount of noise in robot position: $\rho = 0m, 0.5m, 1m$ (robot positions in its paths are randomly distributed within ρ radius of its original (first round) path); number of changes in the map: between 0 and 100, rounds without change - with change: both varying between 2 and 10; and we also varied the loop closure threshold of the algorithm from very low to very high. Loop closure threshold is proposed to set a medium value (default Matlab value is appropriate). If it is too low, obviously, too much error will be gathered, but if it is too high loop closure edges of changes will be dropped. (Decreasing loop closure number per round can indicate map change too, but it will not tell its location without further ado.) High errors in robot position, or too much change resulted in incorrect maps (or dropped loop closures for high matching thresholds). These will be left out of the following evaluation because in our earlier assumption we know, the robot circles, so too much deviation from the original path can be filtered. Also, too much change (more than 100% of original object area appeared - about 35–40 objects) cases indicated by a significant drop of loop closure edges per round are left out. In the final evaluation there were about 20.000 loop closures in 330 rounds.

As we have mentioned previously we evaluated the change of the residual error of the pose graph and its normalized with the number of loop closures. Table 1 shows the detection results based on different thresholds, xc, and xuc indices mean the function value at the position of change and unchanged, F-rate$_{uc}$ refer to the detection of "unchangingness".

We found that, when the areas of the new objects less than 3% of the areas of the original objects, the change does not influence much the matching and thus the residual error respect to loop closure number. These tests with little changes on the map cause the recall values not equal to 1.0 in Table 1.

Another conclusion we can draw from our test is that increasing the offset to the original path can lead to variation in residual error trend because of the completely new route. While the proposed threshold gave satisfactory results for the 'same-path' cases (first row of Table 1), for the 'noisy-path' cases not so (third row). Here, we increased this threshold value to deal with the possible higher deviation (fourth row). This can be done, because the traversed path is known, but the effect of the new route planned to be considered in a more appropriate way in the future.

5.2 Real-Life Experiments

We did real-life tests with an automatized forklift equipped with Sick NAV350 sensor for navigation purposes. The equipment was provided and the test was conducted in the VVRL[1]. The 2D point clouds acquired by the laser scanner was used for SLAM and the environment change localization. The AGV has done 5–5 rounds in the original and the changed environment (just as in some simulated test cases). The measurements in the different laps were executed approximately with the same position and orientation (as the vehicle circles on the same route in the production line). The results are presented in Fig. 5.

Table 1. Results based on different thresholds

Noise - threshold values	Precision	Recall	F-rate	Average $\frac{\frac{d}{dn}F(p^*)_{xc}}{th}$	Maximum $\frac{\frac{d}{dn}F(p^*)_{xuc}}{th}$	F-rate$_{uc}$
0 m–th	1.0	0.83	0.91	22.14	0.57	0.99
0 m–2.5 th	1.0	0.5	0.66	0.23	0.3325	0.99
0.5 m–th	0.46	1.0	0.63	14.29	2.22	0.96
0.5 m–2.5 th	1.0	0.83	0.91	5.71	0.88	0.99

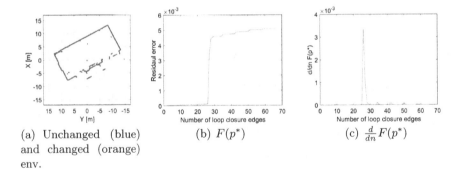

(a) Unchanged (blue) and changed (orange) env.

(b) $F(p^*)$

(c) $\frac{d}{dn}F(p^*)$

Fig. 5. Real-life experiment (Color figure online)

The predicted behaviour in the residual error function can be well-observed in both diagrams of Fig. 5. The extremum in the differential where the change in the environment perceived first, easily can be located as it exceeds the calculated

---

[1] Vehicle Vision Research Laboratory of the Faculty of Transportation Engineering and Vehicle Engineering's Department of Material Handling and Logistics Systems of Budapest University of Technology and Economics.

threshold for this 'same-path' case. After the change has been localized, if one would like to find the exact change, it is enough to examine the corresponding loop closure edge and the two scans of its nodes.

6   Conclusion

In the paper, a methodology is proposed to automatically localize changes in SLAM generated maps and so avoid the brute force based feature matching. This is extremely useful in cases where mobile robots frequently traverse approximately the same path (e.g. AGVs in a production line, vehicles with a fixed path as a railway) or in case of more vehicles with an information network. Analysis of the optimization residual can be useful in many cases. The success of the proposed method was proven with many 2D test cases and real-life experiments. The method can be easily extended to 3D or adapt to other sensors. Some proof of concept tests of the extension was already made, in the future, we would like to develop this extension and consider the effect of completely different paths.

Acknowledgment. The publication was supported by the European Commission through the Centre of Excellence in Production Informatics and Control (EPIC), grant No. 739592. The research was further supported by the Hungarian Scientific Research Fund No. OTKA/NKFIH K_120499.

References

1. Alcantarilla, P.F., Stent, S., Ros, G., Arroyo, R., Gherardi, R.: Street-viewchange detection with deconvolutional networks. Auton. Rob. **42**(7), 1301–1322 (2018). https://doi.org/10.1007/s10514-018-9734-5
2. Dissanayake, M.W.M.G., Newman, P., Clark, S., Durrant-Whyte, H.F., Csorba, M.: A solution to the simultaneous localization and map building (SLAM) problem. IEEE Trans. Robot. Autom. **17**(3), 229–241 (2001). https://doi.org/10.1109/70.938381
3. Droeschel, D., Behnke, S.: Efficient continuous-time SLAM for 3D lidar-based online mapping. In: 2018 IEEE International Conference on Robotics and Automation (ICRA), pp. 1–9 (2018). https://doi.org/10.1109/icra.2018.8461000
4. Durrant-Whyte, H., Roy, N., Abbeel, P.: A linear approximation for graph-based simultaneous localization and mapping. In: MITP (2012). https://doi.org/10.7551/mitpress/9481.003.0011
5. Engel, J., Schöps, T., Cremers, D.: LSD-SLAM: large-scale direct monocular SLAM. In: Fleet, D., Pajdla, T., Schiele, B., Tuytelaars, T. (eds.) ECCV 2014. LNCS, vol. 8690, pp. 834–849. Springer, Cham (2014). https://doi.org/10.1007/978-3-319-10605-2_54
6. Grisetti, G., Kümmerle, R., Stachniss, C., Burgard, W.: A tutorial on graph-based SLAM. IEEE Intell. Transp. Syst. Mag. **2**, 31–43 (2010). https://doi.org/10.1109/mits.2010.939925
7. Hess, W., Kohler, D., Rapp, H., Andor, D.: Real-time loop closure in 2D LIDAR SLAM. In: 2016 IEEE International Conference on Robotics and Automation (ICRA), pp. 1271–1278, May 2016. https://doi.org/10.1109/ICRA.2016.7487258

8. Khosoussi, K., Huang, S., Dissanayake, G.: Exploiting the separable structure of SLAM. In: Robotics: Science and Systems (2015). https://doi.org/10.15607/rss. 2015.xi.023

9. Lajoie, P.Y., Hu, S., Beltrame, G., Carlone, L.: Modeling perceptual aliasing in SLAM via discrete-continuous graphical models. IEEE Rob. Autom. Lett. **PP**, 1 (2019). https://doi.org/10.1109/LRA.2019.2894852

10. Mendes, E., Koch, P., Lacroix, S.: ICP-based pose-graph SLAM. In: 2016 IEEE International Symposium on Safety, Security, and Rescue Robotics (SSRR), pp. 195–200, October 2016. https://doi.org/10.1109/SSRR.2016.7784298

11. Mu, B., Liu, S.Y., Paull, L., Leonard, J.J., How, J.P.: SLAM with objects using a nonparametric pose graph. In: 2016 IEEE/RSJ International Conference on Intelligent Robots and Systems (IROS), pp. 4602–4609 (2016). https://doi.org/10.1109/iros.2016.7759677

12. Mur-Artal, R., Tards, J.D.: ORB-SLAM2: an open-source SLAM system for monocular, stereo, and RGB-D cameras. IEEE Trans. Rob. **33**(5), 1255–1262 (2017). https://doi.org/10.1109/TRO.2017.2705103

13. Newcombe, R.A., Lovegrove, S.J., Davison, A.J.: DTAM: dense tracking and mapping in real-time. In: 2011 International Conference on Computer Vision, pp. 2320–2327, November 2011. https://doi.org/10.1109/ICCV.2011.6126513

14. Oh, J.H., Jeon, J.D., Lee, B.H.: Place recognition for visual loop-closures using similarities of object graphs. Electron. Lett. **51**(1), 44–46 (2015). https://doi.org/10.1049/el.2014.3996

15. Sakurada, K., Okatani, T., Deguchi, K.: Detecting changes in 3D structure of a scene from multi-view images captured by a vehicle-mounted camera. In: 2013 IEEE Conference on Computer Vision and Pattern Recognition, pp. 137–144, June 2013. https://doi.org/10.1109/CVPR.2013.25

16. Sakurada, K., Okatani, T.: Change detection from a street image pair using CNN features and superpixel segmentation. In: BMVC (2015). https://doi.org/10.5244/c.29.61

17. Salas-Moreno, R.F., Newcombe, R.A., Strasdat, H., Kelly, P.H.J., Davison, A.J.: SLAM++: simultaneous localisation and mapping at the level of objects. In: 2013 IEEE Conference on Computer Vision and Pattern Recognition, pp. 1352–1359, June 2013. https://doi.org/10.1109/CVPR.2013.178

18. Saputra, M.R.U., Markham, A., Trigoni, N.: Visual SLAM and structure from motion in dynamic environments: a survey. ACM Comput. Surv. **51**(2), 37:1–37:36 (2018). https://doi.org/10.1145/3177853

19. Torr, P.H.S., Zisserman, A.: MLESAC: a new robust estimator with application to estimating image geometry. Comput. Vis. Image Underst. **78**, 138–156 (2000). https://doi.org/10.1006/cviu.1999.0832

20. Ulusoy, A.O., Mundy, J.L.: Image-Based 4-d reconstruction using 3-d change detection. In: Fleet, D., Pajdla, T., Schiele, B., Tuytelaars, T. (eds.) ECCV 2014. LNCS, vol. 8691, pp. 31–45. Springer, Cham (2014). https://doi.org/10.1007/978-3-319-10578-9_3

21. Underwood, J.P., Gillsj, D., Bailey, T., Vlaskine, V.: Explicit 3D change detection using ray-tracing in spherical coordinates. In: 2013 IEEE International Conference on Robotics and Automation, pp. 4735–4741, May 2013. https://doi.org/10.1109/ICRA.2013.6631251

22. Wang, H., Huang, S., Frese, U., Dissanayake, G.: The nonlinearity structure of point feature SLAM problems with spherical covariance matrices. Automatica **49**(10), 3112–3119 (2013). https://doi.org/10.1016/j.automatica.2013.07.025

23. Wang, H., Huang, S., Khosoussi, K., Frese, U., Dissanayake, G., Liu, B.: Dimensionality reduction for point feature slam problems with spherical covariance matrices. Automatica **51**, 149–157 (2015). https://doi.org/10.1016/j.automatica.2014.10.114

24. Xiao, W., Vallet, B., Schindler, K., Paparoditis, N.: Street-side vehicle detection, classification and change detection using mobile laser scanning data. ISPRS J. Photogrammetry Remote Sens. **114**, 166–178 (2016). https://doi.org/10.1016/j.isprsjprs.2016.02.007

25. Yu, C., et al.: DS-SLAM: a semantic visual SLAM towards dynamic environments. In: 2018 IEEE/RSJ International Conference on Intelligent Robots and Systems (IROS), pp. 1168–1174 (2018). https://doi.org/10.1109/iros.2018.8593691

Initial Pose Estimation of 3D Object with Severe Occlusion Using Deep Learning

Jean-Pierre Lomaliza[1] and Hanhoon Park[1,2]([✉]) [iD]

[1] Department of Electronic Engineering, Pukyong National University, Busan 48513, South Korea
hanhoon.park@pknu.ac.kr
[2] School of ITMS, University of South Australia, Adelaide, SA 5001, Australia

Abstract. During the last decade, augmented reality (AR) has gained explosive attention and demonstrated high potential on educational and training applications. As a core technique, AR requires a tracking method to get 3D poses of a camera or an object. Hence, providing fast, accurate, robust, and consistent tracking methods have been a main research topic in the AR field. Fortunately, tracking the camera pose using a relatively small and less-textured known object placed on the scene has been successfully mastered through various types of model-based tracking (MBT) methods. However, MBT methods requires a good initial camera pose estimator and estimating an initial camera pose from partially visible objects remains an open problem. Moreover, severe occlusions are also challenging problems for initial camera pose estimation. Thus, in this paper, we propose a deep learning method to estimate an initial camera pose from a partially visible object that may also be severely occluded. The proposed method handles such challenging scenarios by relying on the information of detected subparts of a target object to be tracked. Specifically, we first detect subparts of the target object using a state-of-the-art convolutional neural networks (CNN). The object detector returns two dimensional bounding boxes, associated classes, and confidence scores. We then use the bounding boxes and classes information to train a deep neural network (DNN) that regresses to camera's 6-DoF pose. After initial pose estimation, we attempt to use a tweaked version of an existing MBT method to keep tracking the target object in real time on mobile platform. Experimental results demonstrate that the proposed method can estimate accurately initial camera poses from objects that are partially visible or/and severely occluded. Finally, we analyze the performance of the proposed method in more detail by comparing the estimation errors when different number of subparts are detected.

Keywords: Initial camera pose estimation · Severe occlusion · Partially visible object · Deep learning · Subpart detection · Model-based tracking · Mobile augmented reality

This work was supported by a Research Grant of Pukyong National University (2019).

J. Blanc-Talon et al. (Eds.): ACIVS 2020, LNCS 12002, pp. 325–336, 2020.
https://doi.org/10.1007/978-3-030-40605-9_28

1   Introduction

Augmented reality (AR) has been shown to be a very good tool for interactive education or training [1–3]. AR can be divided into several categories by the method used to track camera pose. Marker-based AR [4,5] is usually the most robust way to track the camera pose, which uses a marker with known pattern and geometry; however, it is usually prone to failure of tracking in case of occlusion of the marker. Moreover, it is also limited in the range of movements since the marker should always be visible on each frame. On the other hand, markerless AR [6,7] offers more liberty of movements to the users, but is not known to be as robust as the marker-based AR. A good alternative to enabling robust tracking without using marker is model-based AR [8–11] where known 3D mesh model of a less-textured object (a target object placed on the scene) is used to track camera pose. Model-based AR can face the same problems as marker-based AR, i.e., the object should be always perfectly visible; however, for the training or educational purposes, teaching through interactive virtual contents that are directly added to the object being studied seems to be the best way to learn. Thus, model-based AR are more suitable in the training or educational AR scenarios, where the target object is the one being studied and the user or learner will always focus the camera on the object. In addition, note that 3D mesh model of real objects can be obtained very accurately in terms of dimensions, by using various 3D scanning technologies.

In the era of blooming of learning-based algorithms that outperform conventional algorithms in many research fields, the ideal scenario to tracking the camera pose accurately and robustly would be to use machine learning through powerful convolutional neural network (CNN) architectures, as done in [12–14]. As learning-based techniques estimate camera pose independently at each frame, they are known to be more reliable. However, they work well with relatively small objects and at the moment are not fast enough to be used to track objects in each frame on mobile platform in real time.

On the other hand, model-based tracking methods [8–11] can be adapted to run on mobile platform in real time (it will also be demonstrated in this paper). Moreover, model-based tracking methods can track the camera pose robustly even when the target object is partially visible in the camera frame. However, model-based tracking methods basically work in such a way that the camera pose in the current frame is obtained by computing a differential pose from the camera pose in the previous frame. Thus, model-based tracking methods are prone to be less reliable when the camera pose in the previous frame is inaccurate. Moreover, due to the dependency for camera pose in the previous frame, model-based tracking methods usually require the user to manually initialize the camera pose in the first frame. To avoid such inconveniences, this paper proposes a learning-based method that can estimate initial camera pose in cases where the target object is partially visible or/and severely occluded on mobile platform. Our initial pose estimator can also be used to recover the camera pose when tracking failure occurs. The proposed method basically detects subparts of a target object and use the geometric information of the detected subparts to estimate camera pose. Therefore, we also offer an analysis of the impact of the numbers detected subparts over the accuracy of the pose estimation.

The remaining parts of this paper are organized as follows. In Sect. 2, the proposed method for initial camera pose estimation will be explained in detail. In Sect. 3, the performance of the proposed initial pose estimator will be analyzed experimentally. In Sect. 4, conclusion and future studies will be presented.

2 Proposed Method

This paper focuses on the estimation of initial camera pose in the case where the target object to be tracked is partially visible in the camera frame or/and severely occluded. Therefore, rather than trying to detect the entire target object, we first propose to detect subparts of the target object using a state-of-the-art CNN, Google's SSD-MobileNet [15], that is also fast enough to be used for pose recovery when the main camera tracking method (in this paper, a model-based tracking method, Optimal Local Searching (OLS) [8]) fails to track the object. After detection of such subparts, the initial camera pose is then estimated using a deep neural network (DNN) that takes as input bounding box information of detected subparts. The process flowchart of the proposed method is shown in Fig. 1. The remaining body of this section will focus on explaining each step and contribution of this paper.

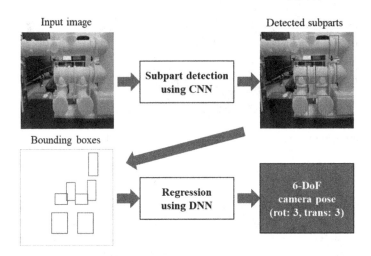

Fig. 1. Process flow of the proposed method.

2.1 Subpart Detection Using CNN

CNNs have demonstrated incredible performance in various applications due to the progress done in the machine learning field and considerably increased availability of data. One of the applications in which CNNs perform well is object detection and recognition, where the accuracy better than humans was achieved [16]. Thus, there are a variety of available CNN architectures that can be used directly on our method to achieve subpart detection with high accuracy. However, even though those CNNs work on a computer in real time, they are still very slow on mobile platform, such as mobile phone, which is the target platform of the proposed method. Hence, to our knowledge, the modified version of single shot multibox detector (SSD) [17], SSD-MobileNet [15], is the best architecture having the best tradeoff between speed and accuracy on mobile platform. The CNN-based object detector takes as input a three channels 300 × 300 image and returns all detected objects together with corresponding locations and confidence scores. In our case, we trained the CNN object detector to detect seven subparts of four different classes visible on one side of the target object, as shown in Fig. 2. The dataset for training the CNN-based object detector was collected using an annotation program shown in Fig. 2.

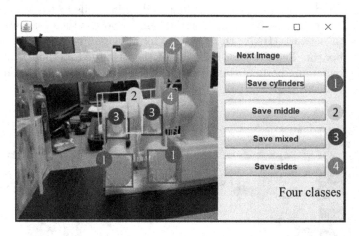

Fig. 2. Annotation program used to collect training dataset for the CNN-based object detector.

2.2   Camera Pose Estimation Using DNN

After detecting subparts of the target object, their locations are used as input to a DNN to regress to the initial camera pose parameters. Note that the CNN-based object detector does not always detect all subparts from the camera frame or some of those subparts may not be visible in the camera frame. Due to that fact, we trained different DNNs accordingly to numbers of detected subparts. One of outputs of the CNN-based object detector is the bounding box indicating location and boundaries of detected subparts. Those bounding boxes are described by height, width, top-left point, and bottom-right point. We added the aspect ratio as additional information. Hence, for each subpart, we obtain seven parameters that are fed into the regression DNN. As shown in Fig. 2, the target object has seven different subparts of four different classes.

In our experiments, using information of only one bounding box did not achieve an accuracy good enough to be used. Hence, we consider cases where the number of detected subparts varies between two and seven.

It is also important to mention that, we handle all possible arranged combinations of subparts. In other words, our trained DNNs take as additional input a code corresponding to subparts that are being combined. For example, for cases where only two subparts are detected, we have 21 possible combinations obtained by taking the seven subparts by two in order. Hence, the code corresponding to combinations will range from 0 to 20. The same logic is applied to cases where the other numbers of subparts are detected.

The overall architecture of a regression DNN is shown in Fig. 3. The number of inputs k varies accordingly to the number of detected subparts. In other words, different DNNs with different numbers of inputs (two to seven) are used, as mentioned previously. Besides exploiting multiple networks accordingly to numbers of inputs, we also attempt to estimate gradually refined pose through multiple phases. Our system has in total four phases where the first two are for initial estimations and the others are for refinements. In the first phase, phase_1, we train six different DNNs to estimate three translation parameters of the camera pose. Each of the six DNNs has different number of inputs and the input number is computed by Eq. 1. In the second phase, phase_2, we train other six DNNs to estimate rotation parameters of the camera pose. As most mobile devices have integrated orientation sensors, we use those sensors to our advantage. Here, as the gravity vector always points to the same direction, we use the angle between the gravity vector and the normal vector to the device's screen, as one of camera rotation parameters. That corresponds to the orientation around the x-axis. Thus, networks in phase_2 only estimate two rotation parameters. Moreover, the orientation acquired from the sensors data is used as an additional input to networks of phase_2. Outputs from phase_1 are also used as additional inputs to phase_2, as described in Eq. 2. The following phase_3 and phase_4 take charge of refinements for translation and rotation parameters estimated in phase_1 and phase_2, respectively. Those last two phases take as additional inputs, outputs from the previous phases, as described in Eqs. 2 and 3.

$$k_{phase\_1} = (N_{BB} \times 7) + 1, \tag{1}$$

$$k_{phase\_2}, k_{phase\_4} = (N_{BB} \times 7) + 1 + 1 + 3, \tag{2}$$

$$k_{phase\_3} = (N_{BB} \times 7) + 1 + 1 + 2. \tag{3}$$

Here, N_{BB} represents the number of detected bounding boxes. The number '7' that appears in each equation corresponds to the bounding box parameters that was described before. The first additional '1' in each equation is the parameter for identifying which subparts have been detected (or combined). The second '1' in Eqs. 2 and 3 is the additional input coming from the orientation sensor. Finally, the numbers '3' and '2' in Eqs. 2 and 3 represent the additional inputs coming from the outputs of the previous phases.

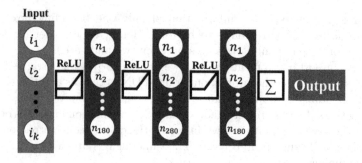

Fig. 3. Structure of regression DNN used for camera pose estimation from bounding box information. i and n represent input and hidden nodes, respectively.

2.3   Camera Pose Tracking and Recovery

After initial camera pose estimation, we use the OLS method [8] to keep tracking the camera pose in subsequent frames, frame by frame. Here, as the original OLS method was meant for desktop platforms, we tweaked several parameters to make it run on mobile platform in real time. Note that the OLS method can often fail to track the camera pose when the tracking conditions get poor (e.g., when quick and large camera motion occurs or when the target object is severely occluded). Thus, when the tracking failure happens (i.e., the tracking error is larger than a threshold), the proposed initial camera pose estimation method is performed again to recover the camera pose. With the recovered pose, the OLS method restarts tracking.

3   Experimental Results and Discussion

To create dataset used to train our regression DNN, we wrote a program in C++ using OpenGL. The program generated images of a target object with different camera poses, backgrounds, and lighting conditions. We also simulated scenarios where the object was partially visible and/or severely occluded. After generating those synthetic images, we used the trained CNN to detect subparts in each of those images and saved the information of detected subparts to text files. Finally, we used the saved information to train DNNs that estimate the camera pose. We used Python programming language and Google's Tensorflow library [18] for our DNN implementations. After training the DNNs, we generated other synthetic images in order to get data that were then used to evaluate accuracy of trained DNNs (Fig. 4). Camera poses were estimated from the synthetic images and estimation errors in each estimation phase for different input sizes are shown in Table 1. The errors are averaged differences between ground truth poses and estimated poses. As the number of detected subparts (bounding boxes) increased, the estimation errors tended to decreased. However, the errors were already small enough in the first two phases and the improvement (refinement) in the last two phases were not significant. Actually, the proposed method performed well without the refinement phases when the input size is bigger than three. Hence, in real scenarios, the refinement phases were used only when two or three subparts were detected.

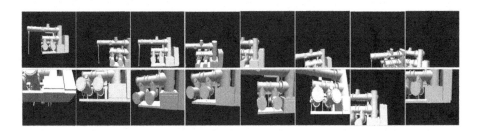

Fig. 4. Examples of synthetic images used for evaluating the accuracy of trained DNNs. The background was colored in black to facilitate the subpart detection by the CNN-based object detector, which reduces the effect of incorrectly detected subparts.

Table 1. Average estimation errors (in mm for translation parameters and in degrees for rotation parameters) of our DNN for different input sizes (= numbers of bounding boxes) and phases.

Input size	phase_1			phase_2		EVM	phase_3			phase_4		EVM
	e_{tx}	e_{ty}	e_{tz}	e_{ry}	e_{rz}		e_{tx}	e_{ty}	e_{tz}	e_{ry}	e_{rz}	
2	2.19	2.09	4.26	3.34	3.99	7.3747	1.87	1.59	3.93	3.24	3.52	6.6601
3	1.43	1.25	2.90	2.12	2.41	4.7200	1.13	1.07	2.57	2.05	2.16	4.2302
4	1.20	0.92	2.27	1.77	1.65	3.6461	1.05	1.20	2.18	1.56	2.17	3.7996
5	0.95	0.81	2.14	1.28	1.42	3.1200	0.88	0.83	2.18	1.42	1.52	3.2469
6	1.01	0.76	1.77	1.91	1.19	3.1200	1.01	0.79	3.59	1.34	1.39	4.2731
7	0.76	0.56	2.03	0.85	1.42	2.7840	0.59	0.81	1.86	1.13	1.10	2.6360

To analyze the errors in more detail, we computed the error vector magnitude (EVM), which is the norm of the 5D vector representing errors in translation and rotation parameters, i.e.,

$$EVM = \sqrt{e_{tx}^2 + e_{ty}^2 + e_{tz}^2 + e_{ry}^2 + e_{rz}^2}. \qquad (4)$$

Here, e_t and e_r represent the estimation errors in translation and rotation parameters, respectively. Using the EVM values, we can evaluate if the initial camera pose estimated by each input size is so accurate that the OLS algorithm would be able to start/keep tracking with the estimated initial camera pose. To analyze how much the OLS method is influenced by the initial camera pose, we increasingly added random noise to the vector representing the ground truth initial camera pose and investigated if the OLS algorithm can start/keep tracking with the noise-added camera poses, as shown in Fig. 5. Finally, based on the results of Fig. 5, we can then evaluate the estimation errors in Table 1 using the EVM values. Except for the input size of 2 where the probability of successful tracking would be around 85%, all other bigger input sizes had small EVM values (smaller than 5), such that the OLS method would always succeed to start/keep tracking.

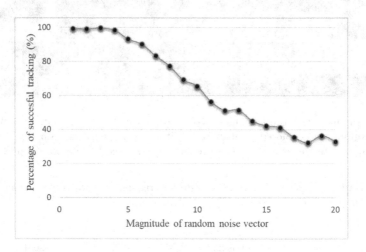

Fig. 5. Probability of successful tracking of the OLS algorithm according to the magnitude of noise added to the ground truth initial camera pose.

Figure 6 shows mobile AR results using the proposed method in real scenarios, including challenging ones where the target object is partially visible and severely occluded. When the target object was mostly visible and not occluded, the proposed method could consistently and accurately estimate the initial camera pose, as shown in Fig. 6a. Then, even if the degree of occlusion and visibility got increased, the proposed method could robustly estimate the initial camera pose using the information of small numbers of detected subparts, as shown in Figs. 6b–e. Hence, experimental results demonstrate that the proposed method, by exploiting locally detected subparts, can handle the challenging cases where the target object is partially visible or/and severely occluded. In all the scenarios, the OLS method could start/keep tracking with the initial camera pose estimated by the proposed method.

Besides initial pose estimation, the proposed method could be used also for recovery of camera pose in case of camera tracking failure. Our implementation works in such a way that, when the OLS method fails to track (i.e., the tracking error does not converge within the threshold), the proposed method runs and, once there are at least two subparts detected, the camera pose is recovered. The process of camera pose recovery is shown in Fig. 7. In the first frame, an initial pose was estimated and refined by the proposed method. Then, the camera pose was successfully tracked using the OLS method over frames. However, the

Fig. 6. Mobile AR results using the proposed method for different scenarios with varying numbers of detected subparts due to different degrees of occlusion and visibility. The first column shows subparts detected by the CNN. The second column shows 3D graphic models (in red) rendered with the initial camera pose estimated by the proposed method. The last column shows 3D graphic models (in green) rendered with the camera pose tracked using the OLS method. (Color figure online)

tracking failure occurred in the 135th frame and the camera pose was recovered immediately by the proposed method. Then, the OLS method restarted tracking and kept tracking in the subsequent frames.

Table 2 shows the processing time in the mobile AR experiments. The CNN-based object detection was still time-consuming on mobile environments and the initial pose estimation took approximately 90–110 ms depending on the devices. However, since the process only needed to be performed once at the beginning or recovery, the camera pose tracking could be done in real time.

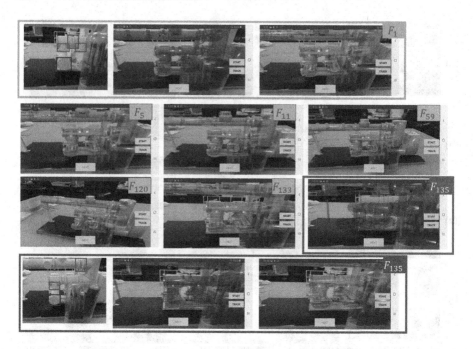

Fig. 7. Pose initialization, tracking, and recovery processes. The first row shows initial pose estimation and refinement in the first frame (frame #1). The second and third rows show tracking and AR results using the OLS method from frame #5 to frame #133. The right image of the third row shows a tracking failure by the OLS method at frame #135. Finally, the last row shows the camera pose recovery using the proposed method.

Table 2. Average processing time in ms on two smartphones.

	Subpart detection	DNN regression	OLS refinement	Total
Samsung Galaxy S8	65	28	35	128
LG V30	80	32	40	152

4   Conclusion and Future Work

In this paper, we proposed a method that estimates the initial camera pose on mobile platform using deep learning. The proposed method first locally detected subparts of a target object using a state-of-the-art CNN architecture and then fed the geometric information extracted from the detected subparts into a DNN that regressed to the initial camera pose. Using information of locally detected subparts made the proposed method very robust against challenging cases where the target object is partially visible in the camera frame or/and severely occluded by other objects. Hence, the proposed method is shown to be more practical to be used in cases where the target object is large in scale and would not always fit in a single camera frame.

In this paper, we did not care much about the lighting conditions. In future, it will be necessary to analyze the performance of the proposed method under poor or challenging lighting conditions. Also in this paper, we limited the camera viewpoint to one side of the target object for the experimental convenience. Thus, we need to extend the viewpoints to the other sides and analyze the performance of the proposed method.

References

1. Augmented Reality in Education. http://k3hamilton.com/AR/AR-Home.html. Accessed 29 Aug 2019
2. Billinghurst, M., Kato, H., Poupyrev, I.: The MagicBook - moving seamlessly between reality and virtuality. IEEE Comput. Graph. Appl. **21**(3), 6–8 (2001)
3. Dias, A.: Technology enhanced learning and augmented reality: an application on multimedia interactive books. Int. Bus. Econ. Rev. **1**(1), 69–79 (2009)
4. Kato, H., Billinghurst, M.: Marker tracking and HMD calibration for a video-based augmented reality conferencing system. In: 2nd IEEE and ACM International Workshop on Augmented Reality, pp. 85–94 (1999)
5. Wagner, D., Schmalstieg, D.: First steps towards handheld augmented reality. In: Seventh IEEE International Symposium on Wearable Computers, pp. 127–135 (2003)
6. Comport, A.I., Marchand, É., Chaumette, F.: A real-time tracker for markerless augmented reality. In: 2nd IEEE/ACM International Symposium on Mixed and Augmented Reality, p. 36 (2003)
7. Comport, A.I., Marchand, E., Pressigout, M., Chaumette, F.: Real-time markerless tracking for augmented reality: the virtual visual servoing framework. IEEE Trans. Vis. Comput. Graph. **12**(4), 615–628 (2006)
8. Seo, B.-K., Park, H., Park, J.-I., Hinterstoisser, S., Ilic, S.: Optimal local searching for fast and robust textureless 3D object tracking in highly cluttered backgrounds. TVCG **20**(1), 99–110 (2014)
9. Prisacariu, V.A., Reid, I.D.: PWP3D: real-time segmentation and tracking of 3D objects. IJCV **98**(3), 335–354 (2012)
10. Henning, T., Ulrich, S., Elmar, S.: Real-time monocular pose estimation of 3D objects using temporally consistent local color histograms. In: ICCV, pp. 124–132 (2017)
11. Tjaden, H., Schwanecke, U., Schömer, E.: Real-time monocular segmentation and pose tracking of multiple objects. In: Leibe, B., Matas, J., Sebe, N., Welling, M. (eds.) ECCV 2016, Part IV. LNCS, vol. 9908, pp. 423–438. Springer, Cham (2016). https://doi.org/10.1007/978-3-319-46493-0_26
12. Tekin, B., Sinha, S.N., Fua, P.: Real-time seamless single shot 6D object pose prediction. In: CVPR (2018)
13. Akgul, O., Penekli, H.I., Genc, Y.: Applying deep learning in augmented reality tracking. In: SITIS, pp. 47–54 (2016)
14. Xiang, Y., Schmidt, T., Narayanan, V., Fox, D.: PoseCNN: a convolutional neural network for 6D object pose estimation in cluttered scenes. In: RSS (2018)
15. Sandler, M., Howard, A., Zhu, M., Zhmoginov, A., Chen, L.-C.: Mobilenetv2: inverted residuals and linear bottlenecks. In: CVPR (2018)
16. He, K., Zhang, X., Ren, S., Sun, J.: Deep residual learning for image recognition. In: CVPR (2016)

17. Liu, W., et al.: SSD: single shot multibox detector. In: Leibe, B., Matas, J., Sebe, N., Welling, M. (eds.) ECCV 2016, Part I. LNCS, vol. 9905, pp. 21–37. Springer, Cham (2016). https://doi.org/10.1007/978-3-319-46448-0_2
18. Tensorflow: Machine Learning Library. https://www.tensorflow.org/. Accessed 29 Aug 2019

Automatic Focal Blur Segmentation Based on Difference of Blur Feature Using Theoretical Thresholding and Graphcuts

Natsuki Takayama[1]($\boxtimes$) and Hiroki Takahashi[1,2]

[1] Graduate School of Informatics and Engineering,
The University of Electro-Communications, Tokyo 182-8585, Japan
takayaman@uec.ac.jp, rocky@inf.uec.ac.jp
[2] Artificial Intelligence Exploration Research Center, Tokyo 182-0026, Japan

Abstract. Focal blur segmentation is one of the interesting topics in computer vision. With recent improvements of camera devices, multiple focal blur images of different focal settings can be obtained by a single shooting. Utilizing the information of multiple focal blur images is expected to improve the segmentation performance. We propose one of the automatic focal blur segmentation using a pair of two focal blur images with different focal settings. Difference of blur features can be obtained from an image pair which are focused on an object and background, respectively. A theoretical threshold identifies the object and background in the difference of blur feature space. The proposed method consists of (i) the theoretical thresholding in the blur feature space; and (ii) energy minimization based on Graphcuts using color and blur features. We evaluate the proposed method using 12 and 48 image pairs, including single objects and flowers, respectively. As results of the evaluation, the averaged Informedness of the initial and the final segmentation are 0.897 and 0.972 for the single object images, and 0.730 and 0.827 for the flower images, respectively.

Keywords: Difference of blur features · Focal blur segmentation · Graphcuts · Theoretical thresholding

1 Introduction

Images often include partial blur which is caused by focal settings of a camera and distances to objects in a scene. Such images are called focal blur images, and analysis of the blur information is one of the important research topics of computer vision. The focal blur image includes several useful information, for example, the relative depth of a scene and region of interest of a photographer. Focal blur segmentation, which segments an image into sharp and blur regions, is an essential technique to extract such information and contributes to various applications.

© Springer Nature Switzerland AG 2020
J. Blanc-Talon et al. (Eds.): ACIVS 2020, LNCS 12002, pp. 337–347, 2020.
https://doi.org/10.1007/978-3-030-40605-9_29

(a) Object-focused image.   (b) Background-focused image.

Fig. 1. Example of image pair with different focus settings.

Automatic focal blur segmentation using a single image, however, is one of the difficult topics. In spite of the much effort in this direction [2, 6–10, 12–14], the further improvement of segmentation performance is required for the practical applications.

On the other hand, multiple focal blur images of different focal settings can be obtained by a single shooting with recent improvements in camera devices. Such cameras are becoming standard for recent smartphones. In addition, it is possible to obtain multiple focal blur images with conventional cameras by using a video during a focus operation [3]. Utilizing the information of multiple focal blur images is expected to improve the segmentation performance.

In this paper, we propose an automatic focal blur segmentation using a focal blur image pair with different focal settings. The proposed method utilizes difference of blur feature from the focal blur image pair which focused on an object and background as shown in Fig. 1. The difference of blur feature has a benefit that a theoretical threshold can be defined in the difference of blur feature space. This characteristic allows automatic thresholding without seeking a threshold under strict assumptions of feature distribution.

The proposed segmentation employs a stepwise strategy. First, the difference of blur feature map is estimated from the focal image pair, and it segmented by the theoretical thresholding. Next, mathematical morphology which consists of erosion and dilation is applied to estimate erroneous segmented regions of the initial segmentation. Finally, Graphcuts [11] using color and blur features are applied to correct the initial segmentation.

We build a focal blur image database to evaluate the proposed method. The database includes 12 and 48 focal blur image pairs which include single objects and flowers, respectively. This paper reports the segmentation performance of the proposed method using the database.

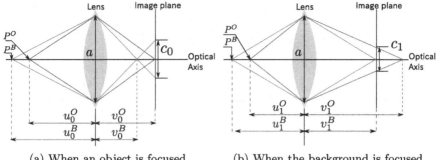

(a) When an object is focused.   (b) When the background is focused.

Fig. 2. Geometrical relationships under the thin lens approximation.

2 Difference of Blur Feature Using Different Focal Settings

Generally, camera focusing is conducted by moving a lens or camera along its optical axis. Figure 2 depicts geometrical relationships between an object, background, lens, and image plane under the thin lens approximation. Figure 2(a) is a geometrical relationship when an object is focused. Figure 2(b) is that of the background. We note that the object, background, and image plane are fixed in Fig. 2. a in Fig. 2 indicates an aperture of the lens. u denotes distances between the lens and the object or that of the lens and background. v denotes distances between a focal point of the object and the lens or that of a focal point of background and the lens. c is a diameter of Circle of Confusion (CoC). P^O and P^B are positions of the object and background, respectively. The object and background are sharply and blurry projected in Fig. 2(a), respectively. In Fig. 2(b), it is vice versa.

Let f denotes a focal length of the lens. The geometrical relationships in Fig. 2(a) and (b), and the lens formula $1/f = 1/u + 1/v$ lead the diameters of CoC c_0 and c_1 as follows:

$$c_0 = af \frac{u_0^B - u_0^O}{u_0^B(u_0^O - f)} \tag{1}$$

$$c_1 = af \frac{u_1^B - u_1^O}{u_1^O(u_1^B - f)} \tag{2}$$

Let c^O and c^B indicate the CoC of the object and background, respectively. While the diameters are $c_0^O \simeq 0$ and $c_0^B \simeq c_0$ in Fig. 2(a), they are $c_1^O \simeq c_1$ and $c_1^B \simeq 0$ in Fig. 2(b). At this time, $c_0^O - c_1^O \leq 0$ and $c_0^B - c_1^B \geq 0$ are held because $c_0^O \leq c_1^O$ and $c_0^B \geq c_1^B$. This means that the difference of blur feature of the focal blur image pair has opposite signs in the object and background regions, respectively. Therefore, the theoretical threshold in the difference of blur feature space to segment the object and background is defined as 0.

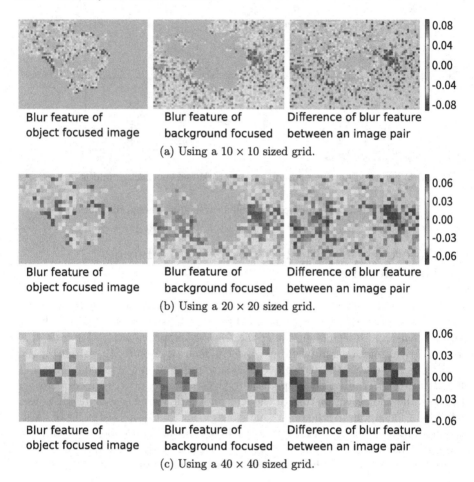

Fig. 3. Examples of the difference of blur feature maps.

3   Difference of Blur Feature Estimation

The difference of blur feature is estimated in local regions. We apply the author's blur map generation scheme [8] in this paper.

First, the focal blur image pair is divided into local regions using multiple-sized grids of 10×10, 20×20, and 40×40 pixels. Next, the blur feature of each local region of the object and background focused images are estimated using ANGHS (Amplitude Normalized Gradient Histogram Span). ANGHS is an extension of GHS [6] which evaluates variation of normalized gradient in a local region, and it is robust to variation of amplitude and size of local regions. The local gradient is normalized by the local intensity amplitude. We use common intensity amplitude $A = \max(A_0, A_1)$ for the corresponding local regions of the object-focused and background-focused images to calculate the difference of blur feature. A_0 and A_1 indicate local intensity amplitude of a local region in the

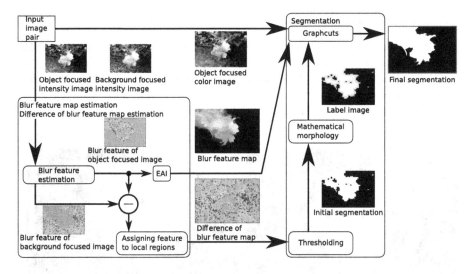

Fig. 4. Process overview. (Color figure online)

object and background focused images, respectively. A_0 and A_1 are estimated using the 95% confidence interval of the local intensities under the Gaussian assumption. Finally, the difference of blur feature is estimated by subtracting ANGHS of the background focused image from that of the object focused image for each corresponding local region.

Figure 3 shows examples of the difference of blur feature maps. Figure 3(a), (b), and (c) show the blur and difference of blur feature when 10×10, 20×20, and 40×40 sized grids are used, respectively. As shown in the difference of blur feature, we can find the local regions of the object and background has positive and negative values in the difference of blur feature space.

4   Automatic Blur Segmentation

4.1   Process Overview

Figure 4 shows the process overview of the proposed method. First, the input color images are converted to intensity images using $RGB \rightarrow YUV$ color conversion. Next, the difference of blur feature map is estimated as explained in Sect. 3. At the same time, the blur feature map of the object-focused image is estimated based on the author's method [8] using Edge Aware Interpolation (EAI) [4]. After that, the theoretical thresholding is applied to the difference of blur feature map as the initial segmentation. And then, mathematical morphology is applied to the initial segmentation result to estimate the erroneous regions. Finally, Graphcuts using the color image and blur feature map of the object-focused image corrects the initial segmentation.

4.2   Theoretical Thresholding

The difference of blur feature maps of each grid size are segmented by thresholding with theoretical threshold 0. After that, we apply a voting process to decide the label value of the initial segmentation. Each pixel of the initial segmentation is labeled as the object region when two of the three thresholded results vote to the object region, and it is labeled as the background region in other cases.

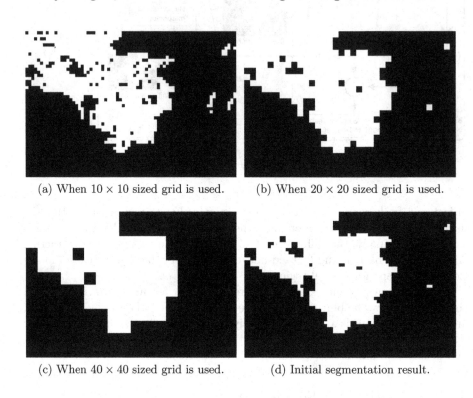

(a) When 10 × 10 sized grid is used.   (b) When 20 × 20 sized grid is used.

(c) When 40 × 40 sized grid is used.   (d) Initial segmentation result.

Fig. 5. Example of initial segmentation.

The thresholded results of 10 × 10, 20 × 20, and 40 × 40 sized grids, and the initial segmentation are shown in Fig. 5. As shown in Fig. 5(a), the error due to the artificial block shape is relatively small, but there is many small isolated regions and holes when a small-sized grid is used. On the other hand, the error due to the block shape increases, and the isolated regions are merged when a large-sized grid is used as shown in Fig. 5(c). The voting process balances the number of isolated regions and the artificial block shapes as shown in Fig. 5(d).

The almost remaining errors of the initial segmentation are near from boundaries of the object and background. Therefore, mathematical morphology including erosion and dilation is conducted to estimate the erroneous regions. 41 × 41 matrix with all of the entries are 1 is used as a structuring element in this paper.

Figure 6 shows the result of the mathematical morphology of the initial segmentation. The light gray color in Fig. 6 indicates the label changed pixels after the erosion. The dark gray color in Fig. 6 indicates those of the dilation.

Fig. 6. Example of label image after the mathematical morphology.

4.3   Graphcuts Using Color and Blur Features

We apply Graphcuts using color and blur features as the final segmentation. In order to apply Graphcuts, we formulate the final segmentation as an energy minimization problem.

$$E(\boldsymbol{X}) = \Sigma_i E_1(X_i) + \gamma \Sigma_{i,j} E_2(X_i, X_j), \tag{3}$$

where $\boldsymbol{X}(X_i \in \{0, 1, 2, 3\})$ is the label image after the mathematical morphology. $X_i = 0$ and $X_i = 3$ indicate that the i_{th} pixel is in the object and background regions, respectively. The labels of these pixels are not changed after the mathematical morphology. $X_i = 1$ and $X_i = 2$ indicate that the i_{th} pixel is likely in the object and background regions, respectively. The labels of these pixels are changed after the erosion and dilation, respectively. The first and second terms are referenced as the data and smoothness terms, respectively. The data term evaluates the validity of the applied label of each pixel. The smoothness term evaluates the continuity of labels between adjacent pixels. γ is a constant that adjusts the effect of the data and smoothness terms. The data and smoothness terms are mapped as the cost of a graph. Then, Graphcuts solves Eq. (3) by cutting the graph into the two sub-graphs which indicate the object and background regions using min-cut/max-flow algorithm [1].

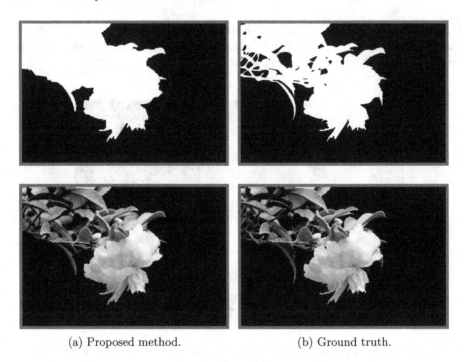

(a) Proposed method.   (b) Ground truth.

Fig. 7. Example of final segmentation and object extraction.

We suppose the pixels of $X_i = 0$ and $X_i = 3$ are correctly segmented in the initial segmentation. Therefore, we assign the fixed high cost to the data term of $X_i = 0$ and $X_i = 3$ not to change those labels by Graphcuts.

On the other hand, the likelihood under a color and blur feature models are assigned to the pixels of $X_i = 1$ and $X_i = 2$. The color and blur feature models are learned based on 5 components Gaussian Mixture Models. The feature models are learned for the object and background regions individually according to $\boldsymbol{X}_{X_i=0\vee1}$ and $\boldsymbol{X}_{X_i=2\vee3}$ of the label image.

The cost of the data term is calculated as follows:

$$E_1(X_i) = \sum_k -\log d^C(k) + \sum_k -\log d^B(k), \tag{4}$$

where $d^C(k)$ and $d^B(k)$ indicate the likelihood of the color and blur feature models, respectively, and these are calculated as follows:

$$d^C(k) = \frac{w_k^C}{|\boldsymbol{\sigma}_k^C|^{1/2}} \exp\left(-\frac{(\boldsymbol{I}_i - \boldsymbol{\mu}_k^C)^T (\boldsymbol{\sigma}_k^C)^{-1}(\boldsymbol{I}_i - \boldsymbol{\mu}_k^C)}{2}\right), \tag{5}$$

$$d^B(k) = \frac{w_k^B}{(\sigma_k^B)^{1/2}} \exp\left(-\frac{(b_i - \mu_k^B)^2}{2(\sigma_k^B)^2}\right). \tag{6}$$

Table 1. Summarized performances.

	Maximum	Minimum	Average	SD
Initial-Single object	0.937	0.828	0.897	0.030
Final-Single object	0.997	0.893	0.972	0.036
Initial-Flower	0.908	0.433	0.730	0.103
Final-Flower	0.984	0.402	0.827	0.109

$I_i = [I_i^r, I_i^g, I_i^b]$ and b_i indicate the pixel values of the color image and blur feature map, respectively. w_k indicates a relative weight of the k_{th} Gaussian component. This paper uses the relative number of pixels belongings to each component. $|\sigma_k^C|$ indicates a determinant of a covariance matrix.

The cost of the smoothness term is calculated as follows:

$$E_2(X_i, X_j) = \frac{|X_i - X_j|}{\text{dist}(i, j)} \exp\left(-\frac{(I_i^4 - I_j^4)^T(I_i^4 - I_j^4)}{2 < (I_i^4 - I_j^4)^T(I_i^4 - I_j^4) >}\right), \qquad (7)$$

where $\text{dist}(i, j)$ is Euclidean distance between adjacent pixels; $I_i^4 = [I_i^r, I_i^g, I_i^b, b_i]^T$.

Figure 7 shows an example of the final segmentation and object extraction according to the result. $\gamma = 30$ was used as an experimental value in this example. A manually defined ground truth is also shown as a comparison. Graphcuts corrects the errors of the initial segmentation as shown in Fig. 7.

5   Performance Evaluation

This paper reports the performance of the proposed method using 12 and 48 focal blur image pairs which include single objects and flowers. We used Informedness [5] to evaluate the performance.

The summarized performances of the initial and final segmentation are described in Table 1. The performance of the single objects and flowers are separately evaluated. "SD" in Table 1 is standard deviation. Moreover, the examples of the focal image pairs and segmented results are shown in Fig. 8. We can find the effectiveness of the stepwise segmentation from the comparison of the initial and final segmentation in Table 1 and Fig. 8. Some flower images, however, degrades the performance in the final segmentation. The current proposed method has difficulties in handling thin structures, as shown in the right-most column in Fig. 8. In such cases, Graphcuts tends to merge the object regions, and the errors are increased.

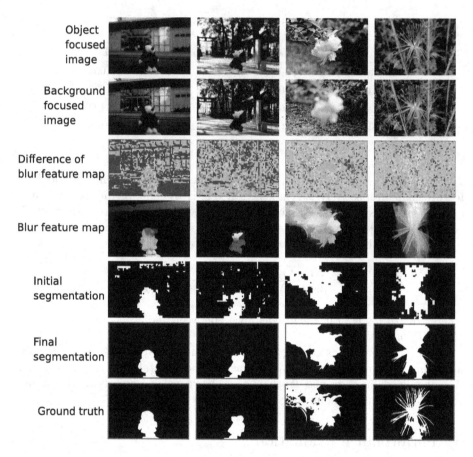

Fig. 8. Examples of segmented results.

6   Conclusions and Future Work

In this paper, we proposed an automatic focal blur segmentation using the pair of two focal blur images with different focal settings. The difference of blur feature can be obtained from the image pair which are focused on the object and background, respectively. We have shown that the theoretical threshold is defined in the difference of blur feature space. The effectiveness of the proposed method which consists of the theoretical thresholding and Graphcuts is presented experimentally.

In future research, we aim to extend the proposed method to a multiple-class segmentation using more focal images. Moreover, improvements in the handling of thin structures must be addressed.

References

1. Boykov, Y., Kolmogorov, V.: An experimental comparison of min-cut/max- flow algorithms for energy minimization in vision. IEEE Trans. Pattern Anal. Mach. Intell. **26**(9), 1124–1137 (2004)
2. Chakrabarti, A., Zickler, T., Freeman, W.: Analyzing spatially-varying blur. In: Proceedings of IEEE Conference on Computer Vision and Pattern Recognition, pp. 2512–2519, June 2010
3. Hyeongwoo, K., Christian, R., Christian, T.: Video depth-from-defocus. In: Proceedings of Fourth International Conference on 3D Vision, pp. 370–379, October 2016
4. Levin, A., Lischinski, D., Weiss, Y.: A closed-form solution to natural image matting. IEEE Trans. Pattern Anal. Mach. Intell. **30**(2), 228–242 (2008)
5. Powers, D.M.W.: Evaluation: from precision, recall and f-measure to ROC, informedness, markedness and correlation. Int. J. Mach. Learn. Technol. **2**(1), 37–63 (2011)
6. Renting, L., Zhaorong, L., Jiaya, J.: Image partial blur detection and classification. In: Proceedings of IEEE Conference on Computer Vision and Pattern Recognition, pp. 1–8, June 2008
7. Shi, J., Xu, L., Jia, J.: Discriminative blur detection features. In: Proceedings of IEEE Conference on Computer Vision and Pattern Recognition, pp. 2965–2972, June 2014
8. Takayama, N., Takahashi, H.: Blur map generation based on local natural image statistics for partial blur segmentation. IEICE Trans. Inf. Syst. **E100–D**(12), 2984–2992 (2017)
9. Zhang, W., Cham, W.K.: Single image focus editing. In: Proceedings of IEEE International Conference on Computer Vision Workshops, pp. 1947–1954, September 2009
10. Yuan, L., Chun, Y.: Automatic segmentation of background defocused nature image. In: Proceedings of 2nd International Congress on Image and Signal Processing, pp. 1–5, October 2009
11. Yuri, Y.B., Marie-Pierre, J.: Interactive graph cuts for optimal boundary and region segmentation of objects in N-D images. In: Proceedings of IEEE International Conference on Computer Vision, vol. 1, pp. 105–112, July 2001
12. Zhi, L., Weiwei, L., Liquan, S., Zhongmin, H., Zhaoyang, Z.: Automatic segmentation of focused objects from images with low depth of field. Pattern Recogn. Lett. **31**(7), 572–581 (2010)
13. Zhu, X., Cohen, S., Schiller, S., Milanfar, P.: Estimating spatially varying defocus blur from a single image. IEEE Trans. Image Process. **22**(12), 4879–4891 (2013)
14. Zhuo, S., Sim, T.: Defocus map estimation from a single image. Pattern Recogn. **44**(9), 1852–1858 (2011)

Feature Map Augmentation to Improve Rotation Invariance in Convolutional Neural Networks

Dinesh Kumar$^{(\boxtimes)}$ (iD), Dharmendra Sharma(iD), and Roland Goecke(iD)

University of Canberra, Bruce, ACT 2617, Australia
{dinesh.kumar,dharmendra.sharma,roland.goecke}@canberra.edu.au
https://www.canberra.edu.au/

Abstract. Whilst it is a trivial task for a human vision system to recognize and detect objects with good accuracy, making computer vision algorithms achieve the same feat remains an active area of research. For a human vision system, objects seen once are recognized with high accuracy despite alterations to its appearance by various transformations such as rotations, translations, scale, distortions and occlusion making it a state-of-the-art spatially invariant biological vision system. To make computer algorithms such as Convolutional Neural Networks (CNNs) spatially invariant one popular practice is to introduce variations in the data set through data augmentation. This achieves good results but comes with increased computation cost. In this paper, we address rotation transformation and instead of using data augmentation we propose a novel method that allows CNNs to improve rotation invariance by augmentation of feature maps. This is achieved by creating a rotation transformer layer called Rotation Invariance Transformer (RiT) that can be placed at the output end of a convolution layer. Incoming features are rotated by a given set of rotation parameters which are then passed to the next layer. We test our technique on benchmark CIFAR10 and MNIST datasets in a setting where our RiT layer is placed between the feature extraction and classification layers of the CNN. Our results show promising improvements in the networks ability to be rotation invariant across classes with no increase in model parameters.

Keywords: Convolutional Neural Network · Rotation invariance · Data augmentation · Deep learning · Feature maps

1 Introduction

Classifying and recognising objects in the visual world in a trivial task for humans, but is still a great challenge for researchers to accomplish the same level of accuracy using computers in the Computer Vision field. While human vision system can easily filter information from the environment, the same variability in natural environments is an obstacle for effective computer vision tasks [14]). Inspired by how the visual cortex of mammals detects features through

© Springer Nature Switzerland AG 2020
J. Blanc-Talon et al. (Eds.): ACIVS 2020, LNCS 12002, pp. 348–359, 2020.
https://doi.org/10.1007/978-3-030-40605-9_30

studies by Hubel and Wiesel [9,10], algorithms such as the CNN have achieved great success in various computer vision tasks such as image classification, object detection, visual concept discovery, semantic segmentation and boundary detection. Algorithms purely based on CNNs or used as a basis of other complicated algorithms are applied in various practical domains such as in self driving cars, defence and security, mobile devices, medical image processing and quality assurance in manufacturing industries.

While CNNs are accepted as the state-of-art method for solving image classification problems, they still have certain limitations such as their inability to handle different variations of transformations on the input such as transformation of rotations. Some research show max pooling layers achieve some translation and rotation invariance [20]. However the exact level of invariance achieved is unknown and usually dependent on the number of max pooling layers used in the model. A popular technique used to make CNNs invariant is to apply data augmentation on the training data. Various random transformations are applied on the input data which is then trained on the CNN. Though this gives good classification results, it is seldom practical particularly when working with large datasets. This technique also contradicts with how the natural vision system works which does not need to be exposed to variations of the same scene to learn invariance. Dicarlo et al. [3] reports detailed architecture of the visual pathways and suggesting invariance encoding happens within the ventral stream. We use these findings to suggest feature map augmentation in the deeper end of the CNN model, allowing the classifier to learn variations of the input data in terms of its transformed features. We initially trial our method on one form of transformation namely rotation.

There is a genuine need for CNNs to learn invariant features for several applications where the pose or orientation of the examples cannot be determined in real time or be similar to the learnt examples. One such case is the application of CNNs in skin cancer research, where models are trained on skin lesion images and deployed on mobile devices that can be used by people as an early warning system. Not only the lesion images may be different in appearance to the learnt example, the angle used to take an image of a suspected skin lesion by individuals may vary drastically.

In this paper, we introduce a simple technique to improve rotation invariance in CNNs by augmentation of feature maps. This is achieved by creating a rotation invariant transformer (RiT) layer that can be placed at the output end of a convolution layer. In RiT features are rotated by a given set of rotation parameters which are then passed to the next layer. Unlike data augmentation methods, our technique does not involve any manipulation of the input data. We conduct extensive experiments to evaluate rotation invariance performance of CNNs combined with RiT on widely used CNN architectures [8] and benchmark datasets. In our experiments, RiT layer is placed at the tail end of the feature extraction pipeline before the classification layer (Fig. 1). The location of the RiT layer to be at the end of all CNN layers is to allow final extracted features to be fed into RiT for transformation prior to being sent to the classification layer.

The results are presented for LeNet5 [15] and modified VGG networks [19] VGG16-5 (*MVGG5*) and VGG16-7 (*MVGG7*) trained on MNIST [16] and CIFAR10 [13] datasets. First the datasets are trained on models LetNet5, MVGG5 and MVGG7 to establish benchmark results and for comparison. Then, *RiT* layer is added to the models and retrained on MNIST and CIFAR10 datasets. We study the effect of transforming feature maps (by applying rotations) on the network's ability to classify test images subjected to varying rotation angles and compare with our benchmark. Here we wish to evaluate the network's performance as we increase the rotation angles of test images. We also study the performance of the CNN+*RiT* models by sampling images from each dataset class, applying rotations on them and evaluating the accuracy. Finally, we study CNN+*RiT* network on color and grey-scale images. In all our case studies, results of CNN+*RiT* are compared with our benchmarks.

Our test results show overall improvements in classification of test images over several distinct rotation angles in comparison to benchmark results. We are also able to demonstrate higher classification accuracy for several classes based on test images sampled from each class. Our results also indicate better performance of CNN+*RiT* network on color images than on grey-scale images.

The rest of the paper is organised as follows: Sect. 2 reviews related work while Sect. 3 introduces our model. Section 4 describes our experiment design and results obtained followed by a summary of outcomes and conclusion in Sect. 5.

2   Recent Advancements

Numerous literature reports a fundamental problem in CNNs in its lack of ability to be spatially invariant to input data that has been subjected to common transformations such as translation, scaling, rotation and small deformations [11,12,17]. Reasons are mainly attributed to the restriction of the receptive field (kernel) in a CNN to address small patches of the image at a time allowing excellent local feature extraction but failing to extract spatial global features such as contours. To solve the problem of invariance in CNNs research work has mostly been focused on finding solutions for individual invariance problems separately such as rotation invariance, translation invariance and scale invariance. This is mainly due to application domain needs that requires the network to perform better in one invariance aspect rather than for all so as to avoid making the network complex which either affects computation time or increases network parameters.

Several papers have proposed various architectures to address rotation invariance problems in CNN networks. In the area of object detection Cheng et al. [2] proposed a rotation-invariant and Fisher discriminative layer within R-CNN framework [6] which they trained by introducing a regularization constraint on the cost function. Though reported to have achieved state-of-the-art results, their model is still noted to have used data augmentation as the first step of training. In this paper our aim is to show rotation-invariance without input data augmentation. In another work [1] the authors developed a new layer for

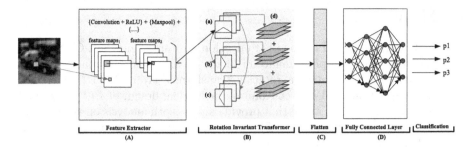

Fig. 1. Architecture of CNN+RiT which comprises of (A) standard convolution layers, (B) RiT layer, (C) flatten layer and (D) fully connected classifier layer. Features extracted by the convolution layer (a) are fed to the RiT, rotated using predefined rotation parameters which in our case are 90° clockwise (b) & 90° anti-clockwise (c). The input features (a) and the rotated features (b) and (c) are stacked and returned as output. These are then reshaped into a vector form by the flatten layer (C) and forwarded to the classifier for learning (D).

CNNs called Rotation Invariant CNN (RICNN) to detect objects on remote sensing images. Their model is also trained by its own objective function and regularization constraint. In contrast to our proposed method of augmentation by rotation of feature maps, Marcos et al. [18] in their work reported encoding rotation invariance in the CNN model by rotating filters. Their paper describes several advantages of their method in accounting for rotation invariance for texture classification. Further, their work was trialled on shallow CNNs. In domains where embedding rotation invariance is crucial such as for galaxy morphology prediction and classification, Dieleman et al. [4] propose a deep CNN model that works by exploiting translational and rotational symmetry in the annotated images from the Galaxy Zoo project. They achieved this by applying the same set of feature detectors (filters) to various rotated versions of the input image resulting in a form of data augmentation guided by rotational symmetry.

Manipulating feature maps to achieve rotation invariance has been studied to some extent. For example Follmann et al. [5] describe work where they use rotational pooling layers in conjunction with rotational convolutions to achieve rotation invariance. Finally an influential piece of work in this domain by Jaderberg et al. [11] introduced an end-to-end trainable module called the *Spatial Transformer* that spatially transforms feature maps by passing them through the transformer's localisation network, grid generator and sampler in succession. The heart of the *Spatial Transformer* module is the localisation network that contains a feed-forward network which generates and learns the parameters of the spatial transformation that should be applied to the input feature map. This architecture is reported to learn several invariances such as translation, scale, rotation and generic warping. However the limitation of this technique is that it limits the number of objects that can be modelled in a feed-forward network. Whilst much progress and state-of-the-art results and models are shown there is still a lot of scope in the area of learning invariance in CNN models.

3   Model

The goal of our proposed method is to develop a rotation-invariant CNN model in order to improve the performance of image classification for rotated images. This is achieved by creating a rotation transformer layer (RiT) that can be placed at the output end of a convolution layer. The design of RiT is inspired by the work of Dicarlo et al. [3] that provide an in-depth analysis on the internal architecture of the visual system. Their studies suggest that invariance is learnt automatically within the vision system rather than being exposed to variations of the same image. The combined CNN+RiT model comprises of four main parts as shown and described in Fig. 1.

3.1   Rotation Invariant Transformer (RiT) Layer

The RiT forms an integral part of our model. This layer does not require any trainable parameters and its main operations are to apply rotation transformations to the input feature maps and stack them for forward passing. In addition to the input feature maps RiT accepts a list of rotation transformation parameters to apply to the input feature maps. These parameters are set at compile time of the model. In our work we specify two rotations from the list $rot\_list = [90°, 270°]$ where 90° is a clockwise rotation and 270° represents 90° anti-clockwise rotation. The main objective of RiT is to provide feature map augmentation for rotation. The motivation behind choosing these rotation parameters to be in multiples of 90° is to allow a full rotation without losing any parts of a feature map that would otherwise be truncated from the edges. Since there are no parameters to be learnt in this layer, it executes fast and does not add significant computation time to the network. However given that RiT makes copies of the input feature maps increases the output depth dimension of the layer multiplied by a factor of n which is equal to the number of transformation supplied in rot_list. This increases the input dimension for the feed forward classifier but the number of parameters in the hidden layer remains unchanged.

Forward Propagation. Our goal is to create variations of the final feature map from the convolution layer for learning by the classifier. To achieve the forward pass function, RiT accepts a set of input feature maps. The incoming input feature maps are retained while copies of it are rotated according to the rotation parameters supplied (Fig. 1(a), (b) and (c)). The input feature maps plus rotated feature maps are then concatenated to form the final output map for the next layer (Fig. 1(d)). This stacked output is returned as augmented feature maps exiting RiT.

Backward Propagation. There are no trainable parameters in RiT. This makes the implementation of the backward function simple and straight forward. The backward function receives gradients from the network and *unstacks* or *slices* the gradients in the exact same dimensions of the input feature maps it received during the forward pass. It finally returns the gradient slice corresponding to the input feature map.

3.2 Description of Other Layers in Our Model (Fig. 1)

(A) Feature Extractor. The convolutional network of our model is built with the standard convolution, ReLU and max pooling layers. For our work we experimented with LeNet5, MVGG5 and MGG7 networks. The primary role of the convolutional network is to extract features using convolution by small filters. Having extracted local features through earlier convolution layers, the model finally outputs global features with the help of network's deeper convolution layers. It is these global features that becomes the main input for the RiT layer.

(B) Flatten Layer. The role of the flatten layer is unchanged from standard convolutional models. This layer operates on the incoming $2D$ feature maps and reshapes to form a single dimension feature vector.

(C) Fully Connected Layer. Here we use a fully connected neural network (NN). In our work the architecture of the NN is same as those defined in the LeNet5, MVGG5 and MVGG7 [8] models respectively. The learning of rotated features are highly dependent on the structure of this NN hence the design of this layer is important with respect to the number of hidden layers and neurons in those hidden layers. All our experimental models LeNet5, MVGG5 and MVGG7 comprise of two hidden layers which is in line with suggestions by Heaton [7] that two hidden layers are capable of representing functions with any kind of shape and that the optimal size of the hidden layer is recommended to be between the size of its input data and size of its output. LeNet5 satisfies the later criteria which is clearly evident in our results performing better than MVGG5 and MVGG7 networks.

4 Experiments and Results

Our experiments were conducted on MNIST [16] and CIFAR10 [13] datasets. For the feature extraction part of our work we used CNN architectures LeNet5 [15] and modified VGG networks MVGG5 and MVGG7 [19]. We describe the datasets, CNN architectures and our experimental results in the following sections.

4.1 Dataset Description

MNIST: The MNIST dataset contains 70,000 sample images of handwritten digits. This is divided into 50,000 training samples, 10,000 validation and 10,000 test samples. The sample images are grey-scale (1-channel) and of size 28×28 pixels. There are 10 classes for this dataset representing the digits from 0 to 9. The training and test batches have unequal distribution of the number of samples from each class.

CIFAR10: CIFAR10 dataset consists of 60,000 color images of size 32×32 pixels with 3-channels. The dataset is divided into 50,000 training samples and 10,000 test samples. There are no validation samples but this can be drawn at random from the train dataset. The samples are divided into 10 mutually exclusive classes defining various objects. This dataset is divided into five training batches and one test batch.

4.2   CNN Architectures

LeNet5 Network: Proposed by LeCun et al. [15], the LeNet5 network in our work comprises of three sets of convolution layers and two max pooling layers. All convolution layers use a filter size of 5×5 with the number of filters of 6, 16 and 120 respectively. The architecture is described in Table 1. Since we are using two datasets with different dimensions for the input images (32×32 for CIFAR10 and 28×28 for MNIST), the hyper-parameter *padding* for the second convolution layer in LeNet5 network trained on CIFAR10 is set to 1. For LeNet5 model trained on MNIST, *padding* for the first and second convolution layers are set to 2 and 1 respectively. This is done to allow the final feature maps dimensions (f_h, f_w) to be >1 so that rotations of these maps is possible in RiT layer.

Modified VGG Networks: We use the modified VGG networks used by Hosseini et al. in [8] in which the networks were trained on CIFAR10 and MNIST datasets and tested on samples of negative images from the same dataset. In the MVGG7 network the hyper-parameter *padding* is set to 1 for the forth convolution layer to allow the final feature map size (f_h, f_w) to be >1.

4.3   Experimental Setup

End-to-end training was performed for all architectures given in Table 1 on CIFAR10 and MNIST datasets respectively. We first train the benchmark CNNs - LeNet5, MVGG5 and MVGG7 on CIFAR10 and MNIST datasets. This establishes our benchmark results against which we compare results of

Table 1. Architecture of LeNet5 and MVGG networks used in our experiments [8].

Model	Layers
LeNet5	(conv $5 \times 5 \times 6$) → (maxpool 2×2) → (conv $5 \times 5 \times 16$) → (maxpool 2×2) → (conv $5 \times 5 \times 120$) → (fc 84) → (fc 10) → softmax
MVGG5	(conv $3 \times 3 \times 16$) → (conv $3 \times 3 \times 16$) → (maxpool 2×2) → (conv $3 \times 3 \times 48$) → (maxpool 2×2) → (fc 128) → (fc 10) → softmax
MVGG7	(conv $3 \times 3 \times 16$) → (conv $3 \times 3 \times 16$) → (maxpool 2×2) → (conv $3 \times 3 \times 32$) → (conv $3 \times 3 \times 32$) → (maxpool 2×2) → (conv $3 \times 3 \times 48$) → (maxpool 2×2) → (fc 128) → (fc 10) → softmax

Table 2. Sample sizes and number of images in each sample batch per rotation.

Number of classes (CIFAR10/MNIST)	10	
Number of rotations	26	
Rotations ($r°$)	[0, 5, 10, 30, 45, 60, 80, 90, 100, 120, 135, 150, 170, 180, 190, 210, 225, 240, 260, 270, 280, 300, 315, 330, 350, 355]	
Sample size	Images per class	Images per sample, per rotation
10	10	100
100	100	1000
300	300	3000
500	500	5000

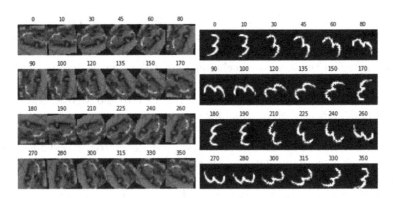

Fig. 2. An example of a rotated test image from CIFAR10 (*left*) and MNIST (*right*) dataset. Numbers indicate the degree of rotation applied on the image. Best viewed in color. (Color figure online)

CNN+RiT networks. Then we train the datasets on ensemble CNN+RiT networks. We repeat training for 50 epochs with learning rate of 10^{-3}, momentum of 0.5 and $L2$ regularization of 10^{-3}. We use batch size 8 for both training and testing. Stochastic gradient decent and cross-entropy was used as learning and loss function respectively. We implemented our models in Python v3.6.8 using accelerated numba library for parallelisation of python code on a Dell Latitude i7 laptop with 16 GB of RAM. Table 3 compares the training losses for all our networks on the two datasets. Whilst the final training loss for MVGG5+RiT is higher than the benchmark MVGG5 CNN, final losses for LeNet5+RiT and MVGG7+RiT are lower by 4.7% and 3.3% respectively than their corresponding benchmark results.

4.4   Accuracy of CNN+RiT network on rotated images

To test CNN+RiT networks on rotated images we generated a random set of images sampled from each class from the CIFAR10 and MNIST datasets and applied various degrees of rotations on these images. Table 2 describes the sample sizes and the number of images selected per class and per rotation for

Table 3. Train losses for all models used in our experiments.

Model	Loss-CIFAR10	Loss-MNIST
LeNet5	0.7945	0.0889
LeNet5+ *RiT*	**0.7569**	**0.0858**
MVGG5	**0.4847**	**0.0742**
MVGG5+*RiT*	0.59	0.0745
MVGG7	0.7173	0.0712
MVGG7+*RiT*	**0.6939**	**0.0665**

that sample size batch. The sampling of various sizes allowed us to observe the consistency of the networks in classifying rotated images drawn from small and large sample sizes. Rotation values were chosen within the range $[0° - 360°)$ to evaluate the effectiveness of our technique particularly for images rotated closer to the rotation degrees of the feature maps (in our case $(90°, -90°)$). Figure 2 shows a sample test image rotated from both CIFAR10 and MNIST test dataset. Rotation of $0°$ indicates no rotation for that image and serves as the base image for comparison.

For illustration, we choose the accuracy graph generated on 500 images sampled from each class in our datasets for each rotation. Figure 3 shows the performance of CNN+RiT networks compared with the respective benchmark models. On CIFAR10 rotated images indicated by the accuracy graphs, a high number of rotated images were classified correctly by the CNN+RiT network compared to

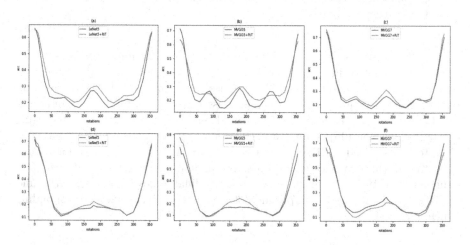

Fig. 3. Shows the performance of CNN+RiT networks compared with the respective benchmark models. Accuracy of the models are tested on 500 random samples drawn for each class from the CIFAR10 and MNIST test datasets per rotation. The x-axis identifies the accuracy at respective rotation points given in Table 2. Graphs (a), (b) and (c) show accuracy on CIFAR10 samples, while (c), (d) and (e) show accuracy on MNIST dataset.

the benchmark models. For example the overall average classification accuracy is better by ≈3% and ≈2% respectively for LeNet5+RiT and MVGG5+RiT networks. On MNIST rotated images, we observe the same trend, however showing greater accuracy when rotation angles approach 0° and 180° than for other rotations for MVGG5+RiT network. The difference between the performances on these datasets indicate CNN+RiT shows better results on color images than on grey-scale images. Overall our results indicate feature map augmentation has an influence on network accuracy and helps improve rotation invariance within CNN networks without increasing network parameters.

4.5   Performance of the CNN+RiT network on dataset classes

Finally we evaluate the consistency and performance of the CNN+RiT models on our dataset classes. We wish to investigate improvements in individual class prediction accuracies and evaluate the consistency of our results with results from benchmark models. Here, for each rotation (Table 2) we select 500 random images from each class and apply the selected rotation. This means for each rotation 5000 rotated images are evaluated for each class by the models. We repeat this method for both datasets. Tables 4 and 5 show the average performance of the proposed models on CIFAR10 and MNIST datasets. For each class the values are calculated by taking the sum of accuracies over all rotations divided by the number of rotations. This method is repeated for all models in our work.

Overall, using RiT with LeNet5, MVGG5 and MVGG7 networks on CIFAR10 dataset has shown superior performance compared to the corresponding benchmark networks (Table 4). Higher average classification accuracies were obtained for **70%** of the classes using the LeNet5+RiT network than the benchmark LeNet5 network (indicated by the bold-face values). In addition, average classification accuracies were higher for **60%** of the classes using MVGG5+RiT network and **50%** of classes using MVGG7+RiT network. The highest overall classification improvement is on class *dog* on the MVGG7+RiT network by 13% (from 46% to 59%) whilst the poorest performance is on class *cat* on the same MVGG7+RiT network. A similar performance is shown by the MVGG5+RiT network confirming that the fully connected layers influence the learning ability of the features. This is due to the same hidden layer parameters used for both the MVGG5 and MVGG7 networks.

Table 4. Average performance on 500 rotated samples for each class from the CIFAR10 test dataset across all rotations.

Model/Class	air- plane	auto-mobile	bird	cat	deer	dog	frog	horse	ship	truck
LeNet5	0.23	0.25	0.31	0.34	0.43	0.42	0.53	0.15	0.21	0.17
LeNet5+RiT	**0.36**	**0.37**	**0.34**	0.26	0.39	**0.43**	0.45	**0.27**	**0.26**	**0.23**
MVGG5	0.25	0.26	0.19	0.51	0.25	0.33	0.61	0.21	0.20	0.26
MVGG5+RiT	**0.32**	**0.29**	**0.28**	0.33	**0.32**	0.32	**0.64**	0.20	**0.32**	0.21
MVGG7	0.27	0.17	0.35	0.48	0.27	0.46	0.55	0.24	0.23	0.21
MVGG7+RiT	**0.31**	**0.27**	**0.40**	0.29	0.27	**0.59**	**0.65**	0.18	0.22	0.20

Table 5. Average performance on 500 rotated samples for each class from the MNIST test dataset across all rotations.

Model/Class	0	1	2	3	4	5	6	7	8	9
LeNet5	0.46	0.09	0.45	0.19	0.13	0.33	0.22	0.20	0.53	0.18
LeNet5+RiT	0.46	**0.14**	0.43	**0.22**	**0.19**	**0.39**	0.14	0.18	0.46	**0.29**
MVGG5	0.26	0.07	0.43	0.30	0.14	0.39	0.17	0.17	0.52	0.18
MVGG5+RiT	**0.43**	**0.21**	0.41	0.20	**0.29**	**0.43**	**0.20**	**0.23**	0.43	0.14
MVGG7	0.48	0.17	0.36	0.24	0.15	0.51	0.19	0.29	0.41	0.18
MVGG7+RiT	0.28	**0.18**	**0.45**	0.21	**0.23**	0.48	0.18	0.15	0.35	**0.20**

Test results on rotated samples from MNIST dataset show similar trends (Table 5). Higher average classification accuracies were obtained for **50%** and **60%** of the classes using LeNet5+RiT and MVGG5+RiT networks respectively. On MGG7+RiT, the model produced higher classification accuracies for **40%** of the classes. Here our results indicate CNN+RiT combination network behaves differently on grey-scale images but overall shows promising improvements in classification accuracies on individual classes with rotated examples. Hence showing that transforming feature maps with controlled rotations as in our RiT layer helps CNN networks become tolerant to rotation of test images.

5   Conclusion

In this work we propose a method to learn rotation invariance in CNNs by introducing a new technique of feature map augmentation within CNN networks. The proposed method uses a new layer called RiT that takes input feature maps from convolution layers and performs rotation on them. The final output from RiT is then learnt by the fully connected network.

Our test results show overall improvements in classification of test data subjected to rotation transformations in comparison to benchmark results on CIFAR10 and MNIST datasets. Further we are able to demonstrate improved average performances across classes for color images than on grey-scale images. From our experimental results we conclude the ensemble CNN+RiT network is able to learn rotation invariance by feature map augmentation with no increase in network parameters. Our method can easily be integrated into CNN based applications requiring rotation invariance such as in skin lesion classification.

Problems and opportunities that require further investigation are to apply more rotations on the feature maps, extend this technique of feature map augmentation to other forms of transformations such as translations and scaling and to apply CNN+RiT to larger and more complex datasets.

References

1. Cheng, G., Zhou, P., Han, J.: Learning rotation-invariant convolutional neural networks for object detection in VHR optical remote sensing images. IEEE Trans. Geosci. Remote Sens. **54**(12), 7405–7415 (2016)

2. Cheng, G., Zhou, P., Han, J.: RIFD-CNN: rotation-invariant and fisher discriminative convolutional neural networks for object detection. In: Proceedings of the IEEE Conference on Computer Vision and Pattern Recognition, pp. 2884–2893 (2016)
3. Dicarlo, J., Zoccolan, D., Rust, N.C.: How does the brain solve visual object recognition? Neuron **73**, 415–434 (2012). https://doi.org/10.1016/j.neuron.2012.01.010
4. Dieleman, S., Willett, K.W., Dambre, J.: Rotation-invariant convolutional neural networks for galaxy morphology prediction. Mon. Not. Roy. Astron. Soc. **450**(2), 1441–1459 (2015)
5. Follmann, P., Bottger, T.: A rotationally-invariant convolution module by feature map back-rotation. In: 2018 IEEE Winter Conference on Applications of Computer Vision (WACV), pp. 784–792. IEEE (2018)
6. Girshick, R., Donahue, J., Darrell, T., Malik, J.: Rich feature hierarchies for accurate object detection and semantic segmentation. In: Proceedings of the IEEE Conference on Computer Vision and Pattern Recognition, pp. 580–587 (2014)
7. Heaton, J.: Introduction to Neural Networks for Java, 2nd edn. Heaton Research Inc., Chesterfield (2008)
8. Hosseini, H., Xiao, B., Jaiswal, M., Poovendran, R.: On the limitation of convolutional neural networks in recognizing negative images. In: 2017 16th IEEE International Conference on Machine Learning and Applications (ICMLA), pp. 352–358. IEEE (2017)
9. Hubel, D.H., Wiesel, T.N.: Receptive fields of single neurons in the cat's striate cortex. J. Physiol. **148**, 574–591 (1959)
10. Hubel, D.H., Wiesel, T.N.: Receptive fields and functional architecture of monkey striate cortex. J. Physiol. **195**, 215–243 (1968)
11. Jaderberg, M., Simonyan, K., Zisserman, A., Kavukcuoglu, K.: Spatial transformer networks. In: Cortes, C., Lawrence, N.D., Lee, D.D., Sugiyama, M., Garnett, R. (eds.) Advances in Neural Information Processing Systems 28, pp. 2017–2025. Curran Associates Inc., New York (2015)
12. Kauderer-Abrams, E.: Quantifying translation-invariance in convolutional neural networks. arXiv preprint arXiv:1801.01450 (2017)
13. Krizhevsky, A., Hinton, G., et al.: Learning multiple layers of features from tiny images. Technical report. Citeseer (2009)
14. Kyrki, V.: Local and global feature extraction for invariant object recognition. Ph.D. thesis, Lappeenrannan University of Technology, Finland (2002)
15. LeCun, Y., Bottou, L., Bengio, Y., Haffner, P., et al.: Gradient-based learning applied to document recognition. Proc. IEEE **86**(11), 2278–2324 (1998)
16. LeCun, Y., Cortes, C., Burges, C.: The MNIST database of handwritten digits. Technical report (1998)
17. Lenc, K., Vedaldi, A.: Understanding image representations by measuring their equivariance and equivalence. In: CVPR (2015)
18. Marcos, D., Volpi, M., Tuia, D.: Learning rotation invariant convolutional filters for texture classification. In: 2016 23rd International Conference on Pattern Recognition (ICPR), pp. 2012–2017. IEEE (2016)
19. Simonyan, K., Zisserman, A.: Very deep convolutional networks for large-scale image recognition. arXiv preprint arXiv:1409.1556 (2014)
20. Xu, Y., Xiao, T., Zhang, J., Yang, K., Zhang, Z.: Scale-invariant convolutional neural networks. CoRR abs/1411.6369 (2014). http://arxiv.org/abs/1411.6369

Automatic Optical Inspection for Millimeter Scale Probe Surface Stripping Defects Using Convolutional Neural Network

Yu-Chieh Ting[1] [iD], Daw-Tung Lin[1]([✉])[iD], Chih-Feng Chen[2] [iD], and Bor-Chen Tsai[2] [iD]

[1] Department of Computer Science and Information Engineering,
National Taipei University,
151, University Road, San-Shia, New Taipei City, Taiwan
fish20160101@gmail.com, dalton@mail.ntpu.edu.tw
[2] CCP Contact Probes CO.,
5F, No. 8 Lane 24, Ho Ping Road, Panchiao District, New Taipei City, Taiwan
{cf_chen,eddie_tsai}@pccp.com.tw

Abstract. Surface defect inspection is a crucial step during the production process of IC probe. The traditional way of identifying defective IC probes mostly relies on the human visual examination through the microscope screen. However, this approach will be affected by some subjective factors or misjudgments of inspectors, and the accuracy and efficiency are not sufficiently stable. Therefore, we propose an automatic optical inspection system by incorporating the ResNet-101 deep learning architecture into the faster region-based convolutional neural network (Faster R-CNN) to detect the stripping-gold defect on the IC probe surface. The training samples were collected through our designed multi-function investigation platform IMSLAB. To circumvent the challenge of insufficient images in our datasets, we introduce data augmentation using cycle generative adversarial networks (CycleGAN). The proposed system was evaluated using 133 probes. The experimental results revealed our method performed high accuracy in stripping defect detection. The overall mean average precision (mAP) was 0.732, and the defect IC probe classification accuracy rate was 97.74%.

Keywords: Automatic optical inspection · IC probe · Surface defect detection · Object detection · Deep learning

1 Introduction

In recent years, technology in the semiconductor manufacturing process has under-gone rapid advancements. Integrated circuits (IC) are getting smaller, the number of pins is increasing and therefore, high-end packaging methods are becoming expensive. Testing before packaging can improve product quality and

© Springer Nature Switzerland AG 2020
J. Blanc-Talon et al. (Eds.): ACIVS 2020, LNCS 12002, pp. 360–369, 2020.
https://doi.org/10.1007/978-3-030-40605-9_31

greatly reduce consumption and costs. Testing dies with a probe card is a crucial process in IC testing. It involves inputting a metal bond pad through certain pins of the probe card, and then flowing electrical signals into the CMOS in the die. After millions of CMOS operations, the resultant signals are outputted through other pins. From these outputted electrical signals, we can determine if the die is working properly. A defect in any IC probe in the probe card can affect the test result of the IC. Therefore, surface defect inspection is another important step during the manufacturing process of the IC probe.

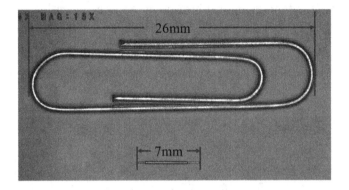

Fig. 1. Size comparison of the paper clip (top) and the IC probe (bottom).

An IC probe can exhibit various defects during the production process, such as stripping gold, acupuncture, tube sputum, and needle tip bluntness. These defects will decline the product quality, and increase materials consumption and costs greatly. Traditional IC probe inspection relies mainly on manual examination by the human eyes through a microscope. However, following long hours of work, the inspection efficiency and accuracy might be diminished. Furthermore, employees will be prone to vision impairment resulting in great loss to the company. The goal of this study is to automatically detect the defects of these tiny IC probes, specifically, gold stripping. The IC probe samples we test in this study are only about 5 to 7 mm long with a diameter of between 0.26 mm and 0.31 mm. Figure 1 compares the sizes of paper clip and an IC probe. As can be observed from Fig. 1, the length of the IC probe is only one forth of the paper clip (7 mm v.s. 26 mm). Based on the appearance and cause of the defect, we classified the stripping-gold defect into two types (Defect01 and Defect02), which are shown in Fig. 2. Defect01 is linked to machine extrusion or human factors, whereby the gold layer on the edge of the probe tube peels off, and the silver color of the inner metal is clearly visible. Defect02 stems from a crack materializing along the metal surface during the manufacturing process; in most cases, there are multiple black line cracks that are often hardly visible. Defect01 is easier to identify and classify, because it can be detected more accurately by sight. However, due to the influence of light and the texture of the IC probe, Defect02 is more difficult to detect.

Recently, the deep learning network has been effectively deployed for object classification and detection. It is also gaining wide use in industrial automated optical inspection. At the same time, in performing many computer vision tasks, traditional image processing algorithms and machine learning algorithms have limitations such as semantic segmentation, object detection, and transfer learning [4]. Therefore, we use faster region-convolutional neural network (Faster R-CNN) [7], which offers advanced performance for object detection tasks, and effectively detects defects on the surface of IC probes.

The remainder of this paper is organized as follows. Section 2 presents the related works of automatic optical inspection. Section 3 describes the proposed method. Training the deep learning model requires a large amount of image dataset. To achieve that, we established a training dataset consisting of original defect samples and data augmentation images. Section 4 presents the experimental results and discussion. The conclusion is presented in Sect. 5.

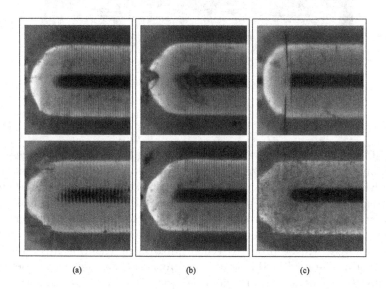

Fig. 2. (a) Normal sample; (b) Defect01 sample; (c) Defect02 sample. (Color figure online)

2 Related Work

Recently, there have been extensive studies on automated optical inspection using computer vision. Compared to detecting objects of average size, detecting tiny objects is more difficult. The traditional automatic optical inspection methods are applying mean filtering and dynamic threshold segmentation to detect large defects; small defects are detected using Gaussian filtering and global threshold segmentation. Yu *et al.* [11] introduce the improved Butterworth high-pass filter for removing noise, and locating small defects on small magnets. Senthikumar *et al.* [8] propose a method for detecting cracks and shrinkage on metal surface images using the iterative thresholding technique. However, the above methods are not suited for our purposes.

The deep neural network is commonly used to detect and classify images of people, animals, vehicles, among other things; in recent times, it is being applied to automatic optical detection. Tao *et al.* [9] propose a method of using an autoencoder to detect small defects on metal surfaces, such as scratches and dust, which are then classified using the convolution neural network (CNN). Chen *et al.* introduce OverFeat [2], a variant of the CNN that can be used directly for the image-based surface defect classification task. Zhu *et al.* [14] propose a method based on the faster R-CNN to detect defects in the X-ray image of a tire; comparison with the traditional method it proves deep learning improves the accuracy rate. In [4], the faster R-CNN is used to inspect the wheel hub, and the results of using VGG-16 and Resnet-101 as the backbone were compared. It was proven that the latter had better detection result.

3   Proposed Method

This section describes the proposed method. The process is divided into several important stages: IC probe image capturing, training dataset collection, data augmentation, and training the faster R-CNN model incorporating Resnet-101. Finally, the trained detection model is used for testing and classifying IC probe defect.

3.1   IC Probe Image Capturing

The diameters of the IC probes in this study range from 0.26 mm to 0.31 mm. When such small objects are concerned, capturing the defect images is a major challenge. After considering parameters, such as working distance, field of view, depth of field, sensor size, and resolution, we settled on a 10.7M pixel high-resolution color camera with a 1X–10X zoom lens. Because the IC probe is cylindrical with a metal surface, we used a hemispherical diffused light source with a white curved reflector at the bottom for lighting. This can effectively reduce the shadow of the IC probe surface during shooting and the effect of reflection on visibility. Furthermore, to ensure that every side of the IC probe is inspected, we designed a semi-automatic platform that includes an x-axis translation module, an x-axis fine-tuning module, a y-axis fine-tuning module, a theta-axis rotation module, and a IC probe-fixing module that collects the images of the defects. We utilized the software development kit (SDK) provided by the camera manufacture while constructing our AOI system so we can get real time defect images from the platform (Fig. 3).

Fig. 3. The constructed IMSLAB-1 platform for IC probe inspection

3.2   ROI Image Segmentation

While capturing the IC probe image, there will be unavoidable skew. Further-more, the defect area is very small, compared to the original image. To effectively improve the inspection result, it is necessary to define the range of the region of interest (ROI) as gold stripping occurs only at the head of the tube. Figure 4 depicts the procedure of obtaining the ROI image. First, we separate the IC probe from the background using HSV (hue, saturation, value) color segmenta-tion as shown in Fig. 4(b). The image is then horizontally aligned (see Fig. 4(c)). Next, we detect the position of the head of the tube by comparing the difference between the needle's width and the tube. Finally, the ROI image is found and provided for the subsequent data augmentation, and for deep learning training as well as for testing process.

3.3   Data Augmentation

Table 1. Details of our original dataset and samples generated from various data augmentation methods.

Type	Normal	Defect01	Defect02	Total
Original	488	1364	262	**1114**
Augmentor	1044	884	589	**2517**
CycleGAN	0	1445	1445	**2890**
Total	1532	2693	2296	**6521**

In deep learning, sufficient training data is required to avoid overfitting. How-ever, we can only collect a limited number of defective IC probe samples during production. Furthermore, manual shooting is time consuming. To increase the

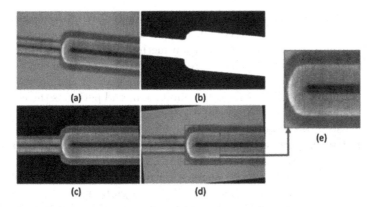

Fig. 4. Illustration of ROI image segmentation procedure: (a) original image, (b) background segmentation, (c) horizontal alignment, (d) ROI detection, (e) final ROI image.

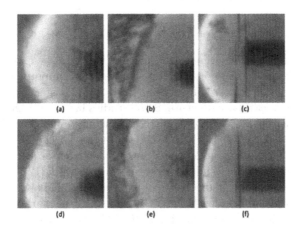

Fig. 5. Examples of training samples. (a), (b), (c): original stripping gold defects samples; (d), (e), (f): artificial defects samples generated by CycleGAN.

number of training samples, we proposed two methods to increase the variety and quantity of dataset.

The first method involves using Augmentor [1], a Python image enhancement library for machine learning. Augmentor uses a random approach that provides majority of standard enhancement practices as well as advanced features such as label protection, random elastic warping, and many additional features for typical enhancement tasks required in machine learning. Augmentor operations contain many parameters that offer precise control over the process of images creation. We morph the defect images to balance the number of each kind of defect images.

The second method is image generation. To increase the variety of the sample dataset, rather than only modifying the existing images, we hope to create new samples to expand our dataset. Recently, Goodfellow *et al.* [3] introduced generative adversarial network (GAN). Isola *et al.* [6] then introduced Pix2Pix, a general-purpose paired image translation method that uses conditional GAN to train a paired dataset to achieve cross domain image conversion. However, for our purposes, it is very difficult to prepare a pair of normal and defect images. CycleGAN [13] circumvents this limitation, and performs the conversion of unpaired image datasets. After obtaining the ROI image of our dataset from the previous step, we establish the normal and defect image datasets. Defect01 and Defect02 are respectively trained into two GANs. The results of the artificial defects generated by the CycleGAN are shown in Fig. 5.

The above-mentioned methods increase the dataset samples. The image numbers are listed in Table 1. We generated 2517 images by using Augmentor and 2890 images by using CycleGAN. After data augmentation, there are total 6521 images in our dataset. Incorporating the augmentative dataset into training improved the accuracy of detecting stripping gold.

3.4   Faster R-CNN ResNet-101

Compared to traditional methods, the feature extraction capabilities of deep learning have proven to be more accurate for various computer vision applications in recent years. Therefore, in this study, we use the faster R-CNN [7], a very common object detection architecture. Some studies have demonstrated that the faster R-CNN has better accuracy and recall rate in object detection. Furthermore, its detection speed is close to real-time detection speed. The faster R-CNN can use different CNNs on convolutional layers to extract object features.

Initially, we utilized the Zeiler-and-Fergus (ZF) network [12], which is lighter than the VGG-16 network, and consists of five convolutional layers and two fully connected layers. Compared with the VGG16, the ZF network with its shorter training time also has a faster detection process. However, to better improve the flaw-detection accuracy, we replace ZF with ResNet-101 [5], which is introduced in [4]. ResNet-101, the winner of the MS COCO 2015 detection and segmentation challenge, introduces skip connection to fit the input from the previous layer to the next layer without any modification to the input [10]. Although ResNet requires longer training time, it offers improved detection accuracy. Figure 6 shows the architecture of the faster R-CNN incorporating ResNet-101.

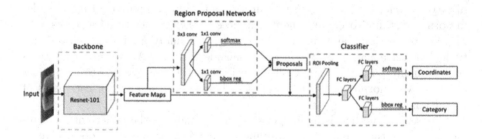

Fig. 6. Network architecture of Faster R-CNN incorporating Resnet-101

4   Experimental Results of IC Probe Inspection

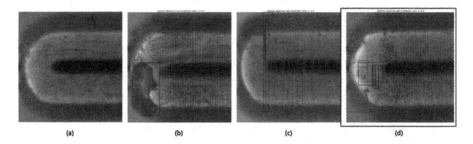

Fig. 7. Examples of detection results: (a) normal case, (b) defect01 case, (c) defect02 case, (d) false alarm case.

Because the IC probe is cylindrical, to completely inspect each surface for defect, we rotate the IC probe, and capture between six to ten images using the proposed detection method. If defect is detected on any of the images, the IC probe is classified as defective. Figure 7 illustrates the examples of defect detection results using the proposed Faster R-CNN incorporating with ResNet-101. During the sampling process, a few of the samples will be misidentified due to disturbances such as dander and dust.

The assessments were performed using four-fold cross-validation method. The existing datasets are randomly divided into four groups; one group is reserved for validation, the others are used as training data. This can provide improve the accuracy rate of the model. Tables 2 and 3 show the detection performance

Table 2. Detection performance (mAP) of the network trained with original ROI dataset.

	Fold-1	Fold-2	Fold-3	Fold-4	Total
AP of Defect01	0.810	0.902	0.895	0.889	**0.874**
AP of Defect02	0.468	0.438	0.440	0.480	**0.457**
mAP	0.639	0.670	0.668	0.685	**0.666**

Table 3. Detection performance (mAP) of the network trained with augmentation dataset.

	Fold-1	Fold-2	Fold-3	Fold-4	Total
AP of Defect01	0.879	0.892	0.909	0.907	**0.897**
AP of Defect02	0.552	0.536	0.612	0.570	**0.568**
mAP	0.715	0.714	0.760	0.739	**0.732**

(mAP) obtained using the proposed Faster R-CNN with ResNet-101 model with the original dataset and the augmentation dataset, respectively. It is apparent that the mAP for Defect01 is higher because it has obvious features that are not affected by the surface texture of the IC probe. The average AP is 0.87, which slightly increases to 0.897 following data augmentation. Compared with Defect01, the mAP of Defect02 demonstrates greater improvement, because the original sample of Defect02 is relatively small. Furthermore, Defect02, due to the influence of light and the texture of the IC probe, is more difficult to detect. The overall average mAP is also improved from 0.666 to 0.732.

As explained in Subsect. 3.4, we will test each image based on the Faster R-CNN with ResNet-101 model, and then classify each IC probe as normal or defective accordingly. The inspection results are shown in Table 4. In our experiment, we tested with 131 IC probes, among which three were misclassified as flawed samples. The accuracy rate is as high as 97.74%.

Table 4. Results of IC probe inspection.

Unit: IC probe		AOI inspection	
		Normal	Defective
Ground Truth	Normal	8	3
	Defective	0	122

5   Conclusion

We proposed a method for detecting the surface stripping-gold defect of the IC probe by using Faster R-CNN with ResNet-101 model. We built a platform for IC probe inspection that can successfully collect clear defect image from the IC probe. We introduce data augmentation by using cycle generative adversarial networks (CycleGAN) to circumvent the challenge of insufficient images in our datasets. Our work also demonstrates that data augmentation datasets can effectively solve the problem of low accuracy arising from insufficient data, since overall mAP has improved from 0.666 to 0.732. Furthermore, our proposed method was successfully applied to the automated optical inspection of IC probe during the production process. The result is verified using the four-fold cross validation, and the IC probe inspection accuracy rate was as high as 97.74%. However, the mAP of Defect02 can still be improved. In the future studies, we aim to integrate our proposed method into real-time systems then apply our system on the production line in order to achieve rapid and efficient detect process.

Acknowledgements. This work is partially supported by the C.C.P. Contact Probes Co., LTD.

References

1. Bloice, M.D., Stocker, C., Holzinger, A.: Augmentor: an image augmentation library for machine learning. arXiv preprint arXiv:1708.04680 (2017)
2. Chen, P.H., Ho, S.S.: Is overfeat useful for image-based surface defect classification tasks? In: 2016 IEEE International Conference on Image Processing (ICIP), pp. 749–753. IEEE (2016)
3. Goodfellow, I., et al.: Generative adversarial nets. In: Advances in Neural Information Processing Systems, pp. 2672–2680 (2014)
4. Han, K., Sun, M., Zhou, X., Zhang, G., Dang, H., Liu, Z.: A new method in wheel hub surface defect detection: object detection algorithm based on deep learning. In: 2017 International Conference on Advanced Mechatronic Systems (ICAMechS), pp. 335–338. IEEE (2017)
5. He, K., Zhang, X., Ren, S., Sun, J.: Deep residual learning for image recognition. In: Proceedings of the IEEE International Conference on Computer Vision and Pattern Recognition, pp. 770–778 (2016)
6. Isola, P., Zhu, J.Y., Zhou, T., Efros, A.A.: Image-to-image translation with conditional adversarial networks. In: Proceedings of the IEEE International Conference on Computer Vision and Pattern Recognition, pp. 1125–1134 (2017)
7. Ren, S., He, K., Girshick, R., Sun, J.: Faster R-CNN: towards real-time object detection with region proposal networks. In: Advances in Neural Information Processing Systems, pp. 91–99 (2015)
8. Senthikumar, M., Palanisamy, V., Jaya, J.: Metal surface defect detection using iterative thresholding technique. In: Second International Conference on Current Trends In Engineering and Technology-ICCTET 2014, pp. 561–564. IEEE (2014)
9. Tao, X., Zhang, D., Ma, W., Liu, X., Xu, D.: Automatic metallic surface defect detection and recognition with convolutional neural networks. Appl. Sci. **8**(9), 1575 (2018)
10. Tsang, S.H.: Review: ResNet–winner of ILSVRC 2015 (image classification, localization, detection) (2018)
11. Yu, Z., Li, X., Yu, H., Xie, D., Liu, A., Lv, H.: Research on surface defect inspection for small magnetic rings. In: 2009 International Conference on Mechatronics and Automation, pp. 994–998. IEEE (2009)
12. Zeiler, M.D., Fergus, R.: Visualizing and understanding convolutional networks. In: Fleet, D., Pajdla, T., Schiele, B., Tuytelaars, T. (eds.) ECCV 2014. LNCS, vol. 8689, pp. 818–833. Springer, Cham (2014). https://doi.org/10.1007/978-3-319-10590-1_53
13. Zhu, J.Y., Park, T., Isola, P., Efros, A.A.: Unpaired image-to-image translation using cycle-consistent adversarial networks. In: Proceedings of the IEEE International Conference on Computer Vision, pp. 2223–2232 (2017)
14. Zhu, Q., Ai, X.: The defect detection algorithm for tire x-ray images based on deep learning. In: 2018 IEEE 3rd International Conference on Image, Vision and Computing (ICIVC), pp. 138–142. IEEE (2018)

Image restauration, Compression and Watermarking

A SVM-Based Zero-Watermarking Technique for 3D Videos Traitor Tracing

Karama Abdelhedi$^{(\boxtimes)}$, Faten Chaabane$^{(\boxtimes)}$, and Chokri Ben Amar$^{(\boxtimes)}$

REsearch Groups in Intelligent Machines, ENIS, 3038 Sfax, Tunisia
{karama.abdelhedi,faten.chaabane,chokri.benamar}@ieee.org

Abstract. The watermarking layer has a crucial role in a collusion-secure fingerprinting framework since the hidden information, or the identifier, directly attached to user identification, is implanted in the media as a watermark. In this paper, we propose a new zero watermarking technique for 3D videos based on Support Vector Machine (SVM) classifier. Hence, the proposed scheme consists of two major contributions. The first one is the protection of both the 2D video frames and the depth maps simultaneously and independently. Robust features are extracted from Temporally Informative Representative Images (TIRIs) of both the 2D video frames and depth maps to construct the master shares. Then, the relationship between the identifier and the extracted master shares is generated by performing an Exclusive OR (XOR) operation. The second contribution uses the SVM and the XOR operation to estimate the watermark. Compared to other zero watermarking techniques, the proposed scheme has proven good results of robustness and transparency even for long size watermarks, which makes it suitable for a tracing traitor framework.

Keywords: Collusion-secure · Fingerprinting · 3D video · SVM · Zero-watermarking · XOR

1 Introduction

The ever growing of the digital era is tightly tied to the diversity of digital tools, the availability of Internet high speed connections and the encroachment of Peer to Peer networks. One key evidence is the affordability and the massive use of digital tools such as personal computers, camcorders and mobile devices: smart phones and tablets [1–4]. Unfortunately, the emerging digital technology phenomenon hides a volte-face. Henceforth, it is obliquely the origin of several unauthorized manipulations of digital content: multiple duplications, illegal redistribution and arbitrary modification. Indeed, the readiness of copying perfectly any type of digital content (image, text, video, audio file) to share it or to redistribute it encourages and leads to a more dangerous phenomenon which is the copyright violation, well-known as the digital piracy. In this context, it was necessary for media holders to find remedies to prevent the digital content from

© Springer Nature Switzerland AG 2020
J. Blanc-Talon et al. (Eds.): ACIVS 2020, LNCS 12002, pp. 373–383, 2020.
https://doi.org/10.1007/978-3-030-40605-9_32

any type of fraud. Henceforth, the digital rights management systems, (DRM), have involved a set of measures to control the use of digital content [5]. One key approach which covers the digital watermarking techniques, belongs to the multimedia forensics and its principle is to use the embedded watermark to identify the media owner. Furthermore, the change performed by the watermark insertion makes the copyright infringement harder but does not alter its quality [6].

Furthermore, one media content which is more and more threatened is the three dimensional(3D) video content. Since this type of video is more attractive and realistic than the traditional 2D video, multimedia makers are striving to produce 3D video directly or convert them from existing 2D video. Moreover, according to [8], 3D video can be stored in one of two major formats. The first one, called the side-by-side format, provides both left and right views for each scene. The second format, consists in attaching the 2D frames to their corresponding depth maps using the image-based rendering (DIBR) technique [9]. In this context, the DIBR format requires less storage and reduced efficiently the costs of bandwidth transmission. Moreover, applying the watermarking to DIBR-based 3D videos generated from 2D frames requires the protection of both, 2D frames and the corresponding depth maps.

As a result, the protection of DIBR-based 3D videos sparked the researchers' interest. In this paper, we focus essentially on watermarking schemes of DIBR-based 3D videos.

Indeed, several watermarking schemes were suggested in the digital right management (DRM) of this video format [10–13]. One classification propose to divide these schemes into two major categories. The first one, called the 2D video frame-based watermarking scheme, embeds the watermarks into 2D video frames. Whereas, the second one is called the depth map-based watermarking one and consists in hiding watermarks in the depth maps. To improve the DRM efficiency, the DIBR algorithm takes the advantage of the fact that the 2D video frames uses the depth maps' characteristics [7,11–15]. To improve more and more the performance of 2D frames-based watermarking schemes, more than work in the literature was striving in using information extracted from depth map to improve one or more requirements. As a first work, a scheme proposed by [10] uses the most important information from visual characteristics provided by depth maps such as the pixels to be embedded by rendering and high-motion regions on the z-axis. In another watermarking scheme [11], authors propose to construct the depth perceptual region of interest, (DP-ROI), by extracting some relevant characteristics as gray contour regions, the foreground, and even the depth-edge, and hence to improve the embedding strength. According to [7,14,15], the main weakness of 2D frames-based watermarking schemes is that they cause distortions to the generated 3D videos and require more enhancement. Besides, when addressing the issue of DRM for DIBR-based 3D videos, it is clear that the 2D video frame-based do not hide any copyright information into the depth maps and still remain insufficient.

A second family of watermarking schemes was consequently proposed by researchers to compensate the lack of copyright for depth maps. The principle of these schemes is to embed watermarks into depth maps which guarantees that no distortions can be seen on synthesized 3D videos [8]. Several proposals have been merged to deal with this requirement, ranging from the Unseen Visible (UVW) schemes [7]; where watermarks embedded after estimating computations and can be perceptible only in particular views, to the Unseen Extractable (UEW) schemes; where watermarks are hidden once DC quantization is performed. Despite the fact that 3D videos are not distorted, the two types suffer from weak robustness results [17]. Similarly, it has been noticed that depth-map based watermarking schemes do not respond to the requirements of DRM for of DIBR-based 3D videos.

According to that, a third family of watermarking schemes, which is zero-watermarking schemes, compensates weaknesses of the two latter families of schemes for robustness and imperceptibility. This family has proven sufficient performance due to the fact that the watermark is not embedded to the signal host. The zero-watermarking schemes are based on computing relationship between video features and the copyright information, which is used then for the copyright step. The main steps of this type of watermarking schemes is a copyright registration step and its identification step [18]. In [8], the main contribution was to improve the traditional zero-watermarking schemes to be suitable for DIBR-based 3D videos, it proposed to protect both 2D frames and depth map to ensure efficient robustness and good imperceptibility. It has more than one key points: it uses Temporally Informative Representative Images, (TIRI), to take the advantage of temporal characteristics, it ensures more robustness to some signal processing attacks, and proposes to use the Visual Secret Sharing, VSS, to identify the copyright information. However, when addressing the high watermark bandwidth, its performance is reduced noticeably. Here, in this paper, the target is to propose a new zero-watermarking technique for DIBR-based 3D videos, which takes the advantage of using a machine learning algorithm, the Support Vector machines algorithm, which improves the identification process and hence the imperceptibility of the scheme, then we propose to use the XOR operation to recover the master share and even the relationship to recover efficiently the watermark. To the best of our knowledge, and according to the requirements of a traitor tracing scheme, the zero-watermarking scheme we propose, is the first watermarking scheme which provides a suitable watermarking layer for an efficient traitor tracing process for 3D videos [19]. It has outperforms the existing watermarking schemes. The paper is arranged as follows: In Sect. 2, we detail the different steps of the general tracing system, essentially the proposed watermarking layer. In Sect. 3, we present the different experimental assessments we carry out to validate the performance the proposed technique. We summarize with a conclusion and future work in Sect. 4.

2   The Generation Traitor Tracing Framework

From a practical standpoint, the whole multimedia fingerprinting system can be considered as a communication chain with a transmitter side, a channel and a receiver side. The transmitter and the receiver are the main parts of the tracing system while the collusion attacks are presented in the transmission channel. As numbered in Fig. 1, the whole fingerprinting framework implies the presence of five essential steps as follows: the fingerprint encoding step, its embedding in the media release, a selection of some releases to participate in a collusion attack and to give rise to a colluded copy, the extraction of the colluded code and the tracing process to trace back the colluders. In this paper, we will detail only the watermarking and the fingerprinting layers.

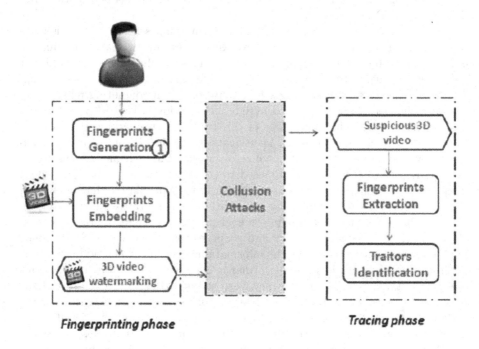

Fig. 1. The general traitor tracing framework.

2.1   The Fingerprinting Layer

The fingerprinting generation step is closely attached to the users of the platform. Labeled by (1) in Fig. 1, it includes the assigning, by the media distributor, of a unique codeword $fingerprint_{i,i\in\{1,\cdots,n\}}$ to each user before using it in the embedding layer to construct the fingerprinted releases $R(f(i)), i \in \{1, \cdots, n\}$. This unique identifier is essentially used to identify each media owner and to protect the media from any illegal treatment [16].

The efficiency of this step depends in a first time on the embedding technique, whose relevance is shown when considered the several attacks the media content should survive during a sharing operation. Moreover, the tracing algorithm is also

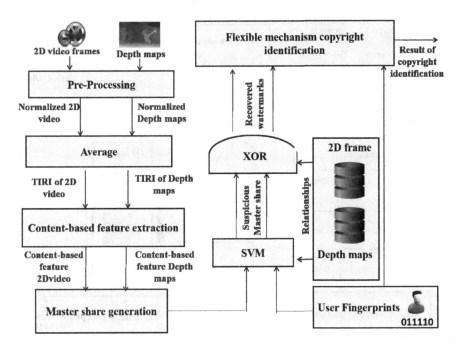

Fig. 2. The copyright registration step.

an important part of the fingerprinting layer, it is a process whose role is to allow the retrieval back of traitorous users when considering the extracted codeword from the colluded release Y. The fingerprinting layer depends on the efficiency of the tracing algorithm and even on its robustness to intentional attacks made by colluders. The diversity of the attacks and the large size of the audience in a multimedia distribution platform make the tracing task more complex. In Sect. 2.2, we detail the proposed SVM-based zero-watermarking layer and its key steps.

2.2 The SVM-Based Zero-Watermarking Layer

A general zero-watermarking scheme consists of two main steps: the copyright registration step and the copyright identification one. The copyright registration phase consists in generating both the relationship of 2D video frames and depth map databases based on relevant extracted content-features and other watermark information. Whereas, in the identification step, the target is to determine the copyright relationship of the released 3D video to detect whether it is a suspicious release.

The Copyright Registration Step. As depicted in Fig. 2 in this phase the relationship of both 2D video frame and depth maps are generated according to a logical operation XOR applied to the identification information and the

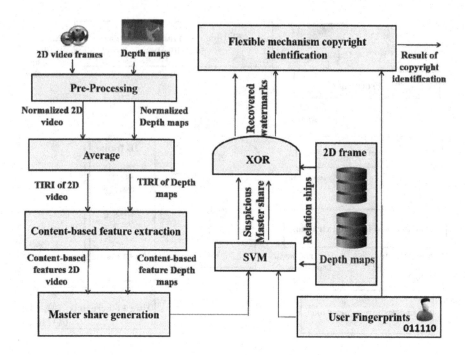

Fig. 3. The copyright identification step.

generated master shares of both 2D videos frames and depth maps. The extraction mechanism of the relationships of the 2D video frames is the same one of the depth maps witch decrease the computation cost. In this proposed technique the two essentials component of the 3D video which are the 2D video frames and depth maps are normalized via spatio-temporal smoothing and sub-simpling and selected the luminance components of both 2D video frames and depth maps. Then the temporally informative representative image (TIRI) [17] of both 2D video and depth maps are generated by using their appropriate luminance components. After affecting 2D-DCT transfers on the TIRIs and selecting the remained low-frequency coefficients, the content features of both 2D video frames and depth maps are determined. Next the Master shares of both 2D video and depth maps are extracted based to their content-feature [17]. Finally the extracted master shares and fingerprints are combined by using XOR operation to generate for each copy of the video a appropriate relations-ships.

The Copyright Identification Step. The copyright identification step is a reverse process of its copyright registration as noticed in Fig. 3. The input of this steps can be a fingerprinting 3D video or a manipulated and fingerprinting 3D video. When 3D video queried, the content-features are extracted from TIRI of both 2D video and depth maps the same as these in copyright registration. Then the master shares of both 2D video and depth maps are generated from their extracted content-features. Using the SVM classifier, each queried video is classified as legal or illegal one. The queried video is classed as suspicious one if one of the 2D video or depth map is illegal. Then for each suspicious video an XOR operation is applied to the recovered master shares and their relevant relations-ships to extract the watermark of the queried video.

3   Experimental Results

In this section, we evaluate the proposed SVM-ZW scheme according to different criteria. In a fist set of tests, we evaluate the robustness of our scheme compared to RZW [8] to show its outperformance, we compute hence the Normalized Correlations, NCs, between the original and the recovered watermark-based features as mentioned in Eqs. 1 and 2 for different types of signal processing attacks defined in [8]. We calculate also the Bit Correction Rate, BCR, which is a similarity measure between the original ans the retrieved watermarks Eq. 3.

The testing database contains 150 different 3D video clips with a total of 3750 different video frames collected from the database [22]. Others are 2D video clips, which are selected from different movies, with their depth maps calibrated using the method in [20]. The video 5 and video 3 contain respectively 100 frames of size 320 * 180 * 25. Moreover, 5 different Tardos identifiers of size 4096 are used in the copyright identification step and in the tracing process.

$$NC_{2d} = \frac{\sum W_{2d}(i,j)W'_{2d}(i,j)}{\sqrt{\sum W_{2d}(i,j)^2}\sqrt{\sum W'_{2d}(i,j)^2}} \tag{1}$$

$$NC_{depth} = \frac{\sum W_{depth}(i,j)W'_{depth}(i,j)}{\sqrt{\sum W_{depth}(i,j)^2}\sqrt{\sum W'_{depth}(i,j)^2}} \tag{2}$$

$$BCR(W,W') = 1 - \frac{1}{2}\sum_{k=1}^{n}\left|W_k - W'_k\right| \tag{3}$$

where $1 \leq i \leq 8, 1 \leq j \leq 8$ and $1 \leq k \leq 8$.

3.1   Watermarking Robustness Results

In this section, we evaluate the robustness and the imperceptibility of the proposed SVM-RZW scheme.

Robustness to Signal Processing Attacks. In Tables 1, 2, compared to RZW [8], NC and BCR values of our scheme are closed to 1 which reveal good robustness results. Which is enhanced by the SVM classifier and the XOR operator.

In Tables 3 and 4, NC and BCR values of our scheme are closed to 1 which reveal good robustness results even against the DIBR related transforms as: warping, dithering, sharpening and cropping [21].

Table 1. NC and BCR values showing imperceptibility results of video 5 for SZW vs RZW.

	NC				BCR			
	SVM-ZW		RZW [8]		SVM-ZW		RZW [8]	
	2D frames	Depth map	2D frames	Depth map	2D frames	Depth maps	2D frames	Depth maps
Gaussian noise 0.005	1	0.977	0.395	0.312	1	0.983	0.789	0.812
Gaussian noise 0.01	1	0.978	0.395	0.312	1	0.983	0.789	0.8125
Gaussian noise 0.05	1	0.973	0.395	0.312	1	0.983	0.789	0.812
Gaussian noise 0.05	1	0.979	0.3953	0.3125	1	0.983	0.789	0.812
Salt & pepper 0.005	1	0.979	0.395	0.312	1	0.983	0.789	0.812
Salt & pepper 0.01	1	0.976	0.395	0.312	1	0.924	0.789	0.8125
Salt & pepper 0.03	1	0.978	0.395	0.312	1	0.983	0.789	0.812
Averagefilter 9	1	0.98	0.3953	0.3125	1	0.9884	0.7891	0.8125
Averagefilter 15	0.97	0.98	0.237	0.312	0.984	0.985	0.789	0.812
Brightness −30%	1	0.97	0.359	0.312	1	0.983	0.789	0.812
Brightness +30%	1	0.98	0.359	0.312	1	0.983	0.789	0.812
Resize 14	1	0.98	0.359	0.312	1	0.985	0.789	0.812
Resize 125	1	0.97	0.359	0.312	1	0.984	0.789	0.812
Resize Crop 1%	0.979	0.978	0.306	0.312	0.98	0.978	0.773	0.812
Resize Crop 2%	0.979	0.976	0.353	0.312	0.98	0.982	0.781	0.812

Table 2. NC and BCR values showing imperceptibility results of video 3 for SZW vs RZW.

	NC				BCR			
	SVM-ZW		RZW [8]		SVM-ZW		RZW [8]	
	2D frames	Depth map	2D frames	Depth map	2D frames	Depth maps	2D frames	Depth maps
Gaussian noise 0.005	1	0.977	0.395	0.312	1	0.983	0.789	0.812
Gaussian noise 0.01	1	0.976	0.3953	0.3125	1	0.983	0.789	0.812
Gaussian noise 0.05	1	0.973	0.3953	0.3125	1	0.983	0.789	0.812
Gaussian noise 0.05	1	0.978	0.395	0.312	1	0.983	0.789	0.812
Salt & pepper 0.005	1	0.972	0.395	0.312	1	0.987 ·	0.789	0.812
Salt & pepper 0.01	1	0.98	0.395	0.312	1	0.927	0.789	0.812
Salt & pepper 0.03	1	0.98	0.395	0.312	1	0.987	0.789	0.812
Averagefilter 9	1	0.98	0.395	0.312	1	0.985	0.789	0.812
Averagefilter 15	0.97	0.98	0.2372	0.312	0.984	0.985	0.789	0.812
Brightness −30%	1	0.972	0.359	0.312	1	0.984	0.789	0.812
Brightness +30%	1	0.972	0.359	0.312	1	0.987	0.789	0.812
Resize 14	1	0.98	0.359	0.312	1	0.987	0.789	0.812
Resize 125	1	0.97	0.359	0.312	1	0.984	0.789	0.812
Resize Crop 1%	0.978	0.972	0.3062	0.312	0.978	0.78	0.789	0.812
Resize Crop 2%	0.979	0.98	0.353	0.312	0.98	0.985	0.77	0.812

Imperceptibility Results. According to [8], it is important to evaluate the watermarking imperceptibility of essentially depth maps because they are more sensitive to perceptible distorsions in case of attacks. Hence, we compute the PSNR value defined in [8], we notice that similar to the RZW scheme, the SVM-RZW has a very high PSNR value, near to the infinity, which demonstrates that there is no distorsions in the depth maps.

Table 3. NC and BCR values showing imperceptibility results of video 3 against some dibr related transforms.

	NC		BCR	
	2D frames	Depth map	2D frames	Depth map
Warping	0.978	0.978	0.984	0.984
Cropping	1	0.968	1	0.976
Dithering	0.979	0.982	0.984	0.986
Sharpening	0.978	0.982	0.984	0.986

Table 4. NC and BCR values showing imperceptibility results of video 5 against some dibr related transforms.

	NC		BCR	
	2D frames	Depth map	2D frames	Depth map
Warping	0.979	0.978	0.984	0.984
Cropping	1	0.968	1	0.976
Dithering	0.979	0.982	0.984	0.986
Sharpening	0.978	0.981	0.984	0.986

3.2 Tracing Results

In order to evaluate the tracing performance of our system against different types of collusion attacks, we use respectively the fingerprint length $m = 4096$, the number of users $n = 100$ and the collusion size $c = 5$. The good detection rate depicted in Fig. 4 demonstrate that the watermarking layer is sufficiently secure and has no impact on the tracing process.

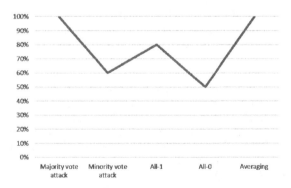

Fig. 4. The Detection rates of Tardos code against some collusion attacks

4 Conclusion

In this paper, we have proposed a new SVM-based zero watermarking scheme for DIBR-based 3D videos. The main contribution of the proposed SVM-RZW is the protection of both the 2D video frames and the depth maps simultaneously and independently. Then, the relationship between the identifier and the extracted master shares is generated by performing an Exclusive OR (XOR) operation. The second contribution uses takes the advantage of using the SVM classifier and the XOR operation to guess efficiently the watermark even under attacks. Compared to other existing zero watermarking scheme, the proposed scheme has proven good results of robustness and transparency even for long size watermark, which makes it represents an original point which makes it suitable for a tracing traitor framework for 3D videos. In a future work, we will focus on evaluating our scheme for more sophisticated attacks of 3D videos.

References

1. Bouchrika, T., Zaied, M., Jemai, O., Ben Amar, C.: Ordering computers by hand gestures recognition based on wavelet networks. In: 2nd International Conference on Communications Computing and Control Applications (CCCA 2012), pp. 1–6 (2012)
2. Teyeb, I., Jemai, O., Zaied, M., Ben Amar, C.: A novel approach for drowsy driver detection using head posture estimation and eyes recognition system based on wavelet network. In: 5th International Conference on Information Intelligence Systems and Applications, pp. 379–384 (2014)
3. Koubaa, M., Elarbi, M., Ben Amar, C., Nicolas, H.: Collusion, MPEG4 compression and frame dropping resistant video watermarking. Multimedia Tools Appl. **56**(2), 281–301 (2012)
4. Charfeddine, M., El'Arbi, M., Ben Amar, C.: A new DCT audio watermarking scheme based on preliminary MP3 study. Multimedia Tools Appl. **70**(3), 1521–1557 (2014)
5. de Rosnay, M.D.: Digital rights management systems and European law: between copyright protection and access control. In: 2nd International Conference on Web Delivering of Music (WEDELMUSIC 2002), Darmstadt, Germany, pp. 117–124 (2002)
6. Thilagavathi, N., Saravanan, D., Kumarakrishnan, S., Punniakodi, S., Amudhavel, J., Prabu, U.: A survey of reversible watermarking techniques, application and attacks. In: International Conference on Advanced Research in Computer Science Engineering Technology (ICARCSET 2015), India, pp. 37:1–37:7 (2015)
7. Pei, SC., Wang, YY.: A new 3D unseen visible watermarking and its applications to multimedia. In: Proceedings of the 3rd Global Conference on Consumer Electronics (GCCE 2014), pp. 140–143 (2014)
8. Liu, X., Zhao, R., Li, F., Liao, S., Ding, Y., Zou, B.: Novel robust zero-watermarking scheme for digital rights management of 3D videos. Sig. Process. Image Commun. **54**, 140–151 (2017)
9. Kauff, P., et al.: Depth map creation and image-based rendering for advanced 3DTV services providing interoperability and scalability. Sig. Process. Image Commun. **22**, 217–234 (2007)

10. Lee, M.J., Lee, J.W., Lee, H.K.: Perceptual watermarking for 3D stereoscopic video using depth information. In: Seventh International Conference on Intelligent Information Hiding and Multimedia Signal Processing (IIH-MSP 2011), pp. 81–84 (2011)

11. Fan, S., Yu, M., Jiang, G., Shao, F., Peng, Z.: A digital watermarking algorithm based on region of interest for 3D image. In: Proceedings of the 8th IEEE International Conference on Computational Intelligence and Security (CIS 2012), pp. 549–552 (2012)

12. Lin, Y.H., Wu, J.L.: A digital blind watermarking for depth-image-based rendering 3D images. IEEE Trans. Broadcast. **57**, 602–611 (2011)

13. Kim, H.D., Lee, J.W., Oh, T.W., Lee, H.K.: Robust DT-CWT watermarking for DIBR 3D images. IEEE Trans. Broadcast. **58**, 533–543 (2012)

14. Pei, S.C., Wang, Y.Y.: Auxiliary metadata delivery in view synthesis using depth no synthesis error model. IEEE Trans. Multimedia **17**, 128–133 (2015)

15. Lin, Y.H., Wu, J.L.: Unseen visible watermarking for color plus depth map 3D images. In: 37th International Conference on Acoustics, Speech and Signal Processing (ICASSP 2012), pp. 1801–1804 (2012)

16. Wagner, R.: Fingerprinting. In: Proceedings of the IEEE Symposium on Security and Privacy, USA, p. 18 (1983)

17. Liu, X., Li, F., Du, J., Guan, Y., Zhu, Y., Zou, B.: A robust and synthesized-unseen watermarking for the DRM of DIBR-based 3D video. Neurocomputing **222**, 155–169 (2017)

18. Gao, G., Jiang, G.: Bessel-Fourier moment-based robust image zero-watermarking. Multimed Tools Appl. **74**, 841–858 (2015)

19. Chaabane, F., Charfeddine, M., Puech, W., Amar, W.: A two-stage traitor tracing strategy for hierarchical fingerprints. Multimedia Tools Appl. (2016)

20. Scharstein, D., Szeliski, R.: A taxonomy and evaluation of dense two-frame stereo correspondence algorithms. Int. J. Comput. Vis. **47**, 7–42 (2002). (April–June)

21. Pereira, S., Voloshynovskiy, S., Madueno, M., Marchand-Maillet, S., Pun, T.: Second generation benchmarking and application oriented evaluation. In: Moskowitz, I.S. (ed.) IH 2001. LNCS, vol. 2137, pp. 340–353. Springer, Heidelberg (2001). https://doi.org/10.1007/3-540-45496-9_25

22. http://www.rmit3dv.com

Design of Perspective Affine Motion Compensation for Versatile Video Coding (VVC)

Young-Ju Choi[1], Young-Woon Lee[2], and Byung-Gyu Kim[1(✉)]

[1] Sookmyung Women's University, Seoul, Korea
{yj.choi,bg.kim}@ivpl.sookmyung.ac.kr
[2] Sunmoon University, A-san, Korea
yw.lee@ivpl.sookmyung.ac.kr
http://ivpl.sookmyung.ac.kr

Abstract. The fundamental motion model of the conventional block-based motion compensation in High Efficiency Video Coding (HEVC) is a translational motion model. However, in the real world, the motion of an object exists in the form of combining many kinds of motions. In Versatile Video Coding (VVC), a block-based 4-parameter and 6-parameter affine motion compensation (AMC) is being applied. The AMC still has a limit to accurate complex motions in the natural video. In this paper, we design a perspective affine motion compensation (PAMC) method which can improve the coding efficiency and maintain low-computational complexity compared with existing AMC. Because the block with the perspective motion model is a rectangle without specific feature, the proposed PAMC shows effective encoding performance for the test sequence containing irregular object distortions or dynamic rapid motions in particular. Our proposed algorithm is implemented on VTM 2.0. The experimental results show that the BD-rate reduction of the proposed technique can be achieved up to 0.30%, 0.76%, and 0.04% for random access (RA) configuration and 0.45%, 1.39%, and 1.87% for low delay P (LDP) configuration on Y, U, and V components, respectively. Meanwhile, the increase of encoding complexity is within an acceptable range.

Keywords: Video coding · Motion estimation · Motion compensation · Affine motion model · Perspective motion model · VVC

1 Introduction

Video compression standard technologies are increasingly becoming more efficient and complex. With continuous development of display resolution and type along with enormous demand for high quality video contents, video coding also plays a key role in display and content industries. After standardizing H.264/AVC [1] and H.265/HEVC [2] successfully, Versatile Video Coding (VVC) [3] is being standardized by the Joint Video Exploration Team (JVET) of ITU-T

ⓒ Springer Nature Switzerland AG 2020
J. Blanc-Talon et al. (Eds.): ACIVS 2020, LNCS 12002, pp. 384–395, 2020.
https://doi.org/10.1007/978-3-030-40605-9_33

Video Coding Experts Group (VCEG) and ISO/IEC Moving Picture Experts Group (MPEG). Obviously, the HEVC is a reliable video compression standard. Nevertheless, more efficient video coding scheme is required for higher-resolution and the newest media services such as UHD and VR.

To develop the video compression technologies beyond HEVC, experts in JVET have been actively conducting numerous researches. VTM provides a reference software model called as VVC Test Model (VTM). At the 11th JVET meeting, VTM2 [4,5] was established with the inclusion of a group of new coding features as well as some of HEVC coding elements.

In motion estimation (ME) and motion compensation (MC), finding precise correlation between consecutive frames is very important to make better coding performance. The fundamental motion model of the conventional block based MC is a translational motion model. In the early researches, a translational motion model-based MC cannot address complex motions in natural videos such as rotation and zooming. Such being the case, during the development of the video coding standards, further elaborate models have been required to handle non-translational motions.

Non-translational motion model-based studies have also been presented in the early research on video coding. Seferidis [6] and Lee [7] proposed deformable block based ME algorithms, in which all motion vectors (MVs) at any position inside a block can be calculated by using control points (CPs). Besides, Cheung and Siu [8] proposed to use the neighboring block's MVs to estimate the affine motion transformation parameters and added an affine mode. After those, affine motion compensation (AMC) has been begun to attract attention.

Later, Narroschke and Swoboda [9] proposed an adjusted AMC to HEVC coding structure by investigates the use of an affine motion model with analyzing variable block size. Huang et al. [10] extended the work in [8] for HEVC and included the affine skip/direct mode to improve coding efficiency. Heithausen [11] and Choi [12] improved the affine motion model to higher order motion model. Li [13] proposed the six-parameter affine motion model and extended by simplifying model to four-parameter and adding gradient-based fast affine ME algorithm in [14]. Because the trade-off between the complexity and coding performance is attractive, the scheme in [14] was proposed to JVET [15] and was accepted as one of the core modules of Joint Exploration Model (JEM) [16]. After that, Zhang [17] proposed a multi model AMC approach. At the 11th JVET meeting in July 2018, modified AMC of JEM has been integrated into VVC and Test Model 2 (VTM2) [4] based on [17].

Although AMC has significantly improved performance over the translational MC, there is still a limit to find complex motion accurately. In natural videos, in the majority of cases, a rigid object moves without any regularity rather than maintains the shape or transform with a certain rate. For this reason, more flexible motion model is desired for new coding tool to raise the encoding quality. In this paper, we propose a perspective affine motion compensation (PAMC) method which improves the coding efficiency compared with existing AMC. The proposed framework is designed to operate adaptively with AMC. In other words, six and four parameter model-based AMC and eight parameter-based perspective ME/MC are performed to select the best coding mode adaptively.

This paper is organized as follows. In Sect. 2, we first present AMC in VVC briefly. The proposed perspective affine motion compensation (PAMC) is introduced in Sect. 3. The experimental results are shown in Sect. 4. Finally, Sect. 5 concludes this paper.

2   Affine Motion Estimation/Compensation in VVC

In HEVC, only translation motion model is utilized for motion compensation module. However, in the real world, the motion of an object exists in the form of combining many kinds of motion. The 4-parameter and 6-parameter motion model for AMC prediction method in the VVC are defined for three motions: translation, rotation and zooming.

As shown in Fig. 1, the affine motion vector field (MVF) of a CU is described by control point motion vectors (CPMVs): (a) two CPs (4-parameter) or (b) three CPs (6-parameter). CP_0, CP_1 and CP_2 are defined as the top-left, top-right and bottom-left corners. For 4-parameter affine motion model, MV at sample position (x, y) in a CU is derived as

$$\begin{cases} mv^h(x,y) = \frac{mv_1^h - mv_0^h}{W}x - \frac{mv_1^v - mv_0^v}{W}y + mv_0^h, \\ mv^v(x,y) = \frac{mv_1^v - mv_0^v}{W}x + \frac{mv_1^h - mv_0^h}{W}y + mv_0^v, \end{cases} \tag{1}$$

and for 6-parameter affine motion model, MV at sample position (x, y) in a CU is derived as

$$\begin{cases} mv^h(x,y) = \frac{mv_1^h - mv_0^h}{W}x + \frac{mv_2^h - mv_0^h}{H}y + mv_0^h, \\ mv^v(x,y) = \frac{mv_1^v - mv_0^v}{W}x + \frac{mv_2^v - mv_0^v}{H}y + mv_0^v, \end{cases} \tag{2}$$

where (mv_0^h, mv_0^v), (mv_1^h, mv_1^v) and (mv_2^h, mv_2^v) are MVs of CP_0, CP_1 and CP_2 respectively. W and H present width and height of the current CU. The $mv^h(x, y)$ and $mv^v(x, y)$ are the horizontal and vertical components of MV for the position (x, y). In order to simplify, block based AMC is applied. The MV at the center position of each 4×4 sub block is derived from CPMVs.

3   Proposed Perspective Affine Motion Estimation/Compensation

Affine motion estimation/compensation in the VVC is applied since it is more efficient than translational motion compensation. The coding gain can be increased by delicately estimating motion on the video sequence in which complex motion is included. However, still it has a limit to accurately find all motions in the natural video.

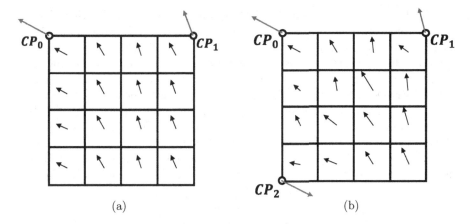

Fig. 1. Affine CPMV and MVF: (a) 4-parameter motion model, (b) 6-parameter motion model.

Affine transformation model has properties to maintain parallelism based on the 2D plane, and thus cannot work efficiently for some sequences containing object distortions or dynamic motions such as shear and 3D affine transformation. In real world, numerous moving objects have irregular motions rather than regular translational, rotation and scaling motions. So, more flexible motion model is desired for future video coding tool to raise the encoding quality.

The basic warping transformation model can estimate motion more accurately, but this method is not suitable because of its high computational complexity and bit overhead by the large number of parameters. For these reasons, we propose a perspective affine motion compensation (PAMC) method which improve coding efficiency compared with the existing AMC method of the VVC. The perspective transformation model-based algorithm adds one more CPMV, which gives degrees of freedom to all four corner vertices of the block for more precise motion vector. Furthermore, the proposed algorithm is integrated while maintaining the AMC structure. Therefore, it is possible to adopt an optimal mode between the existing encoding mode and the proposed encoding mode.

3.1 Perspective Motion Model for Motion Estimation/Compensation

Figure 2 shows that the perspective model with four CPs (b) can estimate motion more flexible compared with the affine model with three CPs (a). Perspective based MVF of a current block is described by four CPs which are matched to $\{mv_0, mv_1, mv_2, mv_3\}$ in illustration. The typical eight-parameter perspective motion model can be described as:

$$\begin{cases} x' = \frac{ax+by+c}{gx+hy+1}, \\ y' = \frac{dx+ey+f}{gx+hy+1}, \end{cases} \tag{3}$$

where a, b, c, d, e, f, g and h are eight perspective model parameters. Among them, parameters g and h serve to give the perspective to motion model. With this characteristic, as though it is a conversion in the 2D plane, it is possible to obtain an effect that the surface on which the object is projected is changed.

Instead of these eight parameters, we used four MVs to equivalently represent the perspective transformation model like the technique applied to AMC of the existing VTM. In video codecs, using MV is more efficient in terms of coding structure and flag bits. In a W x H block as shown in Fig. 2(b), we denote the MVs of $(0,0)$, $(W,0)$, $(0,H)$, and (W,H) pixel as mv_0, mv_1, mv_2 and mv_3. Moreover, we replace $g \cdot W + 1$ and $h \cdot H + 1$ with p and q to simplify the formula. The six parameters a, b, c, d, e and f of model can solved as following Eq. (4):

$$
\begin{cases}
a = \frac{p(mv_1^h - mv_0^h)}{W}, \\
b = \frac{q(mv_2^h - mv_0^h)}{H}, \\
c = mv_0^h, \\
d = \frac{p(mv_1^v - mv_0^v)}{W}, \\
e = \frac{q(mv_2^v - mv_0^v)}{H}, \\
f = mv_0^v.
\end{cases}
\tag{4}
$$

In addition, $g \cdot W$ and $h \cdot H$ can solved as Eq. (5):

$$
\begin{cases}
g \cdot W = \frac{(mv_3^h - mv_2^h)(2mv_0^v - mv_1^v) + (mv_3^v - mv_2^v)(mv_1^h - 2mv_0^h)}{(mv_3^v - mv_2^v)(mv_0^h - mv_1^h) + (mv_3^h - mv_2^h)(mv_3^v - mv_1^v)}, \\
h \cdot H = \frac{(mv_3^h - mv_1^h)(2mv_0^v - mv_2^v) + (mv_3^v - mv_1^v)(mv_2^h - 2mv_0^h)}{(mv_3^v - mv_1^v)(mv_3^h - mv_2^h) + (mv_3^h - mv_1^h)(mv_3^v - mv_2^v)}.
\end{cases}
\tag{5}
$$

Based on Eqs. (4) and (5), we can derive MV at sample position (x, y) in a CU by following Eq. (6):

$$
\begin{cases}
mv^h(x, y) = \frac{\frac{p\left(mv_1^h - mv_0^h\right)}{W}x + \frac{q\left(mv_2^h - mv_0^h\right)}{H}y + mv_0^h}{\frac{p-1}{W}x + \frac{q-1}{H}y + 1}, \\
mv^v(x, y) = \frac{\frac{p\left(mv_1^v - mv_0^v\right)}{W}x + \frac{q(mv_2^v - mv_0^v)}{H}y + mv_0^v}{\frac{p-1}{W}x + \frac{q-1}{H}y + 1}.
\end{cases}
\tag{6}
$$

With the AMC, the designed perspective motion compensation also is also applied by 4×4 sub block-based MV derivation in a CU. Similarly, the motion compensation interpolation filters are used to generate the prediction block.

3.2 Perspective Affine Motion Compensation

Based on the aforementioned perspective motion model, the proposed algorithm is integrated to the existing AMC. A flowchart of the proposed algorithm is shown in Fig. 3.

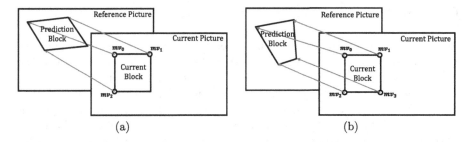

Fig. 2. The motion models: (a) 6-parameter affine model with three CPs, (b) perspective model with four CPs.

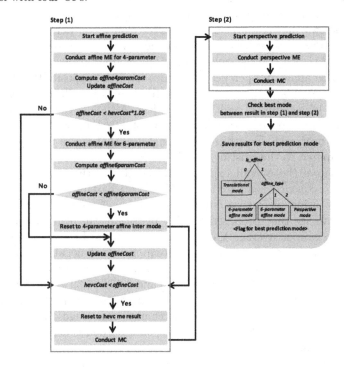

Fig. 3. Flowchart of the proposed overall algorithm.

After performing fundamental translational ME and MC as in HEVC, affine prediction process is conducted first in step (1). Then, the proposed perspective prediction process is performed as step (2). After that, we check the best mode between results from step (1) and step (2) by RD cost check process of the current CU. Once the best mode is determined, the flag for prediction mode are signaled in the bitstream. At this time, two flags are required: affine flag and affine type flag. An affine type flag is signaled for a CU when its affine flag is true. When affine type flag is 2, it means that the current CU is coded in the perspective mode.

4   Experimental Results

To evaluate the performance of the proposed PAMC module, the proposed algorithm was implemented on VTM 2.0 [18]. The 14 test sequences used in the experiments were from class A to class F specified in the JVET common test conditions (CTC) [19]. Experiments are conducted under random access (RA) and low delay P (LDP) configurations and four base layer quantization parameters (QP) values of 22, 27, 32 and 37. According to the JVET CTC documentation, the E class sequences are not used in the RA configuration, and the A class sequences are not used in the LDP configuration. We used 50 frames in each test sequence. The objective coding performance comparison was evaluated by the Bjontegaard-Delta Rate (BD-Rate) measurement [20].

Table 1 shows the comparison of the prediction mode percentage of CU on luminance component for each CTC sequence in the proposed framework. As shown in Table 1, for translational mode, the occupation ratio increases as QP increases, while affine and perspective modes have the opposite trend. Through those results, it can be seen that the proposed perspective affine motion model is most effective in encoding of high quality videos. Although the CU ratio selected in perspective mode occupies a small portion of 4.93% in the RA and 4.35% in the LDP, the proposed algorithm is meaningful in that it can find more delicate motions compared to the existing AMC.

Table 2 shows the overall experimental results of the proposed algorithm for each test sequence compared with VTM 2.0 baseline. The BD-Rate was calculated by using piece-wise cubic interpolation. Compared with the VTM anchor, we can see that proposed PAMC algorithm can bring about 0.07 and 0.12 BD-Rate gain on average Y component in RA and LDP cases respectively, and besides almost tested results on Y component are negative. The BD-rate reduction of the proposed technique can be achieved up to 0.30%, 0.76%, and 0.04% for random access (RA) configuration and 0.45%, 1.39%, and 1.87% for low delay P (LDP) configuration on Y, U and V components, respectively. Moreover, there are only 3% and 6% encoding time change in average, for RA and LDP, respectively. This means that the computational burden is negligible.

In order to improve the coding efficiency through the proposed algorithm, the test sequence have to contain irregular object distortions or dynamic rapid motions. The class A sequence "Campfire" contains a bonfire that moves inconsistently and steadily. Also "CatRobot1" contains a lot of flapping scarves and a flipped book, and class B sequence "BQTerrace" involves the ripples on the surface of the water. All of such moving objects commonly include object distortions whose shape changes. Because of this, those sequences can be compressed more efficiently by the proposed framework.

Figure 4 shows examples of comparing AMC in VVC and the proposed PAMC. Figure 4(a) presents the coded CU with affine mode in VVC baseline and Fig. 4(b) shows the coded CU with affine and perspective mode in the proposed framework. If the unfilled rectangles imply the CUs coded in affine mode and the filled rectangles imply the CUs coded in perspective mode, in Fig. 4(b), some filled rectangles can be found on scarves and on the pages of a book.

Table 1. Comparison of the prediction mode percentage of CU on luminance component.

Class	Sequence	Mode	RA 22	27	32	37	**avg.**	LDP 22	27	32	37	**avg.**
A	Campfire	Trans.	59.22	69.21	68.24	70.35	66.76	-	-	-	-	-
		Aff.	24.37	25.78	27.29	25.72	25.79	-	-	-	-	-
		Persp.	16.41	5.02	4.47	3.93	7.46	-	-	-	-	-
	CatRobot1	Trans.	44.38	40.17	42.79	51.03	44.59	-	-	-	-	-
		Aff.	50.91	57.56	55.03	46.44	52.49	-	-	-	-	-
		Persp.	4.71	2.26	2.18	2.53	2.92	-	-	-	-	-
B	RitualDance	Trans.	60.95	64.52	70.32	69.57	66.34	60.99	66.95	67.21	70.33	66.37
		Aff.	35.78	33.39	27.49	27.97	31.16	35.77	30.80	30.53	27.39	31.12
		Persp.	3.26	2.09	2.19	2.46	2.50	3.24	2.24	2.26	2.28	2.51
	BasketballDrive	Trans.	69.72	76.68	78.99	80.06	76.36	66.19	71.90	76.83	79.12	73.51
		Aff.	26.80	21.72	19.12	18.33	21.49	29.73	25.57	20.91	19.10	23.83
		Persp.	3.48	1.60	1.89	1.61	2.15	4.08	2.53	2.26	1.79	2.66
	BQTerrace	Trans.	51.35	54.71	53.55	54.02	53.41	54.54	51.51	53.15	54.21	53.35
		Aff.	24.04	39.87	41.11	41.96	36.74	34.14	42.70	42.72	41.28	40.21
		Persp.	24.61	5.42	5.34	4.02	9.85	11.31	5.78	4.13	4.52	6.44
C	BasketballDrill	Trans.	53.18	55.43	56.09	55.92	55.16	52.12	54.08	55.47	56.33	54.50
		Aff.	39.42	39.72	39.05	39.30	39.37	41.58	41.16	40.34	39.39	40.62
		Persp.	7.40	4.84	4.86	4.78	5.47	6.30	4.76	4.19	4.28	4.88
	PartyScene	Trans.	57.63	62.45	67.07	68.17	63.83	55.99	59.41	65.32	67.28	62.00
		Aff.	35.76	33.08	29.51	28.50	31.71	38.59	36.08	31.24	29.12	33.76
		Persp.	6.61	4.47	3.41	3.33	4.46	5.43	4.51	3.44	3.59	4.24
D	BQSquare	Trans.	65.54	65.12	66.08	66.71	65.86	65.16	64.90	65.43	66.42	65.48
		Aff.	29.81	31.09	30.43	29.88	30.30	30.60	31.04	30.78	30.01	30.61
		Persp.	4.65	3.79	3.49	3.41	3.84	4.24	4.06	3.79	3.58	3.92
	RaceHorses	Trans.	62.83	62.91	64.00	64.84	63.64	61.22	63.20	64.32	65.08	63.46
		Aff.	33.57	33.79	32.91	31.99	33.06	35.34	33.45	32.55	31.80	33.29
		Persp.	3.60	3.30	3.09	3.18	3.29	3.44	3.35	3.13	3.12	3.26
E	FourPeople	Trans.	-	-	-	-	-	58.39	61.01	62.13	62.53	61.01
		Aff.	-	-	-	-	-	37.57	35.05	34.28	33.64	35.13
		Persp.	-	-	-	-	-	4.04	3.95	3.59	3.83	3.85
	KristenAndSara	Trans.	-	-	-	-	-	53.38	56.98	57.98	58.51	56.72
		Aff.	-	-	-	-	-	41.85	39.22	38.10	37.53	39.18
		Persp.	-	-	-	-	-	4.76	3.80	3.91	3.95	4.11
F	BasketballDrillText	Trans.	53.82	56.15	56.57	57.01	55.89	53.01	55.87	57.81	56.78	55.87
		Aff.	38.45	38.03	37.69	37.12	37.83	40.70	39.25	37.46	37.73	38.78
		Persp.	7.73	5.82	5.74	5.86	6.29	6.30	4.88	4.72	5.49	5.35
	SlideEditing	Trans.	56.47	56.53	56.55	56.58	56.53	56.66	56.73	56.86	56.67	56.73
		Aff.	37.52	37.64	37.58	37.66	37.60	37.79	37.62	37.61	37.72	37.69
		Persp.	6.01	5.83	5.87	5.76	5.87	5.56	5.65	5.53	5.61	5.58
	SlideShow	Trans.	56.77	58.23	57.68	57.58	57.57	56.51	56.63	56.42	56.86	56.61
		Aff.	36.99	35.94	37.15	37.24	36.83	38.16	37.83	38.17	38.00	38.04
		Persp.	6.25	5.82	5.17	5.18	5.60	5.33	5.54	5.41	5.15	5.36
Avg.		Trans.	58.83	61.27	62.69	63.04	**61.46**	57.85	59.93	61.58	62.51	**60.47**
		Aff.	33.81	34.43	33.20	33.00	**33.61**	36.82	35.82	34.56	33.56	**35.19**
		Persp.	7.36	4.30	4.11	3.96	**4.93**	5.33	4.25	3.86	3.93	**4.35**

Table 2. BD-Rate (%) performance and encoder complexity (%) of the proposed algorithm, compared to VTM 2.0 Baseline.

Class	Sequence	RA				LDP			
		Y	U	V	YUV	Y	U	V	YUV
A	Campfire	−0.09%	−0.02%	0.06%	−0.07%	-	-	-	-
	CatRobot1	−0.10%	0.41%	0.33%	−0.03%	-	-	-	-
B	RitualDance	−0.04%	0.15%	0.22%	−0.02%	−0.06%	0.55%	0.01%	−0.03%
	BasketballDrive	−0.14%	−0.05%	0.44%	−0.03%	−0.06%	0.28%	0.47%	−0.06%
	BQTerrace	−0.08%	−0.11%	−0.03%	−0.07%	−0.08%	0.38%	−0.16%	−0.10%
C	BasketballDrill	−0.09%	−0.12%	−0.16%	−0.14%	−0.02%	0.04%	−0.15%	−0.02%
	PartyScene	−0.06%	−0.21%	−0.04%	−0.07%	0.01%	0.46%	0.24%	0.04%
D	BQSquare	0.03%	0.45%	−0.04%	0.04%	−0.03%	−0.42%	−1.87%	−0.09%
	RaceHorses	0.07%	−0.76%	0.09%	0.05%	−0.25%	0.21%	0.49%	−0.17%
E	FourPeople	-	-	-	-	−0.16%	−0.58%	0.15%	−0.18%
	KristenAndSara	-	-	-	-	0.07%	−0.22%	0.03%	0.06%
F	BasketballDrillText	−0.01%	−0.18%	0.25%	−0.01%	−0.07%	−0.57%	−0.28%	−0.13%
	SlideEditing	−0.07%	−0.03%	−0.03%	−0.06%	−0.28%	−0.22%	−0.47%	−0.29%
	SlideShow	−0.30%	1.29%	0.19%	−0.13%	−0.45%	−1.39%	0.08%	−0.46%
Avg.		**−0.07%**	**0.07%**	**0.11%**	**−0.05%**	**−0.12%**	**−0.12%**	**−0.12%**	**−0.12%**
EncT		103%				106%			

In class F which has the screen content sequences, rapid long range motions as large as half a frame often happen like browsing and document editing. Even in this case, the proposed PAMC algorithm can bring BD-Rate gain. However, when the resolution of sequence is too small such as class D, the proposed algorithm could not give good coding gain.

All CTC test sequences used in the experiment have a YUV 4:2:0 format. This means that in the case of the chrominance components, frame images compressed to one quarter are used to encode. The perspective motion model is an advanced model designed to predict accurate movement by using eight parameters, so it can be unstable in color spaces where only a quarter of the pixel information remains. For this reason, even in the same "slideshow" sequence on U component, it can bring about 1.39 BD-Rate gain in the LDP configuration and 1.29 BD-Rate loss in the RA configuration.

From experimental results, the designed PAMC achieved better coding gain compared to VTM 2.0 Baseline. It means that the proposed PAMC scheme can be applied for providing better video quality in terms of a limited network bandwidth environment.

(a) AMC in VVC.

(b) Proposed PAMC.

Fig. 4. Examples of CUs with affine and perspective motion, CatRobot1, RA, QP22, POC24.

5 Conclusions

In this paper, an efficient perspective affine motion compensation framework was designed to estimate further complex motions beyond the affine motion. In the proposed algorithm, an eight-parameter perspective motion model was first defined and analyzed. Then these perspective motion model based motion compensation algorithm was developed. The proposed framework was been implemented in the reference software of VVC. The experimental results showed that the proposed perspective affine motion compensation framework could achieve better BD-Rate performance and negligible encoder complexity compared with the VVC baseline. Especially, the proposed perspective affine motion model can characterize the irregular object distortions or dynamic rapid motions.

Acknowledgement. This research was supported by Basic Science Research Program through the National Research Foundation of Korea (NRF) funded by the Ministry of Education (NRF-2016R1D1A1B04934750).

References

1. Draft, I.T.U.T.: Recommendation and final draft international standard of joint video specification (ITU-T Rec. H. 264—ISO/IEC 14496-10 AVC). Joint Video Team (JVT) of ISO/IEC MPEG and ITU-T VCEG, JVTG050 (2003)
2. Sze, V., Budagavi, M., Sullivan, G.J.: High efficiency video coding (HEVC). In: Integrated Circuit and Systems, Algorithms and Architectures, pp. 1–375 (2014)
3. Bross, B., Chen, J., Liu, S.: Versatile Video Coding (Draft 2), JVET-K1001 (2018)
4. Chen, J., Ye, Y., Kim, S.: Algorithm description for Versatile Video Coding and Test Model 2 (VTM 2), JVET-K1002 (2018)
5. Choi, Y.J., Kim, J.H., Lee, J.H., Kim, B.G.: Performance analysis of future video coding (FVC) standard technology. J. Multimedia Inf. Syst. 4(2), 73–78 (2018)
6. Seferidis, V., Ghanbari, M.: General approach to block-matching motion estimation. Opt. Eng. 32(7), 1464–1474 (1993)
7. Lee, O., Wang, Y.: Motion compensated prediction using nodal based deformable block matching. J. Vis. Commun. Image Represent. 6(1), 26–34 (1995)
8. Cheung, H.-K., Siu, W.-C.: Local affine motion prediction for H.264 without extra overhead. In: IEEE International Symposium on Circuits and Systems (ISCAS), pp. 1555–1558 (2010)
9. Narroschke, M., Swoboda, R.: Extending HEVC by an affine motion model. In: Picture Coding Symposium (PCS), pp. 321–324 (2013)
10. Huang, H., Woods, J.W., Zhao, Y., Bai, H.: Affine SKIP and DIRECT modes for efficient video coding. In: Visual Communications and Image Processing (VCIP), pp. 1–6 (2012)
11. Heithausen, C., Vorwerk, J.H.: Motion compensation with higher order motion models for HEVC. In: IEEE International Conference on Acoustics, Speech and Signal Processing (ICASSP), pp. 1438–1442 (2015)
12. Choi, Y.J., Jun, D.S., Cheong, W.S., Kim, B.G.: Design of efficient perspective affine motion estimation/compensation for versatile video coding (VVC) standard. Electronics 8(9), 993 (2019)
13. Li, L., Li, H., Lv, Z., Yang, H.: An affine motion compensation framework for high efficiency video coding. In: IEEE International Symposium on Circuits and Systems (ISCAS), May 2015
14. Li, L., Li, H., Liu, D., Li, Z., Yang, H., Lin, S., Chen, H., Wu, F.: An efficient four-parameter affine motion model for video coding. IEEE Trans. Circ. Syst. Video Technol. 28(8), 1934–1948 (2018)
15. Lin, S., Chen, H., Zhang, H., Maxim, S., Yang, H., Zhou, J.: Affine transform prediction for next generation video coding, ITU-T SG16 Doc. COM16-C1016 (2015)

16. Chen, J., Qualcomm Inc., Alshina, E., Samsung Electronics, Sullivan, G.J., Microsoft Corp., Jens-Rainer Ohm, RWTH Aachen University, Boyce, J.: Intel: Algorithm Description of Joint Exploration Test Model 1, JVET-A1001 (2015)
17. Zhang, K., Chen, Y.-W., Zhang, L., Chien, W.-J., Karczewicz, M.: An improved framework of affine motion compensation in video coding. IEEE Trans. Image Process. **28**(3), 1456–1469 (2019)
18. Versatile Video Coding (VVC) Test Model 2.0 (VTM 2.0). https://vcgit.hhi. fraunhofer.de/jvet/VVCSoftware_VTM.git. Accessed 1 Dec 2019
19. Bossen, F., Boyce, J., Suehring, K., Li, X., Seregin, V.: JVET common test conditions and software reference configurations for SDR video, JVET-K1010 (2018)
20. Bjøntegaard, G.: Calculation of average PSNR differences between RDcurves, ITU-T SG.16 Q.6, Document VCEG-M33 (2001)

Investigation of Coding Standards Performances on Optically Acquired and Synthetic Holograms

Roberto Corda, Cristian Perra$^{(\boxtimes)}$ (iD), and Daniele Giusto

DIEE, UdR Cnit, University of Cagliari, Cagliari, Italy
cperra@ieee.org

Abstract. Digital holography needs efficient coding tools that facilitate storage and transmission of this type of data in order to reach practical applications. This paper presents an experimental analysis of the performance of different coding tools for the compression of digital holograms. During the experiments, a dedicated compression architecture is employed in order to transform the holographic data in a representation suitable to be provided to the encoders, and for performing an objective quality evaluation of the obtained results. Several state-of-the-art image and video codecs are evaluated on different reference datasets, comprising different types of digital holograms. The evaluation is carried out on the reconstructed images with different metrics, and obtained results are critically analyzed and discussed.

Keywords: Computer generated holography · CGH · Hologram compression · Hologram coding · Objective quality

1 Introduction

The vast majority of 3D displays and headsets are currently based on the stereoscopic principle [21]. However, the lack of correspondence between accommodation distance and convergence distance can cause distress or malaise to the observer [14]. Holography, on the other hand, is distinguished for its unique ability of reproduce all visual cues, giving to the viewer the perception of a three-dimensional image that is exactly the same as the reality [43], and because of this feature, it could revolutionize the field of display applications. Currently, this imaging technique has shown its potential and usefulness in a wide range of application fields, ranging from microscopy [6,34] to interferometry applications [22,35].

Typically, holography is a two-step process, in which the first step is the hologram acquisition, while the second step aims to reproduce the hologram,

This research activity has been partially funded within the Cagliari2020 project (MIUR, PON04a2 00381) and the DigitArch Cluster Top-Down project (POR FESR, 2014–2020).

J. Blanc-Talon et al. (Eds.): ACIVS 2020, LNCS 12002, pp. 396–407, 2020.
https://doi.org/10.1007/978-3-030-40605-9_34

i.e. reconstruct the holographic image from the previously acquired information. There exist several techniques for acquiring the hologram of a real object, but a general setup includes a coherent light source, generally a laser, a beam-splitter, and an acquisition plate, that can also be an electronic device such as a CCD or CMOS sensor [4,38]. The beam generated by the laser is directed towards the beam-splitter, from which two different beams (or "waves") originates: a reference wave and an object wave. The reference wave is so-called because it is headed directly to the acquisition plate, while the object wave first hits the object of interest, and only after that interaction it goes towards the acquisition plate. The interference between reference and object waves is recorded by the acquisition plate and constitutes what is called a hologram. Through this method it is possible to acquire not only information on the amplitude of the electromagnetic wave but also about its phase. The holographic image is finally obtained illuminating the hologram stored in the acquisition plate [38]. The acquisition of holograms with optical setups, although it has the advantage of being able to acquire real objects, is a non-trivial task that moreover requires specialized equipment and environments. In this context, computerized methods represent a viable alternative: they allow to numerically calculate the propagation of the electromagnetic field, simulating the acquisition and reconstruction processes. Typically, the computation of the so-called Computer Generated Hologram (CGH) takes place starting from a three-dimensional representation of a synthetic scene [29]. The three-dimensional scene representation can be of different types, and the generation algorithm typically varies in according to it: computer generated holograms can be originated from point clouds [1,33,42], polygonal models [19,26,41], layer-based representations [7,29,47] or ray-based representations [15,36,45]. Regardless of the method, the computational complexity to create a CGH is typically high, so research in this field continues to be very active [5,11,28,32]. In this work, both optically acquired and computer-generated holograms (mainly related to display applications) have been employed during the experimentation phase.

Appropriate visualization devices are obviously necessary to exploit the great potential of holography. However, despite different types of holographic displays have been proposed so far, high quality multi-user holographic displays are unlikely to be available in the near future [3]. Therefore, during the exploration studies on holography, alternative strategies are employed to circumvent this unavailability [8].

The amount of information required to represent a high quality digital hologram is far greater than conventional image or video contents, and can reach data volumes in the order of the Terasamples [25]. Efficient coding solution are therefore mandatory for storing, transmitting and manage these type of data. In the recent years, different type of proposal have been put forward in order to resolve this great challenge: lossless techniques as Lempel-Ziv, Lempel-Ziv-Welch, Huffman coding, other than DCT and DFT compression [27] and standards codecs for image and video compression [2,30], including also different pre-processing steps before the standard coding stage [39,40]. More holographic-content-aware transforms have been also proposed, as the Gabor wavelet [9] and the Fresnelets [23]. Despite the various contributions, digital holography compression is still an open challenge, in which a standard solution has not been found yet. Within ISO/IEC SC29 WG1, the JPEG PLENO standardization activities are currently exploring, among other activities, the aspect of holographic data compression [37]. In order to develop efficient and effective coding solutions, it is believed that it is necessary to have a good knowledge of the currently state-of-the-art coding solutions on holographic data. The purpose of this research is to perform exploration studies on digital holograms using different coding tools, such as JPEG XT, JPEG XS, JPEG 2000, HEVC. In contrast to previous studies on hologram compression benchmarks as [31], other than include more recent types of codecs, different types of digital holograms (computer generated, optically acquired, monochrome and color) have been tested in this work, with the aim of contributing to the research in the holography compression area and to current standardization activities. This paper is structured as follows. In Sect. 2 an overview of the digital holography main concepts, related to this work, is given. Section 3 describes the employed compression architecture and the metrics used for evaluating the results. Section 4 is dedicated to the description of the holograms under test, the experimental results, and discussion. Finally, the conclusions are drawn in Sect. 5.

2   Digital Holography Basics

During an optical hologram recording, the acquisition device records the interference between the reference wave and the object wave [38]. The digital hologram is often expressed as a complex waveform $U_0(x_0, y_0)$ that can be calculated using several methods, depending on the employed acquisition technique. A common technique is the Phase Shifting, that consists in recording multiple interferograms, varying the phase difference between reference and object wave at each acquisition. A basic case can involve the acquisition of four interferograms, increasing the phase angle at each acquisition of $\frac{\pi}{2}$ radians, for example. In this scenario, the complex waveform can be obtained with the following expression [44]:

$$U_0(x_0, y_0) = \frac{1}{4U_r^*} \left\{ \left[I(x_0, y_0; 0) - I(x_0, y_0; \pi) \right] \right.$$
$$\left. - j \left[I\left(x_0, y_0; \frac{\pi}{2}\right) - I\left(x_0, y_0; \frac{3\pi}{2}\right) \right] \right\} \tag{1}$$

where U_r represents the reference wave, the symbol $*$ the complex conjugate, I is an interferogram and j is the imaginary unit. There exist also improvements of the base Phase Shifting technique, that allow the acquisition of only three or two interferograms [24]. Computer Generated Holograms are instead obtained simulating the light propagation, that can be described through the use of the scalar diffraction theory, by which a scene point (x_0, y_0) can be related to a point in the hologram plane (x, y) placed at distance z [12]:

$$U_r(x, y) = \frac{z}{j\lambda} \iint U_0(x_0, y_0) \frac{\exp\left(j\frac{2\pi}{\lambda}r\right)}{r^2} dx_0 dy_0 \qquad (2)$$

where λ is the wavelength, and $r = \sqrt{(x - x_0)^2 + (y - y_0)^2 + z^2}$. The Eq. 2 is often simplified, and two widely used simplifications are the Angular Spectrum Method (ASM) and the Fresnel Method [20].

The complex waveform that constitutes the hologram is often described with the two standard forms of the complex numbers: the algebraic and the polar form. In the algebraic form, the complex wavefield can be identified with a real and a imaginary part:

$$U_0(x_0, y_0) = real(U_0(x_0, y_0)) + j\, imag(U_0(x_0, y_0)) \qquad (3)$$

while in the polar representation, it can be divided in the amplitude and the phase component:

$$U_0(x_0, y_0) = A(x_0, y_0)e^{j\phi(x_0, y_0)}. \qquad (4)$$

3   Proposed Evaluation Scheme

The holograms employed during the compression studies belong to the B-Com and the EmergImg-HoloGrail v2 databases. The former is composed of different CGH generated with the algorithms proposed in [10,11], while the latter includes optically-recorded holograms acquired using the Phase Shifting technique [2]. In both cases, the digital holograms are represented with a matrix of complex values (or three matrices, if the hologram is RGB), therefore some pre-processing and post-processing steps are required in order to compress these data with standard codecs and evaluate the compression performances. In this respect, the architecture showed in Fig. 1 has been employed.

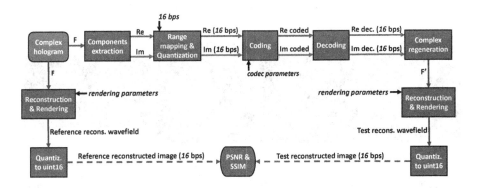

Fig. 1. Proposed evaluation scheme.

The first step is to extract the two components from the matrix F that represents the complex hologram, using the Components extraction block. In these experiments, the algebraic representation is used, since previous studies showed that it provides better compression performance compared to the polar [2,31]. The two extracted components real (Re) and imaginary (Im) are two matrices of real numbers, whose values have an arbitrary variation range. This range is then mapped in the range [0, 1], and subsequently these data are quantized with 16 bits per sample (bps). The process takes place in parallel for the two components. They are then coded and decoded (in parallel) in the Coding and Decoding blocks. In these experiments, four of the most recent state-of-the-art codecs have been chosen: JPEG 2000 [18], JPEG XT [17], JPEG XS [16] and HEVC [13] (Intra) used in lossy mode.

Once decoded, Re dec. and Im dec. are provided as input to the Complex regeneration block: in order to reconstruct the hologram, it is necessary to regain the complex matrix. In this block, both components are dequantized, restored to their original variation ranges and finally joined in a single complex matrix F'. At this point, the holographic image is reconstructed from the hologram that has been compressed (F') and in parallel the same image is reconstructed from the hologram that has not been compressed (F). The reconstruction process is carried out in the Reconstruction & Rendering block, which provide as output the reconstructed wavefied. This wavefield is represented by a matrix of positive real numbers, which in the Quantiz. to uint16 block are quantized to 16 bit unsigned values. Finally, the image reconstructed from the compressed data (Test Image) is compared to the image reconstructed from data that has not undergone any processing (Reference Image) with the PSNR and SSIM [46] metrics, of common use also in holography [3], in order to evaluate the effect and performance of the codec.

4   Experimental Analysis and Results

The compression experiments have been carried out on two optically recorded grayscale holograms belonging to the EmergImg-HoloGrail v2 dataset, namely Astronaut (Fig. 2a) and Dice2 (Fig. 2b), which have a resolution of 2588 × 2588

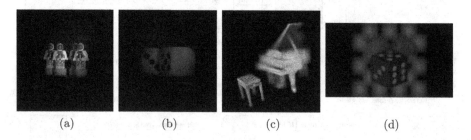

(a)   (b)   (c)   (d)

Fig. 2. Holograms under test (reconstructed reference images): Astronaut (a) Dice2 (b), Piano8k (c) and Dices8k4k (d).

and pixel pitch of 2.2 μm. The other two holograms under test are color computer generated holograms, belonging to the B-Com's dataset: Piano8k (Fig. 2c) with a resolution of 8192 × 8192 and a pixel pitch of 0.4 μm, and Dices8k4k (Fig. 2d) which have resolution of 7680 × 4320 with pixel pitch of 4.8 μm. The results showed in the following, related to the lossy compression, are obtained comparing the image reconstructed from the uncompressed (Reference) hologram and the image reconstructed from the hologram that has been compressed. For each codec, the hologram has been compressed with six different quality levels. Starting from HEVC, the q parameter has been set at $\{33, 25, 20, 15, 10, 7\}$, while for the other codecs the output (compressed) bitrates of HEVC have been used as reference for the quality setting. The "Total Rate" is the total bitrate of the hologram, i.e. the sum of the real and imaginary part bitrates'. In Fig. 3 the results in

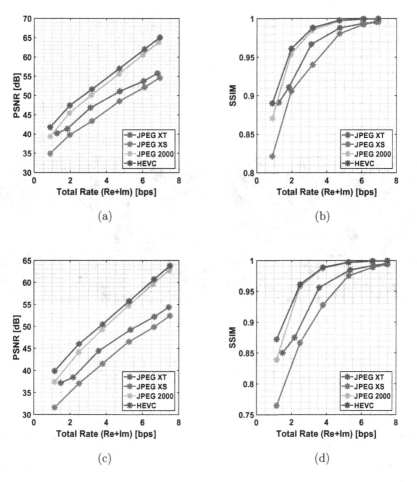

Fig. 3. EmergImg-HoloGrail v2 results: Astronaut PSNR (a) and SSIM (b), Dice2 PSNR (c) and SSIM (d).

terms of PSNR and SSIM of the optically-acquired samples (EmergImg-Holograil v2 dataset) are showed. It can be noted that the codecs perform similary both with Astronaut (Fig. 3a–b) and Dice2 (Fig. 3c–d). The best codec is HEVC, as highlighted by previous works [31], followed by JPEG 2000. These performances can be related to the complexity of these two codecs, higher than JPEG XT and JPEG XS. The performance difference between HEVC and JPEG 2000 tends to taper off as the bitrate increases, both in terms of PSNR and SSIM. JPEG XT and JPEG XS show lower image quality in terms of PSNR at every bitrate, while in terms of SSIM all codecs are nearly 0.99 or above at bitrates of 6 bps or higher. The worst codec is JPEG XS: this fact can be related to the absence of a specific coding profile for grayscale inputs, other than the very low complexity that characterizes this codec. As an example, in Fig. 4 a comparison with Astronaut, between the best (HEVC) and the worst (JPEG XS) codec is showed. The Fig. 4a is the reference image, Fig. 4b is the reconstructed hologram compressed with HEVC, while Fig. 4c compressed with JPEG XS, both at nearly 0.9 bps (lower points in the HEVC/JPEG XS curves, Figs. 3a–b). It can be noted that the HEVC compression doesn't show clear artifacts, as could be expected with PSNR of nearly 40 dB and an SSIM of 0.89. The JPEG XS compression instead, brings up an evident "ghosting effect" in the left, right and bottom of the reconstruction, other than a less sharp image of the four astronauts: this image has a PSNR of nearly 31.5 dB and a SSIM of 0.77.

(a)   (b)   (c)

Fig. 4. EmergImg-HoloGrail v2 Astronaut: reference reconstruction (a); reconstruction from HEVC compression at 0.9 bps, PSNR = 40 dB, SSIM = 0.89 (b); reconstruction from JPEG XS compression at 0.9 bps, PSNR = 31.5 dB, SSIM = 0.77 (c).

In Fig. 5 the compression results of the color computer-generated holograms of B-Com's dataset are showed. It can be noted that HEVC is still the best codec, but with some differences depending on the hologram: in Dices8k4k (Fig. 5a–b) the difference with respect to JPEG 2000 becomes negligible over nearly 6.5 bps, both in terms of PSNR and SSIM, while in Piano8k (Fig. 5c–d) there is a more evident difference, especially in terms of PSNR. JPEG XS shows better performances compared to JPEG XT on these color holograms. This fact could

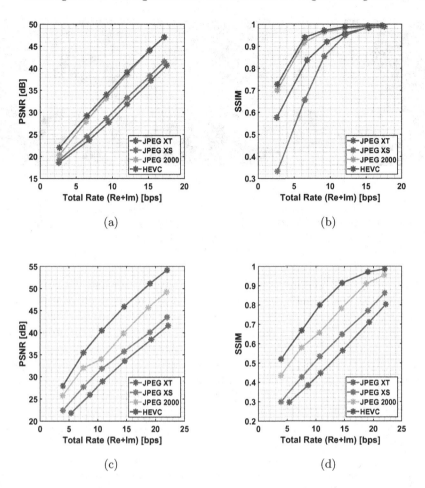

Fig. 5. B-Com results: Dices8k4k PSNR (a) and SSIM (b), Piano8k PSNR (c) and SSIM (d).

be related to the presence of a more specific coding profile (XS main) that is not available for grayscale inputs. The overall SSIM values, for a fixed bitrate, are lower for Piano8k if compared to Dices8k4k: this difference can be related to the stronger speckle noise that affect Piano8k compared to Dices8k4k as can also be noted in Fig. 2. As an example, in Fig. 6 a comparison with Piano8k, between the best (HEVC) codec and in this case JPEG XT (the worst codec) is showed. The Fig. 6a is the reference image, Fig. 6b is the reconstructed holo-gram compressed with HEVC at 3.9 bps, while Fig. 6c compressed with JPEG XT at 5.3 bps. These images correspond to the lower points in the HEVC/JPEG XT curves showed in Fig. 5c–d). It can be noted that, despite the bitrate, the HEVC compression shows clearly visible alternations with a sort of "halo effect" that surrounds the scene, other than a stronger speckle noise and a focus varia-tion, if compared with the reference image. The distortion is even more visible

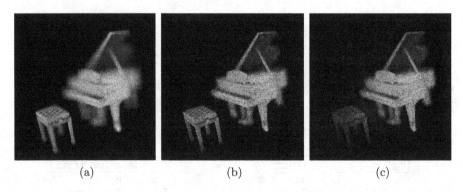

(a)   (b)   (c)

Fig. 6. B-Com Piano8k: reference reconstruction (a); reconstruction from HEVC compression at 3.9 bps, PSNR = 27.9 dB, SSIM = 0.52 (b); reconstruction from JPEG XT compression at 5.3 bps, PSNR = 21.8 dB, SSIM = 0.29 (c).

with the JPEG XT compression, that shows strong aberrations and artifacts. In both cases the image quality can hardly be considered satisfactory, especially if related to the bitrate: these values could be considered relatively high even for a standard image, and even more for a digital hologram, that in order to produce high-quality holographic images may consists of 10^{12} samples or more, as already pointed out. This poor performances, other than the more complex data pattern if compared to standard image contents, could be also related to the speckle noise that characterizes this hologram, that in addition to decrease the reconstructed image "pleasantness", affect also the compression performances. This aspect should be then considered during the compression stage, or better during the acquisition or generation processes.

As a final remark, a non-linear behaviour of the curves (i.e. an unusual relation between the bitrate and the estimated image quality) can be noted in some cases, as the JPEG XT compression of Astronaut and Dice2 (Fig. 3) or the JPEG 2000 compression of Piano8k (Fig. 5c–d). The content-unawareness of the codecs has definitely a non-negligible role in this phenomenon: during the coding stage important parts of the hologram could be affected in a strong (or weak) way, and the impact of these alteration on the reconstructed image cannot be easily predicted, as theoretically every hologram sample could have information of the whole holographic image. These relations are, at the best of the authors' knowledge, still poorly known at present and will need more in-depth investigation during the design of new coding solutions for holographic contents.

5   Conclusions

The compression performances of different state-of-the-art standard codecs have been evaluated on optically acquired and computer-generated digital holograms. In particular, the PSNR and SSIM metrics have been evaluated on the reconstructed images, showing that HEVC and JPEG 2000 are the best codecs.

Although they are designed for standard image contents, PSNR and SSIM can provide useful insights on holographic image reconstructions quality. Nevertheless, a strong validation of image quality metrics will be necessary in order to provide more accurate results. Another important aspect to generate efficient coding solutions is the understating of the main parameters that characterize the holographic image, as the relations between the hologram samples and the reconstructed images, along with disturbing elements as the speckle noise that arises during the acquisition process. Future works will be oriented to further explore these aspects with the aim to develop efficient coding solutions for digital holograms.

References

1. Arai, D., Shimobaba, T., Nishitsuji, T., Kakue, T., Masuda, N., Ito, T.: An accelerated hologram calculation using the wavefront recording plane method and wavelet transform. Opt. Commun. **393**, 107–112 (2017)
2. Bernardo, M.V., et al.: Holographic representation: hologram plane vs. object plane. Signal Process. Image Commun. **68**, 193–206 (2018)
3. Blinder, D., et al.: Signal processing challenges for digital holographic video display systems. Signal Process. Image Commun. **70**, 114–130 (2019)
4. Blinder, D., et al.: Open access database for experimental validations of holographic compression engines. In: 2015 Seventh International Workshop on Quality of Multimedia Experience (QoMEX), pp. 1–6. IEEE (2015)
5. Blinder, D., Schelkens, P.: Accelerated computer generated holography using sparse bases in the STFT domain. Opt. Express **26**(2), 1461–1473 (2018)
6. Cazac, V., et al.: Surface relief and refractive index gratings patterned in chalcogenide glasses and studied by off-axis digital holography. Appl. Opt. **57**(3), 507–513 (2018)
7. Chen, J.S., Chu, D.: Improved layer-based method for rapid hologram generation and real-time interactive holographic display applications. Opt. Express **23**(14), 18143–18155 (2015)
8. Corda, R., Perra, C.: A dataset of hologram reconstructions at different distances and viewpoints for quality evaluation. In: 2019 Eleventh International Conference on Quality of Multimedia Experience (QoMEX), pp. 1–3. IEEE (2019)
9. El Rhammad, A., Gioia, P., Gilles, A., Cagnazzo, M., Pesquet-Popescu, B.: Color digital hologram compression based on matching pursuit. Appl. Opt. **57**(17), 4930–4942 (2018)
10. Gilles, A., Gioia, P., Cozot, R., Morin, L.: Computer generated hologram from multiview-plus-depth data considering specular reflections. In: 2016 IEEE International Conference on Multimedia & Expo Workshops (ICMEW), pp. 1–6. IEEE (2016)
11. Gilles, A., Gioia, P., Cozot, R., Morin, L.: Hybrid approach for fast occlusion processing in computer-generated hologram calculation. Appl. Opt. **55**(20), 5459–5470 (2016)
12. Goodman, J.W.: Introduction to Fourier Optics. Roberts and Company Publishers, Englewood (2005)
13. HEVC Reference Software. https://hevc.hhi.fraunhofer.de/

14. Hoffman, D.M., Girshick, A.R., Akeley, K., Banks, M.S.: Vergence-accommodation conflicts hinder visual performance and cause visual fatigue. J. Vis. **8**(3), 33–33 (2008)

15. Igarashi, S., Nakamura, T., Matsushima, K., Yamaguchi, M.: Efficient tiled calculation of over-10-gigapixel holograms using ray-wavefront conversion. Opt. Express **26**(8), 10773–10786 (2018)

16. Overview of JPEG XS. https://jpeg.org/jpegxs/index.html

17. JPEG XT Reference Software. https://jpeg.org/jpegxt/software.html

18. Kakadu software. http://kakadusoftware.com/

19. Kim, H., Kwon, J., Hahn, J.: Accelerated synthesis of wide-viewing angle polygon computer-generated holograms using the interocular affine similarity of three-dimensional scenes. Opt. Express **26**(13), 16853–16874 (2018)

20. Kim, M.K.: Principles and techniques of digital holographic microscopy. SPIE Rev. **1**(1), 018005 (2010)

21. Konrad, J., Halle, M.: 3-D displays and signal processing. IEEE Signal Process. Mag. **24**(6), 97–111 (2007)

22. Kumar, M., Shakher, C.: Experimental characterization of the hygroscopic properties of wood during convective drying using digital holographic interferometry. Appl. Opt. **55**(5), 960–968 (2016)

23. Liebling, M., Blu, T., Unser, M.: Fresnelets: new multiresolution wavelet bases for digital holography. IEEE Trans. Image Process. **12**(1), 29–43 (2003)

24. Liu, J.P., Poon, T.C., Jhou, G.S., Chen, P.J.: Comparison of two-, three-, and four-exposure quadrature phase-shifting holography. Appl. Opt. **50**(16), 2443–2450 (2011)

25. Lucente, M.: The first 20 years of holographic video-and the next 20. In: SMPTE 2nd Annual International Conference on Stereoscopic 3D for Media and Entertainment, pp. 21–23 (2011)

26. Matsushima, K., Nakahara, S.: Extremely high-definition full-parallax computer-generated hologram created by the polygon-based method. Appl. Opt. **48**(34), H54–H63 (2009)

27. Naughton, T.J., Frauel, Y., Javidi, B., Tajahuerce, E.: Compression of digital holograms for three-dimensional object recognition. In: Algorithms and Systems for Optical Information Processing V, vol. 4471, pp. 280–290. International Society for Optics and Photonics (2001)

28. Nishitsuji, T., Shimobaba, T., Kakue, T., Ito, T.: Review of fast calculation techniques for computer-generated holograms with the point-light-source-based model. IEEE Trans. Ind. Inf. **13**(5), 2447–2454 (2017)

29. Park, J.H.: Recent progress in computer-generated holography for three-dimensional scenes. J. Inf. Disp. **18**(1), 1–12 (2017)

30. Peixeiro, J., Brites, C., Ascenso, J., Pereira, F.: Digital holography: benchmarking coding standards and representation formats. In: 2016 IEEE International Conference on Multimedia and Expo (ICME), pp. 1–6. IEEE (2016)

31. Peixeiro, J.P., Brites, C., Ascenso, J., Pereira, F.: Holographic data coding: benchmarking and extending HEVC with adapted transforms. IEEE Trans. Multimedia **20**(2), 282–297 (2017)

32. Perra, C.: Assessing the quality of experience in viewing rendered decompressed-light fields. Multimedia Tools Appl. **77**(16), 21771–21790 (2018)

33. Phan, A.H., Alam, M.A., Jeon, S.H., Lee, J.H., Kim, N.: Fast hologram generation of long-depth object using multiple wavefront recording planes. In: Practical Holography XXVIII: Materials and Applications, vol. 9006, p. 900612. International Society for Optics and Photonics (2014)

34. Quan, X., et al.: Three-dimensional stimulation and imaging-based functional optical microscopy of biological cells. Opt. Lett. **43**(21), 5447–5450 (2018)
35. Ruiz, C.G.T., Manuel, H., Flores-Moreno, J., Frausto-Reyes, C., Santoyo, F.M.: Cortical bone quality affectations and their strength impact analysis using holographic interferometry. Biomed. Opt. Express **9**(10), 4818–4833 (2018)
36. Sato, H., et al.: Real-time colour hologram generation based on ray-sampling plane with multi-GPU acceleration. Sci. Rep. **8**(1), 1500 (2018)
37. Schelkens, P., et al.: JPEG Pleno: Providing representation interoperability for holographic applications and devices. ETRI J. **41**(1), 93–108 (2019)
38. Schnars, U., Jüptner, W.P.: Digital recording and numerical reconstruction of holograms. Meas. Sci. Technol. **13**(9), R85 (2002)
39. Seo, Y.H., Choi, H.J., Kim, D.W.: 3D scanning-based compression technique for digital hologram video. Signal Process. Image Commun. **22**(2), 144–156 (2007)
40. Seo, Y.H., et al.: Digital hologram compression technique by eliminating spatial correlations based on MCTF. Opt. Commun. **283**(21), 4261–4270 (2010)
41. Shimobaba, T., Kakue, T., Ito, T.: Review of fast algorithms and hardware implementations on computer holography. IEEE Trans. Ind. Inf. **12**(4), 1611–1622 (2015)
42. Shimobaba, T., Masuda, N., Ito, T.: Simple and fast calculation algorithm for computer-generated hologram with wavefront recording plane. Opt. Lett. **34**(20), 3133–3135 (2009)
43. Slinger, C., Cameron, C., Stanley, M.: Computer-generated holography as a generic display technology. Computer **38**(8), 46–53 (2005)
44. Tsang, P.W.M., Poon, T.C.: Review on the state-of-the-art technologies for acquisition and display of digital holograms. IEEE Trans. Ind. Inf. **12**(3), 886–901 (2016)
45. Wakunami, K., Yamaguchi, M.: Calculation for computer generated hologram using ray-sampling plane. Opt. Express **19**(10), 9086–9101 (2011)
46. Wang, Z., Bovik, A.C., Sheikh, H.R., Simoncelli, E.P., et al.: Image quality assessment: from error visibility to structural similarity. IEEE Trans. Image Process. **13**(4), 600–612 (2004)
47. Zhao, Y., Cao, L., Zhang, H., Kong, D., Jin, G.: Accurate calculation of computer-generated holograms using angular-spectrum layer-oriented method. Opt. Express **23**(20), 25440–25449 (2015)

Natural Images Enhancement Using Structure Extraction and Retinex

Xiaoyu Du and Youshen Xia[✉]

Department of Mathematics and Computer Science, Fuzhou University,
Fuzhou, China
ysxia@fzu.edu.cn

Abstract. Variational Retinex model-based methods for low-light image enhancement have been popularly studied in recent years. In this paper, we present an enhanced variational Retinex method for low-light natural image enhancement, based on the initial smoother illumination component with a structure extraction technique. The Bergman splitting algorithm is then introduced to estimate the illuminance component and reflectance component. The de-block processing and illuminance component correction are used for the enhanced reflectance as the ultimate enhanced image. Moreover, the estimated smoother illumination component can make enhanced images preserve edge details. Experimental results with a comparison demonstrate the present variational Retinex method can effectively enhance image quality and maintain image color.

Keywords: Image enhancement · Variational Retinex · Structure extraction

1 Introduction

Nowadays, it is very needed to enhance low-light images because of the poor lighting condition. The image enhancement technique has been widely used in many fields such as underwater exploration and remote sensing, and so on. Many methods for image enhancement were presented in recent decades. They are roughly divided into four categories: histogram equalization algorithm [1,2], wavelet transform algorithm [3], partial differential equation algorithm [4], and Retinex algorithm [5–7]. Among them, Wu [8] proposed a contrast enhancement methods based on linear programming model. This method has a fast computation speed but it can enlarge the noise in the image to be enhanced. Huang et al. [9] proposed an adaptive gamma correction method to prevent excessive enhancement. The Retinex-based methods are based on human color perception system. Various variational Retinex models-based method for low-light image enhancement were presented. In particular, Kimmel et al. [7] gave a general variational framework for Retinex, where the different regularization terms of illuminance were introduced to establish the objective energy function. In recent years, several Retinex model-based methods for enhancing low-light images were presented by using different initial illuminance components [10–15]. However, they suffer from severe halo artifacts.

J. Blanc-Talon et al. (Eds.): ACIVS 2020, LNCS 12002, pp. 408–420, 2020.
https://doi.org/10.1007/978-3-030-40605-9_35

This paper first analyzes the Retinex model-based methods with initial illuminance component estimation, and then presented an enhanced variational retinex method for low-light natural image enhancement based on the initial smoother illumination component estimated by a structure extraction technique [16]. The Bergman splitting algorithm is used to estimate the illuminance component and reflectance component. The de-block processing and illuminance component correction are used for the enhanced reflectance as the ultimate enhanced image. Moreover, the introduced low-light natural image enhancement method can enhance the low contrast images with compression artifacts and preserve image edge details.

2   Retinex Model and Related Works

A classic Retinex model can be described as:

$$I = R \circ L \tag{1}$$

where $\circ$ is the operator of pixel-wise multiplication, I is the input image, R represents the reflectance component, and L represents the illuminance component. It is our goal to estimate the illuminance component and the reflectance component by using the input image.

2.1   Related Works

The estimation of the illuminance component can significantly influence on the reflectance component estimation. In recent years, several Retinex model-based methods for enhancing low-light images were presented by using different initial illuminance components [11–15].

Fu et al. [10] assumed that the gradients of illuminance component approximates the Gaussian distribution, the gradients of reflectance component is formulated with Laplacian distribution. The objective function is expressed as:

$$\min_{L,R} E(L,R) = \|R \circ L - I\|_F^2 + \alpha\|\nabla L\|_F^2 + \beta\|\nabla R\|_1 + \gamma\|L - L_0\|_F^2 \quad s.t. I \leq L \tag{2}$$

where $\|R \circ L - I\|_F^2$ represents the data fidelity term, and $\|\nabla R\|_1$ corresponds to TV reflectance sparsity, enforces piece-wise continuous on the reflectance component R, and $\|\nabla L\|_F^2$ and $\|L - L_0\|_F^2$ are used to ensure the smoothness of illuminance component and avoid the scaling problem. The initial illuminance component L_0 is obtained by averaging the input images. Because this method can't reveal the details of dark image areas, Fu et al. [11] presented using Gaussian low-pass filtering to get the initial illuminance component L_0.

Park et al. [12] presented the minimization of the following objective function:

$$\min_{L,R} E(L,R) = \lambda\|R \circ L - I\|_F^2 + \alpha\|\nabla L\|_F^2 + \beta\|\nabla R\|_1 + \gamma\|L - L_0\|_F^2 \tag{3}$$

where $\|R \circ L - I\|_F^2$ represents the data fidelity term, and $\|\nabla R\|_1$ and $\|\nabla L\|_F^2$ respectively the smoothness terms on the reflectance and illuminance component, and $\|L - L_0\|_F^2$ are used to penalizes the over-enhancement of the reflectance component. The initial illuminance component L_0 is got by the gamma correction. This method can reveal the details of dark images, but the image color is not preserved well.

Guo [13] estimated the illuminance component by using weighted l_1-*norm* to enhanced the contrast of low-light images, and the objective function to be minimized is given by:

$$\min_{L,R} E(L, R) = \lambda \|\hat{L} - L\|_F^2 + \alpha \|W \nabla L\|_1 \tag{4}$$

where $\hat{L}$ represents the initial illuminance, W is the weight matrix, and $\|\hat{L} - L\|_F^2$ represent data fidelity term, and $\|W \nabla L\|_1$ ensure the smoothness of illumination. This method may over enhance images.

Li et al. [14] studied the minimization of the following objective function:

$$\min_{L,R} E(L) = \lambda \|R \circ L + N - I\|_F^2 + \beta \|\nabla L\|_1 + \omega \|\nabla R - G\|_F^2 + \delta \|N\|_F^2 \tag{5}$$

where $\|R \circ L + N - I\|_F^2$ is data fidelity term, where N is the estimated noise map and $\|N\|_F^2$ constrains the overall intensity of the noise. G is the guide map of the gradient of the input I, the term of $\|\nabla R - G\|_F^2$ avoid the noise of reflectance component R, and $\|\nabla L\|_1$ considers the smoothness of the illumination. This method can work well for high noise images, but cannot preserve image details.

Rao et al. [15] studied the minimzation of a general objective function:

$$\min_{L,R} E(L, R) = \|I - R \circ L\|_F^2 + \alpha \|I - R \circ L\|_1 + \beta w \|\nabla L\|_1$$
$$+ \gamma_1 v \|\nabla R\|_1 + \gamma_2 \|L - L_0\|_F^2 \tag{6}$$
$$s.t. \ I \leq L$$

where $\|I - R \circ L\|_F^2$ is data fidelity term, $\|\nabla L\|_1$ and $\|L - L_0\|_F^2$ to estimate illuminance component, and $\|\nabla R\|_1$ is used to estimate reflectance. For the item the $\|I - R \circ L\|_1$, they assumed the error of $R \circ L$ and I is Laplacian distribution. The initial illuminance component L_0 is obtained by averaging the input images.

2.2 Computed Results of Different Initial Illuminance Components

From Sect. 2.1, we see that the Eqs. (2) and (6) use L_0 to ensure the illuminance smoothness, Eq. (3) uses L_0 to enforce the over-enhance of reflectance component. Equation (4) uses the estimated structure to replace the illuminance component. These initial illuminance components used in [10, 13, 15] may cause halo artifacts. Therefore, a good alternative L_0 should be considered for smoothing and preserving edge image details. In this paper, we propose an enhanced variational retinex method for low-light natural image enhancement based on the initial smoother illuminance component with a structure extraction technique.

As an illustrative example shown in Fig. 1, we see that the edges of Figs. 1(a) and (b) are not as well as the edges of Fig. 1(c). Figure 1(d)–(e) is the illuminance component obtained by using the same model but different L_0. It is seen from Fig. 1(d)–(e) that the edge of the finally obtained illuminance component has a halo at the edge portion of the little girl's hand. From Fig. 1(f) that we see that our illuminance component is smooth while maintaining the edge of the image. So from Fig. 1(j) t we see that our enhanced image doesn't have halo artifacts at the edge portion of the little girl's hand. But for other methods shown in Fig. 1(h)–(i), there is a halo in the edge of the little girl's hand. In addition, it is worth noting that Fig. 1(h)–(j) are enhanced images with using same processing steps except for different L_0.

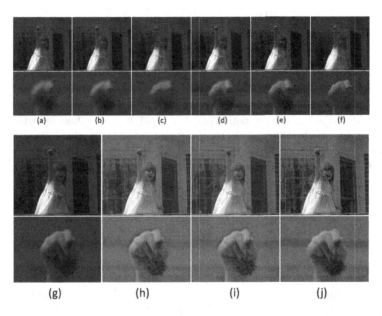

Fig. 1. Comparative results based on different initial illumination components. (a)–(c): L_0 obtained by Gaussian filter, mean filter, and ours; (d)–(f): illumination estimated by (a)–(c); (g): input image; (h)–(j): enhanced images using L_0 from (a)–(c).

3   Proposed Method

3.1   Our Retinex Model

For the balance of contrast, edge details, and color preservation, we present an enhanced variational Retinex model-based method which solves:

$$\min_{L,R} E(L, R) = \|R \circ L - I\|_F^2 + \alpha_1 \|\nabla L\|_1 + \alpha_2 \|\nabla R\|_1 + \alpha_3 \|L - L_*\|_F^2 \quad s.t.\, I \leq L \tag{7}$$

where α_1, α_2 and α_3 are regular term parameters, $\|R \circ L - I\|_F^2$ represents the data fidelity to ensure the fidelity between the input image and the enhanced image; the term of $\|\nabla L\|_1$ and $\|L - L_*\|_F^2$ are for smoothness of the illumination L, and avoid the illumination over-estimate. The term of $\|\nabla R\|_1$ is used to enable piece-wise continuous on the reflectance R. Compared with Rao et al. [15], our method does not contain the term of $\|I - R \circ L\|_1$. This is because the error between $R \circ L$ and I may not satisfy Laplacian distribution. Experimental results in Sect. 4 show that adding the term of $\|I - R \circ L\|_1$ does not improve the quality of the enhanced image. Furthermore, differently from other initial illumination components from papers [10,11,15], we take the initial illuminance component given by [16]:

$$L_* = arg \min_L \|L - I\|_F^2 + \lambda \sum_{i \in \Omega} w_i (\nabla L)_i^2 \tag{8}$$

where the item of $\|L - I\|_F^2$ is the data fidelity, and where w_i is the weight of ∇L, and i refers to the pixels in the window. The Ω was local patches and we set the size of it to 5×5 in this paper.

Since many images are subject to compression artifacts, the reflectance component that we get will have compressed artifact blocks. Then we will remove artifact blocks on the reflectance components. And the method of de-blocking is inspired by [17], but different from [17], we have selected three sizes 4×4, 6×6 and 8×8 local artifact block to smooth the edges. Experiments show that the effect of our method of removing artifact blocks is significant. Then the de-blocking method can be expressed as:

$$E(R^r) = \min_{R^r} \|R^r - R\|_F^2 + \beta \Phi(R) \tag{9}$$

where R is the reflectance component, R^r represent the reflectance that de-blocked, where the term $\|R^r - R\|_F^2$ guarantees the similarity between the reflectance component after the artifact block and the input reflectance component, where $\Phi(R) = \sum_{i \in \Omega} (R_i - R_i^r)^2$ and Ω is the edge of the local block, and this term can smooth the edges of the artifact block. And Fig. 2 shows the overview of our proposed method.

3.2   Solution

We use a Bregman splitting method [18] to solve the Eq. (7). Equation (7) can be rewritten as:

$$\min_{L,R,Z_1,Z_2} \|R \circ L - I\|_F^2 + \beta_1 \|\nabla L + b_1 - Z_1\|_F^2 + \beta_2 \|\nabla R + b_2 - Z_2\|_F^2$$
$$+ \alpha_3 \|L - L_*\|_F^2 + \alpha_1 \|Z_1\|_1 + \alpha_2 \|Z_2\|_1 \tag{10}$$

where β_1, β_2 are penalization parameter, where Z_1, Z_2 are auxiliary variables, where b_1, b_2 are Bregman variables. Now we can split the Eq. (7) into four sub-problems to solve.

The first subproblem is to estimate the reflectance component R, which can be written as:

$$E(R^i) = \min_{R^i} \|R^i \circ L^{i-1} - I\|_F^2 + \beta_2 \|\nabla R^i + b_2^{i-1} - Z_2^{i-1}\|_F^2 \qquad (11)$$

where i represents the ith iteration, then it become easily to get the reflectance component R:

$$R^i = F^{-1}\left\{ \frac{F\left(\frac{I}{L^{i-1}+\xi} + \beta_2 \nabla^T \left(Z_2^{i-1} - b_2^{i-1}\right)\right)}{F\left(1 + \beta_2 \nabla^T \nabla\right)} \right\} \qquad (12)$$

where ∇ represents derivative operator, and it conclude horizontal gradient operator and vertical gradient operator, where ξ represent a small positive number to avoid zero pixels. F represents the fast Fourier transform (FFT) operator, and F^{-1} represents its inverse operator.

The second subproblem is to estimate the illuminance component L, which can be expressed as:

$$E(L^i) = \min_{L^i} \|R^i \circ L^i - I\|_F^2 + \beta_1 \|\nabla L + b_1^{i-1} - Z_1^{i-1}\|_F^2 + \alpha_3 \|L - L_*\|_F^2 \qquad (13)$$

and the solution of illuminance component L is similar to reflectance component R:

$$L^i = F^{-1}\left\{ \frac{F(\frac{I}{R^i+\xi} + \beta_1 \nabla^T(Z_1^{i-1} - b_1^{i-1}) + \alpha_3 L_*)}{F\left(1 + \alpha_3 + \beta_1 \nabla^T \nabla\right)} \right\} \qquad (14)$$

The termination criterion for the L, R iteration set $err_R = \frac{\|R^i - R^{i-1}\|}{\|R^i\|} < 0.001$ and $err_L = \frac{\|L^i - L^{i-1}\|}{\|L^i\|} < 0.001$ or the iteration number greater than 20. Then the third and fourth subproblem of Z_1, Z_2 can be expressed as:

$$E(Z_1^i) = \arg\min_{Z_1} \alpha_1 \|Z_1\|_1 + \beta_1 \|\nabla L^i + b_1^{i-1} - Z_1\|_F^2 \qquad (15)$$

$$E(Z_2^i) = \arg\min_{Z_2} \alpha_2 \|Z_2\|_1 + \beta_2 \|\nabla R^i + b_2^{i-1} - Z_2\|_F^2 \qquad (16)$$

then we can use R^i, L^i to update Z_1, Z_2 as follows:

$$Z_1^i = Shrinkage(\nabla L^i + b_1^{i-1}, \frac{\alpha_1}{2\beta_1}) \qquad (17)$$

$$Z_2^i = Shrinkage(\nabla R^i + b_2^{i-1}, \frac{\alpha_2}{2\beta_2}) \qquad (18)$$

where soft threshold operator $Shrinkage(x, \varepsilon) = \frac{x}{|x|} * max(|x| - \varepsilon, 0)$, and the b_1, b_2 update as:

$$b_1^i = b_1^{i-1} + \nabla L^i - Z_1^i \qquad (19)$$

$$b_2^i = \nabla R^i + b_2^{i-1} - Z_2^i \qquad (20)$$

Algorithm 1

Input: observed image I, parameters $\alpha_1, \alpha_2, \alpha_3, \beta_1, \beta_2$, and ε, and ϵ

Initialization: Estimate L_* by solving equation (8), $tex \leftarrow I - L_*$, $R^0 = 0$, $L^0 = I$, $Z_1^0 = 0$, $Z_2^0 = 0$, $b_1^0 = 0$, $b_2^0 = 0$, set $i = 1$;

do while$(err_R < \epsilon$ **and** $err_L < \epsilon)$

step 1: update R^i by solving equation(11);
step 2: update L^i by solving equation(13);
step 3: update Z_1^i by solving equation(17);
step 4: update Z_2^i by solving equation(18);
step 5: update b_1^i by solving equation(19);
step 6: update b_2^i by solving equation(20);
step 7: $i = i + 1$;

Return R, L

Postprocess: $\hat{L} \leftarrow adjust\ L\ by\ gamma\ correction;$
        $\hat{R} \leftarrow RR\ deblocking\ by\ equation\ (9), where\ RR = tex + R;$

Output: $I_{final} = \hat{R} \circ \hat{L}.$

In this paper, the application of fast Fourier transform (FFT) avoids the inversion process of the matrix and reduces the amount of calculation. For the de-artifact block part of the reflectance, we can easily and quickly get the $\hat{R}$ that after the artifact block by the formula (7). And the postprocess of illuminance is gamma correction: $\hat{L} = L^{\frac{1}{\gamma}}$, where $\hat{L}$ is the gamma corrected illuminance component, where $\gamma = 2.2$; finally, we can summarize the algorithm as Algorithm 1.

4   Experiments

In this section, we conduct experimental comparative analysis from two aspects: simulation experiment comparison, real low-light image experiment comparison. And the experimental comparison methods are Fu's method [10], Guo's method [13], Park's method [12] Li's method [14] and Rao's method [15], and the result of Fu's method, Guo's method and Li's method are generated by the code which downloaded from author's web site, with recommended experiment settings. And we simulated Park's method and the Rao's method, with recommended experiment settings.

4.1   Experiment with Simulation Images

In this section, we select 40 well-exposure images and darkening them, which are from image dataset General-100 [19] and Kodak Lossless True Color Image Suite [20]. And we have dealt with the three channels of RGB separately. And the parameters of this section of the experiment are set $\alpha_1 = 0.01, \alpha_2 = 0.001, \alpha_3 = 5, \beta_1 = 0.1$, and $\beta_2 = 0.01$, respectively. And $\lambda = 0.01$ in Eq. (8), $\beta = 0.05$ in Eq. (9). It is worth noting that the post-processing of this section of the experiment is the same.

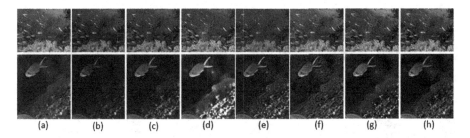

Fig. 2. Comparation of fish image: (a) is the original image; (b) is the input image; (c) is the result of [10]; (d) is the result of [13]; (e) is the result of [12]; (f) is the result of [15]; (g) is the result of [14]; (h) is the result of proposed method

From Fig. 1, we can see the comparison of different methods. In addition, we use peak signal-to-noise ratio (PSNR), structural similarity (SSIM) and multi-layer structural similarity (MS-SSIM)[21] to evaluate enhanced images quality. The larger the values of PSNR, SSIM and MS-SSIM, the higher the similarity between the image and the original image, and the better quality of enhanced images. From Table 1, we can know that PSNR, SSIM and MS-SSIM values of our method are much higher than other methods.

Table 1. Quantitative measurement results of PSNR, SSIM and MS-SSIM

Image	Metrics	[10]	[13]	[12]	[15]	[14]	The proposed
Fish	PSNR	16.7490	17.4717	14.9528	23.2807	18.5220	28.9795
	SSIM	0.8125	0.8609	0.8135	0.9178	0.8742	0.9735
	MS-SSIM	0.9467	0.9236	0.8849	0.9426	0.9548	0.9897
Average40	PSNR	18.5022	17.5708	17.3984	27.4508	20.8975	31.4919
	SSIM	0.7955	0.7834	0.7634	0.9328	0.8294	0.9691
	MS-SSIM	0.9537	0.8858	0.9442	0.9770	0.9640	0.9868

4.2   Experiments with Real Images

In this section, we use experiments to prove the effectiveness of the method that we proposed. So we collect 42 low-contrast, non-uniformly illuminance images from [10, 22, 23] and then enhanced and compared them to other methods. Considering that the color of the non-uniform illumination images, we convert the image to "HSV" channel and only process the brightness channel V to preserve the color. In our experiment, the parameters are set $\alpha_1 = 0.005, \alpha_2 = 0.001, \alpha_3 = 10, \beta_1 = 0.05$, and $\beta_2 = 0.01$, respectively. And $\lambda = 0.005$ in Eq. (8), $\beta = 0.05$ in Eq. (9).

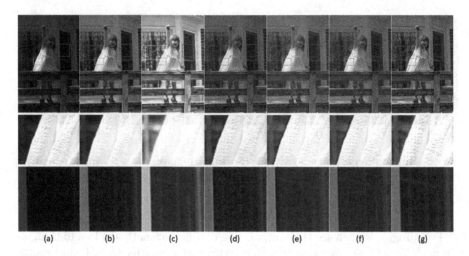

Fig. 3. Comparation of girl image: (a) is the input image; (b) is the result of [10]; (c) is the result of [13]; (d) is the result of [12]; (e) is the result of [14]; (f) is the result of [15]; (g) is the result of proposed method (Color figure online)

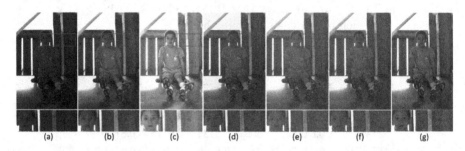

Fig. 4. Comparation of boy image: (a) is original image; (b) is the result of [10]; (c) is the result of [13]; (d) is the result of [12]; (e) is the result of [14]; (f) is the result of [15]; (g) is the result of proposed method (Color figure online)

Figures 3, 4, 5 and 6 show the enhanced images with different methods. From Fig. 3(b)–(c), we can see that the enhanced images are not well in light area, and the brightness and contrast of the image are not high enough. Although Fig. 3(d) and (f) preserve the details of light area, the noise of dark area is enlarged. And from Fig. 3(d)–(f), the edge of the little girl's skirt have a black halo artifact. And our method not only reveals the details of the dark areas, but also does not enlarge noise in the dark area. From Figs. 4 and 5, we can see the brightness of Fu's method [10] enhanced image insufficiently, Guo's method [13] over-enhanced. From Fig. 5(c), we can easily observe Guo's method [13] can't work well between leaves. Park's method [12] remains fine in details, but causes noise amplification and color distortion. From Figs. 3, 4, 5 and 6(e), Li's method [14] suppresses noise but also loses some details. From Figs. 4(f)–(g)

Fig. 5. Comparation of house image: (a) is the input image; (b) is the result of [10]; (c) is the result of [13]; (d) is the result of [12]; (e) is the result of [14]; (f) is the result of [15]; (g) is the result of proposed method (Color figure online)

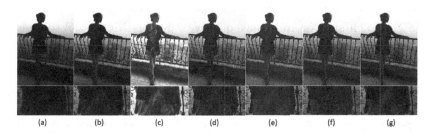

Fig. 6. Comparation of woman image: (a) is the input image; (b) is the result of [10]; (c) is the result of [13]; (d) is the result of [12]; (e) is the result of [14]; (f) is the result of [15]; (g) is the result of the proposed method (Color figure online)

Table 2. Quantitative measurement results of NIQE and BRISQUE

Image	Metrics	[10]	[13]	[12]	[14]	[15]	The proposed
Girl	NIQE	3.0375	3.6283	3.2090	3.4149	5.4157	3.0924
	BRISQUE	21.3127	32.8396	20.0556	28.3397	21.9579	17.9514
Boy	NIQE	3.6307	3.7944	3.9262	4.0041	5.0119	3.5610
	BRISQUE	16.6225	24.2441	10.4890	14.1155	11.0530	8.5796
House	NIQE	2.9440	2.5369	3.4593	3.5704	4.5474	2.7373
	BRISQUE	21.0452	30.6284	22.7221	32.9814	27.9994	21.2879
Woman	NIQE	2.7917	3.6555	2.6747	3.5263	3.8633	2.3239
	BRISQUE	19.9024	29.2165	16.8267	21.7640	15.8443	13.4006
Average(42)	NIQE	3.6490	4.1484	3.6090	4.3127	4.7145	3.4891
	BRISQUE	17.3667	25.5603	17.6936	25.2486	17.2187	15.2268

and 5(f)–(g), we can see the details and color of our method are better than the Rao's method [15]. From Fig. 6(f), we can see that our method effectively removes the artifact blocks and preserves the details. Figure 6(d) and (f) contain the compressed artifact block. Figure 6(d) does not hold the color very well.

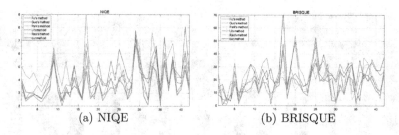

Fig. 7. The plot of BRISQUE and NIQE values for 42 enhanced images (Color figure online)

Comprehensively, our method is effective to eliminate block artifacts and does not reduce the details.

Objectively, we chose Blind/Referenceless Image Spatial Quality Evaluator (BRISQUE) [24] and Natural image quality evaluator (NIQE) [25] to evaluate the enhanced images. The lower the metrics value of NIQE and BRISQUE, the better quality of enhanced images. Table 2 shows the NIQE and BRISQUE indicator value for Figs. 3, 4, 5 and 6 and 42 images averaged, were red represents the lowest value and blue represents the second lowest value. And Fig. 7 shows a curve plot of the NIQE and BRISQUE values for 42 enhanced images using different methods. As shown the Table 2 and Fig. 7, our method have a lower value than other methods. The average value of the NIQE and BRISQUE are more convincing.

5   Conclusion

This paper has presented an enhanced variational retinex-based method for low-light natural image enhancement, based on the initial smoother illumination component with a structure extraction technique. The Bergman splitting algorithm is used to estimate the illuminance component and reflectance component. The de-block processing and illuminance component correction are used for the enhanced reflectance as the ultimate enhanced image. Moreover, The estimated smoother illumination component can make enhanced images preserve image edge details. Experimental results with a comparsion demonstrate the present variational retinex method can effectively enhance image quality and maintain image color than related variational retinex methods.

References

1. Wang, Y., Chen, Q., Zhang, B.: Image enhancement based on equal area dualistic sub-image histogram equalization method. IEEE Trans. Consum. Electron. **45**(1), 1–8 (1999)
2. Kim, Y.T.: Contrast enhancement using brightness preserving bi-histogram equalization. IEEE Trans. Consum. Electron. **43**(1), 1–8 (1997)

3. Loza, A., Bull, D.R., Hill, P.R., Achim, A.M.: Automatic contrast enhancement of low-light images based on local statistics of wavelet coefficients. Digit. Sig. Process. **23**(6), 1856–1866 (2013)
4. Kim, J.H., Kim, J.-H., Jung, S.W., Noh, C.K., Ko, S.J.: Novel contrast enhancement scheme for infrared image using detail-preserving stretching. Opt. Eng. **50**(7), 1–11 (2011)
5. Jobson, D.J., Rahman, Z., Woodell, G.A.: Properties and performance of a center/surround retinex. IEEE Trans. Image Process. **6**(3), 451–462 (1997). A Publication of the IEEE Signal Processing Society
6. Jobson, D.J., Rahman, Z., Woodell, G.A.: A multiscale retinex for bridging the gap between color images and the human observation of scenes. IEEE Trans. Image Process. **6**(7), 965–976 (1997)
7. Kimmel, R., Elad, M., Shaked, D., Keshet, R., Sobel, I.: A variational framework for retinex. Int. J. Comput. Vis. **52**(1), 7–23 (2003)
8. Wu, X.: A linear programming approach for optimal contrast-tone mapping. IEEE Trans. Image Process. **20**(5), 1262–1272 (2011)
9. Huang, S.C., Cheng, F.C., Chiu, Y.S.: Efficient contrast enhancement using adaptive gamma correction with weighting distribution. IEEE Trans. Image Process. **22**(3), 1032–1041 (2013)
10. Fu, X., Liao, Y., Zeng, D., Huang, Y., Zhang, X., Ding, X.: A probabilistic method for image enhancement with simultaneous illumination and reflectance estimation. IEEE Trans. Image Process. **24**(12), 4965–4977 (2015)
11. Fu, X., Sun, Y., LiWang, M., Huang, Y., Zhang, X.P., Ding, X.: A novel Retinex based approach for image enhancement with illumination adjustment. In: 2014 IEEE International Conference on Acoustics, Speech and Signal Processing (ICASSP), Florence, pp. 1190–1194 (2014). https://doi.org/10.1109/ICASSP.2014.6853785
12. Park, S., Moon, B., Ko, S., Yu, S., Paik, J.: Low-light image enhancement using variational optimization-based Retinex model. In: IEEE International Conference on Consumer Electronics (ICCE), Las Vegas, NV, pp. 70–71 (2017). https://doi.org/10.1109/ICCE.2017.7889233
13. Guo, X.: Lime: a method for low-light image enhancement. In: Proceedings of MM International Multimedia Conference 2016, MM 2016. Proceedings of the 24th ACM International Conference on Multimedia, Amsterdam, The Netherlands, pp. 87–91 (2016). https://doi.org/10.1145/2964284.2967188
14. Li, M., Liu, J., Yang, W., Sun, X., Guo, Z.: Structure-revealing low-light image enhancement via robust Retinex model. IEEE Trans. Image Process. **27**(6), 2828–2841 (2018)
15. Rao, Z., Xu, T., Luo, J., Guo, J., Shi, G., Wang, H.: Non-uniform illumination endoscopic imaging enhancement via anti-degraded model and L_1L_2-based variational retinex. EURASIP J. Wirel. Commun. Network. **2017**(1), 1–11 (2017)
16. Xu, L., Yan, Q., Xia, Y., Jia, J.: Structure extraction from texture via relative total variation. ACM Trans. Graph. **31**(6), 1–10 (2012)
17. Li, Y., Guo, F., Tan, R.T., Brown, M.S.: A contrast enhancement framework with JPEG artifacts suppression. In: Fleet, D., Pajdla, T., Schiele, B., Tuytelaars, T. (eds.) ECCV 2014. LNCS, vol. 8690, pp. 174–188. Springer, Cham (2014). https://doi.org/10.1007/978-3-319-10605-2_12
18. Goldstein, T., Osher, S.: The split Bregman method for L1 regularized problems. SIAM J. Imaging Sci. **2**(2), 323–343 (2009)

19. Dong, C., Loy, C.C., Tang, X.: Accelerating the super-resolution convolutional neural network. In: Leibe, B., Matas, J., Sebe, N., Welling, M. (eds.) ECCV 2016. LNCS, vol. 9906, pp. 391–407. Springer, Cham (2016). https://doi.org/10.1007/978-3-319-46475-6_25

20. Kodak Lossless True Color Image Suite. http://r0k.us/graphics/kodak/. Accessed 11 Aug 2019

21. Wang Z., Simoncelli, E.P., Bovik, A.C.: Multi-scale structural similarity for image quality assessment. In: Conference Record of the Thirty-Seventh Asilomar Conference on Signals, Systems and Computers (2003). https://doi.org/10.1109/ACSSC.2003.1292216

22. Fu, X., Zeng, D., Huang, Y., Liao, Y., Ding, X., Paisley, J.: A fusion-based enhancing method for weakly illuminated images. Sig. Process. **129**, 82–96 (2016)

23. Retinex Image Processing. https://dragon.larc.nasa.gov/retinex/pao/news/. Accessed 11 Aug 2019

24. Mittal, A., Moorthy, A.K., Bovik, A.C.: No-reference image quality assessment in the spatial domain. IEEE Trans. Image Process. **21**(12), 4695–4708 (2012). A Publication of the IEEE Signal Processing Society

25. Mittal, A., Soundararajan, R., Bovik, A.C.: Making a 'completely blind' image quality analyzer. IEEE Sig. Process. Lett. **20**(3), 209–212 (2013)

Unsupervised Desmoking of Laparoscopy Images Using Multi-scale DesmokeNet

V. Vishal[1], Varun Venkatesh[2], Kshetrimayum Lochan[1], Neeraj Sharma[3],
and Munendra Singh[1(✉)]

[1] Department of Mechatronics Engineering, Manipal Institute of Technology,
Manipal Academy of Higher Education, Manipal 576104, Karnataka, India
vishalv7197@gmail.com, lochan.nits@gmail.com, munendra107@gmail.com
[2] Sapthagiri Institute of Medical Sciences & Research Center,
Bangalore 560090, India
drvarunv0103@gmail.com
[3] School of Biomedical Engineering,
Indian Institute of Technology (Banaras Hindu University),
Varanasi 221005, India
neeraj.bme@iitbhu.ac.in

Abstract. The presence of surgical smoke in laparoscopic surgery reduces the visibility of the operative field. In order to ensure better visualization, the present paper proposes an unsupervised deep learning approach for the task of desmoking of the laparoscopic images. This network builds upon generative adversarial networks (GANs) and converts laparoscopic images from smoke domain to smoke-free domain. The network comprises a new generator architecture that has an encoder-decoder structure composed of multi-scale feature extraction (MSFE) blocks at each encoder block. The MSFE blocks of the generator capture features at multiple scales to obtain a robust deep representation map and help to reduce the smoke component in the image. Further, a structure-consistency loss has been introduced to preserve the structure in the desmoked images. The proposed network is called Multi-scale DesmokeNet, which has been evaluated on the laparoscopic images obtain from Cholec80dataset. The quantitative and qualitative results shows the efficacy of the proposed Multi-scale DesmokeNet in comparison with other state-of-the-art desmoking methods.

Keywords: Smoke removal · Image enhancement · Laparoscopic surgery

1 Introduction

In laparoscopic surgery, the generation of artefacts like specular reflections, surgical smoke [2], blood and inadequate illumination degrades the image quality, which affects the efficiency of the computer vision algorithms for assistive help at tasks like tracking and detection [14]. The robust performance of the tracking and detection algorithms cannot be realized without removing the artefacts

© Springer Nature Switzerland AG 2020
J. Blanc-Talon et al. (Eds.): ACIVS 2020, LNCS 12002, pp. 421–432, 2020.
https://doi.org/10.1007/978-3-030-40605-9_36

from the images. This study particularly discusses the smoke in the laparoscopic surgery, which blocks surgeon's operative field and affects his ability to carry out various procedures specially in case of image-guided surgical systems. Hence, the removal of the smoke is an essential task in order to realize the full potential and benefits of laparoscopic surgery. Hardware based techniques [23] and other computer vision methods [13,24] have been developed to remove the smoke, which are bound by certain limitations. A digital solution can serve as a useful tool to better visualize the operative field and produce high quality images for the computer vision pipeline of the system. In recent times, the advancements in computer vision has been greatly influenced by Deep learning. The rise of new deep learning networks, objective functions and strategies has enabled to achieve impressive results and breakthroughs with problems that were difficult to be solved by traditional and non-learning based approaches. Further, Goodfellow et al. formulated the Generative Adversarial networks (GANs) [8], which has shown significant progress in the field of image synthesis. The competitive learning between the generator and discriminator networks drives the GANs to realise an implicit loss function that helps produce images that are more sharper and of higher perceptual quality. The generator and discriminator networks both learn the distribution of the training images. The generator tries to produces fake images, while the discriminator network tries to distinguish the fake images from the real images from the training dataset. The min-max optimization between the generator and the discriminator networks, i.e generator minimizes the discriminator's ability to distinguish the fake images from real images and discriminator maximizes its ability to identify real images. GANs has also depicted outstanding results at task of Image-to-Image translation, which requires either paired or unpaired data.

Previously, Isola et al. proposed the Pix2Pix framework [10] that utilizes a conditional generative adversarial network to learn a mapping between the input and output images using paired data. Various tasks like generation of photographs from semantic layouts [11], sketches [21] and edges can be seen using similar approach. The extended work of the Pix2Pix framework resulted in CycleGAN [31] that aimed to tackle the task of image translation using unpaired data. It learns a mapping between two data domains: X and Y rather than just pair of images. This enables it to be unsupervised and allows the generation of images even in the absence of ground truth. Similar work can be seen in case of UNIT [17], DiscoGAN [12] and DualGAN [30].

In the present study, we focus on the task of smoke removal as a single image desmoking that translates a smoke image to an image resembling smoke-free images. The translation of the image from the smoke domain to smoke-free domain depicts considerable decrease in the smoke component and results in the enhancement of the image. The main contributions of the paper are as follows:

1. An unsupervised method that translates smoke images to desmoke images without the need for synthetic ground truths in an unpaired manner. This method helps to remove the need for paired data and simulation to real world domain adaptation.

2. A new generator architecture that consists of multi-scale feature extraction (MSFE) blocks. The MSFE blocks perform robust feature extraction and help capture the smoke features at multiple scales at each encoder level.

3. A new loss function called structure-consistency loss has been proposed. This loss ensures the structure in the image is maintained in the translation framework, hence resulting in the reconstruction of the desmoked images with structure and edges identical to the smoke image.

2 Related Work

2.1 Conventional Methods

Previously, Tchaka et al. [24] utilized the dark channel prior dehazing method (DCP) [9] for the task of smoke removal. They have improved upon the DCP by performing two modifications: decreasing the emphasis on pixel values in the range where smoke is expected to be present and thresholding the dark channel by a constant value. Further, to enhance the contrast they have performed histogram equalization. In [13], Kotwal et al. framed the problem as Bayesian inference problem and performed joint de-smoking and de-noising using a probabilistic graphical, where the uncorrupted image is represented as a Markov random field (MRF) and maximum-a-posteriori (MAP) is used to obtain the enhanced images. The method was extended in [1] to perform specular removal, along with de-smoking and de-noising. Previously proposed methods for desmoking also focused on applying solutions relevant to the problem of dehazing. The mathematical model for the atmospheric scattering is given as follows:

$$I(x) = J(x)t(x) + A(1 - t(x)) \tag{1}$$

where $I(x)$ represents the hazy image, $J(x)$ is the haze-free image, $t(x)$ is the transmission map and A is the global atmospheric light on each x pixel coordinates.

Wang et al. in [27], devised a variational method to estimate the smoke veil for smoke removal assuming smoke component to have low contrast and low inter-channel differences. The results of the method yielded considerable enhancement in the image and higher visual quality. On the other hand, Luo et al. [18] introduced a visibility-driven defogging framework that recovers the atmospheric veil solution as bilateral of bilateral grid and finally obtains the defogged image by Eq. 1. The work also proposed no-reference image quality assessment metrics to quantify the naturalness and sharpness in an image.

2.2 Deep Learning Methods

Recently, Sabri et al. produced a synthetic smoke images dataset using Perlin noise and utilized AOD-net model to transfer learn the task of smoke removal [4]. In [28], Wang et al. trained a encoder-decoder architecture with Laplacian image pyramid decomposition input strategy on a synthetic smoke images and evaluated the performance on real smoke. An improvement of the CycleGAN framework called the Cycle-Desmoke was proposed in [26] for desmoking of laparoscopic images, this work on relied on atrous convolutions to perform multi-scale feature extraction and consisted of a new loss function called the guided-unsharp upsample loss to help the network at the operation of upsampling.

3 Method

This section describes the proposed method for single image desmoking of laparoscopic images. Figure 1 depicts the overview of the Image-to-Image translation framework, which consists of two generator networks G_{DS} and G_S that help generate synthetic desmoked and smoke images respectively, two discriminator networks D_{DS} and D_S that help distinguish the synthesized desmoked images from real smoke-free images and synthesized smoke images from real smoke images. The two mapping functions are desmoking (DS) and re-smoking (RS) that map smoke to desmoke images and smoke-free to smoke images respectively. In addition to the adversarial and cycle-consistency losses, we include structure consistency loss as well, this enables to maintain and preserve the structure and edge information during the mapping operations.

3.1 Adversarial Loss

For both the mapping functions DS and RS, the adversarial loss has been applied. The generator minimizes the discriminator's ability to distinguish the synthetic images from the real images, while the discriminator maximizes its ability to identify the real images. This min-max optimization follows the game theoretic approach and helps realize a loss function. The adversarial losses for the two mapping functions are given as follows:

$$min_{G_{DS}}max_{D_{DS}}L_{GAN}(G_{DS}, D_{DS}, S, SF) = E_{sf \sim p_{data}(sf)}[logD_{DS}(sf)] \\ + E_{s \sim p_{data}(s)}[log(1 - D_{DS}(G_{DS}(s)))] \tag{2}$$

$$min_{G_S}max_{D_S}L_{GAN}(G_S, D_S, SF, S) = E_{s \sim p_{data}(s)}[logD_S(s)] \\ + E_{sf \sim p_{data}(sf)}[log(1 - D_S(G_S(sf)))] \tag{3}$$

3.2 Cycle Consistency Loss

Only adversarial loss cannot ensure the learned mapping function to map an image from input domain to the target domain. The addition of cycle-consistency loss helps to learn mapping functions more accurately. Given an input smoke image s, G_{DS} is the desmoked image and $G_S(G_{DS}(s)) \approx s$ the forward cycle consistency and similarly in the backward cycle consistency: given an input smoke-free image sf, $G_S(sf)$ is the smoke image and $G_{DS}(G_S(sf)) \approx sf$. This loss is applicable only to the generator networks. The loss is given as follows:

$$L_{cyclic}(G_{DS}, G_S) = ||G_{DS}(G_S(sf)) - sf|| + ||G_S(G_{DS}(s)) - s|| \tag{4}$$

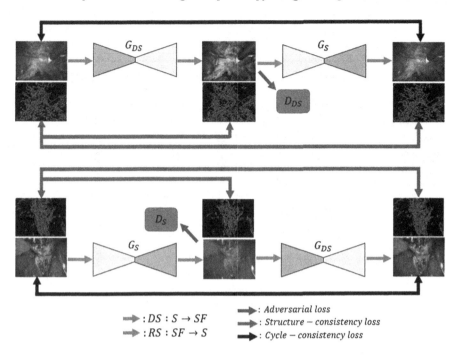

Fig. 1. Representation of the framework of the proposed method.

3.3 Structure Consistency Loss

In order to ensure the structure between the images of the mapping operations is preserved, we include the structure consistency loss. Such a loss is essential at reconstructing the structure and edges in an image similar to the input images. We make use of the Canny edge detection to obtain edge information in the image. The image gradients of the input image serve as the ground truth and through L1 norm we try to generate the image gradients of the translated images to be similar to the input image. This forces the generator networks to produce images that resemble the input images in both edge and structural information and also helps in reduction of artefacts, distortions etc. The structure consistency loss is given as follows:

$$L_{SC}(G_{DS}.G_S) = L_{SF} + L_{SB} \tag{5}$$

$$L_{SF} = \alpha_1 \sum ||I_{canny}(s) - I_{canny}(G_{DS}(s))|| \\ + \alpha_2 \sum ||I_{canny}(s) - I_{canny}(G_S(G_{DS}(s)))|| \tag{6}$$

$$L_{SB} = \alpha_1 \sum ||I_{canny}(sf) - I_{canny}(G_S(sf))|| \\ + \alpha_2 \sum ||I_{canny}(sf) - I_{canny}(G_DS(G_S(sf)))|| \tag{7}$$

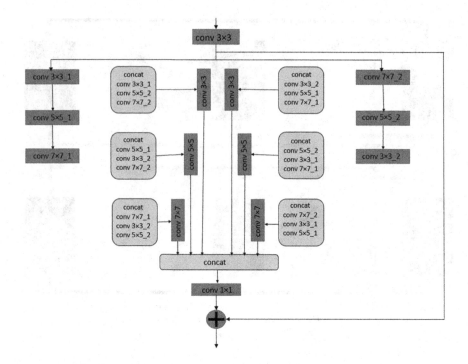

Fig. 2. The structure of the multi-scale feature extraction block.

where L_{SF} and L_{SB} are losses for the forward and backward translations respectively. The terms α_1 and α_2 are set as 0.5 and 1 respectively in order to control the amount of edge information reconstructed. α_1 is set lower than α_2 as the edge information in the input image approximately resembles the translated image, while the edge information of the reconstructed input image has complete resemblance with respect to that of input image.

3.4   Complete Objective Function

The combination of adversarial, cycle-consistency and structure-consistency loss forms the complete objective function is given as follows:

$$L(G_{DS}, G_S, D_{DS}, D_S) = L_{GAN}(G_{DS}, G_S) + L_{GAN}(G_S, D_S)$$
$$+ \lambda_1 L_{cyclic}(G_{DS}, G_S) + \lambda_2 L_{SC}(G_{DS}, G_S) \tag{8}$$

The terms λ_1 and λ_2 are positive weighted terms that help control the training procedure.

3.5   Network Architecture

Multi-scale Feature Extraction Blocks: The detection of features at multiple scales is useful for a wide range of computer vision problems like semantic segmentation [7], object detection [6] and image super-resolution [16]. We propose similar technique for the task of desmoking. Our multi-scale feature extraction blocks consists of different kernel sizes. The input feature map passes sequentially through conv 3×3, conv 5×5, conv 7×7 branch in an incremental kernel dimension pattern and conv 7×7, conv 5×5, conv 3×3 branch in a decremental kernel dimension pattern respectively. By doing so, we are able to vary the receptive field effectively and capture distinctive features from the input feature map. The output feature maps from each convolution operation is concatenated, as depicted in Fig. 2. The concatenated blocks are further convolved by particular kernel sizes and the output feature maps are concatenated and later convolved by 1×1 kernel to maintain the desired channel dimension. If the channel dimension in the input feature map is M, then the channel dimension after every convolution operation is maintained as M. We also include residual learning in order making the learning more efficient and allow the information to flow from the input layer to the output layer. This is brought about a skip connection and element-wise addition.

Generator Architecture: We adopt an encoder-decoder architecture design for the generator. Each encoder layer consists of a multi-scale feature extraction block and is downsampled by a convolution operation with stride 2. The bottleneck consists of 6 residual blocks similar to [31]. Each decoder layer consists of a pixel shuffle [22] operation that receives input feature maps from the respective encoder and decoder layers as depicted in Fig. 3. Once the desired spatial resolution is realized, convolution operations are performed to obtain the output image with same dimensions of that of the input image. This architectural design for the generator network allows it reconstruct the images.

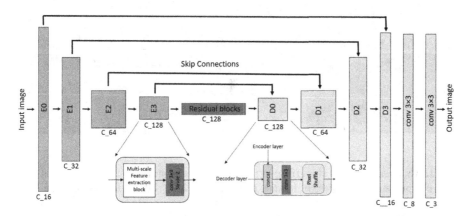

Fig. 3. Representation of the generator architecture. It consists of MSFE blocks at each encoder block and pixel shuffle at each decoder block for upsampling along the corresponding encoder-decoder levels. The C_XX denotes the channel dimension of the output feature map of the particular layer.

Discriminator Architecture: The discriminator D_{DS} and D_S adopt the architecture as in [31]. They consist of five convolutional layers that identify whether the 70×70 overlapping image patches are synthetic or real.

3.6   Experimental Results

This section evaluates the proposed approach on the ITEC Smoke_Cholec80 real laparoscopic image dataset, which is composed of 80 different cholecystectomy procedures. The present study first introduces the dataset and discusses the implementation details. Each component of the proposed approach is evaluated and the results are analyzed. Finally, the study provides the comparison results with other state-of-the-art de-smoking methods.

ITEC Smoke_Cholec80 Image Dataset: The public dataset [15] consists of 100K smoke/non-smoke images extracted from the Cholec80 dataset [25]. We have selected 1200 images each for smoke domain and smoke-free domain as our training set, random 100 images from the training set as the validation set and 200 images as the test set. As there is black corners due to the camera arrangement, the images have been center cropped to 240×320, which will avoid network to learn unnecessary information. The selected images have varying levels of smoke at different depths, to ensure the network is to learn on a diverse set of images.

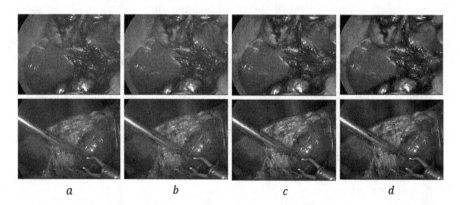

Fig. 4. Ablation study on the utility of the MSFE blocks and the structure consistency loss. a: input smoke images, b: images generated without MSFE blocks but just convolution blocks, c: images generated without structure consistency loss in the overall objective function, d: images from the proposed method

Implementation Details: In order to stabilize the training procedure, the least-squared loss is used instead of the log likelihood loss [19] network is trained with a learning rate of 0.0001 and is linearly decayed to zero for the next 100

Table 1. Quantitative evaluation of ablation study on the validation dataset. The terms a, b, c, d are denoted in Fig. 4.

Method	a		b		c		d	
Image quality	BRISQE	CEIQ	BRISQE	CEIQ	BRISQE	CEIQ	BRISQE	CEIQ
Mean	19.14	3.345	17.95	3.350	17.50	3.317	17.43	3.352

epochs with ADAM as the optimizer. The batch size is fixed to one. Experimentally, the λ_1 and λ_2 terms in the complete objective function are set to 10 and 0.1 respectively. Tensorflow was used to train the network on the NVidia K80 GPU.

Experimental Results: The use of multi-scale feature extraction block in comparison to just convolution layers help to capture features at different scales and this is essential at detecting and reducing the smoke component in the image and also leads to better contrast. The addition of the structure-consistency loss along with adversarial and cycle-consistency loss results in images that are sharper and with enhanced structural information as shown in Fig. 4. Further, the quantitative analysis of ablation study on the validation dataset in terms of image quality measures (i) Blind/Referenceless Image Spatial Quality Evaluator (BRISQE) [20] and (ii) Quality Assessment of Contrast-Distorted Images (CEIQ) [29] is shown in Table 1. The lower values for BRISQE and higher values for CEIQ indicates better perceptual quality.

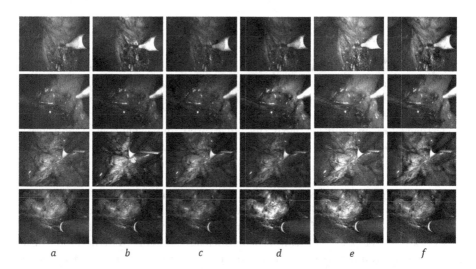

Fig. 5. Qualitative evaluation of smoke removal on randomly selected images from test set. a: input smoke images, b: Non-local Dehaze, c: DehazeNet, d: DCP, e: CycleGAN, f: Proposed Multi-scale DesmokeNet

The proposed method is compared with state-of-the-art methods i.e. Non-local Dehaze [3], Single image haze removal using dark channel prior (DCP) [9], Dehazenet [5] and CycleGAN [31]. Figure 5 shows the qualitative results on random four images. It is quite evident that the desmoked images generated from other comparison methods are low in contrast, dark and result in lesser smoke removal in comparison to the proposed method. The images from proposed method looks more promising and of higher quality. Further, the quantitative results in Table 2 also suggest the better performance of the proposed method over the other state-of-the-art methods.

Table 2. Quantitative evaluation on the test dataset. The values denote the mean of the image quality.

Method	Image quality	
	BRISQE	CEIQ
Test Input	21.14	3.302
Non-local Dehaze	19.05	3.148
DehazeNet	22.04	3.101
DCP	20.949	2.948
CycleGAN	17.12	3.373
Proposed method	**15.39**	**3.382**

3.7 Conclusion

This paper introduced a new unsupervised learning method for removal of smoke in laparoscopic images. The proposed method consists of a new generator architecture of comprising of novel multi-scale feature extraction blocks that help to alleviate the smoke component at different scales. Further, the new structure-consistency loss in addition with the adversarial and cycle-consistency losses results in preserving the structure of the image effectively. The proposed method is qualitatively and quantitatively compared and has shown the edge over other state-of-the-art desmoking methods.

As the surgical smoke removal relies on mechanical solutions that still have lag time, having a digital visualization of the frames that automatically removes smoke will be of great use for the practitioners and surgeons, we plan to extend this work to develop an algorithm that performs in real time. Further, looking at the spatial-temporal consistency between the sequences of frames and we would try to ensure that the level of smoke removal is not dependent on its amount but just on its presence in that frame.

References

1. Baid, A., Kotwal, A., Bhalodia, R., Merchant, S., Awate, S.P.: Joint desmoking, specularity removal, and denoising of laparoscopy images via graphical models and Bayesian inference. In: 2017 IEEE 14th International Symposium on Biomedical Imaging, ISBI 2017, pp. 732–736. IEEE (2017)
2. Barrett, W.L., Garber, S.M.: Surgical smoke: a review of the literature. Surg. Endosc. **17**(6), 979–987 (2003)
3. Berman, D., Avidan, S., et al.: Non-local image dehazing. In: Proceedings of the IEEE Conference on Computer Vision and Pattern Recognition, pp. 1674–1682 (2016)
4. Bolkar, S., Wang, C., Cheikh, F.A., Yildirim, S.: Deep smoke removal from minimally invasive surgery videos. In: 2018 25th IEEE International Conference on Image Processing (ICIP), pp. 3403–3407. IEEE (2018)
5. Cai, B., Xu, X., Jia, K., Qing, C., Tao, D.: DehazeNet: an end-to-end system for single image haze removal. IEEE Trans. Image Process. **25**(11), 5187–5198 (2016)
6. Cai, Z., Fan, Q., Feris, R.S., Vasconcelos, N.: A unified multi-scale deep convolutional neural network for fast object detection. In: Leibe, B., Matas, J., Sebe, N., Welling, M. (eds.) ECCV 2016. LNCS, vol. 9908, pp. 354–370. Springer, Cham (2016). https://doi.org/10.1007/978-3-319-46493-0_22
7. Chen, L.-C., Zhu, Y., Papandreou, G., Schroff, F., Adam, H.: Encoder-decoder with atrous separable convolution for semantic image segmentation. In: Ferrari, V., Hebert, M., Sminchisescu, C., Weiss, Y. (eds.) ECCV 2018. LNCS, vol. 11211, pp. 833–851. Springer, Cham (2018). https://doi.org/10.1007/978-3-030-01234-2_49
8. Goodfellow, I., et al.: Generative adversarial nets. In: Advances in Neural Information Processing systems, pp. 2672–2680 (2014)
9. He, K., Sun, J., Tang, X.: Single image haze removal using dark channel prior. IEEE Trans. Pattern Anal. Mach. Intell. **33**(12), 2341–2353 (2010)
10. Isola, P., Zhu, J.Y., Zhou, T., Efros, A.A.: Image-to-image translation with conditional adversarial networks. In: Proceedings of the IEEE Conference on Computer Vision and Pattern Recognition, pp. 1125–1134 (2017)
11. Karacan, L., Akata, Z., Erdem, A., Erdem, E.: Learning to generate images of outdoor scenes from attributes and semantic layouts. arXiv preprint arXiv:1612.00215 (2016)
12. Kim, T., Cha, M., Kim, H., Lee, J.K., Kim, J.: Learning to discover cross-domain relations with generative adversarial networks. In: Proceedings of the 34th International Conference on Machine Learning-Volume 70, pp. 1857–1865. JMLR. org (2017)
13. Kotwal, A., Bhalodia, R., Awate, S.P.: Joint desmoking and denoising of laparoscopy images. In: 2016 IEEE 13th International Symposium on Biomedical Imaging (ISBI), pp. 1050–1054. IEEE (2016)
14. Lawrentschuk, N., Fleshner, N.E., Bolton, D.M.: Laparoscopic lens fogging: a review of etiology and methods to maintain a clear visual field. J. Endourol. **24**(6), 905–913 (2010)
15. Leibetseder, A., Primus, M.J., Petscharnig, S., Schoeffmann, K.: Real-time image-based smoke detection in endoscopic videos. In: Proceedings of the on Thematic Workshops of ACM Multimedia 2017, pp. 296–304. ACM (2017)
16. Lim, B., Son, S., Kim, H., Nah, S., Mu Lee, K.: Enhanced deep residual networks for single image super-resolution. In: Proceedings of the IEEE Conference on Computer Vision and Pattern Recognition Workshops, pp. 136–144 (2017)

17. Liu, M.Y., Breuel, T., Kautz, J.: Unsupervised image-to-image translation networks. In: Advances in Neural Information Processing Systems, pp. 700–708 (2017)
18. Luo, X., McLeod, A.J., Pautler, S.E., Schlachta, C.M., Peters, T.M.: Vision-based surgical field defogging. IEEE Trans. Med. Imaging 36(10), 2021–2030 (2017)
19. Mao, X., Li, Q., Xie, H., Lau, R.Y., Wang, Z., Paul Smolley, S.: Least squares generative adversarial networks. In: Proceedings of the IEEE International Conference on Computer Vision, pp. 2794–2802 (2017)
20. Mittal, A., Moorthy, A.K., Bovik, A.C.: No-reference image quality assessment in the spatial domain. IEEE Trans. Image Process. 21(12), 4695–4708 (2012)
21. Sangkloy, P., Lu, J., Fang, C., Yu, F., Hays, J.: Scribbler: controlling deep image synthesis with sketch and color. In: Proceedings of the IEEE Conference on Computer Vision and Pattern Recognition, pp. 5400–5409 (2017)
22. Shi, W., et al.: Real-time single image and video super-resolution using an efficient sub-pixel convolutional neural network. In: Proceedings of the IEEE Conference on Computer Vision and Pattern Recognition, pp. 1874–1883 (2016)
23. Takahashi, H., et al.: Automatic smoke evacuation in laparoscopic surgery: a simplified method for objective evaluation. Surg. Endosc. 27(8), 2980–2987 (2013)
24. Tchaka, K., Pawar, V.M., Stoyanov, D.: Chromaticity based smoke removal in endoscopic images. In: Medical Imaging 2017: Image Processing, vol. 10133, p. 101331M. International Society for Optics and Photonics (2017)
25. Twinanda, A.P., Shehata, S., Mutter, D., Marescaux, J., De Mathelin, M., Padoy, N.: EndoNet: a deep architecture for recognition tasks on laparoscopic videos. IEEE Trans. Med. Imaging 36(1), 86–97 (2016)
26. Vishal, V., Sharma, N., Singh, M.: Guided unsupervised desmoking of laparoscopic images using cycle-desmoke. In: Zhou, L., et al. (eds.) OR 2.0/MLCN -2019. LNCS, vol. 11796, pp. 21–28. Springer, Cham (2019). https://doi.org/10.1007/978-3-030-32695-1_3
27. Wang, C., Cheikh, F.A., Kaaniche, M., Beghdadi, A., Elle, O.J.: Variational based smoke removal in laparoscopic images. Biomed. Eng. Online 17(1), 139 (2018)
28. Wang, C., Mohammed, A.K., Cheikh, F.A., Beghdadi, A., Elle, O.J.: Multiscale deep desmoking for laparoscopic surgery. In: Medical Imaging 2019: Image Processing, vol. 10949, p. 109491Y. International Society for Optics and Photonics (2019)
29. Yan, J., Li, J., Fu, X.: No-reference quality assessment of contrast-distorted images using contrast enhancement. arXiv preprint arXiv:1904.08879 (2019)
30. Yi, Z., Zhang, H., Tan, P., Gong, M.: DualGAN: unsupervised dual learning for image-to-image translation. In: Proceedings of the IEEE International Conference on Computer Vision, pp. 2849–2857 (2017)
31. Zhu, J.Y., Park, T., Isola, P., Efros, A.A.: Unpaired image-to-image translation using cycle-consistent adversarial networks. In: Proceedings of the IEEE International Conference on Computer Vision, pp. 2223–2232 (2017)

VLW-Net: A Very Light-Weight Convolutional Neural Network (CNN) for Single Image Dehazing

Chenguang Liu[⊠], Li Tao, and Yeong-Taeg Kim

Digital Media Solutions Lab, Samsung Research America, 18500 Von Karman Ave.
Suite 700, Irvine, CA 92612, USA
cheng.liu1@samsung.com
https://www.sra.samsung.com/

Abstract. Camera imaging is one of the most important application areas of computer image and video processing. However, computational cost is usually the main reason preventing many state of the art image processing algorithms from being applied to practical applications including camera imaging. This paper proposes a very light-weight end-to-end CNN network (VLW-Net) for single image haze removal. We proposed a new Inception structure. By combining it with a reformulated atmospheric scattering model, our proposed network is at least 6 times more light-weight than the state-of-the-arts. We conduct the experiments on both synthesized and realistic hazy image dataset, and the results demonstrate our superior performance in terms of network size, PSNR, SSIM and the subjective image quality. Moreover, the proposed network can be seamlessly applied to underwater image enhancement, and we witness obvious improvement by comparing with the state-of-the-arts.

Keywords: Light-weight · Convolutional neural network · Deep learning · Dehazing · Inception structure

1 Introduction

Capturing outdoor scene images often suffers from complicated, nonlinear and data-dependent noise due to the existence of haze caused by aerosols, which makes dehazing a challenging image restoration and enhancement problem. Dehazing has become an increasingly desirable technique and attracted many research interests in computational photography and computer vision.

Single image dehazing has gained more attentions amongst the existing methods since it is more practical for realistic settings [9]. Most of the state-of-art dehazing methods exploit the atmospheric scattering model (Eq. 1) including the handcrafted feature extraction methods [5,7,11,12] and the deep learning based methods [6,13,16].

$$I(x) = J(x)t(x) + A(1 - t(x)) \tag{1}$$

© Springer Nature Switzerland AG 2020
J. Blanc-Talon et al. (Eds.): ACIVS 2020, LNCS 12002, pp. 433–442, 2020.
https://doi.org/10.1007/978-3-030-40605-9_37

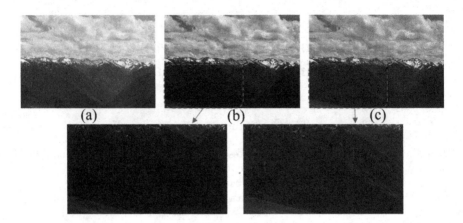

Fig. 1. Visual quality comparison between AOD-Net and our dehazing network. (a) Real world hazy image. (b) Result of AOD-Net. (c) Result of the proposed VLW-Net.

where $I(x)$ is the observed hazy image, $J(x)$ is the haze free image to be recovered, and A and $t(x)$ are two critical parameters which denote the global atmospheric light and the transmission matrix, respectively. To solve Eq. 1 for clear image $J(x)$, most dehazing methods firstly estimate the transmission matrix $t(x)$ from the hazy image $I(x)$, then estimate A using empirical methods.

As the key to achieve reliable haze removal, estimating the transmission matrix $t(x)$ based on the hazy image $I(x)$ attracts the majority of attention. The methods exploiting natural image priors and depth statistics [9,10,19] often encounter with issues of inaccurate color or depth estimation related with unconstrained scene depth structure, and unreliable prior when scene objects are similar to the atmospheric light.

To avoid inaccurate estimation of physical parameters, there have been some algorithms directly learning the transmission matrix $t(x)$ relying on various CNNs. [6] proposed a trainable model for transmission matrix estimation, and [16] put forward a new CNN model that provides coarse to fine transmission matrix generation. Despite their promising results, they need to estimate an accurate medium transmission matrix to recover clean image from haze, which may cause accumulated or amplified errors over each separate estimation step. Besides, CNN based methods, even though have gained prevailing success in many computer vision tasks, always suffer from high complexity because of the high-weight networks.

In AOD-Net [13], the authors reformulated the physical model in a "more end-to-end" fashion and estimated all its parameters in one unified model. By far, it is the most light-weight dehazing network, and outperforms the previous state-of-the-arts in terms of running time [14]. As illustrated in Fig. 1, the proposed dehazing model, even though **6 times** lighter than AOD-Net, maintains more details in dark area without losing the image clearness (Fig. 1).

In this paper, we adopted the reformulated physical model proposed in [13] and designed the dehazing network with a new Inception structure. By integrating the proposed Inception structure with an end-to-end physical model, our dehazing network outperforms the AOD-Net on both model size and picture quality. We trained our model on synthesized haze indoor image dataset and tested it on public indoor and outdoor real world image dataset. The experiments and comparison with the state-of-the-arts demonstrate the effectiveness of the proposed VLW-Net.

2   Modeling

2.1   Transformed Scattering Model

As indicated in [13], in order to achieve fully end-to-end solution, the two parameters $t(x)$ and A are unified into one formula $K(x)$, such that the optimization is directly applied to the reconstruction errors in the image pixel domain. Equation 2 is then used to design the AOD-Net instead of solving 1 for $J(x)$.

$$J(x) = K(x)I(x) - K(x) + b \tag{2}$$

$$K(x) = \frac{\dfrac{1}{t(x)}(I(x) - A) + A - b}{I(x) - 1} \tag{3}$$

where $K(x)$ is a new only variable integrated by $t(x)$ and A, and b is the constant bias with the default value 1. The advantage of jointly learning $t(x)$ and A using Eq. 3 has been proved in [13] that the two factors $1/t(x)$ and A mutually refine each other. We adopted it in designing the proposed dehazing network, which makes our model a fully end-to-end clear image generation model.

2.2   Inception Module

The success of Inception network has gained tremendous attentions in the variety of research fields. As illustrated in [18], the basic idea of the Inception architecture is (1) to approximate and cover an optimal local sparse structure in a convolutional vision network by readily available dense components, and (2) to judiciously apply dimension reductions and projections wherever the computational requirements would increase too much (Fig. 2a). The convolutions using different size of filters applied on the previous layer cover different levels of spatially spread out clusters, and the 1×1 convolutions are used to compute reductions before the expensive 3×3 and 5×5 convolutions.

2.3   Network Design

Inspired by the inception module from [18], we proposed a new Inception module (Fig. 2b) where there are 3×3, 5×5 and 7×7 convolutions after 1×1 convolutions.

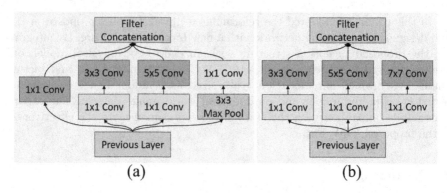

Fig. 2. Inception modules. (a) Inception module with dimension reductions in [18] (b) New inception module proposed in our dehazing network.

In Fig. 2a, the layer of 1×1 convolutions covers the correlated units concentrated in local regions of the layers close to the input layer, and the max pooling layer with stride 2 is to halve the resolution of the grid. However, in the proposed dehazing network, there is already a 1×1 convolution layer right after the input layer, which fullfiled the task of local region concentration of the 1×1 layer in the Inception module. Moreover, we use only one filter for each of the convolutions and the resolution of the convolution output is kept the same. Therefore, we do not have the layer of 1×1 convolutions and the layer of max pooling in the proposed Inception module. To cover more local sparse structure and obtain larger inception field, we introduced 7×7 convolution filter through experiments.

The proposed VLW-Net dehazing network is responsible for estimating $K(x)$ which is applied to the clean image generation module (Eq. 3) to generate clean image $J(x)$. As depicted in Fig. 3, it is composed of three convolutional layers and the new Inception structure. By fusing filters with different size, multi-scale features are extracted in each of the convolutional layers. The application of 1×1 convolutions before each of the convolutions with larger size filters greatly reduced the computational cost.

The input layer of the proposed Inception structure is a concatenation of the outputs of the first two convolutional layers, i.e. the 1×1 and 3×3 convolutional layers. Similar to the algorithm proposed in [16], such multi-scale network design is able to capture multi-scale features, and the concatenation of convolution outputs compensates for the information loss during convolutions. The last layer is a 3×3 convolutional layer applied to the concatenation layer of the proposed Inception module output.

Notably, each of the three convolutional layers outside the has only three filters, and each of the convolutional layers inside the proposed Inception module has only one filter. This leads to a very light-weight design. The proposed VLW-Net dehazing network has only **287** trainable parameters. As a comparison, AOD-Net, the currently most light-weight state-of-the-art [14], has **1761** trainable parameters. The VLW-Net is only **1/6** of AOD-Net in size.

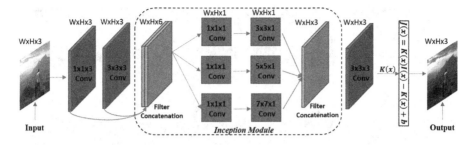

Fig. 3. The proposed dehazing network design and configuration. The "$w \times h \times n$ Conv" represents the convolutions with filters of width w and height h, and n depicts the number of filters.

3   Experiments

3.1   Datasets and Implementation

For training, we utilize the synthesized hazy images which are created using the indoor NYU2 Depth Database [17] with the same method introduced in [13]. For testing, we use real world hazy images and haze-free images in I-HAZE [1] and O-HAZE [2] datasets for quantitative test, and use a number of widely adopted challenging natural hazy images in the dehazing research field for subjective visual comparison. The I-HAZE and O-HAZE datasets are the most recent public benchmarking datasets for evaluating state-of-the-art algorithms quantitatively. There are 35 indoor and 46 outdoor real hazy images, as well as the corresponding haze-free images.

We initialize the weights with Gaussian random variables during training process. The per-pixel Mean Square Error (MSE) loss function is adopted for optimization, and the momentum and decay parameter are set to 0.9 and 0.0001, respectively. We utilize ReLU activation function due to its capability of boosting sparsity and its effectiveness on dealing with vanishing gradient problem. We found the training image size may affect the performance of the proposed dehazing model, and we set the training image size to 400×400 through experiments.

3.2   Experimental Results and Comparison

We compared the proposed model with several representative state-of-the-art dehazing methods: Dark-Channel Prior (DCP) [10], Boundary Constrained Context Regularization (BCCR) [15], DehazeNet [6], Multi-scale CNN (MSCNN) [16] and All-in-One Dehazing Network (AOD-Net) [13]. Thanks to the I-HAZE and O-HAZE dataset which provide haze-free ground-truth when testing on real hazy images, we are able to compare those dehazing results quantitatively in terms of PSNR and SSIM. We also compare with the state-of-the-arts in terms of running time to demonstrate the efficiency of the proposed model.

We first compare the dehazed results on I-HAZE and O-HAZE dataset using full-references PSNR and SSIM, the matrices widely adopted for evaluating and

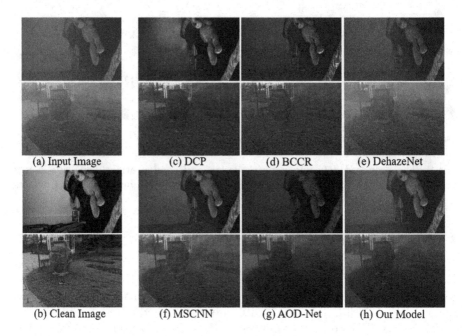

(a) Input Image   (c) DCP   (d) BCCR   (e) DehazeNet

(b) Clean Image   (f) MSCNN   (g) AOD-Net   (h) Our Model

Fig. 4. Examples of dehazed results on I-HAZE and O-HAZE dataset.

comparing image dehazing algorithms these days. The selected indoor and outdoor images and the corresponding dehazed results are shown in Fig. 4. We observe that the classical prior-based methods DCP and BCCR tend to remove more haze but suffer from unrealistic artifacts or unnatural color due to prior estimation noise. Deep learning methods including DehazeNet, MSCNN, AOD-Net and the proposed model tend to smooth visual details in results and generate artifact-free images. However, the result of DehazeNet is still too hazy compared with other methods, MSCNN sometimes generates unnatural color (also observed in Fig. 5), and AOD-Net results in too dark outputs where many dark details are disappeared).

Table 1. Average full-evaluation results of dehazed results on I-HAZE and O-HAZE dataset with running time comparison

		DCP [10]	BCCR [15]	DehazeNet [6]	MSCNN [16]	AOD-Net [13]	Ours
I-HAZE	PSNR	15.285	14.574	16.983	*17.280*	**17.407**	*17.443*
	SSIM	0.711	0.750	**0.771**	*0.791*	0.735	*0.769*
O-HAZE	PSNR	16.586	17.443	16.207	*19.068*	**17.742**	*17.658*
	SSIM	0.735	**0.753**	0.666	*0.765*	0.663	*0.750*
	Time	*1.62 s*	3.85 s	2.51 s	2.60 s	**0.65 s**	*0.26 s*

In Table 1, we compare the PSNR and SSIM score of each algorithm. The learning-based methods outperform natural or statistical priors based methods

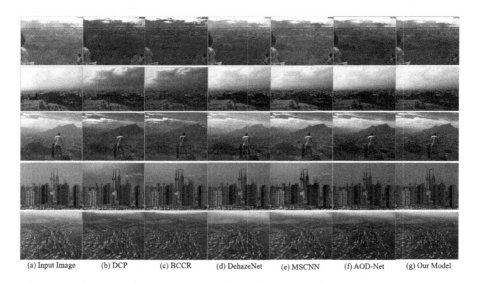

(a) Input Image   (b) DCP   (c) BCCR   (d) DehazeNet   (e) MSCNN   (f) AOD-Net   (g) Our Model

Fig. 5. Visual comparison among dehazing algorithms using real hazy photos. (Color figure online)

in most of the cases. In terms of PSNR, the proposed model ranks the first on I-HAZE indoor dataset and the third on O-HAZE outdoor dataset with minor difference to the second ranking algorithm AOD-Net. In terms of SSIM, the proposed model ranks the third on both of the datasets with minor difference to the second ranking algorithms DehazeNet and BCCR, respectively. However, the proposed VLW-Net achieves the fastest running time which is 10 times faster than MSCNN.

We also report the per-image running time of each algorithm in Table 1. The data of the other methods is from [14]. AOD-Net shows a significant advantage over the others in efficiency due to its lightweight feed-forward structure. In this paper, we run the proposed model and AOD-Net over OTS dataset [14] (image resolution: 550×413) using a machine with 3.20 GHz CPU and 64G RAM. By averaging the running time of the two methods, we get 0.186 s and 0.458 s per image for our model and AOD-Net, respectively. This depicts that our model is 2.5 times faster than AOD-Net in terms of running time. The size of the proposed VLW-Net is 6 times smaller than AOD-Net, hence our model is more hardware friendly. To compare with the other methods under the same benchmark, we convert the running time of our model according to the test in [14] and get 0.26 s per image (0.65/2.5), given the running time of AOD-Net in [14] is 0.65 s per image.

The current metrics PSNR and SSIM, though widely adopted for evaluating and comparing image dehazing algorithms, turn out to be insufficient for characterizing either human perception quality or machine vision effectiveness [14]. We thus qualitatively compare the visual results on selected challenging real hazy photos in Fig. 5. We observe that the methods including DCP, BCCR,

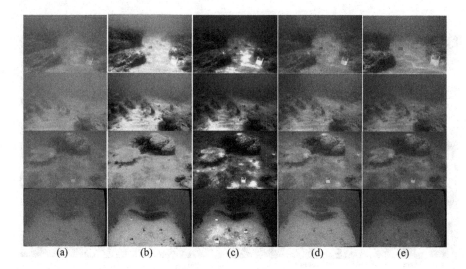

(a)   (b)   (c)   (d)   (e)

Fig. 6. Visual comparison among the underwater dehazing algorithms using real underwater dataset [4]. (a) Input image, and outputs with (b) Emberton et al. [8], (c) Ancuti et al. [3], (d) Berman [4], and (e) Our model.

DehazeNet, MSCNN, and AOD-Net sometimes generate unrealistic noises, artifacts, distorted color, or sometimes crush too many dark details. While, the clean images generated by the proposed method are free of unnatural noises and artifacts.

The algorithm MSCNN, even though scores the best PSNR and SSIM on I-HAZE and O-HAZE dataset during the qualitative test, has the following issues by comparing with the proposed VLW-Net (Fig. 5): (1) less clear image (the 1st, 3rd and 5th rows); (2) color distortion (sky color of the 4th row); (3) crushed dark details (bush area in the 2nd row). The DehazeNet has similar issues to MSCNN comparing with our model. By comparing the results of AOD-Net and our model, our model preserves the dark details much better than AOD-Net (the 2nd and 5th rows).

Without any fine tuning and further training procedure needed, the proposed model can be seamlessly applied to underwater hazy images. To demonstrate the strong adaptiveness of the proposed method, we qualitatively compare it with state-of-the-art underwater dehazing algorithms which are developed specifically for underwater images. In Fig. 6, we demonstrate the underwater image dehazing results with the methods introduced in [3,4,8] and our model. Method in [8] sometimes produces overexposed images with crushed highlight details (1st and 3rd rows), method in [3] distorted the color drastically and sometimes overexposed the highlight area(1st and 2nd rows), and the results with method in [4] have less contrast compared with the proposed method.

4 Conclusion

The paper proposes a very light-weight end-to-end CNN model (VLW-Net) for single image haze removal. We compare the proposed model with a variety of state-of-the-art methods on both indoor and outdoor real haze images, using both objective (PSNR, SSIM) and subjective measurements. The experiments exhibit that our model outperforms the current state-of-the-art methods in both efficiency and accuracy, thanks to the very light-weight network design by integrating a new Inception structure.

References

1. Ancuti, C.O., Ancuti, C., Timofte, R., Vleeschouwer, C.D.: I-HAZE: a dehazing benchmark with real hazy and haze-free indoor images. CoRR. arXiv:1804.05091 (2018)
2. Ancuti, C.O., Ancuti, C., Timofte, R., Vleeschouwer, C.D.: O-HAZE: a dehazing benchmark with real hazy and haze-free outdoor images. In: CVPR Workshops (2018)
3. Ancuti, C.O., Ancuti, C., Vleeschouwer, C.D., Neumann, L., Garcia, R.: Color transfer for underwater dehazing and depth estimation, pp. 695–699 (2017)
4. Berman, D., Levy, D., Avidan, S., Treibitz, T.: Underwater single image color restoration using haze-lines and a new quantitative dataset. arXiv preprint. arXiv:1811.01343 (2018)
5. Berman, D., Treibitz, T., Avidan, S.: Non-local image dehazing. In: IEEE Conference on Computer Vision and Pattern Recognition, vol. 9906, pp. 1674–1682 (2016)
6. Cai, B., Xu, X., Jia, K., Qing, C., Tao, D.: DehazeNet: an end-to-end system for single image haze removal. IEEE Trans. Image Process. **25**, 5187–5198 (2016)
7. Chen, C., Do, M.N., Wang, J.: Robust image and video dehazing with visual artifact suppression via gradient residual minimization. Eur. Conf. Comput. Vis. **9906**, 576–591 (2016)
8. Emberton, S., Chittka, L., Cavallaro, A.: Underwater image and video dehazing with pure haze region segmentation. Comput. Vis. Image Underst. **168**, 145–156 (2018)
9. Fattal, R.: Single image dehazing. ACM Trans. Graph. (TOG) **27**, 2–9 (2008)
10. He, K., Sun, J., Tang, X.: Single image haze removal using dark channel prior. In: 2009 IEEE Conference on Computer Vision and Pattern Recognition, pp. 1956–1963 (2009)
11. Jiang, Y., Sun, C., Zhao, Y., Yang, L.: Image dehazing using adaptive bi-channel priors on superpixels. Comput. Vis. Image Underst. **165**, 17–32 (2017)
12. Ju, M., Zhang, D., Wang, X.: Single image dehazing via an improved atmospheric scattering model. Vis. Comput. **33**, 1613–1625 (2016)
13. Li, B., Peng, X., Wang, Z., Xu, J., Feng, D.: AOD-Net: all-in-one dehazing network. In: 2017 IEEE International Conference on Computer Vision (ICCV), pp. 4780–4788 (2017)
14. Li, B., et al.: Benchmarking single-image dehazing and beyond. IEEE Trans. Image Process. **28**, 492–505 (2018)

15. Meng, G., Wang, Y., Duan, J., Xiang, S., Pan, C.: Efficient image dehazing with boundary constraint and contextual regularization. In: 2013 IEEE International Conference on Computer Vision, pp. 617–624 (2013)
16. Ren, W., Liu, S., Zhang, H., Pan, J., Cao, X., Yang, M.-H.: Single image dehazing via multi-scale convolutional neural networks. In: Leibe, B., Matas, J., Sebe, N., Welling, M. (eds.) ECCV 2016. LNCS, vol. 9906, pp. 154–169. Springer, Cham (2016). https://doi.org/10.1007/978-3-319-46475-6_10
17. Silberman, N., Hoiem, D., Kohli, P., Fergus, R.: Indoor segmentation and support inference from RGBD images. In: Fitzgibbon, A., Lazebnik, S., Perona, P., Sato, Y., Schmid, C. (eds.) ECCV 2012. LNCS, vol. 7576, pp. 746–760. Springer, Heidelberg (2012). https://doi.org/10.1007/978-3-642-33715-4_54
18. Szegedy, C., et al.: Going deeper with convolutions. In: 2015 IEEE Conference on Computer Vision and Pattern Recognition (CVPR), pp. 1–9 (2015)
19. Tang, K., Yang, J., Wang, J.: Investigating haze-relevant features in a learning framework for image dehazing. In: 2014 IEEE Conference on Computer Vision and Pattern Recognition, pp. 2995–3002 (2014)

An Improved GAN Semantic Image Inpainting

Panagiotis-Rikarnto Siavelis, Nefeli Lamprinou, and Emmanouil Z. Psarakis[(✉)]

Department of Computer Engineering and Informatics,
University of Patras, Rion Patras, Greece
{siavelis,lamprinou,psarakis}@ceid.upatras.gr

Abstract. Image inpainting is used to fill in missing regions based on remaining image data. Although the existing methods, that use deep generative models to infer the missing content, produce realistic images, sometimes the results are unsatisfactory due to arithmetical issues caused by the use of unbalanced ingredients of the proposed cost functions. In this paper, we propose a loss that generates more plausible results. Experiments on two datasets show that our method predicts information in large missing regions and achieves pixel-level photorealism, significantly outperforming state-of-the-art methods [24] and [25]. Having improved the semantic image inpainting we focus on applying the method to laparoscopic images that suffer from glares. The modified technique again outperforms its rivals. Moreover, it is faster than classical PDE based inpainting techniques and, more importantly, its running time is almost independent on the size of missing area, both critical issues in medical image processing.

Keywords: Semantic image inpainting · GANs · Laparoscopic images

1 Introduction

Semantic inpainting [19] refers to the task of inferring arbitrary large missing regions in images based on image semantics. This task is more challenging than classical inpainting since we need to predict high frequency content. Numerous applications such as automated image analysis systems that suffer from deteriorated data - e.g. due to highly reflectable surfaces - benefit from efficient semantic inpainting methods if large regions are missing.

Inpainting [8] can be used to enhance image quality in a variety of applications, from the removal of image artifacts to restoration of damaged paintings.

Based on the effective size of the missing area, image inpainting methods can be broadly classified in the following two categories:

This research is implemented through the Operational Program "Human Resources Development, Education and Lifelong Learning" and is co-financed by the European Union (European Social Fund) and Greek national funds., MIS: 5006336.

© Springer Nature Switzerland AG 2020
J. Blanc-Talon et al. (Eds.): ACIVS 2020, LNCS 12002, pp. 443–454, 2020.
https://doi.org/10.1007/978-3-030-40605-9_38

- blind inpainting and
- non-blind inpainting.

Techniques belonging into the first category can be used to detract objects and areas with low range of noise. For instance, total variation [1] takes into account the smoothness property of natural images, which is useful to fill small missing regions or remove spurious noise. On the other hand techniques of the second category are mainly used in problems with larger missing areas. Holes in textured images can be filled by finding a similar texture from the same image [11], either assuming redundancy in the undistorted input, using low rank (LR) modeling [13], or using non local information [5].

Many inpainting techniques have also been applied in the field of medical imaging in order to remove specular reflections. In particular, laparoscopic images often suffer from light specular reflections, also called highlights or glare, on the wet tissue surface. They are caused by the inherent frontal illumination and are very distracting for the observer [17]. Glare affects severely image analysis algorithms because they introduce "fake" pixel values and additional edges. This diminishes image feature extraction, which is an essential technique for reconstruction. Hence, a number of approaches have been proposed for their detection and removal.

Most approaches consist of two steps [17]. In the first step glares at each frame are detected. This is rather straightforward and in most cases basic histogram analysis, thresholding and morphological operations are used. Pixels above an intensity threshold are classified as glare. Some authors additionally propose to check for low saturation as a further strong indication for specular highlights [18,23]. In this context, the usage of various color spaces has been proposed. In the second step, the pixels identified as reflections are enhanced, i.e., modified in a way that the resulting image looks as realistic as possible [17]. Another important aspect is that operator should be informed about this image enhancement, because one cannot rule out the possibility that wrong information is introduced, e.g., a modified pit pattern on a polyp that can adversely affect the diagnosis.

For this second step, the following two different approaches can be distinguished [17]. The first is spatial interpolation, in which only the current frame is considered and the pixels that correspond to glare are inpainted. However, there are cases where the real color information for highlight pixels cannot be determined from the current frame. Hence, a temporal interpolation approach has been proposed that tries to find the corresponding position in preceding and subsequent frames and reuse the information from this position.

Recently, deep learning has been applied to image restoration [15] and semantic inpainting was introduced [14,19] that leverages distribution learning capabilities of Deep Generative Adversarial Networks (GANs). In [24] and [25] the missing region is inferred by a convolutional neural network, while in [4] an improved version of the Wasserstein generative adversarial network that accomplishes better trainability was presented.

In this paper, we build upon two recently proposed techniques [24] and [25], where a trained generative model was proposed for finding the closest encoding of the corrupted image in the latent image manifold using a cost function composed by a context based term and a prior one. This encoding was then passed through the generator to infer the missing context. The use of blending as a post-processing step was also proposed. We propose the modification of the prior term such that its contribution to the total cost to be more balanced. The proposed modification results in a better performance of the inpainting technique making the post-processing step unnecessary.

The paper is organized as follows. In Sect. 2, a brief introduction of the GANs as image generator is presented. In Sect. 3, the problem of semantic inpainting by constrained image generation and the proposed modification are presented. Section 4 contains our experiments and finally Sect. 5 contains our conclusions.

2   GANs as Image Generators

Generative Adversarial Network [3,6,7,12] constitutes the most popular learning framework for estimating generative models via an adversarial process. Figure 1 captures the architecture employed during the training phase of GANs. There is a random vector X with unknown probability density function (pdf) $f(X)$, with X playing the role of a "prototype" random vector. The goal is to design a data-synthesis mechanism that generates realizations for the random vector X.

For this goal we employ a nonlinear transformation $G(Z, \theta)$, known as the Generator, that transforms a random vector Z of known pdf $q(Z)$ (e.g. Gaussian or Uniform) into a random vector Y. We would like to estimate the parameters θ of the transformation so that Y is distributed according to $f(\cdot)$. Under general assumptions, such a transformation always exists [2,9] and it can be efficiently approximated [10] by a sufficiently large neural network, with θ summarizing the network parameters. Adversarial approaches, in order to make the proper selection of θ, employ a second nonlinear transformation $D(\cdot, \vartheta)$ that transforms X and Y into suitable scalar statistics $u = D(X, \vartheta)$ and $v = D(Y, \vartheta)$ and then compute a "mismatch" measure (not necessarily a distance) $\mathcal{J}(u, v)$ between the two random scalar quantities u, v. The second transformation $D(\cdot, \vartheta)$ is also implemented with the help of a neural network, known as the Discriminator. We are interested in the average mismatch between u, v, namely $\mathbb{E}_{u,v}[\mathcal{J}(u, v)]$, which, after the substitution, can be written as:

$$J(\theta, \vartheta) = \mathbb{E}_{X,Y}[\mathcal{J}(D(X, \vartheta),\ D(Y, \vartheta))] = \mathbb{E}_{X,Y}[\mathcal{J}(D(X, \vartheta), D(G(Z, \theta), \vartheta))], \quad (1)$$

with $\mathbb{E}[\cdot]$ denoting the expectation operator. For every selection of the generator parameters θ, we would like to select the discriminator parameters ϑ so that the average mismatch between u, v is maximized. In other words, we design the discriminator to differentiate between the synthetic random vector Y and the prototype random vector X, as much as possible.

3   Semantic Inpainting by Constrained Image Generation

The use of adversarial deep generative models for solving the inpainting image problem, i.e. to fill-in large missing regions in an image, is proposed in [24]. After

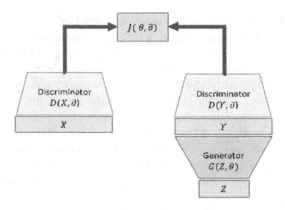

Fig. 1. GAN learning framework

the training phase of a GAN, the generator $G(Z, \theta^\star)$ is able to take a realization Z drawn from the known pdf $q(Z)$ and generate an image mimicking samples from $f(\cdot)$.

Assuming that $G(Z, \theta^\star)$ is efficient in its representation, an image that is not a sample of $f(\cdot)$ (e.g., corrupted data) should not lie on the learned encoding manifold Z. Therefore, the solution is to recover the "closest" encoding $\hat{Z}$ to the corrupted image while being constrained to the manifold. After $\hat{Z}$ is obtained, we can easily generate the missing content by using the trained generative model $G(\hat{Z}, \theta^\star)$.

Fig. 2. Corrupted image (a) and the corresponding binary mask (b)

The process of finding $\hat{Z}$ can be formulated as an optimization problem. To this end, let Y be the corrupted image and M be the binary mask, with size equal to the image, that indicates the missing or corrupted parts of the image. An example of Y and M is shown in Fig. 2. Then, using this notation we define the "closest" encoding $\hat{Z}$ by solving the following optimization problem:

$$\hat{Z} = \arg\min_{Z}\{\mathcal{L}_c(Z|Y, M) + \mathcal{L}_p(Z)\} \tag{2}$$

where $\mathcal{L}_c(\cdot|\cdot)$ and $\mathcal{L}_p(\cdot)$ are the contextual and the prior loss respectively. It is clear that, given the input corrupted image Y and the hole mask M, the contextual loss constrains the generated image, while the prior loss penalizes unrealistic images. The *contextual loss* $\mathcal{L}_c(\cdot)$, being the weighted difference between the recovered image and the uncorrupted portion, is defined as follows:

$$\mathcal{L}_c(Z|Y, M) = \|W \odot (G(Z, \theta^\star) - Y)\|_1 \tag{3}$$

where $\odot$ and $\|X\|_1$ indicate the hadamard operator (or element-wise multiplication) and the l_1 norm of vector X respectively and W the weighting image whose the i-th element is defined as:

$$W_i = \begin{cases} \sum_{j \in N(i)} \dfrac{(1 - M_j)}{|N(i)|}, & \text{if } M = 1 \\ 0, & \text{if } M = 0 \end{cases} \tag{4}$$

where i denotes the pixel index, W_i the weight factor of the i-th pixel, $N(i)$ the set of neighbors of the i-th pixel in a local window, and $|N(i)|$ the cardinality of set $N(i)$. Note that according to Eq. (4), pixels close to a hole are more important, since they are highly correlated with the corrupted pixels while, pixels far away from any holes are not considered in the inpainting process.

On the other hand, the *prior loss* $\mathcal{L}_p(\cdot)$, based on high-level image feature representations instead of pixelwise differences, enforces the recovered image to be similar to the samples drawn from the training set. Recall that in GANs, the discriminator, $D(Y, \vartheta)$, is trained to differentiate generated images from real images. Therefore the prior loss can be chosen to be identical to the GAN loss used for its training [24], i.e.,

$$\mathcal{L}_p(Z) = \lambda \log(1 - D(G(Z, \theta^\star), \vartheta^\star)) \tag{5}$$

or the logit of the trained discriminator [25], i.e.

$$\mathcal{L}_p(Z) = \lambda \operatorname{logit}(D(G(Z, \theta^\star), \vartheta^\star)) \tag{6}$$

where λ a parameter taking positive values to balance between the two loss terms. A large value in this parameter, enforces the results to be more realistic. Note that without $\mathcal{L}_p(\cdot)$, the mapping from Y to Z may result to an unreasonable image [24].

3.1 The Proposed Modified Technique

It is evident from Eq. (2) that the overall inpainting loss is the superposition of the contextual and the prior loss. As we can see from Eq. (3), the contextual loss is nonnegative and bounded from above by some finite positive number. On the other hand, assuming that λ is positive, the prior loss, defined in Eq. (5) and alternatively in Eq. (6), is nonpositive and bounded from above by zero. Thus, the fields of values of the loss functions defined in Eqs. (3) and (5 or 6) have no intersection and they are of opposite signs. This in turn, means that small values in the overall loss do not necessarily mean small values in the subcomponents. In addition, in many cases the discriminator $D(\cdot, \vartheta^\star)$ outputs a very small positive number, which in turn leads to a very small value of the prior loss thus making the prior loss to have no substantial effect in the overall loss. This is the reason why, very often, the results obtained by trained generative model $G(\hat{Z}, \theta^\star)$ are not satisfactory as it is clear from the first and second row of Fig. 3 where are shown the Context and Prior losses obtained from the application of the methods presented in [24] and [25] respectively. Note also that both loss terms are unstable and they do not converge. A post-processing step for ramifying this, is necessary (please see 4-th and 6-th column in Fig. 4). In order to balance the loss terms and avoid this additional step, which must be stressed at this point that it is not permitable in the case of medical images, the use of the following prior loss is proposed instead:

$$\mathcal{L}_p(Z) = \lambda(1 - D(G(Z, \theta^\star), \vartheta^\star)), \tag{7}$$

where λ a positive constant to balance between the two losses.

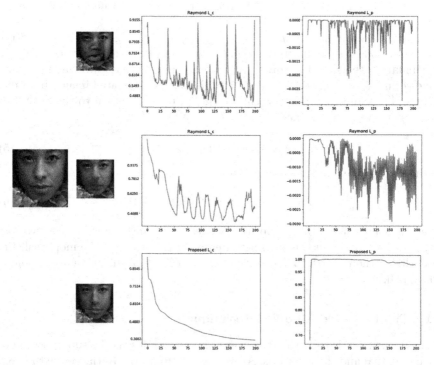

Fig. 3. Original Image (Left) and the inpainted images as well as the corresponding Context and Prior losses obtained by the method in [24] (Top row), the method in [25] (Middle row) and the proposed one (Bottom row)

Note that the new proposed loss function is free from the logarithmic function. This simple modification changes the output range of this loss term to $[0, \lambda]$, where λ some finite positive number. The obtained results from the application of the proposed method in a specific example are shown in the bottom row of Fig. 3. Note the stabilization of the loss terms and their monotone convergence. This is the rule we are going to see in the next section, where we apply the proposed technique in a number of experiments.

4 Experiments

In the following section we compare our method with the one from [24], the state-of-the-art method for semantic inpainting, evaluating both methods results qualitatively and quantitatively, using Peak Signal to Noise Ratio (PSNR) and Structural similarity (SSIM) [22]. The network used is the one in [20], along with the proposed optimal parameters for training. Specifically we set the learning rate $\ell = 0.01$ for both methods and the balancing constant $\lambda = 1$ for our method and $\lambda = 0.003$ for the method in [24], as proposed. In all experiments a window of size 7 is used for the contextual loss (Eq. 3).

For the evaluation of our method two datasets were used: the CelebFaces Attributes Dataset (CelebA) [16] and laparoscopic images from a cholecystectomy.

We used the code implemented in https://github.com/kozistr/Awesome-GANs. The experiments were conducted on a system with an 8core processor, using 32 GB RAM and NVIDIA GeForce GTX 1080 Ti.

4.1 CelebA

The CelebA database contains 202,599 face images with coarse alignment. We use approximately 2000 images for testing. The images are cropped at the center to 64×64, containing faces with various viewpoints and expressions.

We test both methods performance using two types of masks, central and random masks.

In Fig. 4 we can see the resulting reconstructed images in the presence of both types of masks, with and without the post-processing blending step. We can see that the proposed method outputs more realistic images (where blending is not used), with the outlines of the mask pixels being far less visible in both cases. However, the results obtained by methods [24] and [25] with blending in the case of random mask are better.

In Table 1 quantitative results are presented, in respect to PSNR and SSIM metrics and their variations. For both types of masks, the proposed method has higher PSNR and SSIM.

4.2 Laparoscopic Images

The laparoscopic dataset contains 42,592 images of original size 1280×720, from a video provided of a real surgery. This dataset contains high noise from glares. We used patches of size 64×64 in our experiment.

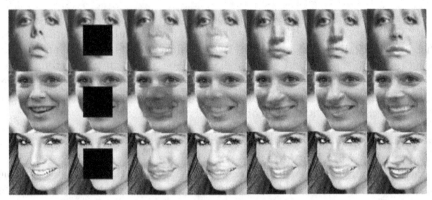

(a) Using central masks.

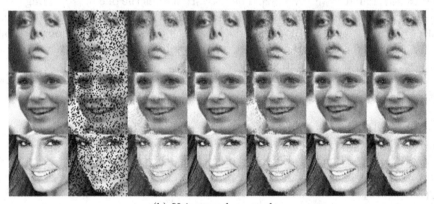

(b) Using random masks.

Fig. 4. (First column) Input image. (Second column) Input image overlayed with mask. (Third column) Inpainting result of method [24] without blending. (Fourth column) Inpainting result of method [24] with blending. (Fifth column) Inpainting result of method [25] without blending. (Sixth column) Inpainting result of method [25] with blending. (Seventh column) Inpainting result of proposed method.

We aim our trained network to be able to remove the glares presenting in those images, due to the camera light, a problem affecting the quality of the video that the surgeon sees. These images present with a problem in evaluating and comparing inpainting methods as we have no ground truth images to use, since all laparoscopy videos obviously have the same glare problem. To bypass this problem, we use as a training set and ground truth, images already inpainted using a classic PDE based inpainting technique [21]. In Figs. 5 and 6 we can see the results of the two compared inpainting methods, as well as the "ground truth" images, inpainted using [21]. The resulting reconstructed images using our method are closer to the images produced using [21], also the color of

Table 1. Quantitative results of the proposed, [24] and [25] for samples from CelebA database. For approximately 2000 images we measure PSNR and SSIM [22]

Method	PSNR (db)	SSIM	Mask type
Proposed	**32.202043 ± 2.2678**	**0.819407 ± 0.045255**	Central
Method [24]	31.593815 ± 2.1429	0.767832 ± 0.046267	Central
Proposed	**31.73239 ± 2.1160**	**0.907708 ± 0.029341**	Random
Method [24]	30.470873 ± 1.4176	0.794872 ± 0.063818	Random
Method [25]	23.60	0.8121	Mean value computed in [25]

the regions is much closer to the surrounding regions, making those regions less distracting to the eye.

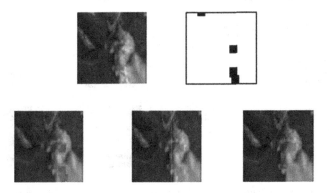

Fig. 5. Top row: (First) Original image. (Second) Glare detection mask. Bottom row: (First) Inpainting result of method [24] without blending. (Second) Inpainting result of method [21]. (Third) Inpainting result of proposed method

In Table 2 we can see that our method, compared to the classic inpainting method [21], achieves a fixed processing time almost independent to the size of the region needing inpainting. Time comparison of our method with methods [24] and [25] are not presented, since all three are tested on the same GAN and as a result the testing times are the same.

Although none of the rivals achieves a running time suitable for real time applications, the running time of the proposed method can be significantly reduced if properly programmed (Fig. 7).

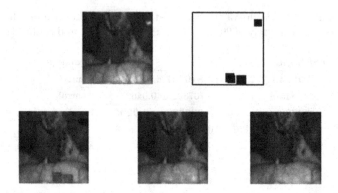

Fig. 6. Top row: (First) Original image. (Second) Glare detection mask. Bottom row: (First) Inpainting result of method [24] without blending. (Second) Inpainting result of method [21]. (Third) Inpainting result of proposed method

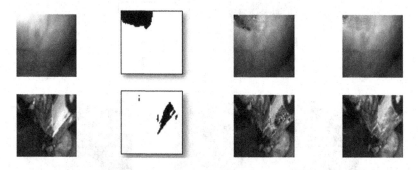

Fig. 7. (First column) Original image. (Second column) Glare detection mask. (Third column) Inpainting result of method [24] without blending. (Fourth column) Inpainting result of proposed method.

Table 2. Mean running times results in seconds for [21] and proposed method

Random mask(% coverage)	Method in [21] (sec)	Proposed (sec)
20	10.68	**9.33**
25	12.51	**9.41**
30	15.54	**9.58**
35	16.81	**9.75**
40	18.92	**9.93**

5   Conclusions

In this paper, an improvement of a recently presented technique that uses a deep generative model for solving the image inpainting problem, was proposed. The proposed modification of the cost function results in more plausible images.

Experiments from two datasets were shown that our method predicts information in large missing regions and achieves pixel-level photorealism. The proposed technique was applied to laparoscopic images that suffer from glares. The modified technique again outperformed its rivals, it was faster than classical PDE based inpainting techniques and, more importantly, its running time was almost independent on the size of missing area.

References

1. Afonso, M.V., Bioucas-Dias, J.M., Figueiredo, M.A.: An augmented lagrangian approach to the constrained optimization formulation of imaging inverse problems. IEEE Trans. Image Process. **20**(3), 681–695 (2010)
2. Andrews, D., Gnanadesikan, R., Warner, J.: Transformations of multivariate data. Biometrics **27**, 825–840 (1971)
3. Arjovsky, M., Chintala, S., Bottou, L.: Wasserstein generative adversarial networks. In: International Conference on Machine Learning, pp. 214–223 (2017)
4. Ballester Nicolau, C.: Semantic image inpainting through improved Wasserstein generative adversarial networks. In: Proceedings of the IEEE Computer Society Conference on Computer Vision and Pattern Recognition, 2019, vol. 4, p. 249–260 (2019)
5. Barnes, C., Shechtman, E., Finkelstein, A., Goldman, D.B.: PatchMatch: a randomized correspondence algorithm for structural image editing. In: ACM Transactions on Graphics (ToG), vol. 28, p. 24. ACM (2009)
6. Basioti, K., Moustakides, G.V., Psarakis, E.Z.: Kernel-based training of generative networks. arXiv preprint. arXiv:1811.09568 (2018)
7. Bengio, Y.: Practical recommendations for gradient-based training of deep architectures. In: Montavon, G., Orr, G.B., Müller, K.-R. (eds.) Neural Networks: Tricks of the Trade. LNCS, vol. 7700, pp. 437–478. Springer, Heidelberg (2012). https://doi.org/10.1007/978-3-642-35289-8_26
8. Bertalmio, M., Sapiro, G., Caselles, V., Ballester, C.: Image inpainting. In: Proceedings of the 27th Annual Conference on Computer Graphics and Interactive Techniques, pp. 417–424. ACM/Addison-Wesley Publishing Co. (2000)
9. Box, G.E., Cox, D.R.: An analysis of transformations revisited, rebutted. Wisconsin Univ-madison Mathematics Research Center, Technical report (1981)
10. Cybenko, G.: Approximation by superpositions of a sigmoidal function. Math. Control Signals Syst. **2**(4), 303–314 (1989)
11. Efros, A.A., Leung, T.K.: Texture synthesis by non-parametric sampling. In: Proceedings of the Seventh IEEE International Conference on Computer Vision, vol. 2, pp. 1033–1038. IEEE (1999)
12. Goodfellow, I., et al.: Generative adversarial nets. In: Advances in Neural Information Processing Systems, pp. 2672–2680 (2014)
13. Hu, Y., Zhang, D., Ye, J., Li, X., He, X.: Fast and accurate matrix completion via truncated nuclear norm regularization. IEEE Trans. Pattern Anal. Mach. Intell. **35**(9), 2117–2130 (2012)
14. Iizuka, S., Simo-Serra, E., Ishikawa, H.: Globally and locally consistent image completion. ACM Trans. Graph. (ToG) **36**(4), 107 (2017)
15. LeCun, Y., Bengio, Y., Hinton, G.: Deep learning. Nature **521**(7553), 436–444 (2015)

16. Liu, Z., Luo, P., Wang, X., Tang, X.: Deep learning face attributes in the wild. In: Proceedings of the IEEE International Conference on Computer Vision, pp. 3730–3738 (2015)
17. Münzer, B., Schoeffmann, K., Böszörmenyi, L.: Content-based processing and analysis of endoscopic images and videos: a survey. Multimedia Tools Appl. **77**(1), 1323–1362 (2018)
18. Oh, J., Hwang, S., Lee, J., Tavanapong, W., Wong, J., de Groen, P.C.: Informative frame classification for endoscopy video. Med. Image Anal. **11**(2), 110–127 (2007)
19. Pathak, D., Krahenbuhl, P., Donahue, J., Darrell, T., Efros, A.A.: Context encoders: feature learning by inpainting. In: Proceedings of the IEEE Conference on Computer Vision and Pattern Recognition, pp. 2536–2544 (2016)
20. Radford, A., Metz, L., Chintala, S.: Unsupervised representation learning with deep convolutional generative adversarial networks. arXiv preprint. arXiv:1511.06434 (2015)
21. Telea, A.: An image inpainting technique based on the fast marching method. J. Graph. Tools **9**(1), 23–34 (2004)
22. Wang, Z., Bovik, A.C., Sheikh, H.R., Simoncelli, E.P., et al.: Image quality assessment: from error visibility to structural similarity. IEEE Trans. Image Process. **13**(4), 600–612 (2004)
23. Yao, R., Wu, Y., Yang, W., Lin, X., Chen, S., Zhang, S.: Specular reflection detection on gastroscopic images. In: 2010 4th International Conference on Bioinformatics and Biomedical Engineering, pp. 1–4. IEEE (2010)
24. Yeh, R.A., Chen, C., Yian Lim, T., Schwing, A.G., Hasegawa-Johnson, M., Do, M.N.: Semantic image inpainting with deep generative models. In: Proceedings of the IEEE Conference on Computer Vision and Pattern Recognition, pp. 5485–5493 (2017)
25. Yeh, R.A., Lim, T.Y., Chen, C., Schwing, A.G., Hasegawa-Johnson, M., Do, M.: Image restoration with deep generative models. In: 2018 IEEE International Conference on Acoustics, Speech and Signal Processing (ICASSP), pp. 6772–6776. IEEE (2018)

Tracking, Mapping and Scene Analysis

CUDA Implementation of a Point Cloud Shape Descriptor Method for Archaeological Studies

David Arturo Soriano Valdez[1]([✉]) [iD], Patrice Delmas[2] [iD], Trevor Gee[2],
Patricio Gutierrez[5], Jose Luis Punzo-Diaz[5], Rachel Ababou[3],
and Alfonso Gastelum Strozzi[4] [iD]

[1] National Autonomous University of Mexico, CDMX, Mexico
`david.soriano@comunidad.unam.mx`
[2] The University of Auckland, Auckland, New Zealand
`p.delmas@auckland.ac.nz`
[3] Centre de Recherches des Écoles de Coëtquidan Saint-Cyr, 56381 Guer, France
`rachel.ababou@st-cyr.terre-net.defense.gouv.fr`
[4] Institute of Applied Science and Technology, CDMX, Mexico
`alfonso.gastelum@icat.unam.mx`
[5] Centro INAH Michoacan, Instituto Nacional de Antropologia e Historia (INAH),
Morelia, Mexico
`http://www.icat.unam.mx/secciones/depar/sub5/unhgm.html`

Abstract. In this work we present a new approach to study shape descriptors of archaeological objects using an implementation of the smoothed-points shape descriptor (SPSD) method that is based on the numerical mesh-free simulation method smoothed-particles hydrodynamics. SPSD can describe the textural or morphological properties of a surface by obtaining a property field descriptor based on the points shape descriptors and a smoothing function over a neighborhood of each point. The neighborhood size depends on a smoothing distance function which drives the field descriptor to either focus on small local details or larger details over big surfaces. SPSD is designed to provide a real-time scientific visualization of cloud points shape descriptors to assist in the field study of archaeological artifacts. It also has the potential to provide quantitative values (e.g. morphological properties) for artifacts analysis and classification (computational and archaeological). Due to the visualisation requirement for a real-time solution, SPSD is implemented in CUDA using an Octree method as the mechanism to solve the neighborhood particles interaction for each point cloud.

Keywords: GPU · Shape-descriptor · SPH

1 Introduction

Due to the increased use of photogrammetry tools and equipment in the area of archaeological sciences [3], there is a renewed interest in the development of

© Springer Nature Switzerland AG 2020
J. Blanc-Talon et al. (Eds.): ACIVS 2020, LNCS 12002, pp. 457–466, 2020.
https://doi.org/10.1007/978-3-030-40605-9_39

non-invasive computation tools that can provide the specialists in the field of archaeology with a method to study the superficial properties of archaeological artifacts. Typical applications include color calibration of the artifacts for museography [12], restoration of artifacts [2], registration of artifacts and their context [13], and textural and morphological analysis of the artifact surface [15], amongst others.

In archaeological studies there is a basic criteria for understanding and characterizing cultural diversity in archaeology that is the concept of style [14]. Understood as the way in which an artifact is made or executed where the agency of the artist, craftsman, teacher or practitioner is inevitably mediated by the norm and social guidelines directly or indirectly, in a given space and time. For example, the form, function, motifs, ornaments, and manufacturing technique are determining factors to characterize stone sculptures [7,8,20].

Computer vision based methods are now used to study superficial properties of the artifacts in an effort to provide information about the methods and tools that were used to create them [5,18]. An approximation to this problem is to obtain the surface mesh of each artifact, and use the associated primitives of representation to measure the morphological and textural descriptions [6]. Another approximation is to work directly with the cloud points obtained from different photogrammetry methods without reconstructing the surface mesh or using the primitives of representation to obtain descriptors of the artifact [1,16]. In this work a novel approach to archaeological artifacts using shape descriptors based on the theory of smoothed-particles hydrodynamics is presented.

Lagrangian mesh-free methods computer simulations are a common practice in the solution of flow behaviors in the area of computational physics [19]. Mesh-free methods, as defined in [17] are "used to establish a system of algebraic equations for the whole problem domain without the use of a predefined mesh for the domain discretisation". One of the most popular Lagrangian method to solve hydrodynamic problems, proposed by Gingold and Monaghan [4], and Lucy [9], is called smoothed-particles hydrodynamics (SPH).

SPH solves the mesh-free modelling problem by dividing the object volume in a set of discrete particles (points on space with local information and the ability to move inside the volume). The local values at each particles are obtained by summing the values of the particles that are inside a kernel centered on each particle using a smoothing function that weights the contribution of each particle according to the distance to their particle of interest. SPH calculates physical properties of the object using a scalar field of points with local physical values at each location. The values are updated at each time step according to the kernel summation at each particle.

Using SPH as a theoretical basis for our work, we developed a smoothed-points shape descriptor (SPSD) method in CUDA that can provide different textural and morphological smoothed descriptors. This can be used as a visual inspection tool for the specialist or as a numerical representation of the artifact or parts of the artifact used in classification or comparison of different artifacts task. Using an implementation of the associated SPSD Octree in CUDA, the calculations are solved realtime, allowing the implementation of the tool in on-site inspection photogrammetry tools.

2   Methodology

The method is applied to a point cloud C represented by the scalar field $\mho(x, y, z)$, were $\mho$ denotes the set of all points P, where P is a tuple (r, A), where $A = \mho(x, y, z)$ is a scalar attribute (feature), and r is the point position in 3D.

SPSD is used on points P of the point cloud C to obtain a property field A_s from the scalar attributes A, the points normals, texture value, or even complex shaders functions like material calculations.

2.1   Property Field

The first step consists in computing a property field A_s from the original scalar feature A_i at each point P_i. The properties field values are calculated using the SPSD implementation (based on the SPH method) [10].

The SPH method adapted as SPSD is used to calculate the property field at each point P_i, $A_s(P_i)$, using the attribute values A_j of the neighbor points P_j (Eq. (1)).

$$A_s(P_i) = \sum_{j=1}^{n} \frac{m_j}{\rho_j} A_s(P_j) W(r_i - r_j, h) \tag{1}$$

Where the current implementation of SPSD considers the physical values mass m and density ρ as equal. n is the number of points in the neighborhood inside the smoothing length h, and W is the smoothing kernel function $W(R, h)$ (Eq. (2)).

$$W(R, h) = \frac{315}{64\pi h^9} \left\{ \begin{matrix} (h^2 - R^2)^3 & 0 \le R \le h \\ 0 & R > h \end{matrix} \right\} \tag{2}$$

Where R is the relative distances between the points i and j.

2.2   Smoothed Normal Field

The first property field used to describe the object surfaces is the smoothed normal field $\overline{n_s}(P_i)$ calculated from the points normal attribute $\hat{n}_i$ (Equ. c).

$$\overline{n_s}(P_i) = \sum_{j=1}^{n} \hat{n}_j W(r, h) \tag{3}$$

The normalized normal value is given by Eq. (4).

$$\hat{n}_s(P_i) = \frac{\overline{n_s}(P_i)}{\|\overline{n_s}(P_i)\|} \tag{4}$$

The absolute difference of smoothed normal field values at each point is used to obtain the smoothed shape descriptor of the object surface. The values are used to provide a visual representation and the absolute differences are solved with respect to the Laplacian of the smoothing kernel function $W(R, h)$, where Eq. (5) shows how SPSD is calculated, while Eq. (6) corresponds to the Laplacian smoothing kernel used.

$$SD(P_i) = \sum_{j=1}^{n} abs(\hat{n_s}(P_i) - \hat{n_s}(p_j))\Delta W(R, h) \tag{5}$$

$$\Delta W = \frac{45}{\pi h^6}(h - R) \tag{6}$$

3   GPU Implementation

The method was implemented on a GPU using Nvidia CUDA. In order to take advantage of the parallel architecture, the radial neighborhood method, which is used to calculate the field properties for the shape descriptor, was optimized by using a Morton Key (a.k.a. MK) [11] (MK) indexed and sorted Octree. By using this radial neighborhood method, it is possible to calculate the neighborhood point set of a point in linear time. The method is highly parallelizable, since every set can be obtained by Octree node.

3.1   Octree

The Octree data structure improves the neighborhood calculation by per-forming spatial division in the set of particles. The 3D spatial division of the work space is symmetrical in the three dimensions. As a result, the initial division provide eight sub-domains, which are referred as node N. The process can be applied on every resulting node for another or more layers of sub-domains division if needed.

The level of every node N is determined by the numbers of its parent nodes. By definition, the work space is the root node, thus it is a parent node for every node. Every node that has a parent node is referred as a child node, while only nodes at the last level have no child, thus they are not parent nodes. The usual approach when spatial division is performed by using an Octree is to isolate every element until they contain only one particle. By using this approach, it is not possible to increase the level indefinitely. Since it is necessary to index and sort the Octree and particles by MK, the structure is limited to a maximum level, the level for a 32 bits value that works as a MK is a 10 depth levels.

The MK for this work is a 32 bits value. Since it is used for 3D data, the MK is composed of segments of three bits per level. The MK is calculated by using the position of the particle which must be normalized to a 10 bits value per dimension. Then, the 3 values are interlaced from the most significant to the least significant bit. The structure of the MK is shown in Fig. 1.

The indexation of the Octree using the above MK coding is possible since it results in a special sequence for the nodes naming, known as z-order. Additional to the indexation, the structure can be sorted to improve efficiency in a parallel architecture.

4 Results

The process was tested by using a model of 1,604,254 particles for a set of archeological artefacts. Particles normals $\hat{n}$ were initially estimated using a commercial software. The process was executed on two different computers:

- Device A. Intel i7-7700HQ @ 2.8 GHz Nvidia GeForce GTX 1070 Laptop GPU (2048 cores at 1443 MHz with 8 Gb memory).
- Device B. Intel i7-7700K @ 4.2 GHz Nvidia GeForce GTX 1080Ti Desktop GPU (3584 cores at 1481 MHz with 11 Gb memory).

The main processes analyzed were the following:

- Smooth Particle Normal Calculation.
- Smooth Particle Shape Descriptor Calculation.

Table 1. Octree level 9, $h = 1.0$.

Process	Device A Time [ms]	Device B Time [ms]
Smooth Particle Normal	94.37	62.68
Smooth Particle Shape Descriptor	107.07	72.04

Table 2. Octree level 8, $h = 1.0$.

Process	Device A Time [ms]	Device B Time [ms]
Smooth Particle Normal	488.99	320.03
Smooth Particle Shape Descriptor	523.18	344.97

Table 3. Octree level 7, $h = 1.0$.

Process	Device A Time [ms]	Device B Time [ms]
Smooth Particle Normal	2573.87	1826.55
Smooth Particle Shape Descriptor	2546.46	1863.71

Table 4. Octree level 6, $h = 1.0$.

Process	Device A Time [ms]	Device B Time [ms]
Smooth Particle Normal	15155.48	11982.87
Smooth Particle Shape Descriptor	15446.44	12433.50

Table 5. Average amount of particles in non-empty nodes by Octree depth level.

Octree depth level	AAPNEN
9	4.35
8	15.39
7	58.44
6	227.74

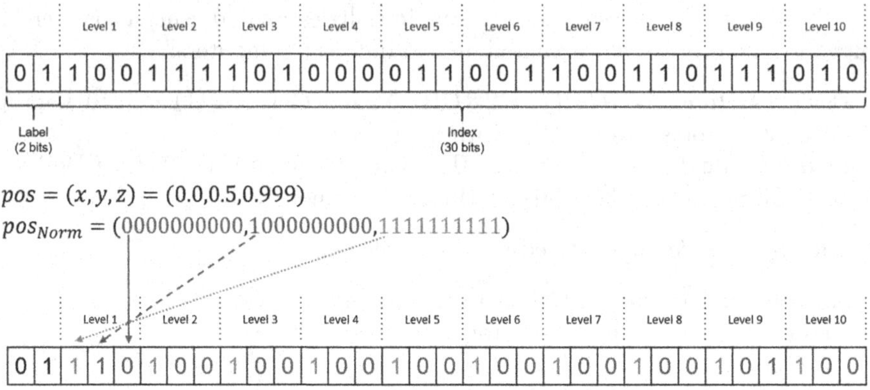

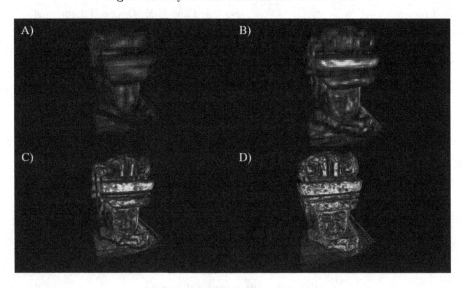

Fig. 1. MK layout and construction for 3D data.

Fig. 2. Smooth Particle Shape Descriptor result at different Octree depth level. (A) Level 6, (B) Level 7, (C) Level 8, (D) Level 9.

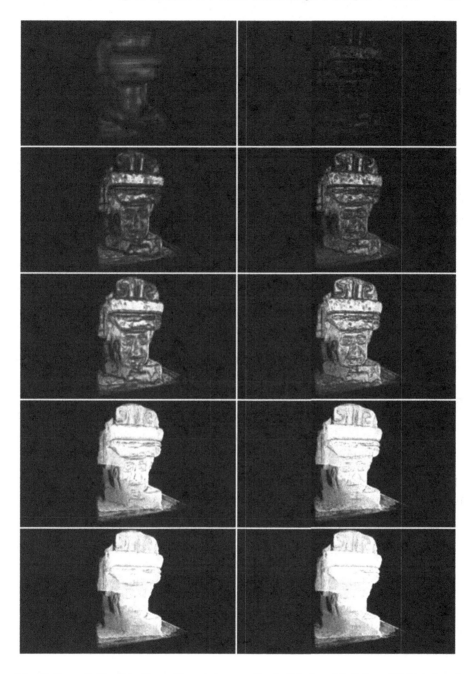

Fig. 3. Smooth Particle Shape Descriptor results. Left is Level 8 Octree. Right is Level 9 Octree. Kernel radius, from top to bottom, $h = (0.5, 1.0, 1.5, 3.0, 4.0)$.

The process was executed using the same kernel radius, $h = 1.0$, but the Octree depth level was changed from level 6 to level 9. Results are shown in Tables 1, 2, 3, and 4.

The resolution of the shape descriptor can be adjusted by increasing or decreasing the kernel radius. The performance of the process is not significantly affected by the kernel radius, it is more affected by the Octree depth level. This is evident since the average amount of particles in non-empty node (AAPNEN) increases as the level depth is reduced, as shown in Table 5.

The results also vary as the depth level changes: Where the depth level increases, the resolution of the shape descriptor also improves. This is helpful where it is necessary to use the shape descriptor at a different resolution alongside the objects. Figure 2 shows a comparison between the result of the shape descriptor as the Octree level depth varies from 6 to 9, while the kernel radius $h = 1.0$ stays the same.

The variation of the kernel radius changes the sensitivity of the SPSD: By increasing the kernel radius only the most significant changes are shown, while by reducing the kernel radius value, almost every changes in shape is recorded. SPSD provides a better resolution as the Octree depth level increases, but it is the kernel size, which is more crucial when it is necessary to focus only in certain shape features. Figure 3 shows results comparison for SPSD: On the left, a level 8 depth Octree is used, while on the right a level 9 depth Octree is used, with from top to bottom the kernel sizes used are: $h = (0.5, 1.0, 1.5, 3.0, 4.0)$.

5 Conclusion

The implementation of a real-time scientific visualization method for archaeological artifacts points cloud exploration allows to introduce a shape descriptor method that make use of the popular cameras that provide real-time point clouds representation of the scene. At the same time, it allows for a new method to obtain the quantification of the shape descriptors as a result of a smoothing function that can, depending on the smoothing kernel, produce field descriptors for different features size.

The methods presented allow the archaeological expert to have a tool that can in real-time provide him with different field descriptors that together can be used to describe different areas of the artifact with different size of features.

SPSD has a tool can provide scientific visualization for the inspection of the artifacts. It can also save each field descriptors as a new scalar attribute in each point tuple. Future work on SPSD will look at introducting color point clouds to explore textural descriptors of the color fields.

References

1. Addison, A.C., Gaiani, M.: Virtualized architectural heritage: new tools and techniques. IEEE MultiMedia **7**(2), 26–31 (2000). https://doi.org/10.1109/93.848422
2. Barreau, J.B., et al.: Ceramics fragments digitization by photogrammetry, reconstructions and applications. In: International Conference on Culturage Heritage, EuroMed 2014, Lemessos, Cyprus, November 2014. https://hal.archives-ouvertes.fr/hal-01090145
3. Brutto, M.L., Meli, P.: Computer vision tools for 3D modelling in archaeology. Int. J. Heritage Dig. Era **1**(1_suppl), 1–6 (2012). https://doi.org/10.1260/2047-4970.1.0.1
4. Gingold, R.A., Monaghan, J.J.: Smoothed particle hydrodynamics - theory and application to non-spherical stars. MNRAS **181**, 375–389 (1977). https://doi.org/10.1093/mnras/181.3.375
5. Gu, Z., Pan, W., Song, G., Qiu, Z., Yang, Y., Wang, C.: Investigating the tool marks of stone reliefs from the Mausoleum of Cao Cao (AD155-AD220) in China. J. Archaeol. Sci. **43**, 31–37 (2014). https://doi.org/10.1016/j.jas.2013.12.005, http://www.sciencedirect.com/science/article/pii/S0305440313004354
6. Hou, M., et al.: A new method of gold foil damage detection in stone carving relics based on multi-temporal 3D LiDAR point clouds. ISPRS Int. J. Geo-Inf. **5**(5) (2016). https://doi.org/10.3390/ijgi5050060, https://www.mdpi.com/2220-9964/5/5/60
7. Joyce, R.A.: Performing the body in pre-hispanic central America. Res.: Anthropol. Aesthetics **33**, 147–165 (1998). https://doi.org/10.1086/RESv33n1ms20167006
8. Lothrop, S.K.: Essays in Pre-Columbian Art and Archaeology. Harvard University Press, Cambridge (1961)
9. Lucy, L.B.: A numerical approach to the testing of the fission hypothesis. Astron. J. **82**, 1013–1024 (1977). https://doi.org/10.1086/112164
10. Monaghan, J.J.: Smoothed particle hydrodynamics. Ann. Rev. Astron. Astrophys. **30**, 543–574 (1992). https://doi.org/10.1146/annurev.aa.30.090192.002551
11. Morton, G.M.: A Computer Oriented Geodetic Data Base and a New Technique in File Sequencing. International Business Machines Co., Ottawa (1966)
12. PARTHENOS (ed.): Digital 3D objects in art and humanities: challenges of creation, interoperability and preservation. White paper. PARTHENOS, Bordeaux, France, May 2017. https://hal.inria.fr/hal-01526713
13. Reu, J.D., et al.: Towards a three-dimensional cost-effective registration of the archaeological heritage. J. Archaeol. Sci. **40**(2), 1108–1121 (2013). https://doi.org/10.1016/j.jas.2012.08.040, http://www.sciencedirect.com/science/article/pii/S0305440312003949
14. Sackett, J.A.: The meaning of style in archaeology: a general model. Am. Antiq. **42**(3), 369–380 (1977). https://doi.org/10.2307/279062
15. Saleri, R., et al.: UAV photogrammetry for archaeological survey: the theaters area of Pompeii. In: 2013 Digital Heritage International Congress (DigitalHeritage), vol. 2, pp. 497–502, October 2013. https://doi.org/10.1109/DigitalHeritage.2013.6744818
16. Setty, S., Mudenagudi, U.: Example-based 3D inpainting of point clouds using metric tensor and Christoffel symbols. Mach. Vis. Appl. **29**(2), 329–343 (2018). https://doi.org/10.1007/s00138-017-0886-7
17. Shapiro, V., Tsukanov, I.G.: Mesh-free method and system for modeling and analysis (2004)

18. Shea, J.J.: Lithic microwear analysis in archeology. Evol. Anthropol.: Issues News Rev. **1**(4), 143–150 (1992). https://doi.org/10.1002/evan.1360010407
19. Tan, J., Yang, X.: Physically-based fluid animation: a survey. Sci. Chin. Ser. F: Inf. Sci. **52**(5), 723–740 (2009). https://doi.org/10.1007/s11432-009-0091-z
20. Wauchope, R., JIKRCL of Congress: Handbook of Middle American Indians. University of Texas Press, Austin (1964)

Red-Green-Blue Augmented Reality Tags
for Retail Stores

Minh Nguyen$^{(\boxtimes)}$, Huy Le, and Wei Qi Yan

School of Engineering, Computer and Mathematical Science,
Auckland University of Technology, Auckland, New Zealand
minh.nguyen@aut.ac.nz

Abstract. In this paper, we introduce a new Augmented Reality (AR)
Tag to enhance detection rates, accuracy and also user experiences in
marker-based AR technologies. The tag is a colour printed card, divided
into three colour channels: red, blue, and green; to label the three components: (1) an oriented marker, (2) a bar-code and (3) a graphic image,
respectively. In this tag, the oriented marker is used for tag detection
and orientation identification, the bar-code is for storing and achieving
numerical information (IDs of the models), and the texture image is
to provide the users with an original sight of what the tag is displaying.
When our new AR tags are placed in front of the camera, the corresponding 3D graphics (models of figures or products) will appear directly on
top of it. Also, we can rotate the tags to rotate the 3D graphics; and
move the camera to zoom in/out or view it from a different angle. The
embedded bar-code could be 1D or 2D bar-codes; the currently popular
QR code could be used. Fortunately, QR codes include position detection patterns that could be used to identify the orientation for the code.
Thus, the oriented marker is not needed for QR code, and one channel is
saved and used for presenting the initially displaying image. Some experiments have been carried out to identify the robustness of the proposed ·
tags. The results show that our tags and its orientations (marker stored
in the blue colour channel) are relatively easy to detect using commodity
webcams. The embedded QR code (painted in blue) is readable in most
test cases. Compared to the ordinary QR tag (black and white), our
embedded QR code has the detection rates of 95%. The image texture
is stored in the red and green channel is relatively visible. However, the
blue channel is missing, which makes it not visually correctly in some
cases. Application-wise, this could be used in many AR applications such
as shopping. Thanks to the large storage of QR Code, this AR Tag is
capable of storing and displaying virtual products of a much wider variety. The user could see its 3D figure, zoom and rotate using intuitive
on-hand controls.

1 Introduction

Shopping retail is one of the essential businesses; and in many cities such as
Singapore, Dubai, and Hong Kong, it is considered the most crucial industry.

© Springer Nature Switzerland AG 2020
J. Blanc-Talon et al. (Eds.): ACIVS 2020, LNCS 12002, pp. 467–479, 2020.
https://doi.org/10.1007/978-3-030-40605-9_40

Today, most people go shopping frequently to satisfy their needs; therefore, shopping can be seen as a must-do activity for many adults, especially women. Shopping malls have been opening everywhere to serve and attract customers. However, with social and technological development, how customers shop have dramatically changed in recent years. Many people alternate their shopping behaviours between in-store, online, and via a mobile device. The growing popularity of online shopping has dramatically put pressure on in-store shopping. Thus, it is vital that shopping retailers offer their customers an in-store experience that they could not receive via another channel. However, there are thousands of products in the world (from different countries, different manufacturers) with the same utilities. These products can vary regarding brand, look, quality, and material. Some critical factors influence whether a customer will purchase a product are: the quality, the look, the current trends, and the brand name. However, the visual look plays a vital role for many people; frequently, how a product looks immediately triggers a customer's buying decision.

In a shopping mall, physical shops cannot display or store all products locally due to available space and cost. The online store typically has more products; however, the visualisation of the online product displayed on the Internet browser is relatively limited. Still, many buyers prefer going to crowded physical stores. For these different kinds of shoppers, there are diverse critical drivers for their shopping behaviour in a store. Customer store preference can be impacted by accessibility, reputation, in-store service and atmosphere. Store atmosphere plays a significant role in customer experience. A positive store experience reinforces customer sanctification and increases sales. Neither channels can provide good enough user shopping experience. Augmented Reality (AR) is one of the possible solutions. AR could be a high-tech facility added to physical shopping stores to increase the availability of products. Moye and Kincade [12] observed that facilities could also affect customer store selection. Customers tend to choose the store with the facilities and infrastructure that eases the shopping process.

AR has been proven to be useful for the retail industry. The report "2016 Retail Perceptions" stated that 40% of customers preferred purchasing an item

(a) View a product catalog   (b) Test a virtual furniture

Fig. 1. Hand-held AR system displays

if they had an AR experience with it. Furthermore, 61% of customers tended to go to a store offering AR experience, and 71% of customers said they would go to the store more frequently if that store offered AR experience. Therefore, according to the report, consumers desire shopping with AR experiences. So, combining AR technology with shopping retail could be a compelling and unique option to increase sales. Figure 1 displays two possible examples of AR, applicable for shopping. As an emerging technology, many fields and industries could benefit from AR. It will generate 120 billion dollars in revenue by 2021 according to current predictions, so it is necessary and vital for retailers to maximise the advantages of this enhanced experience with shoppers. The question is: which AR technology could be applied to the shopping retail industry to enhance the customer experience and also to increase the business turnover? It must be simple, low-cost, and visually attractive. In this paper, we propose a simple type of AR tags with embedded bar-codes, texture pictures and orientation markers all together in its trio-colour Red-Blue-Green printed presentation. On this tag, we can dynamically present 3D graphics of a chosen product efficiently and correctly while some pictorial information still remains. Accordingly, instead of a pair of shoes, the shopper only needs one printed card with the shoes' barcode printed on the card. One shoe-box can fit hundreds of such cards; thus, it could save a significant amount of stocking spaces and cost for the shop.

2 Background of AR Enhancement

AR refers to a technology that combines real and virtual imagery in a real environment. It aligns computer-generated objects with the real world and operates in real-time in three dimensions [1]. The history of AR can be traced back to the 1960s, when the first AR interface was developed by Ivan Sutherland and his students [17] which enabled users to see 3D graphics via a see-through device. By the late 1990s, there was an increased exploration of AR. The systems became more straightforward to create with the development of the ARToolKit. Many symposiums were held to exchange information on the problems and solutions of AR. There are three enabling technologies for AR: tracking techniques, display techniques and interaction techniques. Tracking techniques is a fundamental enabling technology. It can be classified into three groups: sensor-based [7], vision-based [2] and hybrid tracking techniques [16]. Display techniques is the visual display. This is categorized into three groups: projection-based displays [4], hand-held displays [19] and see-through HMD displays [18].

2.1 Marker-Based AR

Marker-based AR is also referred to as "image recognition" or "recognition-based AR". This technology relies on a camera in an AR device to produce a result with a marker acting as a trigger. The marker can be a 2D image with distinct features and patterns that can be detected by the camera [9]. Several common markers are shown in Fig. 2. In this technology, firstly a marker is required to put

(a) ID marker	(b) Barcode	(c) QRcode	(d) Picture

Fig. 2. Common AR markers and their tracking system

in the real world, which is equivalent to determining a 3D plane in reality. At that moment, there is a need to recognise the marker and conduct pose assessment through the camera to determine its location. Marker coordinates are originated by the marker centre. There is a requirement to obtain a transformation to establish a mapping relationship between the marker coordinate and the screen coordinate. The transformation from marker coordinates to screen coordinates requires rotation and translation to the camera coordinates and then referring to the observed screen coordinates.

2.2   Marker-Less AR

Marker-less AR is also called "location-based AR". Marker-less AR is the same as marker-based AR regarding the fundamental principle. However, this technology is not based on authentic markers, which are put in the real environment, so that the AR system could track the position and orientation [11]. Instead, any objects like the face, the hands or the body, which have feature points in the real environment, can be traced to produce a virtual object. In general, the process initially, is to find out feature points such as the corners of an image; then, to describe these points by the feature vector. Finally, it is necessary to match the feature vector from different viewpoints. Popular algorithms used to detect and track local features in an image include Scale Invariant Feature Transform (SIFT), Speed Up Robust Features (SURF), and Simultaneously Localization and Mapping (SLAM) [8]. These algorithms can detect and track local features through specific metrics in an image. In this application, marker-less AR is not useful as we need to have many AR cards with product pictures and barcodes on them to easily keep stocks and quickly search for when needed.

3   Design and Implementation

Overall, we build a system that can read an AR marker and display 3D graphics of commercial products on top of the marker. For ease of usage, there are three requirements of the newly proposed AR marker, as listed below:

1. The marker should be used to detect orientation; e.g. when rotated, the graphics model is also rotated.

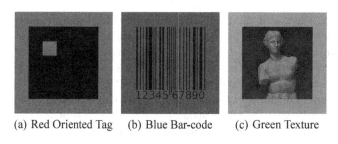

(a) Red Oriented Tag   (b) Blue Bar-code   (c) Green Texture

(d) Result after RGB channels combined

Fig. 3. The overall design of our proposed RGB marker. (Color figure online)

2. The marker should be numerically encrypted and decrypted so that thousands of product models can be stored and displayed efficiently.
3. The marker should hold some visual picture of the displaying product; for instance, the picture of a Nike Air VaporMax shoe. This helps the retailer quickly and easily allocate the correct cards among hundreds of them.

3.1   Design of the Unique AR Marker/Tag

The anaglyphic images motivate the creation of our new AR marker. Anaglyphic images are used to display stereoscopic 3D images, which are seen through dual colour anaglyph glasses (a chromatically opposite colour for each lens). Compared to other 3D displaying techniques, such as auto-stereoscopic display and holographic display, anaglyphic images are the most inexpensive and accessible way for 3D visualisation. In 1853, W. Rollmann first introduced the principle of the 3D anaglyph [14]. This was then applied to practical use by Ducos du Hauron in 1981.

The proposed AR marker applies a similar idea: it collects three unique pictures (1) an oriented marker, (2) a bar-code and (3) a graphic image; and merge these into one by encoding them to three colour channels: green, blue and red. The idea is demonstrated in Fig. 3. The red channel is used to store an orientation robust squared marker (Fig. 3(b)); the blue channel is holding the product bar code (Fig. 3(a)); and the green channel is used to display the original image of the product (Fig. 3(c)). The choice of the colour is decided carefully, we know that our human eyes are most sensitive to yellowish-green colour, then to red, and not very sensitive to blue colour. However, to computer, they are relatively the same. We choose green to present the product photo because it is the most important visual information towards retailers and shoppers. Blue is for the bar-code because it is complex and distracting. Red is then kept for the orientation tag. More details are described in the following sections.

3.2 Red Oriented Tag

To blend between the real and virtual environments harmoniously, it is essential that the 3D graphics are well aligned with the real-world objects. As previously mentioned, the bordered marker is stored in the blue channel for tracking relative position and orientation to the camera. Our marker is made of two parts: solid black square and inner white square. The position of the black square determines the 3D graphics position, while the white square is for orientation. Like other computer vision projects, there is a primary pipeline for marker detection as well: Thresholding, Labeling, Extract contours, Find four corners, and Get information and verification. The converted grayscale image is segmented (the black square is separated from the rest of the image) and a binary image created by thresholding technique. The principle of the thresholding technique is to separate the light and dark regions. Each pixel below a certain threshold is turned to zero, and each pixel above that threshold is transformed to one.

Labelling seeks groups of connected pixels and ultimately identifies the closed area in the image. Each pixel that satisfies the thresholding will obey the following algorithms: First, scan all pixels row by row and assign a preliminary label; Second, merge the equivalent regions that have different labels into one single label. Next is finding the contours of the objects for further determination of the curve, which stands for the solid border of a marker. Future extraction of the corner markers will be by this curve. Also, the edge detection algorithms can draw out the contour. Canny edge detection is a multi-stage edge detection algorithm with noise suppressed at the same time. The first stage is using a Gaussian filter to smooth the image for the noise and unwanted details and textures reduction. The process of smoothing the image I with a Gaussian filter G can be written as

$$G(x) = exp(-x^2/2\sigma^2)/2\pi\sigma^2 \tag{1}$$

$$I(x,y) = [G(x)G(y)] * f(x,y) \tag{2}$$

In the above equation, σ is the Gaussian filter spread, the smoothing degree is determined by it. The second stage is the calculation of the gradient direction and magnitude. A 2×2 neighboring area is often selected in the Canny edge detection algorithm to acquire the corresponding magnitude and direction gradient image. Through the following formulas, the first order partial derivatives in the directions of x and y can be gained:

$$M(x,y) = \sqrt{N_x^2(x,y) + N_y^2(x,y))} \tag{3}$$

$$D(x, y) = arctan[N_y(x,y)/N_x(x,y)] \tag{4}$$

$$N_x = (-S_1 + S_2 - S_3 + S_4)/2 \tag{5}$$

$$N_x = (S_1 + S_2 - S_3 - S_4)/2 \tag{6}$$

Here M(x, y) represents the gradient magnitude of the image, and D(x, y) stands for the gradient direction of the image, also the pixel values of the image $(x, y), (x, y + 1), (x + 1, y), (x + 1, y + 1)$ are stated by S_1, S_2, S_3, S_4 respectively. After gaining the gradient magnitude image M and gradient direction image D, we need to perform non-maximum with the goal to detect the edge and avoid the occurrence of false edges efficiently. A 3×3 adjacent area is chosen in the Canny algorithm to compare a pixel with its two adjacent pixels along the gradient direction. If the magnitude $M(i, j)$ is more significant than the two interpolation results on the gradient direction, a candidate edge point will be arranged to the pixel. Otherwise, a non-edge point will be assigned to it [13]. After applying non-maxima suppression, canny methods employ double thresholds, which consist of low and high thresholds to detect and connect edge points. If a pixel has a gradient magnitude that is bigger than the high threshold, an edge point will be assigned to the pixel, while a pixel with a smaller gradient will be assigned with a non-edge point. For those gradient magnitudes between high and low thresholds, if there is a point around the pixel more significant than the high threshold, then it is the edge point. Following these steps, the edge image is acquired.

After finding the contours of the marker, we need to find further the polygon approximations of the contours for the marker square using the polygon approximation algorithm [20]. This algorithm can reduce the number of points in a curve, which is approximated by a series of points. The functions of this algorithm are to find the distance dimension on the line for each point and conduct the simplified curve reconstruction. Finally, the points of the simplified square can be seen as the marker corners. After identifying the marker corner coordinates, which should be vertical, the 2D and 3D coordinates need to be mapped in the real world so that the 3D objects appear as if they are positioned on top of the marker. The camera is positioned in the origin of the coordinate system of the camera, looking along the Z-axis. Two axes of the coordinate marker system are parallel to the sides of the squared marker. To establish the mapping relationship between the marker coordinates and screen coordinates, we need to get transformation between them.

In the circumstance of coding, all these transformations are a matrix. Also, the transformation is represented by a matrix in linear algebra. Here, we can use homogeneous coordinates [10] to perform the matrix. The scientific name of matrix C is the camera internal reference matrix, and matrix T_m is referred to as the camera external reference matrix, where camera calibration obtains the internal reference matrix in advance, and the external reference matrix is unknown. Thus, we need to define it according to the screen coordinates (x_c, y_c) in advance according to the marker's coordinate system and internal parameter matrix. Then, based on T_m, we can draw the graphic on it. After marker detection and verification, it is possible to display the 3D model on the AR marker.

We want to keep track of the marker's orientation so that the 3D graphics can rotate consequently as the marker rotates from the camera. We will use the small inner square to orient. The detection steps are almost the same as the previous processes to find the large square. Once we finish detecting the solid outer border, next step is to detect the small square that designates the vertex of the large square closest to the small square as "first".

3.3   Blue Product Bar Code

Most products are printed with barcodes for the ease of detecting and calculating the final prices. The commonly used barcode formats in commodity products are EAN-8, EAN-13, and UPC. They are one-dimensional barcodes which can be readable by most Laser scanners used in retails. The same 1D barcode of a particular product can be used for this Blue Product Bar Code. Nowadays, 2D barcodes are more and more frequently used, such as QR code, due to its higher storing capacity and better error correction rates. QR code was introduced by Denso Wave in 1994 [6]. Since then, the QR code has obtained wide popularity in many diverse industries, like construction, marketing, healthcare, tourism, life sciences, and education. The reason for its popularity due to the data stored in QR code is of a higher density than the information in a barcode. Another reason is that it is convenient and straightforward to download and install a code detector onto a mobile, enabling data to be quickly retrieved. Furthermore, it has the advantage that either electronic or static media efficiently distribute data in QR code. Today, QR code can act as either an identifier or a database itself. Likewise, our RGB tag is also capable to store 2D QR code in its blue channel.

3.4   Green Channel Texture Image

The texture image is the photo of the product, which is stored in the marker's green channel so that users can have an essential first sight of what the 3D graphics will be. As stated, human's eyes are most sensitive to green colour. Also, contrast sensitivity of the human visual system plays another important part [3]. If the contrast of the image is high, this provides a better perception of human vision. Contrast refers to the separation of the bright and dark areas present in an image. The difference between the colour and brightness of the object and its background determines the image contrast. Contrast enhancement aims to enhance the visibility of the object by improving the difference between the brightness of the object and its background.

The most widely used technique for increasing the contrast of images is the grey-histogram method. This is a graph displaying the number at which each grey level occurs in an image, plotting the number of pixels for each tonal value [5]. The contrast of an image can be enhanced by changing its grey histogram: distributing the image value to cover a wide range on the graph. Here, we use a method called "Contrast Limited Adaptive Histogram Equalization (CLAHE)", which is optimised based on the adaptive histogram equalisation

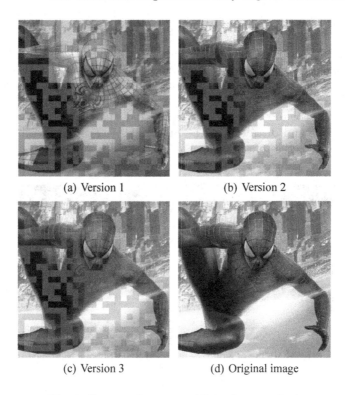

(a) Version 1   (b) Version 2

(c) Version 3   (d) Original image

Fig. 4. Generated new tags (Color figure online)

method (AHE). The difference between AHE and CLAHE is the contrast limit as the CLAHE introduced clipping limit to overcome the noise amplification problem [21]. CLAHE was firstly introduced to process the medical image and has proven to be an effective method for enhancing a low-contrast image. The principle behind this method is that the image is segmented into several non-overlapping regions with nearly equivalent sizes [15]. The steps in this approach are described below:

- First, calculate the histogram of each region.
- Then, based on the required restrictions for contrast expansion, acquire a limit of the clip for clipping histograms.
- Next, for that height, does not exceed the clip restriction, distribute each histogram in the same way again.
- Finally, with the aim of grayscale mapping, determine cumulative distribution functions (CDF) of the generated contrast limited histograms. In the CLAHE technique, pixel mapping can be achieved by integrating the outcomes resulting from the mappings of the four nearest regions linearly.

3.5   Combination of the Three Channels

After identifying the components (1) an oriented marker, (2) a bar-code and (3) a graphic image; we can store them in red, green and blue channels of the final image. Figure 4(a) display such a RGB tag (Version 1) created from the original picture of the Spiderman in Fig. 4(d). As we are not very sensitive to blue colour, the QR code is not too visible; however, the red square of the oriented marker is quite obvious and distracting. The Spiderman is relatively visible with low contrast due to the use of only one channel.

(a) RGB Tag of a ring   (b) RGB Tag of a Nike shoe

Fig. 5. Generated new tags

When we use QR Code as the embedded bar-code, it is known that the QR codes themselves include position detection patterns that could be used to identify the orientation for the tag. Thus, the oriented marker is needed anymore if QR code is used. Thus, one red channel is saved and we may use it for presenting the texture image, to increase the contrast. Figure 4(b) shows the effect of elimination of red orientation marker (Version 2). The contrast definitely increase and the figure of the Spiderman is relatively visible to most people; however, the figure is greyscaled. We slightly improve the look of it by retaining the original red and green colour channels of the Spiderman figure to achieve its version 3 (Fig. 4(c)). Comparing this tag with the original images in Fig. 4(d); we do not see much difference; most pictorial information still remains correctly and the viewer can very quickly recognise that it is a Spiderman figure. Figure 5 displays two other QR tags, Fig. 5(a) displays a ring model from The Lord of the Rings series; and Fig. 5(a) shows a Nike shoe.

4   Results and Performance

Figure 6 displays the 3D model of a Spiderman figure viewed from two angles. The figure is rendered on top of the RGB Tag mentioned above. We conduct a series of experiments to test the performance of this RGB Tag, on the detection

(a) Front Side   (b) Back side

Fig. 6. 3D Spiderman model results

and displaying rates. The steps are summarised as: (1) we choose a new tag; (2) we set the detection distance between the tag and the camera as 20 cm. At this distance, we capture each frame, and scan the marker to detect and decode stored information. If the captured information is correct, we add 1 to the counter. We repeat it until reaching 600 frames and calculate the ratio. (3) We repeat and set the distance to 30 cm and 40 cm. By the end, we record the correctly detected and displaying rates in percentages. Table 1 shows the results. It indicated that our new tag has the 95% detected and displaying rate on average.

Table 1. Our new AR tags detection rate (in percentages)

Models	20 cm away	30 cm away	40cm away
Plane	84.5%	98.5%	90.6%
Spongebob	86.1%	94.8%	89.6%
Astronaut	90.2%	93.1%	87.7%
Tractor	87.9%	96.8%	94.1%
Average rate	87.2%	95.3%	90.5%

5 Conclusions

The objective of this project is to enhance consumers' in-store shopping experience by using AR technology. It aims to boost sales and increase competitiveness among retailed shops. Our solution is to design a system that can display a 3D model of the product on top of the AR tag, which would allow consumers to see the 3D item in the real environment. We designed a Red-Blue-Green AR tag which combines three components. The red channel is to display the orientation marker. The blue channel is used to store a bar-code (or QR Code); and the green channel is used to show the product photo. When our proposed AR system is applied in the retail shopping industry, the physical store could save space and money and instead, consumers could use their phone cameras to scan the AR tag collections to access and view the products themselves. This process is simple and convenient for consumers, especially when the store is busy. Furthermore, our AR tag can attract consumers' attention and provide them with more functions than normal AR tags currently can.

References

1. Azuma, R., Baillot, Y., Behringer, R., Feiner, S., Julier, S., MacIntyre, B.: Recent advances in augmented reality. IEEE Comput. Graph. Appl. **21**(6), 34–47 (2001)
2. Bajura, M., Neumann, U.: Dynamic registration correction in video-based augmented reality systems. IEEE Comput. Graph. Appl. **15**(5), 52–60 (1995)
3. Barten, P.G.: Contrast Sensitivity of the Human Eye and Its Effects on Image Quality, vol. 21. SPIE Optical Engineering Press, Bellingham (1999)
4. Bimber, O., Wetzstein, G., Emmerling, A., Nitschke, C.: Enabling view-dependent stereoscopic projection in real environments. In: 2005 Proceedings of the Fourth IEEE and ACM International Symposium on Mixed and Augmented Reality, pp. 14–23. IEEE (2005)
5. Bora, D.J., Gupta, A.K.: AERASCIS: an efficient and robust approach for satellite color image segmentation. In: International Conference on Electrical Power and Energy Systems (ICEPES), pp. 549–556. IEEE (2016)
6. Kieseberg, P., et al.: QR code security. In: Proceedings of the 8th International Conference on Advances in Mobile Computing and Multimedia, pp. 430–435. ACM (2010)
7. Klinker, G., Reicher, R., Brugge, B.: Distributed user tracking concepts for augmented reality applications. In: 2000 Proceedings of the IEEE and ACM International Symposium on Augmented Reality, (ISAR 2000), pp. 37–44. IEEE (2000)
8. Leonard, J.J., Durrant-Whyte, H.F.: Simultaneous map building and localization for an autonomous mobile robot. In: Proceedings IROS 1991 IEEE/RSJ International Workshop on Intelligent Robots and Systems 1991. Intelligence for Mechanical Systems, pp. 1442–1447. IEEE (1991)
9. Lepetit, V., Fua, P., et al.: Monocular model-based 3D tracking of rigid objects: a survey. Found. Trends® Comput. Graph. Vis. **1**(1), 1–89 (2005)
10. Li, H., Hestenes, D., Rockwood, A.: Generalized homogeneous coordinates for computational geometry. In: Sommer, G. (ed.) Geometric Computing with Clifford Algebras, pp. 27–59. Springer, Heidelberg (2001). https://doi.org/10.1007/978-3-662-04621-0_2
11. Lima, J.P., Simões, F., Figueiredo, L., Kelner, J.: Model based markerless 3D tracking applied to augmented reality. J. 3D Interact. Syst. **1**, 2–15 (2010)
12. Moye, L.N., Kincade, D.H.: Influence of usage situations and consumer shopping orientations on the importance of the retail store environment. Int. Rev. Retail Distrib. Consum. Res. **12**(1), 59–79 (2002)
13. Neubeck, A., Van Gool, L.: Efficient non-maximum suppression. In: 2006 18th International Conference on Pattern Recognition, ICPR 2006, vol. 3, pp. 850–855. IEEE (2006)

14. Rollmann, W.: Zwei neue stereoskopische Methoden (Two new stereoscopic methods). Annalen der Physik (Ann. Phys.) **166**(9), 186–187 (1853)
15. Sasi, N.M., Jayasree, V.: Contrast limited adaptive histogram equalization for qualitative enhancement of myocardial perfusion images. Engineering **5**(10), 326 (2013)
16. Satoh, K., Uchiyama, S., Yamamoto, H., Tamura, H.: Robust vision-based registration utilizing bird's-eye view with user's view. In: Proceedings of the 2nd IEEE/ACM International Symposium on Mixed and Augmented Reality, p. 46. IEEE Computer Society (2003)
17. Sutherland, I.E.: A head-mounted three dimensional display. In: Proceedings of the Fall Joint Computer Conference, part I, 9–11 December 1968, pp. 757–764. ACM (1968)
18. Tuceryan, M., Genc, Y., Nava, N.: Single-point active alignment method (SPAAM) for optical see-through HMD calibration for augmented reality. Presence: Teleoperators Virtual Environ. **11**(3), 259–276 (2002)
19. Wagner, D., Barakonyi, I.: Augmented reality kanji learning. In: Proceedings of the 2nd IEEE/ACM International Symposium on Mixed and Augmented Reality, p. 335. IEEE Computer Society (2003)
20. Wikipedia: Ramer-Douglas-Peucker algorithm, Internet (2017). https://en.wikipedia.org/wiki/Ramer%E2%80%93Douglas%E2%80%93Peucker_algorithm. Accessed 10 May 2018
21. Zuiderveld, K.: Contrast limited adaptive histogram equalization. In: Graphics Gems, pp. 474–485 (1994)

Guided Stereo to Improve Depth Resolution of a Small Baseline Stereo Camera Using an Image Sequence

Trevor Gee[1]($\boxtimes$), Georgy Gimel'farb[1], Alexander Woodward[4], Rachel Ababou[2], Alfonso Gastelum Strozzi[3], and Patrice Delmas[1]

[1] Intelligent Vision Systems Lab, The University of Auckland, Auckland, New Zealand
tgee862@aucklanduni.ac.nz, p.delmas@auckland.ac.nz
[2] Centre de Recherches des Écoles de Coëtquidan Saint-Cyr, 56381 Guer, France
[3] Institute of Applied Science and Technology, Mexico City, Mexico
alfonso.gastelum@icat.unam.mx
[4] RIKEN Center for Brain Science, 2-1 Hirosawa, Wako, Saitoma, Japan
alexander.woodward@riken.jp
http://www.icat.unam.mx/secciones/depar/sub5/unhgm.html

Abstract. Using calibrated synchronised stereo cameras significantly simplifies multi-image 3D reconstruction. This is because they produce point clouds for each frame pair, which reduces multi-image 3D reconstruction to a relatively simple process of pose estimation followed by point cloud merging. There are several synchronized stereo cameras available on the market for this purpose, however a key problem is that they often come as fixed baseline units. This is a problem since the baseline that determines the range and resolution of the acquired 3D. This work deals with the fairly common scenario of trying to acquire a 3D reconstruction from a sequence of images, when the baseline of our camera is too small. Given such a sequence, in many cases it is possible to match each image with another in the sequence that has a more appropriate baseline. However is there still value in having calibrated stereo pairs then? Clearly not using the calibrated stereo pairs reduces the problem to a monocular 3D reconstruction problem, which is more complex with known issues such as scale ambiguity. This work attempts to solve the problem by proposing a guided stereo strategy that refines the coarse depth estimates from calibrated narrow stereo pairs with frames that are further away. Our experimental results are promising, since they show that this problem is solvable provided there are appropriate frames in the sequence to supplement the depth estimates from the original narrow stereo pairs.

Keywords: 3D reconstruction · Binocular stereo · Guided stereo

© Springer Nature Switzerland AG 2020
J. Blanc-Talon et al. (Eds.): ACIVS 2020, LNCS 12002, pp. 480–491, 2020.
https://doi.org/10.1007/978-3-030-40605-9_41

1 Introduction

This work investigates the idea of using guided stereo to improve the resolution of depth measurements from narrow baseline calibrated stereo image pairs, using a selected frame from further away in the image sequence.

Currently there is an increasing need for digital geometric models of real world objects to match the growing innovation in fields such as machine learning, virtual reality and augmented reality. In response to this, there are several different types of sensors on the market that are capable of measuring point clouds. These sensors can roughly be divided into those that interrogate a scene by projecting some sort of emission (called active depth sensors), and those that attempt to infer this information from image data using image processing (called passive depth sensors). While active depth sensors are really effective in some situations, it is still the passive depth sensors that are the most flexible, able to operate in indoor and outdoor environments with a wide variety of different scene sizes. This work focuses on improving the applicability of the many commercial stereo camera systems that are sold with fixed baselines.

The length of the baseline of a stereo camera system directly determines the depth resolution that is measurable by the system. The relationship between depth resolution ΔZ and depth Z, focal length f and the baseline b of a stereo camera system camera is well known and can be expressed as the following equation:

$$\Delta Z = \frac{Z^2}{fb + Z} \tag{1}$$

While ideally it would be great for stereo camera systems to have flexible baselines, this significantly complicates the casing design and makes it more likely that a calibration measurement will be invalidated due to motion between the cameras. Therefore the general trend has been to have fixed baseline cameras. This baseline also tends to be quite narrow since this allows for a more compact product and narrow baseline stereo is easier than wide baseline stereo [16]. See Table 1 for a list of commercially available stereo cameras and their respective fixed baselines.

Table 1. A list of commercially available fixed baseline stereo cameras with their baseline sizes.

Fixed baseline camera	Baseline
FujiFilm FinePix Real W3	75 mm
Minoru 3D Webcam	60 mm
GoPro Hero 3+ Stereo Case	33 mm
Realsenese D435	50 mm
ZED Stereo Camera	120 mm
DUO MC	30 mm

Fig. 1. Two histogram-equalized depth maps of a chair with a wall in the background. The left image shows the original disparity map that was captured using a stereo camera with a narrow baseline. Note that the depth values appear layered. The right image shows the same depth map after applying the proposed technique. Its appearance is much more smooth.

Naturally the main issue with a fixed baseline stereo camera is that it is not possible to customize the baseline of the camera with respect to the scene that is being captured. This means that since manufacturers favour small baselines, that the most common case is that the baseline of the camera is too small for the scene being captured, resulting in a disparity map with large jumps between levels as can be seen on the left image from Fig. 1.

While there is not much that one can do to fix this problem in the case of a single binocular pair, typical stereo cameras are used in conjunction with 3D reconstruction of scenes from sequences of images. In this case it is often possible to have access to overlapping images with a wider baseline, providing the potential to improve upon the initial coarse estimates. The problem is that if we throw the original coarse depth estimates away, then we are no longer benefiting from the fact that we have pairs of calibrated cameras.

The solution investigated in this work is an attempt at a compromise between keeping the original coarse values or using new values between two frames within the sequence that are wider apart. The idea is to use the original values as a basis to guide the search for better values using another frame with a bigger baseline. The advantages to doing this is that the search space for the new values can be made smaller resulting in a quicker search, and that the original perspective can be retained meaning that the sequence of images may still be treated as a sequence of stereo images after the depth refinement approach is complete.

In Sect. 2 the latest relevant literature is discussed. Then the proposed methodology is explained in Sect. 3, followed by how it was tested in Sect. 4, the results in Sect. 5. The work concludes with a discussion of results in Sect. 6.

2 Literature Review

The cornerstone of extracting depth information from a pair of rectified images is the stereo matching algorithm. Stereo matching is an old problem in computer vision, with research dating back to the 1970s [8]. Today it is solved in some contexts, but it is still an active research topic in the general case, with many modern researchers using the latest machine learning techniques [18] in a bid to solve the problem.

In the last 50 years, a vast amount of stereo matching algorithms have been developed. Algorithms like block matching stereo [19] have stuck because they are fast, though not very accurate. More accurate algorithms have stuck too such as Semi-global block matching [10] because they are a reasonable compromise between speed and accuracy and are freely available in popular libraries such as OpenCV [3]. However it is typically the case that one algorithm is more effective in one context, and another algorithm is more effective in another context. This lead to the notion of "guided stereo", the idea that the output from one stereo matching system could be used to influence the output of another.

This idea was used by Nguyen et al. [14] to harness the speed of block matching stereo in combination with the global optimization of Symmetric Dynamic Stereo [7]. It was also used by Chan et al. [4] to merge the benefits of active depth sensing with passive depth sensing. A variant of Guided Stereo is used in the ELAS [6] stereo matching system, that constructs a coarse disparity map using Delaunay triangulation a on sparse set of "reliable" points to guide a more constrained search for accurate correspondences. This work attempts to do something similar in the context of improving a coarse set of disparities, this time using them to guide a search for more accurate results.

There were no actual published work found attempting to solve a problem similar to the one being presented in this publication. Articles on 3D reconstruction from sequences of images tend to be heavily dominated by monocular solutions.

3   Proposed Approach

Given a sequence Φ of n stereo pairs ρ_i indexed by i where $\{i \in \mathbb{Z} \mid 0 \leq i < n\}$. A stereo pair consists of a left image l_i and a right image r_i as $\rho_i = \{l_i, r_i\}$. It is assumed that the entire sequence is captured with the same synchronized stereo system that has been calibrated.

In the case of 3D reconstruction it is useful to convert this sequence of stereo pairs Φ into a sequence of depth frames $\lambda = \{\eta_i\}$ whose elements η_i consist of a color image c_i and a depth image d_i as $\eta_i = \{c_i, d_i\}$. The procedure to achieve this is as follows:

- Calibration parameters are used to remove distortion from the initial stereo pairs ρ_i to produce a distortion free stereo pairs $\rho_{u:i} = \{l_{u:i}, r_{u:i}\}$.
- The distortion free stereo pairs $\rho_{u:i}$ are then warped so that canonical epipolar geometry may be applied to them. This warping process is known as rectification and the approach of Fusiello et al. [5] is an example. The output of this step is a set of distortion free rectified stereo pairs as $\rho_{ur:i} = \{l_{ur:i}, r_{ur:i}\}$. The process of rectification also produces a new set of calibration parameters to

reflect the new alignment of the images. In this work it is presumed that the resultant calibration parameters contain values for focal length f, the optical center $[c_x, c_y]^T$ and the baseline b.

- Each distortion free rectified stereo pair $\rho_{ur:i}$ can be converted into a disparity map Δ_i using an operation known as stereo matching. A stereo matching algorithm finds a dense set of correspondences across left and right images and records these correspondences as a parallax shift between points known as a disparity. An example of a stereo matching algorithm is the Symmetric Dynamic Programming Stereo (SDPS) algorithm proposed by Gimel'farb [7].
- Each disparity map can then be δ_i converted into a depth map d_i. This is straight forward because disparity δ_i is related to depth Z as $Z = fb/\delta$. Note that the focal length f and baseline b are determined during rectification, which in turn gets these values from the original calibration of the camera.
- Finally the depth frame η_i can be formed by renaming the distortion free rectified left image $l_{ur:i}$ to the color image c_i and forming the pair $\eta_i = \{c_i, d_i\}$ with the corresponding depth map d_i calculated from the disparity map Δ_i in the previous step.

Of course at this point, in the case of the narrow baseline (as described in the introduction), d_i within the depth frame η_i is very coarse. The proposed procedure to fix this begins by selecting a suitable match depth frame η_j (Sect. 3.1) for each η_i, then estimating the fundamental matrix F between η_i and η_j (Sect. 3.2), and finally perform guided stereo search between the frames to improve the depth estimates of d_i (Sect. 3.3).

3.1   Find Matching Depth Frame

In this section the goal is, given a depth frame η_i, to find another depth frame η_j with an ideal baseline length b^* to our frame (or as close to it as possible), along with a significant region of overlap between the two frames.

To proceed, the ideal baseline length b^* needs to be decided upon. Choosing the appropriate baseline length is application specific, and assumes that the user knows (perhaps roughly) what depth resolution they require, along with the maximum distance Z_{max} relevant from the camera. While some care is needed in choosing reasonable numbers (the larger the ideal baseline, the smaller the overlap at closer distances), the ideal baseline length b^* can be calculated from a version of Eq. 1 as follows:

$$b^* = \frac{Z_{max}^2 - (\Delta Z)Z_{max}}{f\Delta Z} \tag{2}$$

The next step is to decide which frames η_j are suitable candidate matches. Since it is assumed that both depth frame η_i and matching depth frame η_j come from the same sequence λ, and that the sequence numbers i represent time steps, such that depth frame η_i is expected to be captured 1 time step before depth frame η_{i+1}, it can then be assumed that depth frames η_i with indices close to i

have the most chance of overlapping with η_i. Therefore we only consider depth frames in the range $i - m$ to $i + m$ to be potential good candidates.

From a system user point of view, m is a free parameter that needs to be provided. A good strategy for finding the appropriate value for m can be as follows: If the time step between i and $i + 1$ is on average Δt and the average velocity of the camera is v, then given the ideal baseline length b^*, it is possible to approximate m as $m = b^*/(\Delta t v)$.

After establishing b^* and m, the next step is to assign a baseline length to each candidate frame. This is done using a simple pose estimation based on the coarse depth values.

Given depth frame η_i and potential candidate depth frame η_j, the first step is to extract Fast Features [15] p_k from the color frame c_i found in η_i. These features are then matched to the color frame c_j using Lucas Kanade optical flow with an image pyramid [2]. Each feature p_k in c_i can be converted into a corresponding 3D point using the depth map d_i with the help of the calibration parameters as follows:

$$
\begin{bmatrix} X_k \\ Y_k \\ Z_k \end{bmatrix} = \begin{bmatrix} (x_k - c_x) \times \frac{Z_k}{f} \\ (y_i - c_y) \times \frac{Z_k}{f} \\ Z_k \end{bmatrix} \tag{3}
$$

where $P_k = [X_k\ Y_k,\ Z_k]^\top$ is the coarse 3D location of the feature in the scene, $p_k = [x_k, y_k]$ is the corresponding image coordinate, Z_k is the corresponding depth value from the depth map d_i, $[c_x, c_y]^\top$ is the optical center from the calibration and f is the focal length from the calibration. This essentially sets up a Perspective-n-Point (PnP) problem which can be solved using an algorithm such as Upnp [11]. The baseline is found as translation between depth frame η_i and candidate frame η_j. The frame with the baseline closest to ideal is the one selected. If there is a tie, then the rotation matrix closest to the *identity* matrix resolves the tie.

3.2 Estimate the Fundamental Matrix Between Frames

The previous Sect. 3.1 outlined an approach to finding a best match depth frame η_*. Unfortunately, since the pose estimate was based on coarse depth measurements, it is not good enough for good quality stereo matching. To solve this, the fundamental matrix [13] F is used. The fundamental matrix F defines the epipolar constraint between two images captured with overlap. The epipolar constraint reduces the search space for a corresponding point, across two images, to the epipolar line, which is the projection of the ray (that produced the point in the first image) on the second image. Practically, the fundamental matrix F can be found using the 8-point algorithm [9] in conjunction with RANSAC [1]. The corresponding points used to determine the fundamental matrix can be the same ones used in the previous section.

3.3   The Guided Stereo Algorithm

The depth frame of interest η_i, a candidate depth matching frame η_*, and the fundamental matrix F, are all that is required for the proposed guided stereo depth refinement. As depth frame η_* could be quite far from ideal epipolar geometry, the proposed algorithm searches for new depth values directly along the epipolar lines calculated from the fundamental matrix, rather than using rectification.

Thus, given a pixel, it is a useful start to determine the appropriate search space for stereo matching. Given homogeneous point $\widetilde{u}_l = [x_l, y_l, 1]^\top$ in η_i (the location applies to both the color image c_i and the depth image d_i) it is possible to find this search space as the epipolar line $F\widetilde{u}_l = L_l$ where F is the fundamental matrix and $L_1 = [a, b, c]$ contains the coefficients for the line $ax + by + c = 0$.

Next, it is necessary to determine where to start and stop the search on the line. While we could search the entire line within the image, this is extremely inefficient. It is more efficient to follow a guided search approach, which uses the depth Z_l found in depth map d_i associated with point u_l. Given depth Z_l, Eq. 1 determines the depth resolution ΔZ_l. Using this, we can determine that the search needs to be in the range $Z_l - \Delta Z_l$ to $Z_l + \Delta Z_l$ in terms of depth coordinates. However, it is necessary to find these locations as points on the line $ax + by + c = 0$. To do this, we make use of the well known fact that every epipolar line must contain the epipole. The epipole $\widetilde{e} = [x_0, y_0, 1]$ for image c_* can be found by finding the *nullspace* of the fundamental matrix transposed $F^\top$. The epipolar $\widetilde{e}$ represents the location of zero depth within the image. In order to find an appropriate depth for a given point, it is necessary to travel down the line $ax + by + c = 0$ from the epipole $\widetilde{e}$ for a distance of $\delta_l = fb/Z_l$ (distances along the line are in disparity space). To make this more straightforward we reformulate the line equation $ax + by + c = 0$ as follows:

$$ s_l = \begin{bmatrix} x_0 \\ y_0 \end{bmatrix} + t\frac{1}{\sqrt{a^2 + b^2}} \begin{bmatrix} b \\ -a \end{bmatrix} \tag{4} $$

where s_l is the search location in color image c_* and t is the distance that ranges from $fb/(Z_l - \Delta Z_l)$ to $fb/(Z_l + \Delta Z_l)$.

Now that the search space is defined, performing the actual search is a straightforward traversal of the line from the established beginning to the established end, testing each pixel traversed for an improvement on previous observations. Testing is performed using a Sum of Squared Differences (SSD) Block Matching of greyscale values between images c_i and c_*. The matches with the best score are chosen for the final result.

4   Experimental Methodology

The research presented in this article came about in response to a practical problem. An investigation was being conducted focused on using photogrammetry to estimate the sizes of cabbages and lettuces in an outdoor environment. Image sequences were collected using a GoPro stereo camera system with a fixed 33 mm baseline. This baseline proved to be too small to be of use. The images were discarded from the experiment, however it was decided to see if the ideas presented in this work could be used to extract useful information from these images. It is for this reason that the main strategy for evaluating the presented approach is through the comparison of cabbage 3D reconstruction to actual ground truth values.

4.1 Equipment

Fig. 2. The camera on the left is the Einscanner-Pro+ structured lighting camera with a sub-millimeter RMSE 3D reconstruction accuracy. The cameras on the right are two GoPro Hero 3+ stereo cameras in the provided casing having a baseline of 33 mm.

The main camera system used in the experiment was two GoPro Hero 3+ Black edition in a stereo casing. These cameras have a focal length of about 3 mm and a stereo casing with a baseline of about 33 mm. These cameras come with a physical synchronization cable and in our experiments we captured images in a video at 30 frames per second with the medium field of view setting with an image size of 1920 × 1080 pixels. An image of these cameras is shown in Fig. 2 on the left.

Ground truth models were acquired using the Einscan Pro+ structured lighting camera system. This system features a 100 line scanning system that scans at 550,000 points per second at an Root Mean Squared Error (RMSE) of about 0.3 mm.

4.2 Experiment Description

The experiment took place in an outdoor scene where cabbages and lettuces were placed on the ground and filmed with a GoPro camera. The scene is shown in Fig. 3.

The goal of the experiment was to determine how accurately the cabbage volume could be determined from a sequence of images.

Prior to filming the cabbage and lettuces, ground truth measurements were made using an Einscanner Pro+ which has a sub-millimeter accuracy (substantially smaller than the millimeter accuracy that we hoping to get).

Fig. 3. The scene of our experiments. A row of cabbages and lettuces were placed in an outdoor scene to be measured by a stereo camera system.

Size was quantified by taking the scanned cabbage (or lettuce), segmenting it from the background (using depth and color), fitting a bounding box and then an ellipsoid to the data and then calculating the volume of this ellipsoid which is $4/3 \times \pi \times a \times b \times c$ where a, b and c are axis radii.

In order to apply the ellipsoid technique to the GoPro Sequences, these sequences were converted into 3D models using a simple stereo SLAM technique. This involves initially converting the frames into depth frames as described at the start of Sect. 3, then using the pose estimation as described in Sect. 3.2, before using SSBA bundle adjustment [12] and a moving volume [17] for point cloud merging.

5   Results

An example of a scanned cabbage is shown in the Fig. 7.

The initial sequences from the GoPro had very bad quality depth maps as can be seen from Fig. 4. This lead to them being discarded initially from the experiment. However once the proposed system was developed, improved depth estimates could be constructed from the data as can be seen in Fig. 5.

The ability to generate good quality depth frames from the stereo frames allowed for good construction of 3D models of the data as shown in the Fig. 6.

The Table 2 shows the comparison of the measurements. Acquiring measurements from the original coarse data was hugely difficult as depth was no help in segmenting the data. The results for these measurements tended to be enormous compared to the actual ground truth value. However once the proposed technique was applied, the results were much closer to the ground truth result. Clearly there is still room for improvement in our volume estimation approach, however the focus of this paper was depth improvement, and this has clearly been achieved.

Fig. 4. An example of the left frame and its corresponding depth map as per narrow baseline GoPro stereo camera being used. The depth map has a heavily layered appearance indicating the coarse resolution.

Fig. 5. An example of the left frame and its corresponding depth map after the technique proposed in this work had been applied. While there is a little of the layer artifact remaining, it is far more smoother and more accurate.

Fig. 6. One of the 3D reconstructions created from the GoPro data using the better quality depth maps. The image on the left is an image of the model and the image on the right is a depth map of the model indicating the smooth nature of the blended depth maps.

Fig. 7. An image of the 3D model generated of a cabbage that was scanned using the Einscanner-Pro+.

Table 2. A table showing the measurement comparison for the cabbages and lettuces based on bounding ellipsoid volume. The first column shows the measurements found using the Einscanner-Pro+. The original column shows the approximate values from the coarse data. The final column shows the improved measures once the coarse data has been refined using the proposed approach

No.	Einscanner	Original	Proposed
1	32 157 cm^3	2 148 087 cm^3	32 237 cm^3
2	111 743 cm^3	3 162 327 cm^3	77 880 cm^3
3	115 425 cm^3	2 712 487 cm^3	81 484 cm^3
4	82 730 cm^3	4 715 610 cm^3	67 155 cm^3
5	114 557 cm^3	7 480 572 cm^3	115 980 cm^3
6	24 098 cm^3	22 096 949 cm^3	20 131 cm^3

6 Conclusions

This work focused on the problem of attempting to acquire good quality depth measurements from a fixed baseline stereo camera, when the baseline is too small, however an entire sequence of these frames were available. The described approach selected a candidate frame and performed additional "guided stereo" matching to improve the depth estimates. In the experimental section it was shown that the proposed technique was effective in improving measurements during a photogrammetry experiment to determine the volume of cabbages and lettuce, an important parameter for optimal harvesting strategies.

The research presented in this article is essentially very new and little experimentation was done to determine the precise characteristics of the approach such as when the approach is effective and when the approach fails and what are that limitations in accuracy. This type of experimentation is still active and on going in our lab, however from the perspective of this paper, preliminary results are very encouraging.

References

1. Bolles, R.C., Fischler, M.A.: A RANSAC-based approach to model fitting and its application to finding cylinders in range data. IJCAI **1981**, 637–643 (1981)

2. Bouguet, J.Y., et al.: Pyramidal implementation of the affine lucas kanade feature tracker description of the algorithm. Intel Corporation **5**(1–10), 4 (2001)
3. Bradski, G., Kaehler, A.: Learning OpenCV: Computer Vision with the OpenCV Library. O'Reilly Media, Inc., Sebastopol (2008)
4. Chan, Y.H., Delmas, P., Gimel'farb, G., Valkenburg, R.: Accurate 3D modelling by fusion of potentially reliable active range and passive stereo data. In: Jiang, X., Petkov, N. (eds.) CAIP 2009. LNCS, vol. 5702, pp. 848–855. Springer, Heidelberg (2009). https://doi.org/10.1007/978-3-642-03767-2_103
5. Fusiello, A., Trucco, E., Verri, A.: A compact algorithm for rectification of stereo pairs. Mach. Vis. Appl. **12**(1), 16–22 (2000)
6. Geiger, A., Roser, M., Urtasun, R.: Efficient large-scale stereo matching. In: Kimmel, R., Klette, R., Sugimoto, A. (eds.) ACCV 2010. LNCS, vol. 6492, pp. 25–38. Springer, Heidelberg (2011). https://doi.org/10.1007/978-3-642-19315-6_3
7. Gimel'farb, G.: Probabilistic regularisation and symmetry in binocular dynamic programming stereo. Pattern Recogn. Lett. **23**(4), 431–442 (2002)
8. Hannah, M.J.: Computer matching of areas in stereo images. Technical report, Stanford Univ CA Department of Computer Science (1974)
9. Hartley, R.I.: In defense of the eight-point algorithm. IEEE Trans. Pattern Anal. Mach. Intell. **19**(6), 580–593 (1997)
10. Hirschmuller, H.: Accurate and efficient stereo processing by semi-global matching and mutual information. In: 2005 IEEE Computer Society Conference on Computer Vision and Pattern Recognition (CVPR 2005), vol. 2, pp. 807–814. IEEE (2005)
11. Kneip, L., Li, H., Seo, Y.: UPnP: an optimal $O(n)$ solution to the absolute pose problem with universal applicability. In: Fleet, D., Pajdla, T., Schiele, B., Tuytelaars, T. (eds.) ECCV 2014. LNCS, vol. 8689, pp. 127–142. Springer, Cham (2014). https://doi.org/10.1007/978-3-319-10590-1_9
12. Konolige, K., Garage, W.: Sparse sparse bundle adjustment. In: BMVC, vol. 10, p. 102–1. Citeseer (2010)
13. Luong, Q.T., Faugeras, O.D.: The fundamental matrix: theory, algorithms, and stability analysis. Int. J. Comput. Vis. **17**(1), 43–75 (1996)
14. Nguyen, M., Chan, Y.H., Delmas, P., Gimel'farb, G.: Symmetric dynamic programming stereo using block matching guidance. In: 2013 28th International Conference on Image and Vision Computing New Zealand (IVCNZ 2013), pp. 88–93. IEEE (2013)
15. Rosten, E., Drummond, T.: Machine learning for high-speed corner detection. In: Leonardis, A., Bischof, H., Pinz, A. (eds.) ECCV 2006. LNCS, vol. 3951, pp. 430–443. Springer, Heidelberg (2006). https://doi.org/10.1007/11744023_34
16. Roszkowski, M.: Overview of the major challenges in the wide baseline stereo vision. In: Photonics Applications in Astronomy, Communications, Industry, and High-Energy Physics Experiments 2011, vol. 8008, p. 800818. International Society for Optics and Photonics (2011)
17. Roth, H., Vona, M.: Moving volume kinectfusion. In: BMVC, vol. 20, pp. 1–11 (2012)
18. Taniai, T., Matsushita, Y., Sato, Y., Naemura, T.: Continuous 3D label stereo matching using local expansion moves. IEEE Trans. Pattern Anal. Mach. Intell. **40**(11), 2725–2739 (2017)
19. Tao, T., Koo, J.C., Choi, H.R.: A fast block matching algorithm for stereo correspondence. In: 2008 IEEE Conference on Cybernetics and Intelligent Systems, pp. 38–41. IEEE (2008)

SuperNCN: Neighbourhood Consensus Network for Robust Outdoor Scenes Matching

Grzegorz Kurzejamski[1] , Jacek Komorowski[1(✉)] , Lukasz Dabala[1] ,
Konrad Czarnota[1], Simon Lynen[2] , and Tomasz Trzcinski[1]

[1] Institute of Computer Science, Warsaw University of Technology, Warsaw, Poland
jacek.komorowski@gmail.com
[2] Google, Zurich, Switzerland

Abstract. In this paper, we present a framework for computing dense keypoint correspondences between images under strong scene appearance changes. Traditional methods, based on nearest neighbour search in the feature descriptor space, perform poorly when environmental conditions vary, e.g. when images are taken at different times of the day or seasons. Our method improves finding keypoint correspondences in such difficult conditions. First, we use Neighbourhood Consensus Networks to build spatially consistent matching grid between two images at a coarse scale. Then, we apply Superpoint-like corner detector to achieve pixel-level accuracy. Both parts use features learned with domain adaptation to increase robustness against strong scene appearance variations. The framework has been tested on a RobotCar Seasons dataset, proving large improvement on pose estimation task under challenging environmental conditions.

Keywords: Feature matching · Pose estimation · Domain adaptation

1 Introduction

Finding correspondence between keypoints detected in images is a key step in many computer vision tasks, such as panorama stitching, visual localization or camera pose estimation. Traditional approaches rely on detecting keypoints in images, computing their descriptors and nearest neighbour search in the feature descriptor space. Many works aim at making keypoint detection and description methods robust to changes of viewing conditions. [1,9] investigated keypoint matching capabilities under global affine geometric transformations and image degradation by modelled noise. In practical applications, scene appearance can change significantly due to different atmospheric conditions, seasons or time of the day. An example of image degradation caused by these factors is shown in Fig. 3. Scene appearance changes caused by these factors are complex, non-linear and difficult to model analytically.

J. Blanc-Talon et al. (Eds.): ACIVS 2020, LNCS 12002, pp. 492–503, 2020.
https://doi.org/10.1007/978-3-030-40605-9_42

Existing camera pose estimation approaches are commonly based on finding correspondences between local features detected in images. First, putative matches are found by using nearest neighbour search in the feature descriptor space. Then, matches are filtered using a robust model parameter estimation method, such as RANSAC [5], with an appropriate geometric consistency criteria, such as epipolar constraint.

In this paper, we propose a feature matching method robust to significant changes of scene appearance. The method combines correspondence estimation at a coarse scale, based on Neighbourhood Consensus Networks (NCNet) [15], with pixel-level accuracy, precise keypoint detection. By using domain adaptation technique [6], the proposed algorithm is robust to scene appearance changes caused by time of the day or seasonal differences. Unlike traditional feature matching approaches, our method imposes semi-global spatial consistency of established correspondences and integrates matching and correspondence filtering steps. This boost the performance under extreme scene appearance variations. See Fig. 1 for sample results.

Matches produced by NCNet are dense, spatially consistent but are at a coarse scale and have a limited spatial resolution. Each location in a feature map processed by NCNet corresponds to the block of 16×16 pixels. To allow an accurate camera pose estimation, we further push the accuracy to pixel-level. This is achieved by detecting keypoints in original resolution images and matching keypoints in each pair of grid cells matched at a course scale by NCNet. Architecture of our keypoint detector is similar to work of DeTone *et al.* [4]. The resultant correspondences are spatially consistent at a course scale and have pixel-level accuracy required for precise camera pose estimation.

We evaluate our method on RobotCar Seasons dataset using different traversals in different weather and daytime conditions and compare its performance against state of the art keypoint matching techniques. Our method shows significantly better results under heavy scene appearance changes than traditional techniques based on feature matching using nearest neighbour search.

2   Related Work

From the very beginning of SLAM [8], common methods for data association are based on nearest neighbour search in a feature descriptor space. The traditional approach is to use hand-crafted feature detectors and descriptors, such as SIFT [9], ORB [16] or AKAZE [1]. Recently, neural-network based feature detectors and descriptors, such as Superpoint [4], gain popularity. However, matching sparse features is challenging under strong viewpoint or illumination changes due to large scene appearance changes.

Through the years, the data association step was a bit overlooked in research, which focused more on the improvement of feature detection and descriptors. There are only few works proposing an efficient feature matching method alternative to nearest neighbour search. Recent advances include local neighbourhood consensus based algorithms [2] to add robust filtering stage after nearest neighbours step. In [19] authors propose a discriminative descriptor that includes

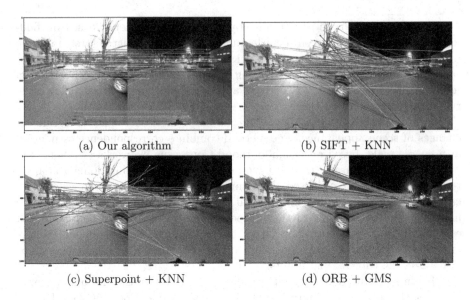

(a) Our algorithm

(b) SIFT + KNN

(c) Superpoint + KNN

(d) ORB + GMS

Fig. 1. Results of different feature descriptor matching strategies. Our method yields consistent correspondences despite strong scene appearance change. For brevity, We show only a random subset of all three thousand correspondences found by our method.

not only local feature representation, but also information about the geometric layout of neighbouring keypoints. There's a number of works related to optical flow and using deep neural nets to estimate the optical flow between a pair of images [13,20]. The optical flow methods give good results for light geometric transformations, but they are not suited for outdoor images, taken at various times of a day and from significantly different viewpoints. Interesting idea of determining dense correspondences between images with strong viewpoint, using hierarchical correlation architecture, is introduced in [14]. Recently, [15] proposes Neighbourhood Consensus Networks (NCNet), a robust method for establishing dense correspondences between two images. The method enforces semi-global spatial consistency between correspondences, and is robust to significant scene appearance changes. However its practical usage is limited to relatively low resolution images, due to usage of computationally-demanding 4D convolution.

One of the key challenges in computer vision is to devise features invariant to scene appearance changes. Large changes in illumination and viewpoint or changes due to seasonal or time-of-the-day variations have a strong impact on scene appearance. Feature descriptors of the same scene captured under different conditions can be significantly different. This adversely impacts performance of feature correspondence estimation process. To increase robustness against scene appearance changes due to seasonal or time-of-the-day differences, we use domain adaptation technique proposed in [6].

3   Network Architecture

High-level architecture of our solution is presented in Fig. 2. It consists of two parts: Domain Adaptation and Neighbourhood Consensus stage, trained independently from each other. The first part is based on Superpoint [4] keypoint detector and descriptor architecture. Encoder is a VGG-style feature extraction network trained with domain adaptation using adversarial loss [6]. It produces a domain-adapted convolutional feature map with spatial resolution decreased by 16 compared to the input image size. The feature map is fed into a Descriptor Decoder network computing dense feature descriptor map. The feature map produced by Encoder is forked into Keypoint Decoder. It uses sub-pixel convolution [18] (pixel shuffle) to create full resolution keypoint response map, indicating precise keypoint locations. Second part, Neighbourhood Consensus stage, is based on Neighbourhood Consensus Networks [15] architecture. It takes as an input two dense feature descriptor maps computed from input images. These maps have reduced spatial resolution, where each spatial location in the feature map corresponds to 16×16 pixel block in the input image. This allows efficient computation of dense correspondences between feature maps taking into account spatial consistency between all matches. Then, to achieve pixel-level accuracy, in each 16×16 pixels block we find four keypoints by taking argmax in the corresponding region of the keypoint response map. The keypoints are matched between corresponding blocks using nearest neighbour search in the descriptor space. For each spatial location in the feature map (corresponding to block of 16×16 pixels) we find four keypoint matches. This is depicted as Merge Data block in Fig. 2. The resultant correspondences are spatially consistent at a course scale and have pixel-level accuracy needed for camera pose estimation.

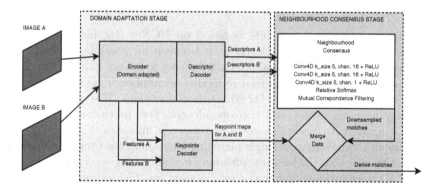

Fig. 2. Architecture overview of our solution.

3.1   Domain Adaptation Stage

The first stage of our solution is based on Superpoint [4] architecture. It uses convolutional neural network for detection and description of corner-like keypoints. It employs non-maximum suppression to enforce even distribution of detected keypoints across the image. In our preliminary experiments, we found that using grid-based detectors, with even distribution of keypoints, improves the results of camera pose estimation. Traditional keypoint detectors, such as SIFT or AKAZE, tend to cluster detected keypoints in some parts of the image, which has adverse effect on accuracy of the pose estimation task.

Encoder uses a VGG-style architecure to compute the feature map from the input image. It consists of convolutional layers, spatial downsampling via pooling and non-linear ReLU activation functions computing feature maps with descreased spatial resolution and increased number of channels. It uses four max-pooling layers with a stride two, each decreasing the spatial dimensional of the processed data by two. The output from the third max-pooling layer, with the spatial dimensions $W/8 \times H/8$ and 128 channels, where W, H is size of the input image, is forked to the Keypoint Decoder head. Keypoint Decoder consists of two convolutional layers with ReLU non-linearity and produces a 65-channel tensor with $W/8 \times H/8$ spatial resolution. The first 64 channels correspond to non-overlapping 8×8 pixel regions and the last channel corresponds to 'no keypoint' dustbin. Thus, each spatial location of the resulting $W/8 \times H/8$ feature map corresponds to 8×8 pixel region and values at different channels correspond to the probability of a keypoint location at each pixel of this region (with the last channel corresponding to the probability of no keypoints in the region). The output from the forth max-pooling layer, with the spatial dimensions $W/16 \times H/16$ and 256 channels, is fed to Descriptor Decoder head. Descriptor Decoder consists of two convolutional layers with ReLU non-linearity and produces a 256-channel descriptor map with $W/16 \times H/16$ spatial resolution. Each spatial location of the descriptor map corresponds to 16×16 pixel region of the input image.

3.2 Neighbourhood Consensus Stage

Neighbourhood Consensus module is based on NCNet [15] method. First it computes four dimensional similarity map between all possible pairs of locations (each being two dimensional point) in two input feature maps. Then, four-dimensional convolutions are applied to modify score of each 4D similarity map location depending on the score of neighbourhood locations. This enforces spatial consistency between established correspondences. Soft mutual neighbourhood filtering is applied, to boost the score of reciprocal matches. Finally, for each location in one feature map, a single matching location in the other feature map is selected by choosing the location with maximal similarity score. Due to application of 4D convolutions, matching score between two 2D locations is influenced by matching scores of neighbourhood locations. This results in more spatially consistent matches compared to the simple feature matching approach based on a nearest neighbour search in the descriptor space. Because our method operates on downscaled feature maps, with resolution reduced by 16 in each dimension, using four dimensional convolution on a Cartesian product of 2D locations in each feature map is computationally feasible.

4   Training

This section describes the training process for our solution. We use RobotCar-Seasons [17] dataset, described in Sect. 4.1. Domain-adapted Encoder, Descriptor Decoder and Keypoints Decoder Detector are trained independently from Neighbourhood Consensus stage. First is described in Sect. 4.2, latter in Sect. 4.3.

4.1   Database

RobotCar-Seasons [17] is a subset of Oxford Robotcar Dataset [10]. It contains sequences of images captured during multiple traversals through the same route in different weather or season conditions (dawn, dusk, night, night-rain, overcast-summer, overcast-winter, rain, snow, sun). Each traversal contains images from three cameras: left, right and rear. One sequence (overcast-reference), captured at favourable conditions, is marked as a reference traversal. The ground truth contains an absolute pose of each image with respect to the world coordinate frame and intrinsic parameters of each camera.

Images have a relatively low resolution and contain all sorts of artifacts present in images acquired by real-world outdoor camera systems. There are overexposured areas, significant blur, light flares and compression artifacts in many images. Exemplary images are shown in Fig. 3.

Fig. 3. Examples from RobotCar Seasons dataset. From the left: night, dusk and overcast samples. We can see blurring, overexposition and low texture content in various images.

4.2   Domain Adapted Feature Descriptor and Detector

The first stage of SuperNCN, Superpoint-based keypoint detector and descriptor, is trained using a Siamese-like [3] approach. In order to achieve domain-invariance, that is to make the resulting descriptors similar whenever the image of the observed scene is taken during the day or at night, the domain adaptation using an adversarial loss [7] technique is used. The architecture of one of the twin modules building the Siamese network used to train the keypoints descriptor is shown in Fig. 4. It includes an encoder, keypoints decoder and a descriptor decoder, which together form a typical feed-forward architecture. Domain adaptation is achieved by adding a domain classifier connected to the last layer of the encoder via a gradient reversal layer. Gradient reversal layer multiplies the gradient by a negative constant during the backpropagation-based training. Gradient reversal forces the feature map produced by the encoder to be domain agnostic (so the domain classifier cannot deduce if it's produced from an image taken during the day, at night, in summer or winter). This is intended to make the resultant descriptors domain-invariant, so descriptors of a same scene point captured at different conditions are as close to each other as possible.

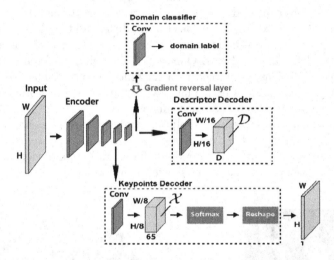

Fig. 4. Architecture of one twin module of the Siamese network used to train keypoint detector and domain invariant descriptor.

The Siamese network is trained by presenting pairs of similar (corresponding to the same scene point) and dissimilar (corresponding to different scene points) image patches. Patches are sampled from pairs of images: an image randomly chosen from one of the Oxford RobotCar traversals (dawn, dusk, night, overcast-summer, overcast-winter, rain, snow) and it's transformed version created by applying a random perspective warp and photometric distortion. Additionally, a numeric domain label (a traversal identifier) is given, identifying the source traversal from which the input image is sampled. During the training, the domain label and domain type inferred by the domain classifier are used to compute cross entropy loss. By adding the gradient reversal layer between the domain classifier and the encoder, the encoder is forced during the training to produce feature maps that are invariant to the image domain (so the domain classifier fails at inferring whether the input image is taken during the day or at night).

4.3   Neighbourhood Consensus

Important characteristic of Neighbourhood Consensus Network is that it can be trained with weakly supervised data. It only needs pairs of images labeled as similar and dissimilar. The ground truth information about corresponding locations in both images is not needed. Loss for training Neighbourhood Consensus Network is defined as:

$$L(I^A, I^B) = -y(mean(s^A) + mean(s^B)) \text{ , where}$$

$$s^A_{ijkl} = \frac{\exp(c_{ijkl})}{\sum_{ab} \exp(c_{abkl})}, \ s^B_{ijkl} = \frac{\exp(c_{ijkl})}{\sum_{cd} \exp(c_{cdkl})}, \text{ and}$$

$$y = \begin{cases} 1 & \text{if images depict the same location} \\ 0 & \text{if images depict different location} \end{cases}$$

I^A and I^B are images from the database with corresponding scaled down (a, b) and (c, d) dimensions. c is a 4D similarity score space computed by neighbourhood consensus of size (a, b, c, d). It is worth mentioning that a, b, c, and d dimensions are interpreted as sizes of scaled down feature space in real implementation. Computing a full resolution 4D hypercube is not feasible with current hardware (memory-wise and computational-wise).

For training, we sample pairs of similar images by taking images from the same traversal with relative translation and rotation below 5 m and 30° thresholds respectively. We chose pairs of dissimilar images by taking images with relative translation or rotation above these thresholds. We used all cameras: right, left and rear for sampling training data. Training is performed on 4 RTX 2080Ti GPUs for 5 epochs and using images downsampled to 512×512 pixel resolution for efficiency. We found that the choice of downsampled resolution has a limited impact on the learning process significantly. Original NCNet produces good results for 400×400 pixel input images.

5   Experimental Setup

We compare performance of our solution with classical and state-of-the-art image matching methods: SIFT+KNN (SIFT [9] features and matching individual features using nearest neighbour search in the descriptor space), AKAZE+KNN (AKAZE [1] features and nearest neighbour-based matching approach), SP+KNN (CNN-based Superpoint [4] feature detector and descriptor and nearest neighbour-based matching approach), DGC (neural network-based Dense Geometric Correspondence Network [11]), ORB+GMS (ORB [16] features and grid-based GMS [2]) matching method.

For evaluation purposes we choose two traversals from RobotCar Seasons dataset acquired during challenging environmental conditions: *rain* and *night*. We match images from these traversals with images in the reference traversal *overcast-reference*, taken at day-time at favourable conditions. See leftmost (night) and rightmost (overcast) images in Fig. 3.

We evaluate the performance using the standard relative pose estimation task. For each image in a non-reference traversal, we find a set of close images in the *overcast-reference* traversal using the ground truth poses provided with the dataset. Close images are defined as images for which distance between camera centres is below 10 m and camera viewing angle differs by less than $\pi/4$. Then, we estimate relative pose between each pair of close images, one from the reference traversal and the other from the non-reference traversal, using a method under evaluation. For methods containing a separate feature detection step (SIFT+KNN, AKAZE+KNN, SP+KNN, ORB+GMS), in each pair of close images we detect N keypoints using a keypoint detector being evaluated. In our experiments, in each image we choose $N = 1,000$ keypoints with the strongest response with exception to evaluation of ORB+GMS method, which by design requires a very large number of keypoints. For ORB+GMS we choose $N = 10,000$ strongest keypoint. Then, we compute keypoint descriptors at the keypoint locations. For methods using KNN matching step (SIFT+KNN, AKAZE+KNN, SP+KNN) we first find putative matches between keypoints by finding the nearest neighbour in the descriptor space. Then, Lowe's ratio test with threshold set to 0.75 is applied to filter putative matches. DGC [11] method operates directly on pairs of images, producing a dense correspondence map. Finally, we use RANSAC [5] with 5-pt Nister algorithm [12] to estimate essential matrix $\mathbf{E}$ from the found correspondences. Rotation matrix $\hat{R}$ and translation vector $\hat{T}$ are computed from the estimated essential matrix $\mathbf{E}$.

Rotation error R_{err} is measured as the rotation angle needed to align ground truth rotation matrix R and estimated rotation matrix $\hat{R}$.

$$R_{err} = \cos^{-1} \frac{\mathrm{Tr}\,(\Delta R) - 1}{2} \quad ,$$

where $\Delta R = R^{-1}\hat{R}$ is the rotation matrix that aligns estimated rotation $\hat{R}$ with the ground truth rotation R and $\mathrm{Tr}(\Delta R)$ is a trace of ΔR. An estimation of relative pose from two images given known intrinsic parameters, is possible only up to a scale factor. To compute translation error, the recovered translation vector $\hat{T}$ is first brought to the same scale as the ground truth vector.

$$T_{err} = \left\| \frac{\|T\|}{\|\hat{T}\|}\hat{T} - T \right\| \quad ,$$

We report the ratio of successful rotation and translation estimation attempts. For rotation, it's calculated as the percentage of rotation estimation attempts with rotation error less then a threshold: $R_{err} < \theta_R$. For translation, it's calculated as the percentage of translation estimation attempts with translation error less then a threshold times ground truth distance between the cameras: $t_{err} < \theta_T \|T\|$.

6   Results

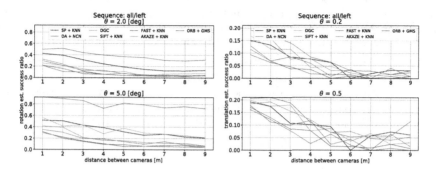

Fig. 5. Pose estimation results averaged on all pairs of sequences from *rain* and *night* traversals matched with a reference traversal.

Evaluation results are presented in Fig. 5. Left plot shows a ratio of successful rotation estimation attempts and right plot shows a ratio of successful translation estimation attempts as a function of a distance between camera centres. Results are averaged for all pairs of close images, one from the reference traversal and one from non-reference traversal: *rain* or *night*.

Our method (denoted DA+NCN) achieves significantly higher rotation estimation success ratio than all other evaluated approaches (see left plot in Fig. 5). Despite very challenging conditions and large scene appearance changes between *overcast-reference* and non-reference (*night, rain*) traversals, it is able to successfully recover relative rotation between cameras. It beats other approaches by a large margin. In terms of translation estimation success ratio (right plot in Fig. 5) our method is better than other compared approaches by a smaller margin. It must be noted that translation estimation success ratio is relatively small for all evaluated methods. For small camera center displacement (below 3 m) it's about 20% for our method but then it sharply drops to about 5–10%. Other methods perform even worse. This can be attributed to the fact, that in images from left and right camera there are a lot of planar structures such as building facades. Planar points create a degenerate configuration for essential matrix estimation task. In such cases all evaluated methods fail to produce correct results.

7   Conclusions and Future Work

The proposed method allows robust matching of local features between two images under a strong scene appearance changes. The success of our method can be attributed to two main factors. First, domain adaptation by injecting adversarial loss during the network training produces features robust to variable environmental conditions. Traditional feature descriptors differ significantly when computed for well lit scene and at night. Using domain adaptation allows learning features more robust to variable environmental conditions. Second, using NCNet promotes spatial smoothness of computed correspondences by boosting the matching score of spatially consistent groups of matches. This further increases performance in challenging environmental conditions, compared to traditional approaches, where features are matched in isolation.

In future work, we'd like to investigate the root cause of the performance difference in the rotation and translation estimation tasks. In rotation estimation task our method significantly outperforms all other evaluated approaches, whereas in translation estimation task it's better by a small margin. We also plan to improve the computational performance of the proposed method. Currently it can efficiently process medium resolution images (1024×1024 pixels) but it's too slow for practical usage to process high resolution FullHD images.

References

1. Alcantarilla, P.F., Solutions, T.: Fast explicit diffusion for accelerated features in nonlinear scale spaces. IEEE Trans. Patt. Anal. Mach. Intell **34**(7), 1281–1298 (2011)
2. Bian, J., Lin, W.Y., Matsushita, Y., Yeung, S.K., Nguyen, T.D., Cheng, M.M.: GMS: grid-based motion statistics for fast, ultra-robust feature correspondence. In: Proceedings of the IEEE Conference on Computer Vision and Pattern Recognition, pp. 4181–4190 (2017)
3. Chopra, S., Hadsell, R., LeCun, Y., et al.: Learning a similarity metric discriminatively, with application to face verification. In: CVPR (1), pp. 539–546 (2005)
4. DeTone, D., Malisiewicz, T., Rabinovich, A.: Superpoint: self-supervised interest point detection and description. In: The IEEE Conference on Computer Vision and Pattern Recognition (CVPR) Workshops (2018)
5. Fischler, M.A., Bolles, R.C.: Random sample consensus: a paradigm for model fitting with applications to image analysis and automated cartography. Commun. ACM **24**(6), 381–395 (1981)
6. Ganin, Y., Lempitsky, V.: Unsupervised domain adaptation by backpropagation. In: Proceedings of the 32nd International Conference on International Conference on Machine Learning, vol. 37, pp. 1180–1189. JMLR.org (2015)
7. Ganin, Y., et al.: Domain-adversarial training of neural networks. J. Mach. Learn. Res. **17**(1), 2030–2096 (2016)
8. Kramer, S.: A bayesian perspective on why the EKF fails in passive tracking. In: Proceedings of the IEEE 1996 National Aerospace and Electronics Conference NAECON 1996, vol. 1, pp. 98–101. IEEE (1996)
9. Lowe, D.G.: Object recognition from local scale-invariant features. In: Proceedings of the International Conference on Computer Vision, ICCV 1999, vol. 2, p. 1150. IEEE Computer Society, Washington, DC (1999)
10. Maddern, W., Pascoe, G., Linegar, C., Newman, P.: 1 Year, 1000km: the Oxford RobotCar dataset. Int. J. Robot. Res. (IJRR) **36**(1), 3–15 (2017)
11. Melekhov, I., Tiulpin, A., Sattler, T., Pollefeys, M., Rahtu, E., Kannala, J.: DGC-net: dense geometric correspondence network. In: 2019 IEEE Winter Conference on Applications of Computer Vision (WACV), pp. 1034–1042. IEEE (2019)
12. Nistér, D.: An efficient solution to the five-point relative pose problem. IEEE Trans. Pattern Anal. Mach. Intell. **26**(6), 0756–777 (2004)

13. Ranjan, A., Black, M.J.: Optical flow estimation using a spatial pyramid network. In: Proceedings of the IEEE Conference on Computer Vision and Pattern Recognition, pp. 4161–4170 (2017)
14. Revaud, J., Weinzaepfel, P., Harchaoui, Z., Schmid, C.: Deepmatching: hierarchical deformable dense matching. Int. J. Comput. Vision **120**(3), 300–323 (2016)
15. Rocco, I., Cimpoi, M., Arandjelović, R., Torii, A., Pajdla, T., Sivic, J.: Neighbourhood consensus networks. In: Proceedings of the 32nd Conference on Neural Information Processing Systems (2018)
16. Rublee, E., Rabaud, V., Konolige, K., Bradski, G.: Orb: an efficient alternative to sift or surf. In: Proceedings of the 2011 International Conference on Computer Vision, ICCV 2011, pp. 2564–2571. IEEE Computer Society, Washington, DC (2011)
17. Sattler, T., et al.: Benchmarking 6dof outdoor visual localization in changing conditions. In: Proceedings of the IEEE Conference on Computer Vision and Pattern Recognition, pp. 8601–8610 (2018)
18. Shi, W., et al.: Real-time single image and video super-resolution using an efficient sub-pixel convolutional neural network. In: Proceedings of the IEEE Conference on Computer Vision and Pattern Recognition, pp. 1874–1883 (2016)
19. Trzcinski, T., Komorowski, J., Dabala, L., Czarnota, K., Kurzejamski, G., Lynen, S.: Scone: siamese constellation embedding descriptor for image matching. In: Proceedings of the European Conference on Computer Vision (ECCV) (2018)
20. Weinzaepfel, P., Revaud, J., Harchaoui, Z., Schmid, C.: Deepflow: large displacement optical flow with deep matching. In: Proceedings of the IEEE International Conference on Computer Vision, pp. 1385–1392 (2013)

Using Normal/Abnormal Video Sequence Categorization to Efficient Facial Expression Recognition in the Wild

Taoufik Ben Abdallah[1(✉)], Radhouane Guermazi[2(✉)],
and Mohamed Hammami[3(✉)]

[1] Faculty of Economics and Management, MIR@CL, University of Sfax, Sfax, Tunisia
taoufik.benabdallah@fsegs.rnu.tn
[2] Saudi Electronic University, Riyadh, Kingdom of Saudi Arabia
r.guermazi@seu.edu.sa
[3] Faculty of Sciences, MIR@CL, University of Sfax, Sfax, Tunisia
mohamed.hammami@fss.rnu.tn

Abstract. The facial expression recognition in real-world conditions, with a large variety of illumination, pose, resolution, and occlusions, is a very challenging task. The majority of the literature approaches, which deal with these challenges, do not take into account the variation of the quality of the different videos. Unlike these approaches, this paper suggests treating the video sequences according to their quality. Using Isolation Forests (IF) algorithm, the video sequences are categorized into two categories: normal videos that visibly express clear illumination and frontal pose of face, and abnormal videos that present poor illumination, different poses of face, occulted face. Two independent facial expression classifiers for the normal and abnormal videos are built using Random Forests (RF) algorithm. The experiments have demonstrated that processing independently normal and abnormal videos can be used to improve the efficiency of the facial expression recognition in the Wild.

Keywords: Facial expression · Isolation Forests · Normal video sequences · Abnormal video sequences · Random Forests · Uncontrolled environment

1 Introduction

Facial expressions are a practical and important means of human communication. They help express the feelings, emotions, attitudes and behavior of humans [21]. Thus, research on facial expressions is a fundamental issue, affecting many areas of science such as psychology, behavioral science, medicine and computer science. One of the increasingly important research fields is the Automatic Facial-Expression Recognition (AFER). It can be useful in a wide range of applications [13] such as intelligent human-computer interfaces, educational software,

© Springer Nature Switzerland AG 2020
J. Blanc-Talon et al. (Eds.): ACIVS 2020, LNCS 12002, pp. 504–516, 2020.
https://doi.org/10.1007/978-3-030-40605-9_43

etc. Considerable attention has been dedicated to AFER from videos. The literature on this field shows a variety of approaches. However, the majority of these approaches consider the facial expression recognition under a lab-controlled environment where the faces are captured in frontal pose with fairly clear illumination. As a matter of fact, the facial expressions are artificial with almost the same degree of intensity.

Recently, Dhall *et al.* have created the Acted Facial Expressions in the Wild (AFEW) dataset [6–8] to promote the transition from lab-controlled to uncontrolled settings [18]. To the authors' best knowledge, AFEW is the largest and the most famous public video dataset proposed to train facial expression recognition models in real-world conditions. In contrast to the lab-controlled environment video datasets like Cohn-Kanade (CK+) [14] and MMI [22], AFEW shows different intensities of expressions due to the high variations of subjects with different ages, gender, and ethnic backgrounds. In addition to the variation of degree of expression, AFEW presents many other challenges like the low resolution of the videos, the face occlusions, the variation of poses, etc. To deal with these challenges, many researchers have proposed various approaches, which can be divided into two main categories: the traditional-based approaches and the deep learning-based approaches.

The traditional-based approaches are based on low-level features and shallow learning as SVM [27] and decision trees [4,24]. The low-level features have intensively studied over the past few decades, and a variety of methods have been proposed based on appearance and/or motion. Several operators have been proposed in the literature to define features that take into consideration not only the spatial information but also the temporal one as well. For example, in [6], the authors handled some AFER's challenges as the variation of illumination, the variation of face poses, the variation of face scales, and the occlusions, using the spatio-temporal extended-variant of LBP, called LBP on Three Orthogonal Planes (LBP-TOP) [34]. Compared to the use of samples captured under a lab-controlled environment, the use of samples in the Wild decreases the performance of AFER. To go further, Huang *et al.* [12] proposed the Spatio-Temporal Local Monogenic Binary Patterns (STLMBP) operator in order to analyze the magnitude, the orientation, and the phase information for facial macro-expression recognition in the Wild. Kaya *et al.* [15] proposed the Local Gabor Binary Patterns on Three Orthogonal Planes (LGBP-TOP) operator for facial macro-expression recognition in the Wild. The authors applied a set of Gabor Filters on the frames of a video sequence in which they used LBP-TOP to detect features. They observed that the use of the LGBP-TOP operator slightly improves the performance of AFER compared to the use of the basic LBP-TOP operator. Guo *et al.* [11] explored other mechanisms so as to describe facial appearance for facial macro-expression recognition in the Wild. They constructed the longitudinal facial expression atlases to obtain salient facial feature changes during an expression process. In [5] and [30], the authors used multiple features to describe facial appearance and motion information. The appearance features were detected through the Histogram of Oriented Gradients from Three

Orthogonal Planes (HOG-TOP) operator. The motion features were derived from the warp transformation of facial points that captures facial configuration changes. Experiments have demonstrated that the second way of combination enhances the discriminative power of features used for AFER in the Wild. The low-level features have been reported to be incapable of addressing the challenges of uncontrolled environment. Indeed, the performance of the majority of these approaches does not exceed 50% [11]. Other modalities such as acoustic (*i.e.,* voice) features have also been used in multimodal systems to improve the facial expression recognition [5], but the enhancement is not important.

The deep learning-based approaches are based on artificial neural networks of multiple layers. Each layer transforms its input data into a slightly more abstract and composite representation [20,31]. These neural networks make it possible to detect high-level features and classify the facial expressions into a unified process. The Convolutional Neutral Network (CNN) and Recurrent Neural Network (RNN) are the two most fundamental network architectures which have shown great superiority in extracting discriminating features and modeling temporal relationship within sequences [18]. Yao *et al.* [32] designed a new CNN architecture, referred to as HoloNet, that reduces redundant filters. It is applied simultaneously to detect high-level features from images and classify facial expressions in the Wild. Sun *et al.* [25] and Li *et al.* [17] proposed the Region-based CNN (R-CNN) to learn features for AFER. Several other CNN-derived architectures have implemented to detect the temporal information from video sequences. In [9,28] and [33], the improved version of the Recurrent Neural Network (RNN), called Long Short Term Memory (LSTM), is applied to link the detected features from each frame to those of the others. Recently, Sun *et al.* [26] have proposed the Multi-channel Deep Spatial-Temporal feature Fusion neural Network (MDSTFN) to exploit the spatial and temporal features detected from an image for AFER. Fan *et al.* [9] used LSTM to model the temporal relation within feature sequences that are extracted by a fine-tuned CNN. The major drawback of the deep learning-based approaches is the need for a large amount of data to avoid the overfitting of the classifier. The related works cited above do not achieve promising results especially for facial expression recognition in the Wild. Contrary to the above-mentioned approaches, some works as [15] and [29] are based on a deep neural architecture to detect features and a shallow learning method to build the classifier.

In our study, through a deep examination of the AFEW dataset, it was observed that the quality of some videos is poor. Thereby, the detection and tracking of faces in these videos is wrong, which influences the efficiency of the model. One possible solution to this problem is to treat this kind of videos separately. Based on the Isolation Forest (IF) method presented in [19], the purpose of this paper is to study if a separate treatment of normal video sequences that visibly express clear illumination and frontal pose of face, and abnormal videos that present poor illumination, different poses of face, occulted face, and poor face detection and tracking, can improve the ability of our low-dimensional feature space $PCA[PTLBP^{u2}]$ [1] to automatically recognize facial expression in real-world conditions.

The remainder of this paper is organized as follows. Section 2 describes the proposed facial expression recognition approach and focuses on the step of normal/abnormal video sequence categorization. Section 3 shows experimental results and discusses the effectiveness of our proposal. Section 4 draws a conclusion and some perspectives.

2 Proposed Approach

We propose to build an approach for an automatic video facial expression recognition. It is based on a low-dimensional temporal feature space in order to produce a classifier, making it possible to recognize facial expression represented by a video sequence that captured under lab-controlled environment or in the Wild. The proposed feature space is called Pyramid of uniform Temporal Local Binary Patterns (PTLBP^{u2}) [1]. Figure 1 shows a flowchart of the proposed approach. It is performed in four main steps.

Step 1: Preprocessing the input video sequences through face detection and tracking.

Step 2: Expressing the video sequences through PCA[PTLBP^{u2}] features [1]. It represents a temporal low-dimensional space through the second and the third levels of a pyramid representation, using the 33 discriminating cuboids selected by applying the method proposed in [1] (8 from level 2 and 25 from level 3).

Step 3: Classifying the video sequences into two categories: normal and abnormal on the basis of the Isolation Forests (IF) algorithm [19]. This classification is particularly valuable for facial expression recognition in the Wild where the quality of video sequences is invariant. Indeed, a good or "normal" video sequence shows a clear illumination and a frontal pose without occlusion of the eyes, nose, and mouth. Contrariwise, a poor or "abnormal" video sequence presents poor illumination, different face poses and scales, and occlusions. Figure 2 shows some examples of abnormal video sequences of cropped faces.

Normal/Abnormal Video Sequence Categorization. To automatically classify video sequences as normal or abnormal, we rely on the Isolation Forests algorithm (IF) [19]. This algorithm separates an instance from other instances. Since the anomalies are few and different, they are more likely to be separated. Unlike other existing anomaly-detection algorithms, the IF takes advantage of the process of scaling up to handle extremely large data size and high-dimensional problems with a large number of irrelevant attributes without defining distances or density measures [19].

Given O the vector of training video sequences, firstly, we calculate the PCA[PTLBP^{u2}] feature matrix $trainV$. Further, we apply IF to build the model that separates normal from abnormal video sequences. This model is formed by nbT binary trees whose leaf-nodes contain only one sample. Each internal-node has exactly two child nodes, $node.left$ and $node.right$.

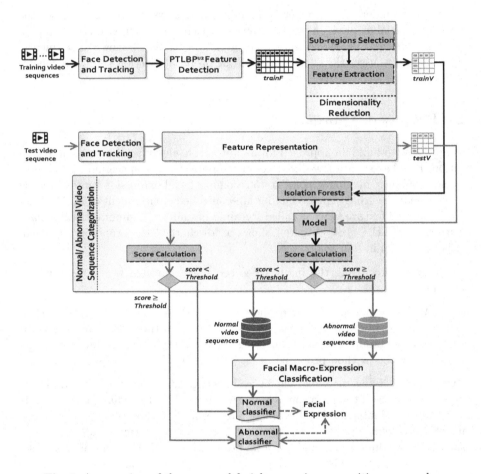

Fig. 1. An overview of the proposed facial expression recognition approach

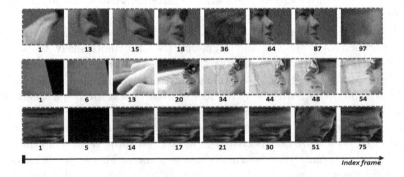

Fig. 2. An example of abnormal video sequences

For each building of a tree t_j ($j \in [1..nbT]$), we proceed as follows:

1. Selecting randomly ψ samples from $trainV$. The result is registered in a ψ-by-nbE matrix, referred to as $subV$.
2. Initializing t_j, by the root node that contains all samples of $subV$.
3. Selecting randomly a feature, noted p, from $subV$ and determine its corresponding index i_p, its minimum value min_p, and its maximum value max_p.
4. Calculating the average value q between min_p and max_p defined as $\frac{(min_p + max_p)}{2}$.
5. $\forall k \in [1..\psi]$, decomposing the current node into sub-nodes: $nodeLeft$ contains samples of $subV$ with $p < q$ (i.e. $subV(k, i_p) < q$); and $nodeRight$ contains samples of $subV$ with $p \geq q$ (i.e. $subV(k, i_p) \geq q$).
   Repeating (1), (2), (3), (4) and (5) for each $nodeLeft$ and $nodeRight$ until each node represent only one instance.

Finally, for a given video sequence, we project it on the obtained model, and we calculate an anomaly score, noted $score$. This score is defined by the Eq. 1:

$$score = 2^{-\frac{avg\_path}{c}} \tag{1}$$

Where $avg\_path = \frac{\sum_{j=1}^{nbT} path^{(j)}}{nbT}$; $path^{(j)}$ is the number of edges of the path between the root node of t_j and the terminating node of the given video sequence; and $c = 2\ln(\psi - 1) + 0.58 - (\frac{2(\psi-1)}{\psi})$ is a variable that used to normalize each path length; It is equivalent to the unsuccessful search measure in the Binary Search Tree (BST) [23]. The abnormal video sequence has a shorter average path than a normal video sequence and resides closer to the root of the tree. So, if $score \geq threshold$, the given video sequence is considered as abnormal; otherwise, it is considered as normal [19].

Step 4: Building the facial expression recognition classifier using a supervised learning technique. There is no one optimal algorithm for all situations. It is strongly dependent on the data nature and user expectations. The performance of our approach has been evaluated using three different supervised learning techniques: the Support Vector Machine SMO [16,27], C4.5 decision tree [24], and the Random Forests (RF) [4]. Indeed, SMO represents one of the widely applied techniques for AFER in the literature; C4.5 decision tree offers simple models, in the form of rules, and easily interpretable; and RF uses the bootstrap aggregating strategy in building trees to correct the decision trees' problem of overfitting.

3 Experiments

To validate the proposed approach, we reserve the following subsections for presenting firstly the facial-expression dataset and the different conducted experiments.

3.1 Facial Expression Dataset

Several datasets accessible to the research communities are available in the literature. The majority of these datasets contains images or video sequences labeled by the universal facial expressions captured or collected either under a lab-controlled environment [14,22]. A limited number of datasets present video sequences captured under real-world conditions where the most famous dataset is the Acted Facial Expressions in the Wild (AFEW) [6]. In our work, we have evaluated the proposed approach using the AFEW dataset.

AFEW dataset [6] contains video sequences collected from different movies with various head poses, occlusions, and illuminations [18]. It provides close-to-real-world conditions of varied subjects in terms of sexes and ages. In our experiments, we have used version 7.0 of AFEW. It is divided into three sets. The training set contains 773 samples, the validation set contains 383 samples, and the test set contains 653 samples. The selection of the different subsets is done in an independent manner in terms of subjects and movie/ TV sources. Compared to the version 6 of AFEW [7], only 90 video sequences are added to the AFEW 7.0's test subset, but the training and the validation sets are not modified. The Mixture of Parts (MoP) model [10] is applied to detect and track the faces. However, after face detection and tracking, Only 756 out of 773 training video sequences, and 371 out of 383 validation video sequences are available which are used in our experiments. These video sequences are labeled by seven expressions: 147 (resp. 63) video sequences from the training set (resp. test set) are labeled *"happiness"*, 113 (resp. 59) video sequences *"sadness"*, 80 (resp. 44) video sequences *"fear"*, 74 (resp. 39) video sequences *"disgust"*, 71 (resp. 46) video sequences *"surprise"*, 133 (resp. 59) video sequences *"anger"*, and 138 (resp. 61) video sequences *"neutral"*. As the test video sequences are not labeled, we have not used them to evaluate the performance of our approach.

3.2 Experimental Results

To investigate the performance of the proposed approach, we have carried out all classifiers generated by SMO, C4.5 or RF.

In all these experiments, for the AFEW 7.0 dataset, we have used the video sequences of the AFEW 7.0's training set to build the proposed classifiers and the video sequences of the AFEW 7.0's validation set to estimate the performance. As a validation metric, we have calculated the accuracy rate. In our work, the pyramid representation requires that the length and the width of the frames must be divisible by 2, 4, and 8. Thus, we have defined a size of frames equals 128×128 pixels.

Facial Expression Recognition Without Normal/Abnormal Video Sequence Categorization. We have evaluate the three classifiers generated using the AFEW 7.0 training set. All experiments are conducted on the validation set without applying the normal/abnormal video sequence categorization (Table 1).

Table 1. Experimental results of the three classifiers based on PCA[PTLBP^{u2}] feature space, using the AFEW 7.0 validation set

	Classification algorithm	Number of sub-regions	Number of features	Accuracy (%)
(1)	SMO	33	39	46.09
(2)	C4.5	33	39	44.74
(3)	RF	33	39	47.71

The result of facial expression recognition in the Wild are very low. The best accuracy is obtained by the RF classifier. It does not exceed 47.71%. In fact, the head movements, the pose, the illumination, and the background variation increase the difficulties of the face detection and the facial expression classification as well.

To sum up all the results, we conclude that RF is better than SMO and C4.5 to recognize facial expressions in the Wild.

Impact of Normal/Abnormal Video Sequence Categorization on Facial Expression Recognition in the Wild. In order to verify the impact of using the normal/abnormal video sequence categorization on the performance of facial expression recognition in the Wild, we have carried out several experiments. As described in Subsect. 2, we apply the IF algorithm to generate a model for normal/abnormal video sequence categorization. Three main parameters of IF can influence the effectiveness of the model built: the size of the sub-sample randomly selected for building each tree in the forest ψ, the number of trees nbT, and the anomaly threshold score $threshold$. Based on the empirical studies, through several datasets at different sizes, proposed in [19], we define $\psi = 256$, $nbT = 100$, and $threshold = 0.5$. As a result, 525 (resp. 288) video sequences from the training set (resp. validation set) are classified as normal, and the 231 (resp. 83) remaining video sequences are classified as abnormal. Table 2 shows the number of video sequences, per facial expression, classified as normal and abnormal for both the training and validation sets of the AFEW 7.0 dataset.

When we visualize the result of the normal/abnormal video sequence categorization based on the IF algorithm, we notice that the most abnormal video sequences present poor illumination, different face poses, and occlusions. So, the obtained results show the effectiveness of the proposed IF model.

According to the previous results, RF is more adequate than SMO and C4.5 for facial expression recognition. So, we will use it as the classification technique based on PCA[PTLBP^{u2}] to generate both normal and abnormal classifiers separately. Using the 525 (resp. 231) normal (resp. abnormal) training video sequences, we build the normal (resp. abnormal) classifier. After that, using the validation set, we calculate the accuracy rate of using normal and abnormal classifiers separately and together (normal+abnormal). Obviously, the normal classifier displays better

Table 2. Normal and abnormal video distribution of the training and validation sets of AFEW 7.0 based on the IF model

Facial expressions	Normal video sequences		Abnormal video sequences	
	Training	Validation	Training	Validation
Neutral (NE)	106	55	32	6
Anger (AN)	82	41	51	18
Disgust (DI)	48	31	26	8
Happiness (HA)	103	51	44	12
Fear (FE)	54	30	26	14
Sadness (SA)	84	49	29	10
Surprise (SU)	48	31	23	15
Total	**525**	**288**	**231**	**83**

performance (63.54% of accuracy) than abnormal classifier (49.40%). Therefore, the performance of the "normal+abnormal" classifier reaches 60.38% of accuracy.

Figure 3 shows the performances of predicting each facial expression with and without normal/abnormal video sequence categorization. In fact, we compare the classifier generated by RF based on PCA[PTLBP^{u2}], using all the video sequences of the training set of AFEW 7.0 with the classifier that separates normal from abnormal video sequences ("normal+abnormal" classifier).

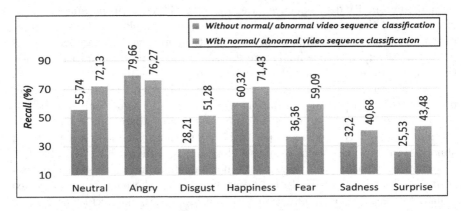

Fig. 3. Impact of the normal/abnormal video sequence categorization on AFER in the Wild using the validation set of the AFEW 7.0 dataset

We observe that using the normal/abnormal video sequence categorization increases the performance of the classifier in the recognition of the majority of facial expressions. The improvement of recall depends on the expression. It varies from 8.48% (the *"sadness"* facial expression) to 23.07% (the *"disgust"* facial expression). The increase in the performance can be assigned to the use of two different classifiers, each of them adequate to deal with some conditions (normal or abnormal). Moreover, using the normal/abnormal video sequence categorization decreases by 3,39% the recall of only one out of seven expressions, which is the *"angry"* facial expression.

To conclude, the use of normal/abnormal video sequence categorization for facial expression recognition in the Wild can improve the accuracy of the basic classifier generated by RF, using the PCA[PTLBP^{u2}] features to represent each video sequence.

3.3   Analysis and Discussions

Considering that RF is based on a random selection of samples and features, we decide to repeat the experiments based on 10 times in order to evaluate the impact of the random selection on the proposed approach. For each iteration, we calculate the accuracy rate. Then, we measure the standard deviation in order to quantify the magnitude of the dispersion of accuracies from their mean. A low standard deviation indicates that random selection does not affect the effectiveness of the proposed approach. It is worth noting that in all the previous experiments based on RF, we consider the results of the iteration that record an accuracy rate close to the mean. Using the "normal+abnormal" classifier, we find that the mean of accuracies equals 60.24 with 1.35 as a standard deviation. Therefore, we deduce that the effectiveness of the proposed classification algorithm is not affected by the random selection of parameters.

Comparative Study. We have conducted a comparison of our approach with other related approaches in the literature in terms of the number of features, the number of video sequences, and the accuracy rate. Table 3 reports the comparison of our proposal with some recent related works using AFEW (version 6.0 or 7.0).

Table 3. The comparison of the proposed approach vis-a-vis the related approaches on the AFEW dataset

	Feature space	Class. Algo	Nb. Samples	Nb. Features	τ (%)
Dhall *et al.* [7]	LBP-TOP	SVM	383	–	38.80
Yao *et al.* [32]	CNN architecture[a]		383	1024	51.96
Chen *et al.* [5]	Hybrid features	SVM	383	3964	40.20
Fan *et al.* [9]	3DCNN + LSTM architecture[a]		383	–	51.96
Afshar *et al.* [3]	Hybrid features	ELM	378	26588	42.86
Vielzeuf *et al.* [28]	3DCNN + LSTM architecture[a]		380	297	52.20
Kaya *et al.* [15]	Hybrid features	ELM	371	–	52.30
Kaya *et al.* [15]	CNN + Hybrid features	ELM	371	–	57.02
Ours	**PCA[PTLBP^{u2}]**	**RF**	**371**	**[24–34]**	**60.24**

[a]Unified process (feature detection + classification)

Several of the related works have proposed traditional-based approaches [3,5,7,15]. Recent works have represented the deep learning-based approaches according to different architectures of deep neural networks [9,15,28,32].

According to Table 3, the proposed approach is the best one according to the accuracy that reaches 60.24% for facial macro-expression recognition in the Wild. Comparing the performance of our approach to that of the literature approaches, the enhancement varies from 3.22% [15] to 21.44% [7]. Although the enhancement is not important compared to [15], we provide a lower number of features. Indeed, the proposed approach needs only 39 features to present a video sequence. Also, it is based only on visual features without incorporating audio features unlike some related works such as [15,32]. Probably, the incorporation of additional features can enhance the performance of the proposed approach.

To conclude, our proposal is considered as a strong competitor not only to the traditional-based approaches but also to the deep learning-based approaches in the literature to recognize facial expressions in the Wild.

4   Conclusion and Perspectives

Automatic Facial Expression Recognition in the Wild is a very challenging problem due to different expressions under arbitrary poses, variability of illumination and occlusions. Different from existing approaches, in this paper, we formulate the AFER by separating the analysis of videos categorized as Normal from videos categorized as Abnormal.

To address this challenge, we have proposed a normal/abnormal video sequence categorization approach based on Isolation Forest (IF) algorithm. Our proposal is based on a low-dimensional feature space called PCA[PTLBP^{u2}]. Furthermore, we compare the SMO, C4.5 and RF classification techniques to create robust AFER classifiers.

The findings of our research are quite convincing, and thus the following conclusions can be drawn:

- Separating normal from abnormal video sequences can improve the facial expression recognition in the Wild. With this categorization, we record the best accuracy rate compared to the most famous literature studies. Thus, we proved that building separate classifiers improves significantly the results.
- Using PCA[PTLBP^{u2}] feature space shows its effectiveness to represent the video sequence for both normal and abnormal video sequences. Despite the fact that the proposed feature space is a low-dimensional space, it represents effectively the video sequences.
- Observing that the traditional-based approaches can be under certain conditions more efficient than the deep-learning-based approaches compared to some state-of-the-art works.
- Using an imbalanced distribution of facial has an important effect on the performance of the classifier to detect a particular expression.

In our future research, we intend to concentrate on studying the impact of the different parameters of the IF algorithm on the performance of the proposed approach. We also plan to test other alternatives to classify normal and abnormal videos like the unsupervised Random Forests [2]. We should also estimate the impact of the normal/abnormal video categorization on other handcrafted features like STLMBP [12] and HOG-TOP [5]. As the distribution of the different expression on AFEW 7.0 is imbalanced, we can also study the impact of applying appropriate learning algorithms to improve the facial expression recognition rate.

References

1. Abdallah, T.B., Guermazi, R., Hammami, M.: Facial-expression recognition based on a low-dimensional temporal feature space. Multimedia Tools Appl. **77**(15), 19455–19479 (2018)
2. Afanador, N.L., Smolinska, A., Tran, T.N., Blanchet, L.: Unsupervised random forests: a tutorial with case studies. Chemometrics **30**(5), 232–241 (2016)
3. Afshar, S., Salah, A.A.: Facial expression recognition in the wild using improved dense trajectories and fisher vector encoding. In: International Conference on Computer Vision and Pattern Recognition Workshops, Las Vegas, NV, USA, pp. 1517–1525. IEEE (2016)
4. Breiman, L.: Random forests. Mach. Learn. **45**(1), 5–32 (2001). https://doi.org/10.1023/A:1010933404324
5. Chen, J., Chen, Z., Chi, Z., Fu, H.: Facial expression recognition in video with multiple feature fusion. IEEE Trans. Affect. Comput. **9**(1), 1–12 (2016)
6. Dhall, A., Goecke, R., Ghosh, S., Joshi, J., Hoey, J., Gedeon, T.: From individual to group-level emotion recognition. In: International Conference on Multimodal Interaction, pp. 524–528. ACM, New York (2017)
7. Dhall, A., Goecke, R., Joshi, J., Hoey, J., Gedeon, T.: Video and group-level emotion recognition challenges. In: International Conference on Multimodal Interaction, pp. 427–432. ACM, New York (2016)
8. Dhall, A., Goecke, R., Lucey, S., Gedeon, T.: Collecting large, richly annotated facial-expression databases from movies. IEEE MultiMedia **19**(3), 34–41 (2012)
9. Fan, Y., Lu, X., Li, D., Liu, Y.: Video-based emotion recognition using CNN-RNN and C3D hybrid networks. In: International Conference on Multimodal Interaction, pp. 445–450. ACM, New York (2016)
10. Felzenszwalb, P.F., Huttenlocher, D.P.: Pictorial structures for object recognition. Comput. Vis. **61**(1), 55–79 (2005)
11. Guo, Y., Zhao, G., Pietikäinen, M.: Dynamic facial expression recognition with Atlas construction and sparse representation. IEEE Trans. Image Process. **25**(5), 1977–1992 (2016)
12. Huang, X., He, Q., Hong, X., Zhao, G., Pietikainen, M.: Improved spatiotemporal local monogenic binary patterns for emotion recognition in the wild. In: International Conference on Multimodal Interaction, Istanbul, Turkey, pp. 514–520. ACM (2014)
13. Imotions: Facial Expression Analysis: The Complete Pocket Guide (2016). https://imotions.com/blog/Facial-Expression-Analysis
14. Kanade, T., Cohn, J.F., Tian, Y.: Comprehensive database for facial expression analysis. In: International Conference on Automatic Face and Gesture Recognition, Grenoble, France, pp. 46–53. IEEE (2000)
15. Kaya, H., Gürpınar, F., Salah, A.A.: Video-based emotion recognition in the wild using deep transfer learning and score fusion. Image Vis. Comput. **65**, 66–75 (2017)
16. Keerthi, S.S., Shevade, S.K., Bhattacharyya, C., Murthy, K.R.: Improvements to Platt's SMO algorithm for SVM classifier design. Neural Comput. **13**(3), 637–649 (2001)

17. Li, J., et al.: Facial expression recognition with faster R-CNN. Procedia Comput. Sci. **107**, 135–140 (2017)
18. Li, S., Deng, W.: Deep facial expression recognition: a survey. Computer Vision and Pattern Recognition (2018, to appear)
19. Liu, F.T., Ting, K.M., Zhou, Z.H.: Isolation forest. In: International Conference on Data Mining, pp. 413–422. IEEE, Washington (2008)
20. Liu, W., Wang, Z., Liu, X., Zeng, N., Liu, Y., Alsaadi, F.E.: A survey of deep neural network architectures and their applications. Neurocomputing **234**, 11–26 (2017)
21. Mehrabian, A.: Communication without words. Psychol. Today **2**(4), 53–56 (1968)
22. Pantic, M., Valstar, M., Rademaker, R., Maat, L.: Web-based database for facial expression analysis. In: International Conference on Multimedia, Amsterdam, Netherlands. IEEE (2005)
23. Preiss, B.R.: Data Structures and Algorithms with Object-Oriented Design Patterns in Java. Wiley, Hoboken, first edn (1999)
24. Quinlan, J.R.: C4.5: Programs for Machine Learning, 1st edn. Morgan Kaufmann, San Francisco (1993)
25. Sun, B., Li, L., Zhou, G., He, J.: Facial expression recognition in the wild based on multimodal texture features. Electron. Imaging **25**(6), 1–8 (2016)
26. Sun, N., Li, Q., Huan, R., Liu, J., Han, G.: Deep spatial-temporal feature fusion for facial expression recognition in static images. Pattern Recogn. Lett. **119**, 49–61 (2019)
27. Vapnik, V.: The Nature of Statistical Learning Theory, 1st edn. Springer, New York (1995). https://doi.org/10.1007/978-1-4757-3264-1
28. Vielzeuf, V., Pateux, S., Jurie, F.: Temporal multimodal fusion for video emotion classification in the wild. In: International Conference on Multimodal Interaction, pp. 569–576. ACM, New York (2017)
29. Wang, F., Lv, J., Ying, G., Chen, S., Zhang, C.: Facial expression recognition from image based on hybrid features understanding. Vis. Commun. Image Represent. **59**, 84–88 (2019)
30. Yan, H.: Collaborative discriminative multi-metric learning for facial expression recognition in video. Pattern Recogn. **75**, 33–40 (2018)
31. Yan, J., Zheng, W., Cui, Z., Tang, C., Zhang, T., Zong, Y.: Multi-cue fusion for emotion recognition in the wild. Neurocomputing **309**, 27–35 (2018)
32. Yao, A., Cai, D., Hu, P., Wang, S., Sha, L., Chen, Y.: HoloNet: towards robust emotion recognition in the wild. In: International Conference on Multimodal Interaction, pp. 472–478. ACM, New York (2016)
33. Yu, Z., Liu, G., Liu, Q., Deng, J.: Spatio-temporal convolutional features with nested LSTM for facial expression recognition. Neurocomputing **317**, 50–57 (2018)
34. Zhao, G., Pietikäinen, M.: Dynamic texture recognition using local binary patterns with an application to facial expressions. IEEE Trans. Pattern Anal. Mach. Intell. **29**(6), 915–928 (2007)

Distributed Multi-class Road User Tracking in Multi-camera Network For Smart Traffic Applications

Nyan Bo Bo[1,2(✉)], Maarten Slembrouck[1,2], Peter Veelaert[1,2],
and Wilfried Philips[1,2]

[1] imec, Kapeldreef 75, 3001 Leuven, Belgium
[2] TELIN-IPI, Ghent University, Sint-Pietersnieuwstraat 41, 9000 Gent, Belgium
{Nyan.BoBo,Maarten.Slembrouck,Peter.Veelaert,Wilfried.Philips}@ugent.be

Abstract. Reliable tracking of road users is one of the important tasks in smart traffic applications. In these applications, a network of cameras is often used to extend the coverage. However, efficient usage of information from cameras which observe the same road user from different view points is seldom explored. In this paper, we present a distributed multi-camera tracker which efficiently uses information from all cameras with overlapping views to accurately track various classes of road users. Our method is designed for deployment on smart camera networks so that most computer vision tasks are executed locally on smart cameras and only concise high-level information is sent to a fusion node for global joint tracking. We evaluate the performance of our tracker on a challenging real-world traffic dataset in an aspect of *Turn Movement Count* (TMC) application and achieves high accuracy of 93% and 83% on vehicles and cyclist respectively. Moreover, performance testing in anomaly detection shows that the proposed method provides reliable detection of abnormal vehicle and pedestrian trajectories.

Keywords: Road user tracking · Smart camera network · Distributed computing · Trajectory analysis · Road traffic statistics · Smart traffic

1 Introduction

Automatic detection and tracking of road users, *i.e.*, vehicles, cyclists and pedestrians is an active research topic in computer vision since it is one of the essential building blocks in smart traffic and intelligent surveillance applications. Traffic features extracted from trajectories of road users are vital for understanding the behavior of road users, modeling of traffic, evaluation of traffic scenes and eventually automatic traffic flow control to obtain optimal efficiency. Many existing trackers use information from a single camera to track one or more classes of road users. To tackle coverage limitation of a single camera view tracking, a network of multiple cameras is used. A considerable number of methods have been proposed on tracking of road users across multiple cameras with overlapping/non-overlapping views. However, all these methods boil down to a single tracking

© Springer Nature Switzerland AG 2020
J. Blanc-Talon et al. (Eds.): ACIVS 2020, LNCS 12002, pp. 517–528, 2020.
https://doi.org/10.1007/978-3-030-40605-9_44

with camera handover or target re-identification to associate local trajectories of each camera view to produce global trajectories.

A relatively small amount of work, for instance [4,7,8,15], has exploited the advantages of using information from cameras observing the same traffic scene with high overlapping view from different angles. By observing the same road user from different view angles, the occlusion problem may be significantly mitigated. Centralized multi-camera tracking methods [4,7,15] require images from all views in order to localize and track road users. In these methods, the number of cameras is limited by computational and communication bottlenecks. Fortunately, the introduction of smart cameras [12] allows the execution computer vision algorithms locally on cameras and only compact high-level information is sent to a decision node for joint trajectory estimation.

In this paper, we propose a distributed multi-camera road user tracking system which is capable of simultaneously tracking pedestrians, cyclists and vehicles with high accuracy. Each smart camera locally tracks road users on its image plane using recursive Bayesian estimation. Local 2D estimates are then sent to a fusion node on which ground plane positions are jointly estimated. For the performance evaluation of our proposed method, we captured a 90 min long multi-camera dataset of a real traffic scene at the intersection of five streets using four cameras. Instead of directly measuring the accuracy of the trajectories, which involves exhaustive manual annotation, we perform an experimental analysis 7 on the performance of high-level tasks such as turning movement counting and anomaly detection. The experimental analysis shows that our method provides trajectories for turning movement count with high accuracy of 93% on vehicles and 83% on cyclists, and reliable anomaly detection.

The rest of this paper is organized as follows: Sect. 2 briefly discusses the related work on tracking of road users. Section 3 describes our proposed tracking method in details. Section 4 explains how do we experimentally measure the performance of the proposed tracker and presents our findings. Section 5 summarizes this paper.

2   Related Work

Numerous road user tracking methods have been proposed over the past two decades. Early methods use feature points [14], foreground blobs [4,16] and histograms [5] in combination with various data association/filtering frameworks to track road users. With an increase in available computation power and the introduction of more efficient classifier algorithms, many methods [6,17] utilize object detector responses as observations in trajectory estimation. This approach is called *tracking-by-detection* in visual tracking literature since an object of interest is first detected and then its trajectory is estimated by associating subsequent detector responses through time.

The popularity of the *tracking-by-detection* approach was boosted by the introduction of more accurate object detectors based on *Convolutional Neural Networks* (CNN). A vehicle tracker proposed by Qui *et al.* [10] deploys the *You*

Only Look Once YOLO object detector [11] and objects detected at two different time instances are associated by finding the optimal matches of *Kanade-Lucas-Tomasi* (KLT) feature points. Their work does not address partial/occlusion of vehicles which is very common in practical traffic surveillance. Ooi *et al.* [9] first detects road users with a *Region-based Fully Convolutional Network* (RFCN) [1]. Then data associating is performed by minimizing a cost function which incorporates object type (car, bicycle, pedestrian, etc.), position, size and color between objects detected at different time instances using the Hungarian algorithm. Missing detection due to occlusion, noise, etc. are handled by the prediction of a Kalman filter. However, if the detector fails to localize an object being tracked for an extended period of time, the prediction of the Kalman filter without measured evidences may drift from an actual trajectory.

Road users can be observed from different viewpoints using a network of cameras to tackle the occlusion problem since it is quite unlikely that a particular road user is occluded in all views. Tang [15] performs an inverse projection of foreground pixels onto the common plane and then deploys an overlap reasoning of the inverse projected blobs for joint position estimation. Instead of an inverse projection, the *Probabilistic Occupancy Mapping* (POM) approach computes the probability of positions on the ground plane being occupied by one or more road users. This is done by projecting a hypothesized volume of road users on the image plane and measuring how well it matches to observations from one or more cameras. The original work of Fluret *et al.* [2] uses foreground blobs as observations in POM computation. A recent work of Nishikawa [7] utilizes responses from the YOLO object detector as observations and *K-Shortest Path* (KSP) optimization to fit trajectories to the computed POM through time. This approach achieves its optimal performance when the KSP trajectory fitting is done on all frames in video, making it an offline tracker, *i.e.*, tracks after the whole video is captured. It is possible to deploy it as an online tracker with small delay by performing trajectory fitting in a batch of a few frames but the performance is then often suboptimal.

3   Proposed Method

Our distributed tracking system consists of K smart cameras observing a traffic scene from different angles and a fusion node as depicted in Fig. 1. Internal parameters such as camera matrix and distortion coefficients as well as external parameters such as rotation matrix and translation vectors are obtained during the camera network installation. In visual tracking literature, an object being tracked is usually denoted as a target. Therefore, the word road user and target will be used interchangeably throughout this paper. Each smart camera locally tracks all targets in its view by using recursive Bayesian state estimation on its image plane and sends local position estimates to the fusion node. Local position estimates of the same target from different viewpoints are fused as a single joint position on the global ground plane. These joint position estimates are associated through time using recursive Bayesian state estimation on the ground plane to produce optimal global trajectories. The following subsections describes both local tracking on smart cameras and global tracking on the fusion node in detail.

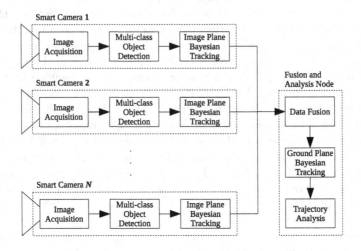

Fig. 1. Building blocks of our proposed tracker.

3.1   Tracking on Smart Camera

At a give time t, a smart camera captures an image observed in its view and feeds it to the YOLO object detector. Output of the object detector is a set of N detections $\boldsymbol{d}_i = (u_i, v_i, w_i, h_i, \alpha_i, \lambda_i) : i \in \{1, 2, \ldots, N\}$ on its image plane where (u_i, v_i), w_i, h_i, α_i and λ_i are the detection center, width, height, object class (pedestrian, cyclist and vehicle) and reliability score respectively. Our tracker keeps the state of M targets, each of which is represented with a vector. Given a set of current detections $D_t = \{\boldsymbol{d}_{1,t}, \boldsymbol{d}_{2,t}, \ldots, \boldsymbol{d}_{N,t}\}$ and a set of previously known states of all targets $K_{t-1} = \{\boldsymbol{k}_{1,t-1}, \boldsymbol{k}_{2,t-1}, \ldots, \boldsymbol{k}_{M,t-1}\}$, the task of a local tracker is to estimate a new state of each target as $K_t = \{\boldsymbol{k}_{1,t}, \boldsymbol{k}_{2,t}, \ldots, \boldsymbol{k}_{M,t}\}$. The state of each target is predicted and updated using recursive Bayesian estimation as:

$$\text{predict}: P(\boldsymbol{k}_t|\boldsymbol{d}_{1:t-1}) = \int P(\boldsymbol{k}_t|\boldsymbol{k}_{t-1})P(\boldsymbol{k}_{t-1}|\boldsymbol{d}_{t-1})d\boldsymbol{k}_{t-1} \tag{1}$$

and

$$\text{update}: P(\boldsymbol{k}_t|\boldsymbol{d}_{1:t}) = \frac{P(\boldsymbol{d}_t|\boldsymbol{k}_t)P(\boldsymbol{k}_t|\boldsymbol{d}_{1:t-1})}{P(\boldsymbol{d}_t|\boldsymbol{d}_{1:t-1})}. \tag{2}$$

Using Eq. (1) together with a constant acceleration motion model, state of targets $\hat{K}_t = \{\hat{\boldsymbol{k}}_{1,t}, \hat{\boldsymbol{k}}_{2,t}, \ldots, \hat{\boldsymbol{k}}_{M,t}\}$ at time t can be predicted. These predicted states have to be associated with corresponding detections in D_t so that Eq. (2)

can be used to update the predicted state using detections as observations. When observing a traffic scene, due to physical constraints, the variation in displacement and size of a target in terms of pixels in two consecutive frames are relatively small. Therefore, a detection d_t corresponding to a target j will be very close to the predicted $\hat{k}_{j,t}$. Moreover, the object class α of the corresponding detection and target's state should be the same. Therefore, we define a function which calculates the association cost between a detection and a target as follows:

$$\delta(d_i, k_j) = \sqrt{(u_i - u_j)^2 + (v_i - v_j)^2} + |w_i - w_j| + |h_i - h_j| + \delta_{type}(\alpha_i, \alpha_j). \quad (3)$$

The first term in Eq. (3) is simply the Euclidean distance while the second and third term compute the absolute differences in height and width respectively. The last term makes sure that $\delta(d_i, k_j)$ is very large when the detected class is different from target's class, formally expressed as:

$$\delta_{type}(\alpha_i, \alpha_j) = \begin{cases} 0 & \alpha_i \text{ and } \alpha_i \text{ are the same road user class} \\ \xi & \text{otherwise} \end{cases} \quad (4)$$

where ξ is a very large constant: $\xi \gg 1$.

The Hungarian method is one of the most popular methods to obtain the assignment of detections to targets with total minimal cost. However, it often occurs that the YOLO detector fails to detect one or more targets while new road users entering the scene are detected. This sometimes causes mismatch errors in the Hungarian method although total matching cost is at its minimum. Therefore, we propose to use a greedy matching algorithm as described in Algorithm 1. Only matched pairs with cost lower than a threshold Λ are added to

---

Algorithm 1. Greedy matching.

1: **Input:** D_t: detections, $\hat{K}_t$: predictions
2: **Output:** *pairs*: matched detection and target pairs
3:
4: Initialize *pairs* as a list
5: **for** $k \leftarrow 1$ to $minimum(N, M)$ **do**
6:     $d, k \leftarrow \underset{d \in D_t, k \in \hat{K}_t}{\arg \min} \; \delta(d, k)$
7:     $D_t \leftarrow D_t \setminus d, \; \hat{K}_t \leftarrow \hat{K}_t \setminus k$
8:     **if** $\delta(d, k) < \Lambda$ **then**
9:         Append tuple (d, k) to the list *pairs*
10:     **end if**
11: **end for**

---

the list *pairs*. Detections without matching target are initialized as new targets. For targets without any matching detection, a correlation-based template matching method similar to Guan *et al.* [3] is used to generate a corresponding detection. Finally, the predicted states of all targets are updated using their corresponding detections. Under an assumption of the Gaussian distribution of noise in motion and observation models, aforementioned recursive Bayesian state estimation simply becomes Kalman filtering. Center points on upper and lower edges of bounding boxes $r^{upper} = (u^{upper}, v^{upper})$ and $r^{lower} = (u^{lower}, v^{lower})$ of all targets are then sent to the fusion node.

3.2   Joint Position Estimation and Tracking

Consider a road user in the scene observed by a set of N smart cameras $C = \{c_1, c_2, \ldots, c_N\}$ from different angles. Only a subset of cameras $C_{vis} \subseteq C$ may be able to track the target. A target may be outside of the view of some cameras, occluded, or the detector may simply fails to detect it. In this case, only cameras in the subset C_{vis} are able to estimate the position of the targets in their own image coordinates. Suppose that a smart camera c accurately estimates the position of a target. A line connecting the two center points r_c^{upper} and r_c^{lower} must be the best approximation of the projection of a hypothetical vertical line, which length is the height h of the target, placed at a true physical position $(x, y, 0)$ of the target. In the ideal situation,

$$L(x, y, h) = |r_c^{lower} - \rho_c(x, y, 0)|^2 + |r_c^{upper} - \rho_c(x, y, h)|^2 \approx 0, \qquad (5)$$

where a projection function $\rho_c(x, y, z)$ projects a point in 3D world coordinates on to the image coordinates of the camera c.

However, due to the presence of uncertainty in the detection and camera calibration, $L(x, y, h)$ will be usually not zero, but will only attain a minimum value greater than zero. When local estimates from a set of cameras C_{vis} are available, the error function $L(x, y, h)$ can be extended to the multi-camera case:

$$L(x, y, h) = \sum_{c \in C} (|r_c^{lower} - \rho_c(x, y, 0)|^2 + |r_c^{upper} - \rho_c(x, y, h)|^2). \qquad (6)$$

The joint estimate of the target's position and height is then found by minimizing the error function $L(x, y, h)$ over all positions (x, y) and possible heights:

$$\hat{x}, \hat{y}, \hat{h} = \arg \max_{x, y, h} L(x, y, h). \qquad (7)$$

Subsequently, using x and y as state variables, a discrete Bayes filter with constant velocity motion model is applied to suppress the noise in the join joint estimation.

4   Experimental Analysis

4.1   Traffic Dataset

For the performance evaluation of our method, we capture a multi-camera video using four *GoPro HERO 4* cameras aiming at an intersection of five streets in the city of Ghent, Belgium. Camera positions and streets layout are depicted in Fig. 2 where top view of the intersection is obtained from *Google Maps*. All streets allow two-way traffic except *Street 4* which does not allow incoming traffic from the junction. Videos are captured at 30 FPS with the HD Ready resolution of 1280×720 pixels. All four video streams are loosely time synchronized and the duration is approximately one and a half hour. All types of usual road users such as cars, trucks, bus, cyclists and pedestrians go trough the intersection causing partial/full occlusion in one or more camera views making it a very challenging dataset.

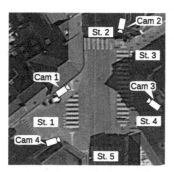

Fig. 2. Camera layout plotted on Google Maps.

4.2   Automatic Turning Movement Count of Vehicles and Cyclists

In smart traffic application, turning movement count (TCM) at intersection is essential information to understand the traffic flow for optimal traffic management. For example, more efficient signal timing plan can be derived from TMC information. Therefore, we measure the performance of our tracker in a TMC application for vehicles and cyclists. First, trajectories of road users denoted as $S = \{s_1, s_2, \ldots, s_N\}$ are generated using the proposed tracker. Then we define regions at the beginning of each streets: Ω_1, Ω_2, Ω_3, Ω_4 and Ω_5 for *Street 1, 2, 3, 4* and *5* respectively. If a trajectory passes through a region Ω_f first and a region Ω_l last, it is classified as a trajectory coming from *Street f* and going into *Street l*. This simple rule-based method produces the TMC as listed in Table 1.

Both tables in Table 1 clearly show that *Street 2* and *5* are the main streets as TMC between two streets for both vehicles and cyclists are much higher than others. Approximately 60% of traffic (both vehicles and cyclists) at the intersection goes trough *Street 2* and *5*. The second most used path is between *Street 1* and *2* constituting approximately 10% of the total traffic flow. The street with the lowest incoming vehicle traffic is *Street 3* since only three incoming vehicles are detected). However, there should be no incoming vehicle in *Street 4* since it is just a one-way street allowing only outgoing traffic. From visual inspection of those trajectories, we learn that all three trajectories are wrong. In fact, two trajectories are false positive trajectories and one is the first part of segmented trajectories of a vehicle. The vehicle turns into the *Street 4* from *Street 5* but the driver realizes that *Street 4* is a one-way street and incoming traffic is not allowed. Therefore, the vehicle turns back and goes into the *Street 3*. This results in two isolated trajectories of the same target causing error in automatic TMC application.

Table 1. Results of automatic turning movement count at five arms intersection.

		Destination								Destination			
		St. 1	St. 2	St. 3	St. 4	St. 5			St. 1	St. 2	St. 3	St. 4	St. 5
Origin	St. 1	0	7	6	0	12	Origin	St. 1	1	25	6	5	7
	St. 2	28	5	15	0	142		St. 2	33	2	2	6	137
	St. 3	8	10	0	0	6		St. 3	9	2	0	2	9
	St. 4	9	8	12	0	8		St. 4	4	24	2	1	5
	St. 5	14	122	8	3	4		St. 5	4	117	10	7	6
		(a) Vehicle trajectories							(b) Cyclist trajectories				

To obtain numerical results of TMC's accuracy, we randomly selected 100 trajectories and visually inspected in the videos if they are correct. We achieve an accuracy as high as 93% on vehicles and 83% on cyclists. The key of achieving high accuracy in turning movement count is to be able to track targets while avoiding tacking loss (resulting segmented trajectories instead of a complete trajectory for a target) and identity switches. Figure 3 illustrates the example of typical trajectories produced by our tracker. Two cyclists are fully occluded by a white van in the view of *Camera 1* and *2*. However, they are visible in the view of *Camera 3* and *4* (partially occluding each other). Although there is no local estimate available from *Camera 1* and *2*, our tracker fuses local estimates from *Camera 3* and *4* to produce accurate joint estimates and tracks two cyclists without any tracking lost.

4.3   Anomaly Detection of Road Users

Road users usually follow similar paths as they move about in traffic. For example, the majority of pedestrians walk on the road side pavement and cross the road along a pedestrian crossing. A trajectory which is very different from the common trajectories is regarded as an abnormal trajectory. For instance, a pedestrian may cross the road without using a pedestrian crossing. If trajectories are clustered based on their similarity, common trajectories form clusters which contain the majority of the trajectories while abnormal trajectories form clusters containing only a few trajectories. In this subsection, we assess the reliability of our tracker in trajectory anomaly detection application. For this purpose, we propose a greed clustering algorithm as described in Algorithm 2. The algorithm uses the Euclidean distance between point pairs from r_j and r_k defined

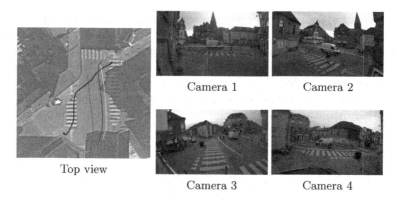

Fig. 3. Example tracking results of a pedestrian (red trajectory), two cyclists (blue and cyan trajectories) and a van (green trajectory). (Color figure online)

by Dynamic Time Warping algorithm [13] as a dissimilarity measure, which is denoted as $\delta_{DWT}(r_j, r_k)$ in Algorithm 2. The operator $|.|$ computes the cardinality of a set.

Algorithm 2. Greedy clustering.

1: **Input:** $S = \{s_1, s_2, \ldots, s_N\}$: a set of trajectories
2: **Output:** $S_1, S_2, \ldots, S_M$: sets of trajectories
3: $i \leftarrow 1$, $S_i \leftarrow \emptyset$
4: **while** $S \neq \emptyset$ **do**
5:    $S_i' \leftarrow S_i$
6:    **for each** $r_j \in S$ **do**
7:       **if** $\frac{\sum_{r_k \in S_i} \delta_{DWT}(r_j, r_k)}{|S_i|} < \Upsilon$ **then**
8:          $S_i \leftarrow S_i \cup \{r_j\}$, $S \leftarrow S \setminus \{r_j\}$
9:       **end if**
10:   **end for**
11:   **if** $S_i' = S_i$ **then**
12:      $i \leftarrow i + 1$, $S_i \leftarrow \emptyset$
13:   **end if**
14: **end while**

All trajectories from *Street 5* to *2* are grouped into a single cluster and no possible anomaly is detected as shown in Fig. 4a. For trajectories coming from *Street 5* and going into *Street 1*, two usual clusters are formed and one anomaly is detected. When there is no vehicle/cyclist waiting at the mouth of *Street 1* to go into other streets, vehicles coming from *Street 5* tend to make smaller turns to go into *Street 1*. These trajectories form a cluster of common trajectories which are shown as green trajectories in Fig. 4b. Vehicles make bigger turns when there are vehicles/cyclists waiting at the mouth of *Street 1* resulting in another cluster of common trajectories shown as red trajectories in Fig. 4b. The abnormal trajectory shown as blue trajectory in Fig. 4b is the trajectories of a car which drives onto a pavement as it turns into *Street 1* and parks for a while before continue driving down the same street. Figure 4c depicts the detected abnormal trajectory which is a results of a car coming out of *Street 3*, which turns into *Street 4* and parks at the mouth of the street for about five minutes. Then the car drives out of *Street 4* and goes into *Street 2*. The other trajectories shown in red are usual trajectories from *Street 3* to *2*.

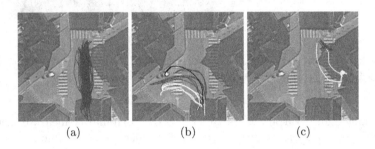

(a)   (b)   (c)

Fig. 4. Examples of anomaly detection in vehicle trajectories. (Color figure online)

Pedestrian trajectories are also clustered using Algorithm 2 and examples of the resulting clusters are shown in Fig. 5. As expected, common trajectories are clustered as big clusters which usually are along pedestrian crossings (as shown

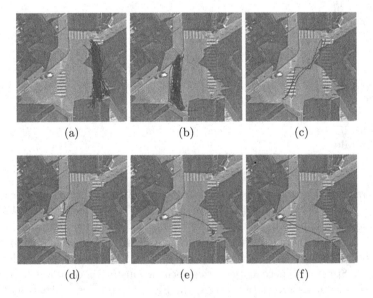

(a)   (b)   (c)

(d)   (e)   (f)

Fig. 5. Examples of anomaly detection in pedestrian trajectories. (Color figure online)

in Fig. 5a and b) and pavements. Three similar abnormal trajectories are formed when pedestrians cross the road from the corner of *Street 1* and *5* to the corner of *Street 2* and *3*, as shown in Fig. 5c. A pedestrian walks to the middle of the junction and comes back while crossing *Street 1* using the pedestrian crossing. This results in a trajectory which is quite different from the other common trajectories as shown in Fig. 5d. A trajectory shown in Fig. 5e is formed by a person getting out of a parked car (the same car which causes the abnormal vehicle trajectory shown in Fig. 4b) and walked straight to a shop at the corner of *Street 4* and *5*. A trajectory of the same person returning from the shop to the car is also detected as abnormal trajectory in another cluster as depicted in Fig. 5f.

5 Conclusion

This paper presented the multi-camera tracking method which simultaneously tracks multiple classes of road users such as pedestrians, cyclists and vehicles. It provides reliable trajectories for subsequent smart traffic applications such as turning movement count and anomaly detection. The distributed design of the proposed tracker allows the deployment on smart camera networks, keeping all local computer vision tasks on smart cameras. Only concise high-level information is exchanged for joint position estimation. Each smart camera locally estimates the image plane position of targets by recursive Bayesian estimation using YOLO detector responses and template matching as observations. Locally estimated image positions of targets are then transmitted to the fusion node where corresponding ground plane positions are jointly estimated by minimizing proposed cost function. Performance of the proposed method was assessed in context of the turning movement count application and achieved an accuracy as high as 93% on vehicles and 83% on cyclists. Abnormal itineraries of vehicles and pedestrians were also detected with high reliability by clustering trajectories produced by the proposed tracker.

Acknowledgment. This research received funding from the Flemish Government under the *"Onderzoeksprogramma Artificiële Intelligentie (AI) Vlaanderen"* programme.

References

1. Dai, J., Li, Y., He, K., Sun, J.: R-FCN: object detection via region-based fully convolutional networks. In: Proceedings of the 30th International Conference on Neural Information Processing Systems. NIPS 2016, pp. 379–387. Curran Associates Inc., USA (2016)
2. Fleuret, F., Lengagne, R., Fua, P.: Fixed point probability field for complex occlusion handling. In: Tenth IEEE International Conference on Computer Vision, ICCV 2005, vol. 1, pp. 694–700, October 2005
3. Guan, J., et al.: Template matching based people tracking using a smart camera network. In: Proceedings of SPIE, vol. 9026, pp. 1–9 (2014)

4. Hu, Z., Wang, C., Uchimura, K.: 3D vehicle extraction and tracking from multiple viewpoints for traffic monitoring by using probability fusion map. In: 2007 IEEE Intelligent Transportation Systems Conference, pp. 30–35, September 2007

5. Lee, S., Baik, H.: Origin-destination (O-D) trip table estimation using traffic movement counts from vehicle tracking system at intersection. In: IECON 2006–32nd Annual Conference on IEEE Industrial Electronics, pp. 3332–3337, November 2006

6. Liu, L., Xing, J., Ai, H.: Multi-view vehicle detection and tracking in crossroads. In: The First Asian Conference on Pattern Recognition, pp. 608–612, November 2011

7. Nishikawa, Y., Sato, H., Ozawa, J.: Multiple sports player tracking system based on graph optimization using low-cost cameras. In: 2018 IEEE International Conference on Consumer Electronics (ICCE), pp. 1–4, January 2018

8. Nyan, B.B., Veelaert, P., Philips, W.: Occlusion robust symbol level fusion for multiple people tracking. In: Proceedings of the 12th International Joint Conference on Computer Vision, Imaging and Computer Graphics Theory and Applications (VISIGRAPP 2017), vol. 6, pp. 216–226 (2017)

9. Ooi, H.-L., Bilodeau, G.-A., Saunier, N., Beaupré, D.-A.: Multiple object tracking in urban traffic scenes with a multiclass object detector. In: Bebis, G., et al. (eds.) ISVC 2018. LNCS, vol. 11241, pp. 727–736. Springer, Cham (2018). https://doi.org/10.1007/978-3-030-03801-4_63

10. Qiu, H., et al.: Kestrel: video analytics for augmented multi-camera vehicle tracking. In: 2018 IEEE/ACM Third International Conference on Internet-of-Things Design and Implementation (IoTDI), pp. 48–59, April 2018

11. Redmon, J., Divvala, S., Girshick, R., Farhadi, A.: You only look once: unified, real-time object detection. In: 2016 IEEE Conference on Computer Vision and Pattern Recognition (CVPR), pp. 779–788, June 2016

12. Rinner, B., Wolf, W.: An introduction to distributed smart cameras. Proc. IEEE 96(10), 1565–1575 (2008)

13. Sakoe, H., Chiba, S.: Dynamic programming algorithm optimization for spoken word recognition. IEEE Trans. Acoust. Speech Signal Process. 26(1), 43–49 (1978)

14. Saunier, N., Sayed, T.: A feature-based tracking algorithm for vehicles in intersections. In: Proceedings of the The 3rd Canadian Conference on Computer and Robot Vision, CRV 2006, p. 59. IEEE Computer Society, Washington (2006)

15. Tang, H.: Development of a multiple-camera tracking system for accurate traffic performance measurements at intersections. Final report, Intelligent Transportation Systems institute, Center for Transaction Studies, University of Minnesota (2013)

16. Hu, W., Xiao, X., Fu, Z., Xie, D., Tan, T., Maybank, S.: A system for learning statistical motion patterns. IEEE Trans. Pattern Anal. Mach. Intell. 28(9), 1450–1464 (2006)

17. Zhang, H., Geiger, A., Urtasun, R.: Understanding high-level semantics by modeling traffic patterns. In: 2013 IEEE International Conference on Computer Vision, pp. 3056–3063, December 2013

Vehicles Tracking by Combining Convolutional Neural Network Based Segmentation and Optical Flow Estimation

Tuan-Hung Vu[1]([✉]), Jacques Boonaert[1], Sebastien Ambellouis[2], and Abdelmalik Taleb Ahmed[3]

[1] IMT Lille Douai, Douai, France
{tuan-hung.vu,jacques.boonaert}@imt-lille-douai.fr
[2] IFSTTAR, Villeneuve d'Ascq, France
sebastien.ambellouis@ifsttar.fr
[3] Université Politechnique Hauts-de-France, Valenciennes, France
abdelmalik.taleb-ahmed@uphf.fr

Abstract. Object tracking is an important proxy task towards action recognition. The recent successful CNN models for detection and segmentation, such as Faster R-CNN and Mask R-CNN lead to an effective approach for tracking problem: tracking-by-detection. This very fast type of tracker takes into account only the Intersection-Over-Union (IOU) between bounding boxes to match objects without any other visual information. In contrast, the lack of visual information of IOU tracker combined with the failure detections of CNNs detectors create fragmented trajectories. Inspired by the work of Luc *et al.* that predicts future segmentations by using Optical flow, we propose an enhanced tracker based on tracking-by-detection and optical flow estimation in vehicle tracking scenario. Our solution generates new detections or segmentations based on translating backward and forward results of CNNs detectors by optical flow vectors. This task can fill in the gaps of trajectories. The qualitative results show that our solution achieved stable performance with different types of flow estimation methods. Then we match generated results with fragmented trajectories by SURF features. DAVIS dataset is used for evaluating the best way to generate new detections. Finally, the entire process is test on DETRAC dataset. The qualitative results show that our methods significantly improve the fragmented trajectories.

Keywords: Vehicles tracking · Instance segmentation · Optical flow

1 Introduction

1.1 Context

Application of achievements in computer vision and machine learning for transportation field, especially autonomous driving car, is a very active research

© Springer Nature Switzerland AG 2020
J. Blanc-Talon et al. (Eds.): ACIVS 2020, LNCS 12002, pp. 529–540, 2020.
https://doi.org/10.1007/978-3-030-40605-9_45

domain. Thanks to the recent successful CNNs model for object detection and segmentation [1,6,7,11], we can achieve almost fully spatial information about all ones and all things involved in traffic networks. Those strong models outperform all the previous traditional methods that use hand-crafted features then learning representation. Naturally, the next step for creating autonomous driving car is actions or activities recognition in transportation, particularly abnormal activities that may cause dangerous situations. Most of the effective methods for normal daily action recognition [13–15] starts with a tracking step for obtaining objects' trajectories. Thus, we observe that object tracking is an important proxy task towards action recognition.

Recently, the rise of CNNs provides us various successful models for object detection and segmentation: Faster R-CNN [6], YOLO [11], FCN [16], Mask R-CNN [1], etc. Among them, Mask R-CNN is our best candidate because we can benefit from various types of information (object class, localization and segmentation at instance level) using only one network. Intuitively, the more information we get from objects, the more effective solution we have for the tracking step. Thanks to the strong performances of Mask R-CNN, we have a simple and effective approach for object tracking: tracking-by-detection. This type of tracker first applies an object detector to each video frame then associates these detections to tracks. One of the most representative and popular tracker of that kind is IOU Tracker [2]. It is the core tracker of winning solutions for multi-objects tracking challenge in AVSS 2018 [5].

1.2   Drawback of Regular Methods: IOU Tracker and Mask RCNN

Instead of using visual information to match the objects locations, IOU Tracker takes into account the IOU between bounding boxes to associate them together. This feature first, makes the performance of the tracker completely dependent from the performance of the detector itself. Second, the lack of visual information lead to the confusing detections between object's classes in overlapped case.

(a)   (b)   (c)   (d)

Fig. 1. Common missing detection of Mask R-CNN in consecutive frames: (a) detection; (b) non-detection; (c) detection; (d) non-detection

On the other hands, Mask R-CNN shows its drawback in the case of false negative error on usual objects (see Fig. 1) although it yields strong performances on COCO dataset [8]. This false negative results or missing detections from Mask R-CNN can lead to cases of failure in the next step of object's tracking because the object trajectories gets "broken". Obviously, confusing detections between object's classes from Mask R-CNN can also produces wrong tracks, for instance when creating new tracking processes instead of maintaining existing ones.

1.3   Objective

Our work aims to build an enhanced tracker that can limit the drawback of fragmented trajectories by Mask R-CNN detector on IOU Tracker while maintain their advantages of fast tracker. The first step is filling in the gaps of fragmented detections using a generative approach. The next step consists in combining the generated results with the current results from the detectors. Then, by applying the idea of IOU trackers, IOUs between the bounding boxes are used to eliminate the overlapping results. The final step is performed by associating the trajectories with SURF feature [3], which is a high performance feature for image matching. Our contributions are the following:

- Introduction of fast and efficient generative approach using optical flow for fill in the discontinuity of object tracking. Our solution is stable with difference type of flow estimation.
- An enhanced tracker integrating IOU tracker based Mask R-CNN with visual information.

2   Background

2.1   Mask R-CNN

Mask R-CNN [1] is a conceptually simple, flexible, and general framework for objects instances segmentation. In principle, Mask R-CNN is an extension of Faster R-CNN [6] that constructs a third branch as FCN [16] for segmentation. Therefore, we find Mask R-CNN is an effective combination of elements from the classical computer vision tasks for object detection and semantic segmentation. Mask R-CNN surpasses all previous state-of-the-art single-model results on the COCO dataset for both instance segmentation and bounding box detection tasks. Though, it has limitation in implementation for other datasets. In particular, the detection can get unstable when the object is rapidly changing its appearance. Such events frequently occur in transportation based video sequences, for instance when a vehicle is turning right or left or when its apparent size increase or decrease owing to its relative move with respect to the camera.

2.2   IOU Object Tracker

Before the rise of CNNs models, almost all successful trackers performed object location based on hand-crafted features: IHTLS [20], H2T [21], CMOT [19], etc. Because of the use of (so called) too simple concepts, these trackers show their limited performances when facing difficult scenarios from recent multi object tracking challenges [5]. Recently, IOU Tracker [2] based on the strong performances of CNN based detector surpassed all previous methods in both easy and difficult scenario. Bochinski *et al.* considered only the overlaps between bounding boxes obtained from detector to associate them. It benefits from the fast and strong performance of CNNs based detector without any other visual information. As a consequence, this can be seen as a bit risky process due to the complete dependence of the tracking task to the accuracy of the detector: as discussed above, both false positive and negative error of detector can cause fragmented trajectories.

2.3   Optical Flow

Optical flow is the apparent motion of brightness patterns in the image. Because the brightness itself is greatly influenced by the moves of the objects within the observed scene, optical flow is an effective way to estimate the motions field generated by the objects' trajectories. Thus, optical flow estimation is an important research domain in computer vision. There are several successful models for estimating optical flow: LDOF [4] and Full Flow [29] estimate large displacements of flow vector thanks to optimization and deep matching, while Flownet2.0 [22] and its lighter version PwC-Net [23] compute flow vector using an encode-decode CNNs architecture. Generally, optical flow is usually a basement for action recognition. On the other hand, Luc *et al.* [9] introduced a new utilization of optical flow for predicting future segmentation. This first work suggests us a promising way to exploit optical flow for generating the missing video information.

3   Related Work

Generative modeling of future RGB video frames has recently been studied using a variety of techniques: prediction of future human pose [24], generative adversarial training model [26], forecasting convolutional features [9]. All their work focus on predicting future information which they never can not achieve. This context is a bit different than our work where we address the problem of generating only the missing information the detector failed to produce. We can freely generate backward and forward segmentation or bounding boxes based on some current results from the detectors. Despite the difference between those contexts, we apply the simple baseline approach from Luc *et al.* [9] that translated segment on the basis of flow vectors.

4   Generating Object Segmentation by Mask R-CNN and Optical Flow

Luc *et al.* [9] proposed two baseline methods to generate future segmentation based on flow vector $F_{t-1 \rightarrow t}$ from frame $t - 1$ to t and the current instance segmentation I_t at frame t. They were called $Shift$ and $Warp$ (Fig. 2a).

- **Warp** approach translates each pixel of instance mask independently using the flow vector at the corresponding position inside this mask. To yield the new mask, the object class and the confident score is copied from the previous one. The predicted mask and flow field are used to make the next prediction, and so on. This approach is suitable for longtime prediction because of the ability of rescaling objects based on flow vector. In contrast, predicted mask suffers from an accumulated error phenomenon and has many holes inside its boundaries. To fill in those gaps, a post-processing step with morphological operator is necessary.
- **Shift** approach, in brief, entirely shifts instance mask using the average flow vector calculated inside the mask. Then the object class and the confident score are also copied from the previous mask. While this approach can avoid accumulating errors and generating holes inside the mask, it is not really suitable for longtime prediction due to the non-rescaling of the mask.

We propose an extension beyond the two original baselines. Instead of only predicting future segmentation (forward), we also generate past segmentation (backward) by projecting the flow vector in opposite direction (Fig. 2b). Intuitively, the missing detection of Mask R-CNN does not last along many frames so we can generate new segmentation in both forward and backward directions.

Furthermore, we also consider the development of flow vector beyond one time step. Given instance results I_t, I_{t+N} of frame t and $t + N$ achieved from Mask R-CNN detector, we have (Fig. 2c):

$$I_{t+j;0<j<N} = \sum p_{t+j} | \mathbf{F}_{t \to t+j}(p_t) = p_{t+j}, \mathbf{F}_{t+j \to t+N}(p_{t+j}) = p_{t+N} \quad (1)$$

where p_i is the pixel of instance I_i, $\mathbf{F}_{i \to l}$ is a translation by flow vectors from frame i to frame l. To avoid the accumulated errors of optical flow estimation methods, we minimize $N = 2$, $j = 1$.

This condition maintain the continuity of trajectories along the flow vectors when we generate new masks. We apply this extension in both forward and backward directions with both *Shift* and *Warp* approaches. Therefore, we have 8 ways to generate object segmentation to fill in the gaps of trajectories.

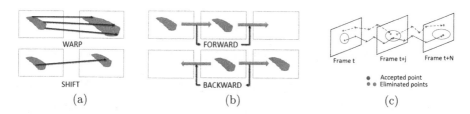

Fig. 2. Generating object segmentation (a) Shift and warp translation; (b) Forward and backward translation; (c) Combined results beyond one time step

T.-H. Vu et al.

5 Improving IOU-Tracker with Generated Informations and SURF Features

Intuitively, the more boxes we have, the more accurate the IOU Tracker is. Despite the strong performances of Mask R-CNN, the missing detections in some frames are inevitable. The technique of generating new segment can be similarly applied for generating new bounding boxes. We consider each box as a special mask which contains only four pixel corresponding to four vertices of box and then the process is repeated. After applying generative approach, we get a situation where we have more overlapping boxes. To eliminate those redundant boxes, we apply the idea of IOU tracker that takes into account the IOU between boxes. Let B_{tG} denote the set of generated boxes b_{tG}^i and B_{tM} denote the set of instance boxes b_{tM}^i directly detected by Mask R-CNN at frame I_t. To combine the two sets, we compare the overlapping region with a threshold σ_{IOU} by running this algorithm:

Algorithm 1. Eliminate overlapping boxes

for $i = 1 \rightarrow ||B_{tG}||$ **do**
   $count \leftarrow 0$
   for $j = 1 \rightarrow ||B_{tM}||$ **do**
      if $IOU(b_{tG}^i, b_{tM}^i) \geq \sigma_{IOU}$ and $class(b_{tG}^i) = class(b_{tG}^i)$ **then**
         $count \leftarrow count + 1$
      end if
   end for
   if $count = 0$ **then**
      add b_{tG}^i to B_{tM}
   else
      discard b_{tG}^i
   end if
end for

It means that we trust more in the results provided by the Mask R-CNN detector (this assertion is based on the analysis of the experimental results described in Sect. 5.1). After this step, we have only one set B_t containing the box b_t^i of frame I_t.

The next stage consists in applying the IOU Tracker. Each box b_t^i is compared with all the boxes from the L previous frames, where L is the length of the tracker. We associate a box to the previous box which obtains the same class and the maximum overlapping region. If we can not find any IOU value greater than the threshold σ_{IOU} along all L frames, the current tracking process is terminated and we start a new track.

To enhance our tracker, we propose to use SURF [3] to verify the matching boxes. This feature have shown its efficiency on image matching problem. We extract SURF points for each bounding boxes. Each box b_t^i is compared with all boxes of L previous frames. If we find two boxes associated to two different classes but with their IOU value and the number of matching SURF points greater than threshold σ_{IOU} and σ_{SURF}, respectively, we associate them together. To avoid the negative effect caused by extracting SURF features from a too small box, we set another threshold σ_S to select the suitable boxes which exhibit a value greater than this threshold. For all the boxes with a value smaller than σ_S, we do not extract SURF and we only take into account IOU value for the association step. The whole process is illustrated by Fig. 3.

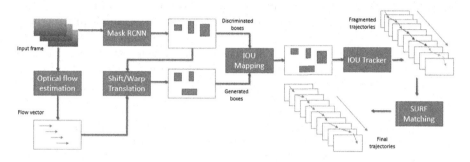

Fig. 3. Improving IOU-Tracker with generated informations and SURF features

6   Experiment

We separately evaluate each stage using different datasets. First, we compare the techniques for generating objects segmentation stages using the DAVIS dataset [10]. Then, we chose the most suitable methods to apply for the tracking stage. The qualitative evaluation regarding the improvement of the IOU Tracker is performed using the UA-DETRAC dataset [5].

6.1   Generating Object Segmentation

DAVIS 2016 consists of fifty high quality, Full HD video sequences, spanning multiple occurrences of common video object segmentation challenges such as occlusions, motion blur and appearance changes. Each video is accompanied by densely annotated, pixel-accurate and per-frame ground truth segmentation. Because the UA-DETRAC only takes into account threes classes (bus, van and car) we restrict the use of the DAVIS sequences to the ones that deal with the same types of vehicles. Unfortunately, only a few sequences within the DAVIS dataset satisfy this constraint. Worst, those sequences are not challenging enough to generate the missing detections from Mask R-CNN we want to cope with. Thus, in order to prepare the situation of false negative errors, we run Mask R-CNN for each frames, then randomly discard some of the generated. The original annotation of discarded frames are utilized as ground-truth to compare with the generated results. The performance is measured with the mean Intersection-over-Union (mIOU) metric, which first computes the IOU for each semantic class and then computes the average over classes. IOU is defined as follows:

$$IOU = \frac{TruePositive}{TruePositve + FalseNegative + FalsePositive} \tag{2}$$

First, Mask R-CNN is applied to each frames. Next, we use LDOF [4] and Full Flow [29] to extract the optical flow vectors from pairs of frames. Then, we do the **Shift** and **Warp** translation in both backward and forward directions. Morphological operators is only executed as post-processing with the **Warp** method. The extension beyond more than one step time that we call "combined results" in Fig. 2c is also evaluated in both directions. Here, we choose the simplest case with $N = 2$ and $j = 1$. Our results with LDOF optical flow is illustrated in Tables 1 and 2; with Full Flow optical flow in Table 3.

Table 1. Results of generating new segmentation on DAVIS 2016 by LDOF with "car" and "bus" classes. S: Shift; W: Warp without morphological post-processing; W-M: Warp with morphological post-processing; R: True masks of Mask R-CNN which are eliminated to make the missing detection context; Combine-B: Extension case in backward direction; Combine-F: Extension case in forward direction

Method	Bus				Car			
	S	W	W-M	R	S	W	W-M	R
Backward	79.31	75.59	**81.76**	89.46	65.13	59.21	**70.92**	92.10
Forward	79.05	75.84	**82.07**	89.46	65.04	60.33	**69.15**	92.10
Combine-B	**79.35**	71.47	77.35	89.46	**66.87**	46.91	57.86	92.10
Combine-F	**79.03**	71.76	77.35	89.46	**65.36**	48.44	56.37	92.10

Table 2. Results of generating new segmentation on DAVIS 2016 by Full Flow with "car" and "bus" classes.

Method	Bus				Car			
	S	W	W-M	R	S	W	W-M	R
Backward	79.24	76.72	**82.36**	89.46	65.66	60.16	**70.55**	92.10
Forward	79.05	76.63	81.96	89.46	65.34	61.00	**68.94**	92.10
Combine-B	**79.31**	72.42	77.57	89.46	**66.41**	48.03	57.33	92.10
Combine-F	**79.06**	72.64	77.48	89.46	**65.63**	49.19	56.66	92.10

We observe that the morphological operators are important with the **Warp** method. After adding this post-processing, the performances are significantly

Table 3. Average performance of generating new segmentation on DAVIS 2016 with "car" and "bus" classes.

Method	LDOF			Full Flow			Mask R-CNN
	Shift	Warp	Warp-M	Shift	Warp	Warp-M	
Backward	72.22	67.40	**76.34**	72.45	68.44	**76.46**	90.78
Forward	72.05	68.08	**75.61**	72.19	68.81	**75.50**	90.78
Combine-B	**73.11**	59.19	67.60	**72.86**	60.23	67.45	90.78
Combine-F	**72.20**	60.10	66.86	**72.35**	60.192	67.07	90.78

increased. While **Shift** works stably for all of the methods, this is not the case with the **Warp** based approach. In a simple case, where we only consider optical flow within two consecutive frames in both direction, **Warp** significantly outperforms **Shift**. Conversely, when we take into account the development of optical flow along many frames, **Shift** gives us better results. This difference can be explained by the accumulated error of optical flow. The more frames we process, the more optical flow error we integrate. **Shift** is chosen to reduce this issue. Furthermore, we did not see any significant differences for each method when we performed LDOF or Full Flow optical flow estimation. Most of average results in Table 3 are similar (Full Flow is slightly better) despite Full Flow significantly outperform LDOF in the original task of optical flow estimation for both MPI Sintel and KITTI Dataset. We draw that the performance of generating new segmentation depends on how we use the flow vector, not the type of flow. Therefore, our methods allow us to benefit regardless of the optical flow estimation methods, then fast methods are highly prioritized. On the other hand, we find that the original masks created by Mask R-CNN always give us better accuracy than the optical flow generated masks. This analysis suggests us to put more confidence in the discriminative results from Mask R-CNN than in the generative results from optical flow when we combine those results for the next tracking stages.

6.2 Enhanced Tracker with IOU-Tracker Based Mask R-CNN and Optical Flow

Based on the results from previous stages, we choose the **Shift** generator to create the new bounding boxes from optical flow. Additionally, tracking only in forward direction is more natural and simpler. Thus, we perform the **Shift** generator in forward direction. Full Flow [29] is used for estimating optical flow vectors. The discriminative boxes of Mask R-CNN and the generative boxes of **Shift** are combined thanks to a IOU mapping (and its σ_{IOU}) to discard all the generative boxes matching a discriminative box w.r.t its location and its object class. As discussed above, we trust more in discriminative boxes provided by Mask R-CNN. For the IOU tracking step, we choose $L = 5$ and use the same parameter σ_{IOU} as for the previous step. All the parameters σ_{IOU}, σ_{SURF}, are determined by experiment. The results show us that $\sigma_{IOU} = 0.5$, $\sigma_{SURF} = 1$ and $\sigma_S = 50 \times 50$ are the best choices. The qualitative results are shown in Fig. 4. We observe that the trajectories are less fragmented after applying our techniques.

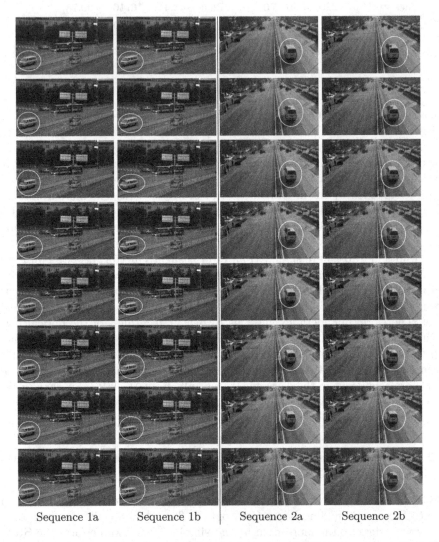

| Sequence 1a | Sequence 1b | Sequence 2a | Sequence 2b |

Fig. 4. Qualitative results of our solution for improving IOU Tracker based Mask R-CNN detector; the trajectories of the car inside the red circle are improved: (1a) (2a) Original IOU Tracker based Mask R-CNN detector, for sequence 1a we can track only one frame over 8 frames and sequence 2a achieves 3 frames over 8 frames; (1b) (2b) Improved IOU Tracker based Mask R-CNN detector with Shift generator and SURF matching, both sequences now achieve 6 frames over 8 frames

7   Future Works and Conclusion

We presented a tracker based generative method to solve the missing detections issue of a very popular deep network called Mask R-CNN. Our new method exploits an optical flow vectors estimation. It gives promising and usable results for the generative step and it produces stable performance regardless types of optical flow. In turn, this generator is applied to improve IOU tracker with adding SURF matching. Qualitative evaluation shows that our method significantly improves the continuity of tracking trajectories.

In the context of our future works, in order to go beyond the use of SURF, we will exploit more useful visual information to link the fragmented trajectories. Furthermore, different type of generator such as GANs will be evaluated to find a competitive process, in line with the performances of a detector such as Mask R-CNN.

References

1. He, K., Gkioxari, G., Dollár, P., Girshick, R.B.: Mask R-CNN. In: IEEE International Conference on Computer Vision, Italy (2017)
2. Bochinski, E., Eiselein, V., Sikora, T.: High-speed tracking-by-detection without using image information. In: International Workshop on Traffic and Street Surveillance for Safety and Security at IEEE AVSS, Italy (2017)
3. Bay, H., Ess, A., Tuytelaars, T.V., Gool, L.: Speeded-up robust features (SURF). Comput. Vis. Image Underst. **110**, 346–359 (2008)
4. Brox, T., Malik, J.: Large displacement optical flow: descriptor matching in variational motion estimation. IEEE Trans. Pattern Anal. Mach. Intell. **33**, 500–513 (2011)
5. Lyu, S., et al.: UA-DETRAC 2017: report of AVSS2017 & IWT4S challenge on advanced traffic monitoring. In: 14th IEEE International Conference on Advanced Video and Signal Based Surveillance AVSS (2017)
6. Ren, S., He, K., Girshick, R.B., Sun, J.: Faster R-CNN: towards real-time object detection with region proposal networks. IEEE Trans. Pattern Anal. Mach. Intell. TPAMI (2017)
7. Wang, L., Lu, Y., Wang, H., Zheng, Y., Ye, H., Xue, X.: Evolving boxes for fast vehicle detection. In: IEEE International Conference on Multimedia and Expo ICME (2017)
8. Lin, T.-Y., et al.: Microsoft COCO: common objects in context. In: Fleet, D., Pajdla, T., Schiele, B., Tuytelaars, T. (eds.) ECCV 2014. LNCS, vol. 8693, pp. 740–755. Springer, Cham (2014). https://doi.org/10.1007/978-3-319-10602-1_48
9. Luc, P., Couprie, C., LeCun, Y., Verbeek, J.: Predicting future instance segmentation by forecasting convolutional features. In: Ferrari, V., Hebert, M., Sminchisescu, C., Weiss, Y. (eds.) ECCV 2018. LNCS, vol. 11213, pp. 593–608. Springer, Cham (2018). https://doi.org/10.1007/978-3-030-01240-3_36
10. Perazzi, F., Pont-Tuset, J., McWilliams, B., Van Gool, L., Gross. M., Sorkine-Hornung, A.: A benchmark dataset and evaluation methodology for video object segmentation. In: Computer Vision and Pattern Recognition CVPR (2016)
11. Redmon, J., Farhadi, A.: YOLO9000: better, faster, stronger. In: The IEEE Conference on Computer Vision and Pattern Recognition CVPR (2017)

12. Wang, H., Klaser, A., Schmid, C., Liu, C.-L.: Dense trajectories and motion boundary descriptors for action recognition. Int. J. Comput. Vis. IJCV **103**, 60–79 (2013)
13. Wang, H., Schmid, C.: Action recognition with improved trajectories. In: IEEE International Conference on Computer Vision ICCV (2013)
14. Simonyan, K., Zisserman, A.: Two-stream convolutional networks for action recognition in videos. In: Advances in Neural Information Processing Systems NIPS (2014)
15. Wang, L., Qiao, Y., Tang, X.: Action recognition with trajectory-pooled deep-convolutional descriptors. In: IEEE Conference on Computer Vision and Pattern Recognition CVPR (2015)
16. Shelhamer, E., Long, J., Darrell, T.: Fully convolutional networks for semantic segmentation. IEEE Trans. Pattern Anal. Mach. Intell. TPAMI **39**, 640–651 (2017)
17. He, K., Zhang, X., Ren, S., Sun, J.: Deep residual learning for image recognition. In: IEEE Conference on Computer Vision and Pattern Recognition CVPR (2016)
18. Roth, S.: Discrete-continuous optimization for multi-target tracking. In: Proceedings of the IEEE Conference on Computer Vision and Pattern Recognition CVPR (2012)
19. Bae, S., Yoon, K.: Robust online multi-object tracking based on tracklet confidence and online discriminative appearance learning. In: IEEE Conference on Computer Vision and Pattern Recognition CVPR (2014)
20. Dicle, C., Camps, O.I., Sznaier, M.: The way they move: tracking multiple targets with similar appearance. In: IEEE International Conference on Computer Vision ICCV (2013)
21. Wen, L., Li, W., Yan, J., Lei, Z., Yi, D., Li, S.Z.: Multiple target tracking based on undirected hierarchical relation hypergraph. In: IEEE Conference on Computer Vision and Pattern Recognition CVPR (2014)
22. Ilg, E., Mayer, N., Saikia, T., Keuper, K., Dosovitskiy, A., Brox, T.: FlowNet 2.0: evolution of optical flow estimation with deep networks. In: IEEE Conference on Computer Vision and Pattern Recognition CVPR (2017)
23. Sun, D., Yang, X., Liu, M.-Y., Kautz, J.: PWC-Net: CNNs for optical flow using pyramid, warping, and cost volume. In: IEEE Conference on Computer Vision and Pattern Recognition CVPR (2018)
24. Villegas, R., Yang, J., Zou, Y., Sohn, S., Lin, X., Lee, H.: Learning to generate long-term future via hierarchical prediction. In: Proceedings of the 34th International Conference on Machine Learning ICML (2017)
25. Walker, J., Doersch, C., Gupta, A., Hebert, M.: An uncertain future: forecasting from static images using variational autoencoders. In: Leibe, B., Matas, J., Sebe, N., Welling, M. (eds.) ECCV 2016. LNCS, vol. 9911, pp. 835–851. Springer, Cham (2016). https://doi.org/10.1007/978-3-319-46478-7_51
26. Mathieu, M., Couprie, C., LeCun, Y.: Deep multi-scale video prediction beyond mean square error. In: International Conference on Learning Representations ICLR (2016)
27. Radford, A., Metz, L., Chintala, S.: Unsupervised representation learning with deep convolutional generative adversarial networks. In: International Conference on Learning Representations ICLR (2016)
28. Vondrick, C., Pirsiavash, H., Torralba, A.: Anticipating the future by watching unlabeled video. In: IEEE Conference on Computer Vision and Pattern Recognition CVPR (2016)
29. Chen, Q., Koltun, V.: Full flow: optical flow estimation by global optimization over regular grids. In: IEEE Conference on Computer Vision and Pattern Recognition CVPR (2016)

Real-Time Embedded Person Detection and Tracking for Shopping Behaviour Analysis

Robin Schrijvers[1,2], Steven Puttemans[1(✉)], T. Callemein[1], and Toon Goedemé[1]

[1] EAVISE, KU Leuven, Jan De Nayerlaan 5, 2860 Sint-Katelijne-Waver, Belgium
{steven.puttemans,timothy.callemein,toon.goedeme}@kuleuven.be
[2] PXL Smart-ICT, Hogeschool PXL, Elfde Liniestraat 24, 35000 Hasselt, Belgium
robin.schrijvers@pxl.be

Abstract. Shopping behaviour analysis through counting and tracking of people in shop-like environments offers valuable information for store operators and provides key insights in the stores layout (e.g. frequently visited spots). Instead of using extra staff for this, automated on-premise solutions are preferred. These automated systems should be cost-effective, preferably on lightweight embedded hardware, work in very challenging situations (e.g. handling occlusions) and preferably work real-time. We solve this challenge by implementing a real-time TensorRT optimized YOLOv3-based pedestrian detector, on a Jetson TX2 hardware platform. By combining the detector with a sparse optical flow tracker we assign a unique ID to each customer and tackle the problem of loosing partially occluded customers. Our detector-tracker based solution achieves an average precision of 81.59% at a processing speed of 10 FPS. Besides valuable statistics, heat maps of frequently visited spots are extracted and used as an overlay on the video stream.

Keywords: Person detection · Person tracking · Embedded · Real-time

1 Introduction

Mapping the flow and deriving statistics (e.g. the amount of visitors or the time spent in as store) of people visiting shop-like environments, holds high value for store operators. To this day, people counting in retail environments is often being accomplished by using cross-line detection systems [24] or algorithms counting people through virtual gates [32].

To accurately count visitors, one can place a computer in the network of the store with access to already available security cameras, deploying software that detects and tracks people in the store automatically and stores the results on a central storage system (e.g. an in-store server or the cloud). In order to run these software solutions, one needs expensive and bulky dedicated computing hardware, frequently covered by a desktop GPU (e.g. NVIDIA RTX 2080). On

© Springer Nature Switzerland AG 2020
J. Blanc-Talon et al. (Eds.): ACIVS 2020, LNCS 12002, pp. 541–553, 2020.
https://doi.org/10.1007/978-3-030-40605-9_46

the other hand, the recent availability of lightweight and affordable embedded GPU solutions, like the NVIDIA Jetson TX2, can be a valid alternative. This is the main motivator to build an embedded and cost-effective people detection and tracking solution. An example of the camera viewpoint from the designed setup can be seen in the left part of Fig. 1.

The remainder of this paper is organized as follows. Section 2 discusses related work on person detection and tracking, along with available embedded hardware solutions. Section 3 provides details about the proposed implementation, while experiments and results are discussed in Sect. 4. Finally, Sect. 5 summarizes this work and discusses useful future research directions.

Fig. 1. Examples of (left) the viewpoint from our camera setup in the store and (right) the generated heat map for the store owner using the combined detector-tracker unit.

2   Related Work

The related work section is subdivided in three subsections, each focusing on a specific subtopic within this manuscript. Subsect. 2.1 starts by discussing literature on person detection, where Subsect. 2.2 continues on specific optimizations in this technology for embedded hardware. Finally Subsect. 2.3 focuses on the person tracking part.

2.1   Person Detection

Robust person detection solutions have been heavily studied in literature. Originally person detection made use of handcrafted features, combined with machine learning to generate an abstract representation of the person [6,9,12,13,15,47]. While these approaches showed promising results, their top-accuracy and flexibility make that these algorithms are not suited for the very challenging conditions in which they will be deployed in this application. Dynamic backgrounds, illumination changes and different store lay-outs are only some of the reasons of a high rate of false positive detections. While re-training the detectors for every store layout would theoretically solve the issue, this isn't a cost-effective solution for companies.

Convolutional neural networks for detecting persons in images offer potentially more robust solutions [17,20,31,38,39]. By automatically selecting the most discriminate feature set based on a very large set of training data, they quickly pushed the traditional approaches into the background. In literature these deep learning based approaches are subdivided in two categories. Single-shot approaches [31,38–40] solve the detection task by classifying and proposing bounding boxes in a single feed-forward step through the network. Multi-stage approaches [16,17,19,41] include separate networks that first generate region proposals before classifying the objects inside the proposals. A more recent and promising multi-stage approach is Trident-Net [30], where scale-specific feature maps are built into the network. Single-stage approaches tend to have a more compact and faster architecture making them preferred in lightweight embedded hardware solutions.

2.2 Optimizing for Embedded Hardware

Solving the task of person detection on embedded platforms has been an active research topic in recent years. A common approach is to enable deep learning on embedded platforms by studying more compact models such as Tiny-YOLOv2 [39]. While these compacter models are able to run at decent speeds on embedded hardware, they tend to lose some percentage in accuracy compared to their full counterparts (e.g. YOLOv2). Besides going compact, several approaches [34,43, 49] optimize the architecture further by looking at the indirect computation complexity and address efficient memory access and platform characteristics. More recent embedded object detection algorithms introduce optimized filter solutions, like depth-wise separable convolutions [22] and inverted residuals [42], to further optimize the performance of these embedded solutions.

With the uprising of FPGA chips for deep learning, several architectures like [23] try to reduce the number of parameters of the models even further, having a 50x reduction in parameters compared to classic models like [28]. These specifically designed hardware chips also allow for fixed-point 16-bit optimizations through OpenCL [48]. While being very promising, these FPGA systems still lack flexibility. They are in most cases designed for a very specific case, and are thus no cost-effective solution for our problem.

While many of these embedded object detectors are explicitly shaped for running on embedded hardware, detection accuracies are still lower compared to their traditional full blown CNN counterparts (eg. YOLOv3). Taking into account that our solution should perform person detection at a minimal processing speed of 10 FPS and a minimal accuracy of 80%, we opt for integrating an optimized embedded implementation of the YOLOv3 object detector [46] in our pipeline. The architecture smartly combines several optimizations based on mobile convolutions in PyTorch and TensorRT compilations.

The introduction of the NVIDIA Jetson embedded GPU enabled balancing of local-processing, low power consumption and throughput in an efficient way. [35] discusses several implementations of deep learning models on the Jetson platform while considering a range of applications, e.g. autonomous driving or traffic surveillance. The work focuses on obtaining low latency, to make detectors useful for providing valuable real-time feedback. The advantages in both FPGA and embedded GPU systems are the fact that they are substantially more space-efficient and less power-consuming than desktop GPUs.

2.3 Person Tracking

Tracking objects in videos has been studied across many fields. Whenever quasi-static environments occur, motion detection through robust background subtraction is used to identify moving objects in images [8,11,27,29]. More challenging cases involve dynamic environments, e.g. tracking objects in autonomous driving applications or drones. [7] solves the tracking task using detection results of a lightweight object detection algorithm combined with euclidean distance equations, GPS-locations and data from an inertial measurement unit. Because research is moving more and more towards highly accurate CNN-based object detectors, a new sort of tracking is introduced, called tracking-by-detection. By calculating an intersection over union (IoU) between the detections of two consecutive frames and applying a threshold, one can decide if we are dealing with the same object.

Where tracking-by-detection works well in many situations, there are also some downsides with these approaches. When people are missed in several subsequent frames by the initiating detector, the person gets lost during tracking. However, in our application of real-time costumer analysis, we want to keep a unique ID for each detected person, and thus need to fill in this gap automatically. [2] solves this issue by integrating a SVM-classifier with an optical-flow-based tracker. With the uprising of CNN-based detectors, CNN-based object trackers have also been proposed. Limb-based detection with tracking-by-detection is proposed in [1], while [4] explains an approach using IoU information from multiple object detectors. In addition, [36] combines location information and similarity measures to perform tracking-by-detection.

Feature-based tracking algorithms are a valid alternative. Sparse optical flow calculates the flow of objects based on a sparse feature set [33,45] and have been successfully deployed for person tracking [11,26,44]. Dense optical flow [14] calculates the flow for every point in the detection and is thus more computationally complex. [37] propose an object tracker based on weakly aligned multiple instance local features and an appearance learning algorithm.

A final range of trackers use online learning, where a tracker learns the representation of the object on the fly through a classifier when initialized with a bounding box [3,18]. [21] learns more efficiently from these overlapping patches by using kernelized correlation filters, while [25] enables tracking failure detection by tracking objects both forward and backward in time, measuring and qualifying the discrepancies between the two trajectories.

[10] gives a clear general overview of state-of-the-art trackers and their accuracy on a public dataset. For this work, we integrated three object trackers: sparse optical flow, kernelized correlation filters and median flow.

This paper proposes a complete off-the-shelf solution, taking into account the fact that we work on a compact embedded system with limited power consumption, achieve real-time performance (e.g. minimally processing at 10 FPS) and obtain acceptable accuracies (e.g. over 75%) in a single setup. While many of these sub-tasks have been discussed already in literature, to our best knowledge, there isn't a single publication that proposes such end-to-end solution for in-store customer behaviour analysis. On top of that we are combining all parts of the pipeline in a batch system that consumes the resources (both CPU and GPU) of the host system as efficient as possible. This is the biggest novelty this work introduces and will be discussed in further detail in subsection 3.1.

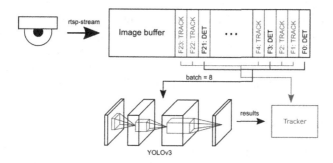

Fig. 2. Pipeline overview with the proposed batch processing approach, passing frames from the image buffer to the detection or tracking unit, based on the batch iterator.

3   Methodology

The goal of this paper is to map the flow of costumers in shop-like environments, with a special focus on detecting and tracking individual costumers using a unique ID. Preferably the system should run minimally at a processing speed of 10 frames per second on top of a compact, easy-to-use, embedded platform (e.g. Jetson TX2). Therefore, we propose a multi-threaded YOLOv3 TensorRT optimized person detector combined with a fast lightweight object tracker.

Figure 2 gives an overview on how our pipeline is implemented on the embedded Jetson TX2 platform. All sub-parts will be discussed in detail in the following subsections.

3.1   Person Detection and Batch Processing

In applications that require real-time performance, one can either choose to maximize throughput or minimize latency. Throughput focuses on the number of images that are processed in a given time slot, while latency focuses on the time needed to process a single image. Given the goal of analyzing the flow of costumers in a given security camera stream, and in the future even in multiple camera streams, throughput is in our case more important than latency. We maximize throughput by performing a batch processing approach that takes advantage parallel GPU processing. As seen in Fig. 2, an image buffer is used to collect all incoming images from the camera stream, at a 1280 × 720 resolution, which are downsampled to a 512 × 288 input resolution before moving to the detector and the tracker unit. We could sent these images directly to our optimized YOLOv3 implementation [46], but since it is only capable of processing data at a maximum speed of 8FPS at this resolution, we skip several frames when collecting a batch that needs to be processed by the YOLOv3 detector.

Take for example a image buffer of 24 images. From those 24 images, only 8 images are selected as a batch that gets passed to the detector unit. The remaining 16 images are sent to the tracker unit, which waits for the processed batch of the detector, so that it can track each detected person throughout the 24 frames series, based on the detections in those 8 reference frames. In order to improve processing speed, both detector and tracker unit are implemented in separate threads, to reduce the processing time delays again as much as possible.

3.2   Person Tracking

Given our hardware setup of a Jetson TX2 embedded platform, we need to carefully consider our hardware resources. Since we're using the on-board GPU to efficiently processes batches of 8 images with the object detector unit, the decision was made to implement a CPU-based tracker to divide the workload as good as possible between the different processing units. We augment the detector unit with a lightweight CPU-based Lukas Kanade Sparse Optical Flow tracker (based on the OpenCV4.1 implementation [5]). This allows us to run the tracker on the CPU, while simultaneously running the GPU-focused YOLOv3 object detector. To achieve the minimally required frame rate of 10 FPS, we downscale the original input frames to a resolution of 512 × 288 pixels.

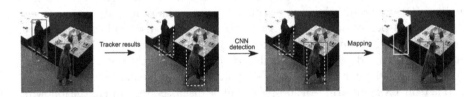

Fig. 3. The detector-tracker pipeline, using tracking-by-detection with interleaved tracker proposals: (solid red) detections (interleaved green) tracker proposals (solid green) tracker proposal accepted as detection in case of a missing detection. (Color figure online)

In between detection frames, two extra frames remain which are only processed by the tracker unit, and thus so far have no knowledge of the location of the detected objects. For each detection inside a detection frame, key feature points are calculated, generating a sparse representation of that detection. Those points are passed into an optical flow mechanism that is used to generate predictions of the locations of these keypoints in the in-between tracker frames. Finally the detections in the next detection frame are used to apply a mapping between the predictions and the detections, possible proposing a slight correction of the bounding box location. Figure 3 illustrates how this detector-tracker combination is working on a sample from the security camera stream. The tracker only produces predictions in the next frame for detections with a detection probability higher than 10%. This allows us to carefully choose the working conditions of the detector and avoid as much false negative detections (persons getting missed) as possible. Detections with a lower confidence will simply be ignored by the tracker and immediately be removed from the tracker memory.

Applying this approach also introduces issues. When customers move outside the field-of-view of the camera stream, their tracking information is kept in memory by the tracker, clogging up the tracker memory and resulting in locally dangling tracks. In order to avoid these dangling tracks, after 5 consecutive detection misses, we force the tracker to remove the tracking information and forget the track altogether, which will end the tracking of that object.

Besides the sparse optical flow approach, two more CPU-based tracking implementations (Kernelized Correlation Filers and Median Flow) are tested to compare tracking robustness in shop-like environments. Both are initialized by the detections of the neural network, just like the sparse optical flow approach. For both trackers, the same input resolution of 512×288 pixels has been used.

3.3   Tracker Memory

Since the selected tracker implementations only takes the information between two consecutive frames into account, we are risking losing a person with a specific ID from tracker memory (e.g. due to exceeding the threshold of missed frames), that can in a later stage, be picked up again by the detector. In order to avoid a new ID being assigned, we introduce a method to keep and match these lost tracks in memory. In case a track is deleted from memory, it first changes its status to lost. A lost track is remembered for maximally 5 s, and is recovered whenever a new detection appears in a location close to the last known location of a deleted track. A match between a lost track and a newly generated track is made by meeting one of the following conditions: the new location lies within a radius of 200 pixels around the last known location, or the new location lies in the quadrant of which the gradient of the lost track points to. In case of multiple new detections in the same quadrant, the detection with the shortest (perpendicular) distance to the gradient is preferred.

3.4   Heat Map Extraction

On top of providing statistics, about the exact location and path followed by customers, users of the system would also like to have some sort of visual confirmation on the activity in their stores. Therefore we propose a heat map based system that gives an overview of the store spots that are most frequently visited within a specific time slot. The heat maps are extracted by mapping the pixels in the detected and tracked bounding boxes on top of a visual layout of the store. The used object detector is known for its jittery bounding boxes, with inconsistent widths and heights. By simply incrementing pixel locations each time a detection box matches a pixel, we end up with a very jittered and visually unpleasing heat map, which we would like to avoid. We solve this by using a weighted increment of the pixels falling within a bounding box. By taking the ratio between the bounding box width and height into account and by giving the center pixel a maximum weight that degrades towards the borders of the bounding box, we end op with a increment value that is less influenced by the jitter, resulting in a visually pleasing heat map. After all frames within the given time slot have been processed, heat maps are normalized to get a meaningful color overlay, as seen in Fig. 1.

4   Experiments and Results

In this section, we evaluate the proposed embedded detector-tracker approach used for in-store customer behaviour analysis by performing four different experiments. We start by evaluating the detector unit and decide on the final deep learned architecture for the detector unit. Next, we add different trackers to the problem to cope with missing detections, optimize parameters and have a look at the influence of changing internal parameters. We then integrate our optimal detector-tracker combination into our batch processing pipeline, to gain processing speed. Finally we take a shot at generating visual heat maps, giving confirmation of frequently visited places in the stores.

4.1   Detector and Tracker Evaluation

A set of 5000 in-store images were collected via the security camera streams inside the store and annotated with ground truth labels. We evaluate both the TensorRT optimized YOLOv2 and YOLOv3 implementations on this set by running them through our NVIDIA Jetson TX2 platform. The left hand side of Fig. 4 shows the difference in obtained average precision (AP). We notice that the YOLOv3 architecture (AP = 84.36%) is performing much better than the YOLOv2 architecture (AP = 64.26%), with a tremendous rise in AP of 20%. This can be explained by the capability of the YOLOv3 architecture to detect persons over a wider variety of scales. Giving the security cameras are mounted to give a single point overview of the store, customers walking further away from the camera are not being detected by the YOLOv2 architecture. On top of that, we also notice that the YOLOv3 architecture is better capable of detecting partially occluded persons.

However, even with the YOLOv3 architecture, we do not yet reach the optimal solution (AP = 100%). After carefully inspecting the missed detections, we notice the detect still drops persons when they are crossing each other or when

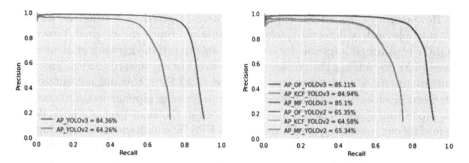

Fig. 4. Precision-recall curves of (left) the embedded and optimized YOLOv2 and YOLOv3 implementation (right) the combined detector-tracker implementations based on optical flow, median flow and kernelized correlation filters.

they are move behind counters and thus get partially occluded. To solve these issues we run our combined detector-tracker units, for which the AP curves are visible in Fig. 4 at the right hand side. We notice that adding the tracker unit slightly helps the detector improve in AP for all the combinations. Based on these experiments we decided to stick with the best performing combination, being YOLOv3 as detector and optical flow as tracker.

One could however argue that the influence of the tracker can also be increased if predicted tracks are kept in memory longer. Until now a track is rejected as soon as the detector unit is unable to find a matching detection in 5 consecutive detection frames. To show the influence of this parameter, we increased this threshold towards 10 missed detection frames and again ran the precision-recall evaluation for the optimal combination again, as seen in Fig. 5 on the left hand side. This change again provides a small boost to the AP of the optimal solution towards 85.35%. However, the influence of varying this parameter even further should be investigated in future research.

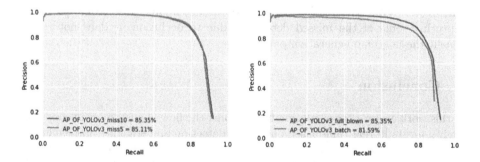

Fig. 5. Influence of (left) increasing the miss threshold for the tracker and (right) applying our batch processing pipeline to gain processing speed.

However, up till now our experiments were targeted at getting the highest AP possible with a detector-tracker combination. Since the original TensorRT YOLOv3 optimized implementation is only able to run the network at an average speed of 3FPS, adding the tracker reduced this further towards 2FPS. Since we want to aim for a minimal processing speed of 10 FPS, this did not suffice. As discussed in subsect. 3.1 we applied a batch processing pipeline to overcome this low processing speed. The obtained AP is seen in the right part of Fig. 5. While increasing our processing speed towards 9.8 FPS by simultaneously reducing the AP towards 81.59%. In order to see if this change might influence the other detector-tracker combinations, we also evaluated those in the batch processing setup. Results of the obtained average tracking speeds can be seen in Table 1, which confirmed that we picked the correct combination before. Median filtering and kernelized correlation filters perform worse due to the linear scale-up in relation to the the amount of detected persons, while this is not true for optical flow. However, we acknowledge that this could depend on the implementation of the tracker used, and that further research on this is necessary.

Table 1. Comparing average detector-tracker processing speeds over the dataset.

Optical Flow	Median Filtering	Kernelized Correlation Filters
9.8 FPS	4.7 FPS	2.1 FPS

4.2 Generating Heat Maps of Frequently Visited Places

Figure 1 illustrates the resulting heat maps that indicate the most frequently visited spots in the store. This gives a clear indication to the shop owners which counters attract customers better than others. Generating heat maps with the detector unit only tends to generate heat maps with visually only very small differences. While the tracker helps in obtaining higher processing speeds in capturing some of the missed detections due to occlusion, it does not really benefit the heatmap generation part.

5 Conclusion

In this work we propose a solution for mapping the flow of customers in shop-like environments in real-time based on person detection and tracking, implemented on an power-efficient and compact embedded platform. In this work we proposed a novel batch processing approach that efficiently uses the power of both GPU and CPU simultaneously, to run a state-of-the-art TensorRT optimized YOLOv3 object detector combined with a sparse optical flow object tracker on the embedded Jetson TX2 platform. We achieve an average precision of 81.59% at a processing speed of 10 FPS on a dataset of challenging real-life imagery acquired in a real shop. While this is not yet the optimal solution, given the challenging conditions (dynamic backgrounds, illumination changes and different store lay-outs), this numbers are quite impressive. On top of that we provide visually pleasing heat maps of the store, giving owners valuable insights to customer behavior. To improve the current results, we are planning to add a person re-identification pipeline to the solution and keep an eye out to new ways of optimizing deep learning pipelines for embedded platforms. This paper creates a proof-of-concept setup that can be further exploited in multiple camera stream setups, exploring how we can use shared memory buffers and shared detection networks between different embedded platforms efficiently.

Acknowledgements. This work is supported by PixelVision, KU Leuven and Flanders Innovation & Entrepreneurship (VLAIO) through a Baekelandt scholarship.

References

1. Andriluka, M., Roth, S., Schiele, B.: People-tracking-by-detection and people-detection-by-tracking. In: Proceedings of CVPR, pp. 1–8 (2008)
2. Avidan, S.: Support vector tracking. In: Proceedings of CVPR, pp. 1064–1072 (2001)
3. Babenko, B., Yang, M.H., et al.: Visual tracking with online multiple instance learning. In: Proceedings of CVPR, pp. 983–990 (2009)
4. Bochinski, E., Eiselein, V., Sikora, T.: High-speed tracking-by-detection without using image information. In: Proceedings of AVSS, pp. 1–6 (2017)
5. Bradski, G., Kaehler, A.: Learning OpenCV: computer vision with the OpenCV library. O'Reilly Media Inc., Sebastopol (2008)
6. Breiman, L.: Random forests. Mach. learn. **45**, 5–32 (2001)
7. Chen, P., Dang, Y., et al.: Real-time object tracking on a drone with multi-inertial sensing data. Trans. ITS **19**(1), 131–139 (2017)
8. Choi, J.W., Moon, D., Yoo, J.H.: Robust multi-person tracking for real-time intelligent video surveillance. Proc. ETRI **37**(3), 551–561 (2015)
9. Dalal, N., Triggs, B.: Histograms of oriented gradients for human detection. In: Proceedings of CVPR, pp. 886–893 (2005)
10. Dendorfer, P., Rezatofighi, H., et al.: CVPR19 tracking and detection challenge: how crowded can it get? arXiv preprint arXiv:1906.04567 (2019)
11. Denman, S., Chandran, V., et al.: Person tracking using motion detection and optical flow. In: Proceedings of DSPCS, pp. 1–6 (2005)
12. Dollár, P., Appel, R., et al.: Fast feature pyramids for object detection. Proc. TPAMI **36**(8), 1532–1545 (2014)
13. Dollár, P., Tu, Z., et al.: Integral channel features. In: Proceedings of BMVC, pp. 91.1-91.11 (2009)
14. Farnebäck, G.: Two-frame motion estimation based on polynomial expansion. In: Scandinavian conference on Image analysis, pp. 363–370 (2003)
15. Felzenszwalb, P.F., McAllester, D.A., et al.: A discriminatively trained, multiscale, deformable part model. In: Proceedings of CVPR, pp. 1–8 (2008)
16. Girshick, R.: Fast R-CNN. In: Proceedings of ICCV, pp. 1440–1448 (2015)
17. Girshick, R., Donahue, J., et al.: Rich feature hierarchies for accurate object detection & semantic segmentation. In: Proceedings of CVPR, pp. 580–587 (2014)

18. Grabner, H., Grabner, M., et al.: Real-time tracking via on-line boosting. In: Proceedings of BMVC, pp. 47–56 (2006)
19. He, K., Gkioxari, G., et al.: Mask R-CNN. In: Proceedings of ICCV, pp. 2961–2969 (2017)
20. He, K., Zhang, X., et al.: Deep residual learning for image recognition. In: Proceedings of CVPR, pp. 770–778 (2016)
21. Henriques, J.F., Caseiro, R., et al.: High-speed tracking with kernelized correlation filters. Proc. TPAMI **37**(3), 583–596 (2014)
22. Howard, A.G., Zhu, M., et al.: Mobilenets: efficient convolutional neural networks for mobile vision applications. arXiv preprint arXiv:1704.04861 (2017)
23. Iandola, F.N., Han, S., et al.: SqueezeNet: AlexNet-level accuracy with 50x fewer parameters and ¡ 0.5 MB model size. arXiv preprint arXiv:1602.07360 (2016)
24. Iguernaissi, R., Merad, D., et al.: People counting based on kinect depth data. In: Proceedings of ICPRAM, pp. 364–370 (2018)
25. Kalal, Z., Mikolajczyk, K., et al.: Forward-backward error: automatic detection of tracking failures. In: Proceedings of CVPR, pp. 2756–2759 (2010)
26. Kanagamalliga, S., Vasuki, S.: Contour-based object tracking in video scenes through optical flow and Gabor features. Optik **157**, 787–797 (2018)
27. Kowcika, A., Sridhar, S.: A literature study on crowd (people) counting with the help of surveillance videos. Int. J. Innovative Technol. Res. 2353–2361 (2016)
28. Krizhevsky, A., Sutskever, I., et al.: Imagenet classification with deep convolutional neural networks. In: Advances of NeurIPS, pp. 1097–1105 (2012)
29. Lefloch, D., Cheikh, F.A., et al.: Real-time people counting system using a single video camera. In: Real-Time Image Processing (2008)
30. Li, Y., Chen, Y., et al.: Scale-aware trident networks for object detection. arXiv preprint arXiv:1901.01892 (2019)
31. Liu, W., Anguelov, D., et al.: SSD: single shot multibox detector. In: Proceedings of ECCV, pp. 21–37 (2016)
32. Liu, X., Tu, P.H., et al.: Detecting and counting people in surveillance applications. In: Advanced Video and Signal Based Surveillance, pp. 306–311 (2005)
33. Lucas, B.D., Kanade, T., et al.: An iterative image registration technique with an application to stereo vision. In: Proceedings of IJCAI, pp. 674–679 (1981)
34. Ma, N., Zhang, X., et al.: ShuffleNet v2: practical guidelines for efficient CNN architecture design. In: Proceedings of ECCV, pp. 116–131 (2018)
35. Mittal, S.: A survey on optimized implementation of deep learning models on the NVIDIA Jetson platform. J. Syst. Architect. **97**, 428–442 (2019)
36. Murray, S.: Real-time multiple object tracking - a study on the importance of speed. arXiv preprint arXiv:1709.03572 (2017)
37. Pernici, F., Del Bimbo, A.: Object tracking by oversampling local features. Proc. TPAMI **36**(12), 2538–2551 (2013)
38. Redmon, J., Divvala, S., et al.: You only look once: unified, real-time object detection. In: Proceedings of CVPR, pp. 779–788 (2016)
39. Redmon, J., Farhadi, A.: YOLO9000: better, faster, stronger. In: Proceedings of CVPR, pp. 7263–7271 (2017)
40. Redmon, J., Farhadi, A.: YOLOv3: an incremental improvement. arXiv preprint arXiv:1804.02767 (2018)
41. Ren, S., He, K., et al.: Faster R-CNN: towards real-time object detection with region proposal networks. In: Advances of NeurIPS, pp. 91–99 (2015)
42. Sandler, M., Howard, A., et al.: Mobilenetv2: inverted residuals and linear bottlenecks. In: Proceedings of CVPR, pp. 4510–4520 (2018)

43. Shafiee, M.J., Chywl, B., et al.: Fast YOLO: a fast you only look once system for real-time embedded object detection in video. arXiv preprint arXiv:1709.05943 (2017)
44. Shashev, D., et al.: Methods and algorithms for detecting objects in video files. In: MATEC Web of Conferences, p. 01016 (2018)
45. Shi, J., Tomasi, C.: Good features to track. In: Proceedings of CVPR, pp. 593–600 (1993)
46. Vandersteegen, M., Vanbeeck, K., et al.: Super accurate low latency object detection on a surveillance UAV. arXiv preprint:1904.02024 (2019)
47. Viola, P., Jones, M., et al.: Rapid object detection using a boosted cascade of simple features. In: Proceedings of CVPR, pp. 511–518 (2001)
48. Wai, Y.J., Mohd Yussof, Z., et al.: Fixed point implementation of Tiny-YOLOv2 using OpenCL on FPGA. In: Proceedings of IJACSA, pp. 506–512 (2018)
49. Zhang, X., Zhou, X., et al.: Shufflenet: an extremely efficient convolutional neural network for mobile devices. In: Proceedings of CVPR, pp. 506–512 (2018)

Learning Target-Specific Response Attention for Siamese Network Based Visual Tracking

Penghui Zhao, Haosheng Chen, Yanjie Liang, Yan Yan, and Hanzi Wang[✉]

Fujian Key Laboratory of Sensing and Computing for Smart City,
School of Informatics, Xiamen University, Xiamen, China
hanzi.wang@xmu.edu.cn

Abstract. Recently, the Siamese network based visual tracking methods have shown great potentials in balancing the tracking accuracy and computational efficiency. These methods use two-branch convolutional neural networks (CNNs) to generate a response map between the target exemplar and each of candidate patches in the search region. However, since these methods have not fully exploit the target-specific information contained in the CNN features during the computation of the response map, they are less effective to cope with target appearance variations and background clutters. In this paper, we propose a Target-Specific Response Attention (TSRA) module to enhance the discriminability of these methods. In TSRA, a channel-wise cross-correlation operation is used to produce a multi-channel response map, where different channels correspond to different semantic information. Then, TSRA uses an attention network to dynamically re-weight the multi-channel response map at every frame. Moreover, we introduce a shortcut connection strategy to generate a residual multi-channel response map for more discriminative tracking. Finally, we integrate the proposed TSRA into the classical Siamese based tracker (i.e., SiamFC) to propose a new tracker (called TSRA-Siam). Experimental results on three popular benchmark datasets show that the proposed TSRA-Siam outperforms the baseline tracker (i.e., SiamFC) by a large margin and obtains competitive performance compared with several state-of-the-art trackers.

Keywords: Siamese networks · Visual tracking · Channel attention

1 Introduction

Visual tracking is a fundamental computer vision task with a variety of applications, such as video surveillance, vehicle navigation and augmented reality. A large number of methods have been proposed for effective visual tracking. Among them, the Siamese network based trackers (e.g., [2,10,17,20,21]) formulate the tracking task as a problem of learning a similarity metric. By using the learned metric, they generate a similarity response map between the target exemplar

© Springer Nature Switzerland AG 2020
J. Blanc-Talon et al. (Eds.): ACIVS 2020, LNCS 12002, pp. 554–566, 2020.
https://doi.org/10.1007/978-3-030-40605-9_47

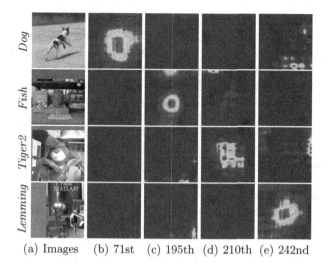

(a) Images   (b) 71st   (c) 195th   (d) 210th   (e) 242nd

Fig. 1. Different channels of the multi-channel response map generated by our tracker on the OTB-100 dataset. There are 256 channels in the multi-channel response map. Different channels correspond to different semantic contents. (a) contains four search regions from video frames in the OTB-100 dataset. We choose the 71st, 195th, 210th and 242nd channels, which are shown from (b) to (e), for demonstration. The 71st channel has high responses on *Dog*, while it has low responses on *Fish*, *Tiger2* and *Lemming*. Similarly, the 195th, 210th and 242nd channels have high responses on *Fish*, *Tiger2* and *Lemming*, respectively.

and each of candidate patches in the search region. The location that has the highest response score is selected as the center of the target. These trackers use a dense and efficient sliding-window strategy to achieve balanced performance between the tracking accuracy and computational efficiency.

Although the Siamese network based trackers have achieved good performance, they are less effective to cope with several challenging scenarios such as fast motion, occlusion, rotation and background clutter. The main reason is that these trackers, especially for SiamFC [2], do not properly exploit the target-specific information containing in the CNN features extracted from the search region. In SiamFC, the CNN features extracted from the target exemplar and the search region are cross-correlated to generate a response map. However, the computation of the response map usually involves considerable amount of target-irrelevant information. As a result, the contribution of target-specific information contained in the CNN features is weakened due to the influence of the target-irrelevant information, leading to the ineffectiveness of the response map for distinguishing the target object from the background.

To address the issue mentioned above, inspired by Xception [6] and CAM [26], we propose a novel attention module, named Target-Specific Response Attention (TSRA). TSRA utilizes a channel-wise cross-correlation operation to generate a multi-channel response map, which is effective to prevent the semantic information contained in each channel of the CNN features from affecting each other. In the multi-channel response map, some channels play a more important role than others in tracking a specific target (see Fig. 1 for some illustrations). Then, TSRA further uses an attention network to enhance the target-specific information by weighing the multi-channel response map.

In practice, the weighted multi-channel response map may be less discriminative since the channels with high responses are occasionally assigned with low weights. To solve this problem, inspired by the residual network [12], we propose a shortcut connection strategy to learn a residual multi-channel response map instead of learning a weighted multi-channel response map. The learned residual multi-channel response map is more discriminative than the weighted response map for robust visual tracking.

Finally, we integrate the proposed TSRA module and the shortcut connection strategy into the classical Siamese based tracker (i.e. SiamFC) to present a new Siamese network based tracker (called TSRA-Siam). The TSRA-Siam tracker is robust to cope with several challenging scenarios in visual tracking, such as fast motion, rotation and occlusion.

The main contributions of this work are three folds:

- We present a Target-Specific Response Attention (TSRA) module, which can be used to enhance the discriminability of the response map for tracking a specific target.
- We propose a shortcut connection strategy, which generates a residual multi-channel response map for discriminative tracking.
- We integrate the proposed TSRA module and the shortcut connect strategy into the framework of SiamFC, and propose a robust Siamese network based tracker (called TSRA-Siam).

The rest of the paper is organized as follows. In Sect. 2, we introduce the related work. In Sect. 3, we describe our TSRA module in detail. In Sect. 4, we present the tracking process. In Sect. 5, we report the experimental results. In Sect. 6, we give the conclusions.

2   Related Work

2.1   Siamese Network Based Tracking

In Siamese network based trackers, the pioneering work is the fully convolutional Siamese network (SiamFC) [2]. Later, many successors have been proposed. For example, SiamFC-tri [10] exploits the underlying relationship of its triplet set to improve the discriminability of SiamFC. CFNet [21] regards the correlation filter as a network layer to compute the similarity between the CNN features generated from the Siamese network. SINT [20] employs the additional optical flow information and it achieves promising performance for tracking. SiamRPN [17] integrates a region proposal module into the backbone network of SiamFC to cope with the changes of target shape and scale. SiamDW [25] is developed to solve the problem of positional bias in training the trackers.

2.2 Attention Mechanisms

Attention mechanisms have been widely used in deep learning, such as image caption [24], machine translation [3] and salient object detection [15]. Recently, attention mechanisms have attracted much attention in visual tracking. For example, DAVT [11] introduces the discriminative spatial attention, which identifies some special regions of the target. ACFN [5] utilizes an attention module to choose a subset of correlation filters for robust visual tracking. CSR-DCF [18] utilizes a foreground spatial reliability map to constrain the correlation filter learning. RTT [7] utilizes a multi-directional Recurrent Neural Network to select reliable regions belonging to the target object. Different from these attention mechanisms, our TSRA learns a set of attention weights to weigh the multi-channel response map in an end-to-end manner.

3 Our Method

3.1 Revisit the Fully Convolutional Siamese Tracking Framework

The fully convolutional Siamese tracking framework (i.e., SiamFC) formulates the tracking problem as an exemplar matching problem in an embedding space. The exemplar matching in SiamFC utilizes a matching metric learned offline on a large number of image pairs to select the best matched patch from candidate patches. This matching metric consists of an embedding function and a similarity function. The embedding function is used to extract the features from the target exemplar $\mathbf{Z} \in \mathcal{R}^{w \times h \times 3}$ and the candidate patches cropped from a search region $\mathbf{X} \in \mathcal{R}^{w_x \times h_x \times 3}$. Here w and h respectively denote the width and height of the target exemplar, and w_x and h_x respectively denote the width and height of the search region. The similarity function is used to compute the similarity score between $\mathbf{Z}$ and $\mathbf{X}$. In SiamFC, the embedding function is the two-branch CNNs, which share the identical parameters p, and the similarity function is the cross-correlation.

The overall process of the fully convolutional Siamese tracking framework can be described as follows: Firstly, the embedding function is used to extract features from $\mathbf{Z}$ and $\mathbf{X}$, respectively. Secondly, the similarity function is used to compute a similarity score between $\mathbf{Z}$ and $\mathbf{X}$. The network outputs a response map $\mathbf{S} \in \mathcal{R}^{w_s \times h_s}$ (where w_s and h_s respectively denote the width and height of the response map). This process can be formulated as follow:

$$\mathbf{S} = \mathrm{Corr}(\varPhi_p(\mathbf{Z}), \varPhi_p(\mathbf{X})) \tag{1}$$

where $\varPhi_p$ denotes the embedding function, and $\mathrm{Corr}(\cdot)$ denotes the cross correlation operator. Finally, the location with the highest response score in $\mathbf{S}$ is selected as the center of the target.

3.2 Target-Specific Response Attention

In a pre-trained convolutional network, some channels of the output CNN features contain the target-specific information, while the other channels contain the target-irrelative information. Thus, only a subset of these channels corresponds to target-specific information during visual tracking. In SiamFC [2], all channels of features extracted from the target exemplar and the search region are used to perform cross-correlation. As a result, the target-irrelative information is involved in the computation of the response map $\mathbf{S}$, which leads to the poor tracking performance. To enhance the target-specific information and suppress the target-irrelative information, we propose the Target-Specific Response Attention (TSRA) module.

Experimental results of Xception [6] suggest that the decouple of cross-channel correlations and spatial correlations will result in more powerful outputs. Inspired by Xception, TSRA separately computes the correlations for each channel of CNN features (i.e., channel-wise correlation) to prevent semantic information contained in each channel of the CNN features from affecting each other. Compared with the single response map in SiamFC, the channel-wise correlation is used to generate a multi-channel response map as follows:

$$\mathbf{S}_{multi} = \mathrm{Corr}_{cw}(\Phi_p(\mathbf{Z}), \Phi_p(\mathbf{X})) \tag{2}$$

where $\mathbf{S}_{multi} \in \mathcal{R}^{w_s \times h_s \times c}$ is the multi-channel response map with c channels, $\mathrm{Corr}_{cw}(\cdot)$ denotes the channel-wise cross-correlation.

Then, an attention network is proposed to focus on the channels corresponding to the target-specific information. Unlike SENet [14] that takes the CNN features as inputs and weighs different channels of features, TSRA takes the CNN features as inputs and weighs the multi-channel response map. The motivation that TSRA weighs the multi-channel response map rather than the multi-channel feature map is that the multi-channel response map is more effective to decouple the cross-channel correlations of feature channels so as to produce a more discriminative response map. By alleviating the influence from other feature channels, the single channel cross-correlation can produce a more discriminative response map, Therefore, the tracker can localize the target more precisely.

More specifically, the proposed attention network consists of a global average pooling (GAP), a multi-layer perceptron (MLP) and a sigmoid function. The GAP is used to generate channel-wise statistics of features, the MLP is used to generate weights of the channels and the sigmoid function is utilized to normalize the weights ranging from 0 to 1. The proposed attention network can be formulated as follows:

$$\boldsymbol{\alpha} = H(\Phi_p(\mathbf{X})), \tag{3}$$

$$\widehat{\mathbf{S}}_{multi} = \boldsymbol{\alpha} \otimes \mathbf{S}_{multi}, \tag{4}$$

where $H(\cdot)$ denotes the proposed attention network, $\otimes$ denotes the channel-wise multiplication, $\boldsymbol{\alpha} \in \mathcal{R}^c$ represents the channel weights, c is the number of feature channels, and $\widehat{\mathbf{S}}_{multi}$ is the weighted multi-channel response map.

3.3   Shortcut Connection on Target-Specific Response Attention

In the work [14], the attention module is implemented by multiplying CNN features with their channel weights. However, by performing such an operation, a small weight may be inappropriately assigned to a channel, where the target has a high response, leading to an inferior tracking performance. To alleviate this problem, we propose to employ a shortcut connection strategy, which is motivated by the structure of residual networks [12]. We reformulate (4) as follows:

$$\widehat{\mathbf{S}}_{final} = (\boldsymbol{\alpha}_{res} + \boldsymbol{I}) \otimes \mathbf{S}_{multi}$$
$$= \widehat{\mathbf{S}}_{res} + \mathbf{S}_{multi} \tag{5}$$

where $\boldsymbol{I} \in \mathcal{R}^c$ denotes the unit vector. $\boldsymbol{\alpha}_{res}$ denotes channel weights generated by the proposed attention network with residual learning. $\widehat{\mathbf{S}}_{res}$ denotes the residual multi-channel response map, and $\widehat{\mathbf{S}}_{final}$ is the output of the shortcut connection.

In (5), we use the proposed shortcut connection strategy to perform residual learning and to generate the residual multi-channel response map, which is much easier to be learned than the weighted multi-channel response map (without residual learning). Based on the shortcut connection strategy, the attention network can be easily trained for generating appropriate attention weights. Thus the proposed TSRA can generate a more discriminative response map for robust visual tracking. Finally, the final response map $\mathbf{S}_{final}$ can be calculated as follows:

$$\mathbf{S}_{final} = \sum_{i=1}^{c} \widehat{S}_{final}^{i} \tag{6}$$

where $\widehat{S}_{final}^{i}$ is the i-th channel of the output of the shortcut connection.

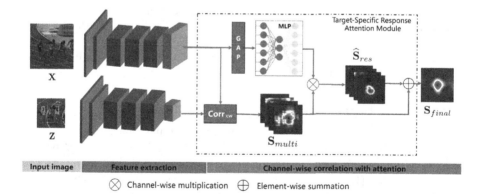

Fig. 2. The overall framework of the proposed TSRA-Siam.

4 Tracking Process

The tracking process of the proposed tracker includes tracker initialization, online detection and scale estimation. Figure 2 shows the framework of the proposed tracker.

Tracker Initialization. Given the first frame with the target location, we crop an image patch centered around the target as the target exemplar. This patch is considered as the input of our framework for feature extraction. The extracted CNN features of the target exemplar are served as the input of the proposed TSRA.

Online Detection. When a new frame comes, we crop the search region from the target center location predicted at the previous frame. Then, the two-branch CNNs are used to extract the features from the search region. After that, we compute the similarity scores between the target exemplar and the search region at the current frame using the features. This is achieved by the proposed TSRA, which outputs a response map. Finally, the maximum of the final response map S_{final} corresponds to the target position.

Scale Estimation. For scale estimation, we adopt the same strategy used in SiamFC [2].

5 Experiments

We implement our tracker in Python with PyTorch. The experiments are conducted on a computer equipped with an Intel 2.4 GHz CPU and an NVIDIA GTX Titan XP GPU. Our tracker runs at a speed of 65.2 FPS.

5.1 Implementation Details

TSRA-Siam is trained offline using the ImageNet Large Scale Visual Recognition Challenge (ILSVRC15) dataset [19]. For each training epoch, we randomly sample 53,200 training pairs from this dataset and set the mini-batch size to 8. We use the stochastic gradient descent (SGD) algorithm with momentum of 0.9 to train our tracker for 50 epochs. The learning rate is exponentially decayed after each epoch from 10^{-2} to 10^{-5}.

5.2 Tracking Benchmarks and Evaluation Metrics

We evaluate the proposed TSRA-Siam on three popular benchmark datasets, including OTB-50 [22], OTB-100 [23] and VOT2017 [16]. For the OTB benchmark datasets, we report the results of one-pass evaluation (OPE) with both the distance precision (DP) and overlap success (OS) plots. We use the DP rates at a threshold of 20 pixels (DPR) in precision plots and the Area Under the Curve (AUC) in success plots to evaluate our tracker. For the VOT benchmark, Accuracy (A), Robustness (R) and Expected Average Overlap (EAO) are used to measure the performance of the competing trackers. When a failure of a tracker is detected, the tracker is re-initialized by the ground truth after five frames.

5.3   Comparison with the State-of-the-Art Trackers

Evaluation on the OTB Datasets. In this paper, we compare our tracker with several state-of-the-art trackers including SiamFC [2], SiamTri [10], CFNet [21], ACFN [5], SRDCF [9], Staple [1], TRACA [4], CSR-DCF [18] and KCF [13]. Among these trackers, SiamFC, SiamTri and CFNet are the Siamese network based trackers, while ACFN, TRACA and CSR-DCF are the trackers using the attention mechanisms.

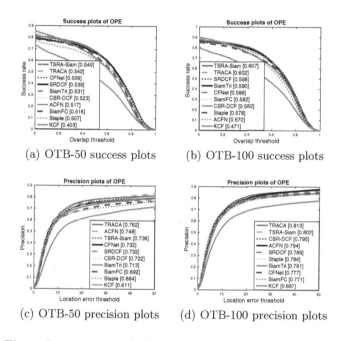

(a) OTB-50 success plots   (b) OTB-100 success plots

(c) OTB-50 precision plots   (d) OTB-100 precision plots

Fig. 3. Comparison on OTB-50 and OTB-100 in terms of OPE.

As shown in Fig. 3, the proposed TSRA-Siam achieves the best performance in terms of the AUC metric on both OTB-50 and OTB-100. TSRA-Siam also achieves competitive results on the AUC metric compared with TRACA on the OTB benchmark. Compared with the baseline tracker (SiamFC), the proposed TSRA-Siam achieves notable improvements on these benchmark datasets. Specifically, the proposed TSRA-Siam obtains 2.0%/2.5% and 2.1%/3.1% improvements in terms of AUC/DPR metrics on OTB-50 and OTB-100, respectively.

To further demonstrate the robustness of the proposed TSRA-Siam, we perform an attribute-based performance analysis. Figure 4 shows the results of 6 major attributes on OTB-100, which are evaluated on the AUC of success plots. These attributes are Fast Motion (FM), In-Place Rotation (IPR), Motion Blur (MB), Out-of-Plane Rotation (OPR), Scale Variation (SV) and Occlusion

Table 1. Experimental results on VOT-2017. The best values are in bold.

Tracker	EAO	Accuracy (A)	Robustness (R)
TSRA-Siam (Ours)	**0.193**	**0.518**	0.582
CSRDCFf	0.158	0.475	0.646
ECOhc	0.177	0.494	**0.571**
SiamFC	0.182	0.502	0.604
Staple	0.170	0.530	0.688
SiamDCF	0.135	0.503	0.988
KCF	0.134	0.445	0.782

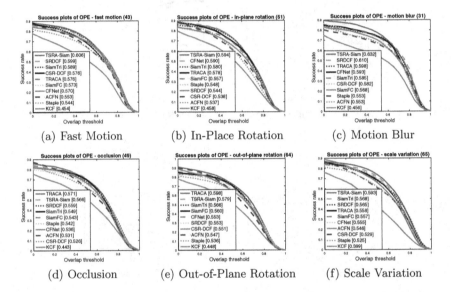

(a) Fast Motion (b) In-Place Rotation (c) Motion Blur

(d) Occlusion (e) Out-of-Plane Rotation (f) Scale Variation

Fig. 4. Success plots for six challenging attributes on OTB-100 in terms of OPE.

(OCC). The results show that our tracker (TSRA-Siam) achieves the best performance on FM, MB, IPR and SV, and it outperforms the baseline tracker (SiamFC) on these challenging attributes. In particular, our tracker achieves a large improvement of 6.4% compared with the baseline tracker on the MB attribute. This is because that the proposed TSRA-Siam effectively enhances the target-specific information and suppresses the target-irrelative information in the multi-channel response map.

Evaluation on the VOT2017 Dataset. On the VOT2017 dataset, we compare our tracker with the baseline tracker, ECOhc [8], SiamDCF, KCF, Staple and CSR-DCFf [18] by using the real-time evaluation. The results are shown in Table 1, from which we can see that. TSRA-Siam achieves the best performance compared with other trackers in terms of EAO and Accuracy.

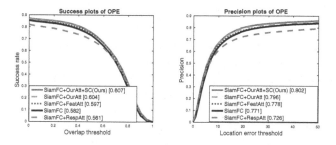

Fig. 5. Ablation study of TSRA-Siam on OTB-100 in terms of OPE.

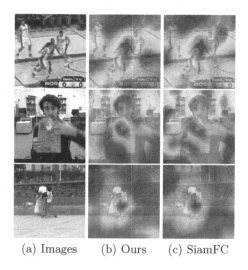

(a) Images   (b) Ours   (c) SiamFC

Fig. 6. Response maps generated by the baseline tracker (SiamFC) and our tracker (TSRA-Siam) on the *Basketball, Clibar* and *Girl2* video sequences of OTB-100. (a) shows the original images, (b) shows the response maps generated by our tracker, and (c) shows the response maps generated by the baseline tracker. As shown in the response maps, our tracker obtains higher responses on the target objects.

5.4   Ablation Study

To illustrate the effectiveness of the proposed TSRA-Siam, we conduct an ablation study by comparing the performance of TSRA-Siam with its three variants on OTB-100. Among these variants, for the proposed TSRA without using the shortcut connection (SC) strategy, we denote it as OurAtt. For the model that takes the CNN features as the input and outputs channel weights for each feature channel, we denote it as FeatAtt. For the model that takes multi-channel response map as the input and outputs channel weights for the channels of the response maps, we denote it as RespAtt. By integrating these variants in the baseline tracker, we obtain three trackers (i.e., SiamFC+OurAtt, SiamFC+FeatAtt and SiamFC+RespAtt). We train these variants with the same parameter settings as in TSRA-Siam. Figure 5 shows the experimental results obtained by these four trackers on OTB-100. Compared with the three variants, the TSRA-Siam using the proposed attention module and the shortcut connection strategy achieves the best performance.

In addition, we visualize the response maps of three search regions of three sequences on OTB-100 to intuitively illustrate the effectiveness of our attention module (as shown in Fig. 6). We can observe that the influence of background is suppressed during the tracking by using the TSRA module, which shows that the proposed TSRA module improves the discriminability of baseline tracker on the background clutter attribute.

6 Conclusions

This paper presents a Target-Specific Response Attention (TSRA) module that performs channel-wise attention on the multi-channel response map for real-time tracking. The proposed TSRA module can enhance the discriminability and adaptability of the Siamese network based tracker in handling target appearance variations. Specifically, we not only exploit target-specific information to weigh the multi-channel response map, but also utilize a shortcut connection strategy to generate residual multi-channel response map for performance improvements. We propose our TSRA-Siam tracker by integrating the TSRA module and the shortcut connection strategy into SiamFC. Experimental results on OTB-50, OTB-100 and VOT2017 demonstrate that our TSRA-Siam tracker can achieve competitive performance at real-time speed compared with several state-of-the-art trackers.

Acknowledgements. This work was supported by the National Natural Science Foundation of China (Grant No. U1605252 and 61872307) and the National Key R&D Program of China (Grant No. 2017YFB1302400).

References

1. Bertinetto, L., Valmadre, J., Golodetz, S., Miksik, O., Torr, P.H.S.: Staple: Complementary learners for real-time tracking. In: Proceedings of Conference on Computer Vision and Pattern Recognition (CVPR), pp. 1401–1409 (2016)
2. Bertinetto, L., Valmadre, J., Henriques, J.F., Vedaldi, A., Torr, P.H.S.: Fully-convolutional siamese networks for object tracking. In: Hua, G., Jégou, H. (eds.) ECCV 2016. LNCS, vol. 9914, pp. 850–865. Springer, Cham (2016). https://doi.org/10.1007/978-3-319-48881-3_56
3. Cho, K., van Merrienboer, B., Gülçehre, Ç., Bahdanau, D., Bougares, F., Schwenk, H., Bengio, Y.: Learning phrase representations using RNN encoder-decoder for statistical machine translation. In: Proceedings of Conference on Empirical Methods in Natural Language Processing (EMNLP), pp. 1724–1734 (2014)
4. Choi, J., Chang, H.J., Fischer, T., Yun, S., Lee, K., Jeong, J., Demiris, Y., Choi, J.Y.: Context-aware deep feature compression for high-speed visual tracking. In: Proceedings of Conference on Computer Vision and Pattern Recognition (CVPR), pp. 479–488 (2018)

5. Choi, J., Chang, H.J., Yun, S., Fischer, T., Demiris, Y., Choi, J.Y.: Attentional correlation filter network for adaptive visual tracking. In: Proceedings of Conference on Computer Vision and Pattern Recognition (CVPR), pp. 4828–4837 (2017)
6. Chollet, F.: Xception: deep learning with depthwise separable convolutions. In: Proceedings of Conference on Computer Vision and Pattern Recognition (CVPR), pp. 1800–1807 (2017)
7. Cui, Z., Xiao, S., Feng, J., Yan, S.: Recurrently target-attending tracking. In: Proceedings of Conference on Computer Vision and Pattern Recognition (CVPR), pp. 1449–1458 (2016)
8. Danelljan, M., Bhat, G., Khan, F.S., Felsberg, M.: ECO: efficient convolution operators for tracking. In: Proceedings of Conference on Computer Vision and Pattern Recognition (CVPR), pp. 6931–6939 (2017)
9. Danelljan, M., Häger, G., Khan, F.S., Felsberg, M.: Learning spatially regularized correlation filters for visual tracking. In: Proceedings of International Conference on Computer Vision (ICCV), pp. 4310–4318 (2015)
10. Dong, X., Shen, J.: Triplet loss in siamese network for object tracking. In: Ferrari, V., Hebert, M., Sminchisescu, C., Weiss, Y. (eds.) ECCV 2018. LNCS, vol. 11217, pp. 472–488. Springer, Cham (2018). https://doi.org/10.1007/978-3-030-01261-8_28
11. Fan, J., Wu, Y., Dai, S.: Discriminative spatial attention for robust tracking. In: Daniilidis, K., Maragos, P., Paragios, N. (eds.) ECCV 2010. LNCS, vol. 6311, pp. 480–493. Springer, Heidelberg (2010). https://doi.org/10.1007/978-3-642-15549-9_35
12. He, K., Zhang, X., Ren, S., Sun, J.: Deep residual learning for image recognition. In: Proceedings of Conference on Computer Vision and Pattern Recognition (CVPR), pp. 770–778 (2016)
13. Henriques, J.F., Caseiro, R., Martins, P., Batista, J.: High-speed tracking with kernelized correlation filters. IEEE Trans. Pattern Anal. Mach. Intell. **37**(3), 583–596 (2015)
14. Hu, J., Shen, L., Sun, G.: Squeeze-and-excitation networks. In: Proceedings of Conference on Computer Vision and Pattern Recognition (CVPR), pp. 7132–7141 (2018)
15. Kalboussi, R., Abdellaoui, M., Douik, A.: Detecting and recognizing salient object in videos. In: Blanc-Talon, J., Helbert, D., Philips, W., Popescu, D., Scheunders, P. (eds.) ACIVS 2018. LNCS, vol. 11182, pp. 62–73. Springer, Cham (2018). https://doi.org/10.1007/978-3-030-01449-0_6
16. Kristan, M., Leonardis, A., et al.: The visual object tracking VOT2017 challenge results. In: Proceedings of International Conference on Computer Vision Workshops (ICCV Workshops), pp. 1949–1972 (2017)
17. Li, B., Yan, J., Wu, W., Zhu, Z., Hu, X.: High performance visual tracking with siamese region proposal network. In: Proceedings of Conference on Computer Vision and Pattern Recognition (CVPR), pp. 8971–8980 (2018)
18. Lukezic, A., Vojir, T., Zajc, L.C., Matas, J., Kristan, M.: Discriminative correlation filter with channel and spatial reliability. In: Proceedings of Conference on Computer Vision and Pattern Recognition (CVPR), pp. 4847–4856 (2017)
19. Russakovsky, O., et al.: ImageNet large scale visual recognition challenge. Int. J. Comput. Vision **115**(3), 211–252 (2015)
20. Tao, R., Gavves, E., Smeulders, A.W.M.: Siamese instance search for tracking. In: Proceedings of Conference on Computer Vision and Pattern Recognition (CVPR), pp. 1420–1429 (2016)

21. Valmadre, J., Bertinetto, L., Henriques, J.F., Vedaldi, A., Torr, P.H.S.: End-to-end representation learning for correlation filter based tracking. In: Proceedings of Conference on Computer Vision and Pattern Recognition (CVPR), pp. 5000–5008 (2017)
22. Wu, Y., Lim, J., Yang, M.: Online object tracking: a benchmark. In: Proceedings of Conference on Computer Vision and Pattern Recognition (CVPR), pp. 2411–2418 (2013)
23. Wu, Y., Lim, J., Yang, M.: Object tracking benchmark. IEEE Trans. Pattern Anal. Mach. Intell. **37**(9), 1834–1848 (2015)
24. Xu, K., Ba, J., et al.: Show, attend and tell: Neural image caption generation with visual attention, pp. 2048–2057 (2015). Computer Science
25. Zhang, Z., Peng, H.: Deeper and wider siamese networks for real-time visual tracking. In: Proceedings of Conference on Computer Vision and Pattern Recognition (CVPR), pp. 4591–4600 (2019)
26. Zhou, B., Khosla, A., Lapedriza, À., Oliva, A., Torralba, A.: Learning deep features for discriminative localization. In: Proceedings of Conference on Computer Vision and Pattern Recognition (CVPR), pp. 2921–2929 (2016)

Author Index

Printed in the United States
By Bookmasters